Neuropsychological Toxicology
Identification and Assessment of Human Neurotoxic Syndromes

Second Edition

CRITICAL ISSUES IN NEUROPSYCHOLOGY

Series Editors

Antonio E. Puente
University of North Carolina, Wilmington

Cecil R. Reynolds
Texas A&M University

Current Volumes in this Series

BEHAVIORAL INTERVENTIONS WITH BRAIN-INJURED CHILDREN
A. MacNeill Horton, Jr.

CLINICAL NEUROPSYCHOLOGICAL ASSESSMENT:
A Cognitive Approach
Edited by Robert L. Mapou and Jack Spector

FAMILY SUPPORT PROGRAMS AND REHABILITATION:
A Cognitive-Behavioral Approach to Traumatic Brain Injury
Louise Margaret Smith and Hamish P. D. Godfrey

NEUROPSYCHOLOGICAL EVALUATION OF THE
SPANISH SPEAKER
Alfredo Ardila, Monica Rosselli, and Antonio E. Puente

NEUROPSYCHOLOGICAL EXPLORATIONS OF
MEMORY AND COGNITION:
Essays in Honor of Nelson Butters
Edited by Laird S. Cermak

NEUROPSYCHOLOGICAL TOXICOLOGY:
Identification and Assessment of Human Neurotoxic Syndromes,
Second Edition
David E. Hartman

THE NEUROPSYCHOLOGY OF ATTENTION
Ronald A. Cohen

A PRACTICAL GUIDE TO HEAD INJURY REHABILITATION:
A Focus on Postacute Residential Treatment
Michael D. Wesolowski and Arnie H. Zencius

PRACTITIONER'S GUIDE TO CLINICAL NEUROPSYCHOLOGY
Robert M. Anderson, Jr.

A Continuation Order Plan is available for this series. A continuation order will bring delivery of each new volume immediately upon publication. Volumes are billed only upon actual shipment. For further information please contact the publisher.

Neuropsychological Toxicology
Identification and Assessment of Human
Neurotoxic Syndromes

Second Edition

David E. Hartman
*Rush Presbyterian–St. Luke's Medical Center
and Chicago Medical School
Chicago, Illinois*

Plenum Press • New York and London

Library of Congress Cataloging-in-Publication Data

Hartman, David E.
  Neuropsychological toxicology : identification and assessment of
human neurotoxic syndromes / David E. Hartman. -- 2nd ed.
    p.   cm. -- (Critical issues in neuropsychology)
  Includes bibliographical references and index.
  ISBN 0-306-44922-6
  1. Neurotoxicology. 2. Behavioral toxicology. I. Title.
II. Series.
  [DNLM: 1. Nervous System--drug effects. 2. Environmental
Pollutants--poisoning. 3. Behavior--drug effects. 4. Cognition-
-drug effects. 5. Emotions--drug effects.   QV 76.5 H333n 1995]
RC347.5.H37 1995
616.8'047--dc20
DNLM/DLC
for Library of Congress   95-39575
                                                                CIP

ISBN 0-306-44922-6

© 1995, Plenum Press, New York
A Division of Plenum Publishing Corporation
233 Spring Street, New York, N.Y. 10013

10 9 8 7 6 5 4

All rights reserved

No part of this book may be reproduced, stored in a retrieval system, or transmitted in any form or by any means, electronic, mechanical, photocopying, microfilming, recording, or otherwise, without written permission from the Publisher

The first edition of this book was published by Pergamon Press, New York, 1988

Printed in the United States of America

To our children, who were born between editions
Sarah Beth (1990) and Adam Maller (1993)
We adore them

Preface to the Second Edition

In the time since the first edition of *Neuropsychological Toxicology* was conceived and published, public health concern about human neurotoxic exposure has steadily increased. Government agencies and professionals in all sectors of health care are increasingly recognizing that workplace neurotoxic exposure effects, like cancer, constitute an important international health problem and require a national agenda of study and treatment.

Non-work-related neurotoxicity has simultaneously become an extremely salient area of concern to society and the scientific community. In particular, pandemic adult cocaine and alcohol abuse with corresponding prenatal neurotoxic sequelae threatens a growing portion of the population with lasting neurological damage. Documentation, rehabilitation, and, ultimately, prevention of these disorders may be contingent on identifying the behavioral and cognitive damage caused by these drugs.

The second edition of *Neuropsychological Toxicology* attempts to address concerns for an expanded survey of a rapidly expanding field as well as to tackle the "wish list" of reviewers and those who have taken time for comment and kind suggestions. I have attempted to expand much of the first edition in both topic and depth. For example, I have rewritten the chapter on testing methods to make clearer the differences in methodologies between experimental and clinical methods and have also included studies where available of PET, SPECT, and BEAM imaging applied to neurotoxic exposure. Two new chapters have been added: Chapter 9, which provides a summary of recent views on multiple chemical sensitivity, sick building syndrome, and psychosomatic disorders, and Chapter 10, which addresses forensic issues. All chapters have been enhanced with new and updated information, including new information on test development, childhood lead exposure, developmental neurotoxicity, fetal alcohol syndrome, prescription drug effects, and solvent-related sequelae. New NIOSH and OSHA exposure levels for industrial toxins were added as available.

Finally, thanks are in order to those researchers and clinicians who discovered errata in the first edition or who have kept me informed of their work. In every case, I have tried to include their information in the new edition.

Preface to the First Edition

This book was originally written from personal need. As Director of Neuropsychology at Cook County Hospital in Chicago, Illinois, I began receiving a steady influx of patients from the hospital's Occupational Medicine Clinic who had been exposed to toxic substances in their place of work. It soon became apparent that conventional theories of neuropsychological dysfunction did not apply to patients whose subjective complaints were often the only available correlates of neurotoxic exposure. Thus, rules of diagnostic interpretation developed for diffuse, lateralized, or localized dysfunction were not appropriate for this population. Neuropsychological interpretation strategies that worked for tumors, closed-head injuries, cerebrovascular disorders, etc., did not appear to fit patients with vaguely defined, diffuse, functional-sounding complaints—patients who had, nevertheless, been exposed to poisons capable of damaging the nervous system. These were patients for whom a conventional neurological examination was not diagnostic. For these patients a neuropsychological examination became the most fine-grained technique available to determine otherwise subclinical effects of neurotoxicity.

After becoming aware of the need for neuropsychological testing with this population, an attempt was made to find books on "neurobehavioral toxicology" that emphasized human clinical applications and diagnostic formulations. There were none written from a neuropsychological perspective. Furthermore, there were few readily available journal articles of any sort on the subject. I had to begin compiling a set of articles and books related to the field that would serve as a personal dictionary to aid in the assessment of these complex and puzzling cases.

My wish to assemble this information in a more accessible form led to the development of this book. It is my hope that *Neuropsychological Toxicology* will now serve others as my filing cabinet full of articles has served me: as a "dictionary" of many human neurotoxic syndromes, as a review of human neurological and neuropsychological effects of common neurotoxins, and as a guide to testing and research.

Neuropsychological Toxicology is written for neuropsychologists, physicians, and other health care providers. It is also intended for representatives of industry, government, and the law with an interest in the cognitive, emotional, behavioral, and neurological effects of neurotoxic substances. The first purpose of this book is to be a reference for clinical practitioners who must evaluate the history, symptoms, behavior, and neuropsychological functioning of individuals whose occupation, hobbies, or voluntary substance abuse exposes them to neurotoxic substances.

It is also written to provide information to members of the multidisciplinary team who seek to unravel the diagnostic, psychological, legal, and industrial complexities inherent to neurotoxic exposure evaluation. Thus, the book is focused on human psychological applications of toxicology research. Many active areas of toxicology could not therefore be included within this domain, including animal research, neurochemical investigations, and nonneuropsychological or neurological effects of toxic substances.

Finally, it is hoped that *Neuropsychological Toxicology* will provide resources and encouragement to those willing to undertake the research that is so greatly needed in this field.

Contents

Chapter 1

Introduction 1
Historical Antecedents of Neuropsychology 3
   Mental Testing 3
   Clinical and Experimental Psychology 4
   Behavioral Neurology 5
   Neuropsychology and Neurotoxicity: Early Start and Long
     Interruption 6
Modern Neurobehavioral Toxicology 7
Neuropsychological Testing Strategies 9
   Standardized (Fixed) Batteries 9
   Flexible Batteries 9
   "Process" Approaches 10
   Performance Testing 10
   Conclusion 11
Historical Views of Neurotoxicity 11
Principles of Neurotoxic Damage 12
   Synaptic Damage from Neurotoxicants 15
   Cellular Damage from Neurotoxicants 15
   Neurochemical Damage from Neurotoxicants 16
Legislative Issues and Neurotoxic Damage 21
Testing Subclinical Neurotoxic Effects 24
Advantages of Neuropsychological Toxicology Assessment 24
Unsolved Challenges in Assessment of Neurotoxic Syndromes 27
   Problems Inherited by the Neuropsychologist 27
   Challenges to the Neuropsychologist 28
Other Issues 30
   Economic Consequences 30
   Legal Consequences 31
   Political Impediments 31

Limited Numbers of Researchers and Clinical Practitioners	31
Barriers to Company, Worker, and Patient Acceptance	31
Neuropsychology in Toxic Substance Research	32
Who Is at Risk?	34
Developmental Neurotoxicity	37
Sex Differences	39
Occupational Risks	40
Other Risks	41
Symptoms of Concern	41
Summary	42

Chapter 2

Evaluation of Neurotoxic Syndromes — 45

Introduction	45
Initial Medical Evaluation	45
Health Questionnaire Construction	47
Areas of Inquiry	48
Neuropsychological Toxicology Test Battery Construction	51
Epidemiological versus Clinical Batteries	52
Clinical Neuropsychological Assessment of Neurotoxic Exposure	53
Clinical Testing Approaches	54
Neuropsychological Toxicology Test Batteries	59
WHO Neurobehavioral Core Test Battery	59
Pittsburgh Occupational Exposures Test Battery	60
California Neuropsychological Screening–Revised	61
Baker, Feldman, White, Harley, Dinse, and Berkey (1983) Battery	61
London School of Hygiene Battery	62
Institute of Occupational Health Battery	62
Tuttle, Wood, and Grether (1976) Battery	62
Valcuikas and Lilis (1980) Battery	63
Putz-Anderson *et al.* (1983) Battery	63
Smith and Langolf (1981) Battery	63
Williamson, Teo, and Sanderson (1982; Williamson, 1990) Battery	63
TUFF Battery	64
Neuropsychological Screening Test Battery	64
IMT Battery	65
SPES Battery	65
NES Battery	66
MANS Battery	67
Psychometric Assessment System/Dementia Screening Battery	67
MTS Battery	68
Naval Biodynamics Laboratory Battery	68

CONTENTS

Walter Reed Performance Assessment Battery	69
Other Computer Batteries	70
Individual Tests	71
Children's Tests	71
Interpretation of Neuropsychological Test Results	72
Summary	72
Testing Cautions	74
Complementary Approaches	75

Chapter 3

Metals	**79**
History	79
Variables That Affect Neurotoxicity	80
Substance	80
Developmental Influences	80
Acute versus Chronic Exposure	80
Individual Variation	81
Neuropsychological Testing of Metal Exposure	81
Aluminum	82
History	82
Individuals at Risk	82
Dialysis Dementia	83
Other Diseases with Possible Links to Aluminum	84
Occupational Exposure	84
Antimony	85
Arsenic	85
History	86
Current Uses and Individuals at Risk	86
Neurological Effects	87
Case Study: Long-Term Consequences of Short-Term Exposure	88
Barium	89
Bismuth	90
Cadmium	91
Chromium	91
Copper	92
Gold	94
History	94
Current Uses and Neurotoxic Effects	94
Iron	94
Lead	95
History	95
Current Uses	96

Intake	97
Neuropathology	98
Exposure	101
Prenatal and Childhood Exposure	102
Adult Lead Exposure	118
Lithium	125
Manganese	125
Neurological and Neuropsychological Aspects	126
Neurophysiological Aspects	127
Mercury	127
History	128
Modern Industrial Risks	130
Occupational Exposure	130
Nonoccupational Exposure	131
Neuropathology—Methylmercury	133
Neuropathology—Inorganic Mercury	135
Neuropsychological Effects	137
Teratogenicity	140
Nickel	140
Platinum	141
Selenium	142
Silicon and Silicone	142
Tellurium	143
Thallium	143
Tin and Organotins	144
Inorganic Tin	144
Organotin	145
Tungsten Carbide	146
Vanadium	147
Zinc	147
Other Metals	148

Chapter 4

Solvents	149
Introduction	149
Exposure	150
Neurophysiological Effects	151
Intake	151
General Toxic Effects	151
Neurotoxicity	152
Teratogenicity	152

Psychiatric Disorder	153
Neuropsychological Symptoms	153
Diagnostic Procedures	154
Neuroradiological and Neurophysiological Techniques	154
International Diagnostic Criteria for Solvent Syndromes	156
Diagnostic Process	158
Solvent Mixtures	159
Acute Effects	159
Chronic Effects	160
Solvent Abuse	186
Acute Effects	187
Chronic Effects	190
Individual Solvents	191
Acetone	192
Carbon Disulfide	192
Carbon Tetrachloride	196
Freon	197
Methyl Chloride	197
n-Hexane and Methyl Butyl Ketone	199
Methyl Isobutyl Ketone	201
Perchloroethylene	201
Styrene	205
Tetrabromoethane	209
Tetrahydrofuran	210
Toluene	210
1,1,1-Trichloroethane	216
Trichloroethylene	217
1,1,1-Trichloromethane	219
Xylene	220
Prognosis for Chronic Solvent Exposure	221
Conclusion	224

Chapter 5

Alcohol   225

Introduction	225
Acute Effects	226
Social Drinking	228
Chronic Effects	229
Prenatal Effects of Alcohol Exposure	229
Chronic Adult Alcoholism without Wernicke–Korsakoff Syndrome	231
Similarities and Interactions with Other Neurotoxic Exposures	233

Disease Pathology 234
Hepatic Dysfunction 235
Other Diseases 236
Evidence of Chronic Effects: Imaging and Electrophysiological Studies 236
Wernicke–Korsakoff Syndrome 242
Comparing Chronic Alcoholics with and without Wernicke–Korsakoff
  Syndrome 244
Chronic Alcoholism without Wernicke–Korsakoff Syndrome 245
Which Tests Show Impairment? 248
Individual Tests 248
Test Batteries 250
Factors Influencing the Neuropsychological Performance of
  Alcoholics 252
Recovery of Function 260
Neurological Recovery 260
Neuropsychological Recovery 262
Isopropyl Alcohol Abuse 265
Conclusions 265

Chapter 6

Drugs 267

Prescription Drugs 267
Anticholinergics 267
Anticonvulsants 268
Antidepressants 273
Antihypertensives 275
Benzodiazepines 275
Chemotherapy and Antineoplastic Drugs 282
Disulfiram (Antabuse) 283
Naloxone 285
Neuroleptics 285
Lithium 290
Muscle Relaxants 294
Steroids 295
Other Drugs 297
Abused and Nonprescription Drugs 297
Amphetamines 297
Amyl Nitrite 298
Caffeine 298
Cocaine 299
Designer Drugs 308
Glue Sniffing 309
LSD and Psychedelic Drugs 309

Marijuana 312
Opiates 320
Phencyclidine (PCP) 321
Polydrug Abuse 324
Sedative-Hypnotics 324
Conclusions—Neuropsychological Toxicology of Drugs 325

Chapter 7

Pesticides 327

Introduction 327
Routes of Exposure and Individuals at Risk 329
Organophosphates 329
   Neurological Effects 332
   Neuropsychological Effects 333
   Effects on Personality and Emotional Functioning 335
Dursban 336
Carbamates 340
Chlorinated Hydrocarbons 340
   DDT 342
Chlorinated Cyclodienes 343
   Chlordane 343
   Agent Orange and Dioxin 344
   Kepone 346
   Telone 346
Other Pesticides 346
   Pyrethroids 346
   Paraquat 347
Fungicides and Fumigants 347
   Chloropicrin 347
   Methyl Bromide 347
   Sulfuryl Fluoride 348
Exposure to Combinations of Pesticides 348
Herbicides 348

Chapter 8

Other Neurotoxins 349

Acrylamide 349
Anesthetic Gases 350
   Introduction 350
   Acute Neuropsychological Effects 351

Developmental Neurotoxicity 353
Nitrous Oxide 353
Arthrinium 354
Aspartame 355
Bromine 355
Butane 356
Carbon Monoxide 356
   Epidemiology 357
   Toxicological Effects 357
   Neurological and Neuropsychological Effects 358
   Neuropsychological Recovery 372
Cassavism 375
Chlorine 375
Cyanide 376
Cycads 378
Extremely Low-Frequency Electromagnetic Radiation 378
Eosinophilia Myalgia Syndrome 379
Ethyl Chloride 386
Ethylene Oxide 387
   Neuropsychological Effects 388
Foods 391
Food Additives 392
Formaldehyde 392
   Neuropsychological Effects 393
Jet Fuels 394
Hydrogen Sulfide 395
Hyperbaric Nitrogen 397
Ionizing Radiation 398
Lathyrus sativus (Neurolathyrism) 398
Ozone 399
Pentaborane 400
Phenol 400
Polychlorinated Biphenyls 401
   Developmental Effects 402
   Neurological Manifestations 403
Propane 404
Silicon and Silicone 404
Toxic Oil Syndrome 405

Chapter 9

Psychosomatic Disorders 407

Multiple Chemical Sensitivity 407
   Patient Profile 408

Etiology of MCS: Multiple Diagnostic Possibilities 409
Treatments 414
Sick Building Syndrome 415
Mass Psychogenic Illness 416
Workplace Stress 417
Posttraumatic Stress Disorder 418
Chronic Technological Disaster 420

Chapter 10

Forensic and Private Practice Issues 421

Toxic Torts 421
Role of the Neuropsychologist as Expert Witness 421
Applying Research Data to Clinical Cases 428

References 431

Index 503

Neuropsychological Toxicology
Identification and Assessment of Human Neurotoxic Syndromes

Second Edition

1

Introduction

There is growing recognition that many industrial and pharmacological agents are *neurotoxic* to the nervous system and its neuropsychological functions. For an agent to be considered neurotoxic it must produce an "adverse change in the structure or function of the nervous system following exposure to [that] agent" (*Neurotoxicity*, 1990). More than nine million individuals come into contact with neurotoxins in the workplace (*Current Intelligence Bulletin*, 1987). However, neurotoxic exposure is not confined to dangerous industrial environments but may also occur at home, in white-collar businesses, during voluntary substance abuse, through prescribed medication usage, or even in the foods or supplements we consume. Neurotoxic substances may be chemically manufactured (e.g., solvents) or naturally occurring (e.g., metals). Neurotoxic substances are as common and well known as lead or mercury, and as obscure as pyrethroids or cycads. There is exposure risk in certain foods, e.g., fugu, the Japanese puffer fish, which, when incorrectly prepared, is fatally neurotoxic to the unfortunate gastronome. Certain common available vitamins are neurotoxic in megadoses, and unregulated folk medicines available in Latin-American botanicas have been shown to contain mercury and other neurotoxic compounds (Wendroff, 1990). L-Tryptophan, an over-the-counter amino acid sleep aid, caused neurotoxic injury in a subset of more than 1500 individuals affected, in the form of *eosinophilia myalgia syndrome*, when a contaminant was apparently introduced into the distillation process, causing central nervous system damage and multiple organ system abnormalities.

The neurotoxicity of other widely distributed substances has not been investigated. For example, the vast majority of food additives and cosmetics remain untested for their neurotoxic potential, this despite development of peripheral neuropathy in users of a white-lead-containing face powder (see Chapter 3) and the discovery that AETT (acetylethyl tetramethyl tetralin), a fragrance component since 1955, was found to cause "degeneration of neurons in the brains of rats and marked behavioral changes ... including irritability and aggressiveness" (*Neurotoxicity*, 1990, p. 54). Withdrawn from the market in 1978, the chronic effects of 23 years of human application will "probably never be known" (p. 54).

Unfortunately, most physicians and other health care professionals have had little training or experience in identifying either the structural or functional components of neurotoxic damage. Spencer (1990b) correctly criticizes the "artificial compartmentalization" that hinders researchers of workplace chemicals, abused drugs, alcohol, and the like from becoming aware of one another's research and addressing the real-world effects of interactive exposures. Most neuropsychologists, until recently, were part of this compartmentalization and not well aware of the behavioral, cognitive, and emotional patterns of nervous system dysfunction caused by exposure to toxic substances. Neuropsychological study of these patterns is so recent that not even its name remains constant among researchers. It has been called *behavioral toxicology, neurobehavioral toxicology, psychological toxicology,* or simply *neurotoxicology,* but these distinctions probably have more to do with the aesthetic preferences of the researchers than objective differences in investigation.

Nevertheless, neuropsychologists have steadily increased their research and clinical involvement in neurotoxic exposure issues. Neuropsychology's unique contribution to neurotoxic exposure investigations is its emphasis on quantified analysis of behavior, cognition, and affect, thereby providing an accurate measure of one part of the chain of "the multidisciplinary understanding of the nervous system and its manifold responses to chemical substances" (Spencer, 1990b, p. 14). Neuropsychologists who participate in such investigations routinely utilize research and clinical data from diverse (and sometimes divergent) specialties, including neurology, occupational medicine, psychiatry, public health administration, epidemiology, industry, and pharmacology. Whether the data are derived from *in vitro* or *in vivo* studies, all neurotoxin research, from the cellular to the sociological, can be considered branches of this field.

Regardless of the name chosen, this area of specialization explores clinical and theoretical aspects of "changes in behavior under the influence of certain natural or artificial substances" (Richelle, 1983, p. 127). When additional inferences are made about the relationship between substance-induced behavior change and nervous system alteration or toxicity, the term "neurobehavioral" toxicology is often employed.

What's in a name? This book has been titled *Neuropsychological Toxicology* not because yet another name for this field is required, but rather to emphasize the relevance of neuropsychological theory and testing procedures to the understanding of overt and subclinical effects of toxicants on the nervous system. Substituting the more comprehensive "neuropsychological" in place of "neurobehavioral" seems to address the wide scope of the field. "Neuro*behavioral*" connotes the restrictive limitations of directly observable action in an organism or the stimulus–response theories of Pavlov or Skinner. "Neuro*psychological*" recognizes the more global and holistic aspects of the human psyche, a domain that includes the concepts of cognition, affect, *and* behavior, both subjectively experienced and objectively measurable.

The psychology of the human organism, its cortical functions and objectively measured correlates of those functions, are well within the purview of clinical neuropsychology. So too is the assessment of toxic damage to those structures and functions. What's in a name, therefore, is the identification of a field where neuropsychological researchers may have a unique and valuable contribution: assessing observable alterations in human cognition, affect, and behavior when the nervous system is exposed to toxic substances. To call that branch of neuropsychology concerned with toxic substances *neuropsychological toxicology* is to gather previously unrelated areas of investigation under a single conceptual banner. Thus, neuropsychological evaluation of alcohol, drugs (abused and prescription), toxic industrial substances, pesticides, and as yet unresearched substances like biotoxins and allergens can all be considered part of the same field.

The value of the neuropsychologist's contribution to the assessment of neurotoxicity has been concisely summarized by Lezak (1984), who states that the "neuropsychological assessment probably offers the most sensitive means of examining the effects of toxic exposure, of monitoring for industrial safety, and of understanding the complaints and psychosocial problems of persons exposed to these toxins" (p. 28).

Definition. To appreciate Lezak's enthusiastic endorsement of neuropsychological methods in toxicological investigations, a brief definition and synopsis of modern neuropsychology may be helpful. Neuropsychology is the study of the relationship between nervous system structure and its corresponding behavioral function. Just as the heart pumps blood, the brain can be conceived of as a "pump" of behavior. Quantified measures of that output form the bulwark of *clinical neuropsychology,* a subdiscipline of psychology that employs specialized tests to correlate neuropathology with cognitive and behavioral abnormality. In clinical diagnosis, then, neuropsychological methods are employed to determine how patterns of cognition, emotion, and behavior covary with brain or nervous system damage.

HISTORICAL ANTECEDENTS OF NEUROPSYCHOLOGY

The actual term "neuropsychology" was apparently first used by Sir William Osler in 1913 at the opening address of the Phipps Psychiatric Clinic of Johns Hopkins Hospital (Bruce, 1985); however, clinical neuropsychological methods developed over the past century as an outgrowth of research in several areas, principally mental testing, clinical and experimental psychology, behavioral neurology, and psychiatry.

Mental Testing

During the 19th century, the mental testing movement sought to develop quantified measures of individual ability, with the original aim of optimally educating French schoolchildren with differing intellectual abilities. Without knowing the specific areas of brain activity measured by these tests, mental testing advocates nevertheless developed a complex set of neurobehavioral tests that assessed a wide variety of mental functions, including memory, motor skills, speech and language, attention and concentration, spatial skills, and many other components of the modern neuropsychological evaluation. The three contributions of the mental testing movement relevant to clinical neuropsychology were (1) the development of neuropsychological measurement apparatus, (2) norm-based assessment of individual differences, and (3) statistical data analysis of those differences (Hartman, 1991).

Clinical and Experimental Psychology

Early 20th century clinical and experimental psychologists developed methodology for psychometric study of abnormal mental states, including neurotoxic exposures; they were no doubt encouraged by an "increasing interest and value in the differentiation of original mental defect from mental deterioration, and the demonstration of special abilities and handicaps" (Woolley & Hall, 1926, p. 412). Psychologists answered this call by describing neurological and psychiatric abnormalities with familiar psychometric tools. Subjects of investigation included natural and artificially induced states of toxic impairment, effects of mental disease on memory, and cognitive effects of physical activity, fasting, and other general health impairments (e.g., Hoch, 1904). Nascent neuropsychological investigations were conducted on diagnostic groups in common use at the time, including dementia praecox, alcoholic psychosis, chorea, epilepsy, syphilis, hysteria, hypopituitarism, and other disorders (Cotton, 1912; Yerkes, Bridges, & Hardwick, 1915).

One of the most important promoters of modern neuropsychological concepts was Shepherd Ivory Franz (1874–1933). Franz, psychologist, honorary physician, and teacher of Lashley, advocated standardized clinical and psychometric examination of mental status using psychological and psychophysiological methods. Beginning in 1910, he conducted a series of lectures and demonstrations of mental examination methods at the Government Hospital for the Insane so that "the mental examination of patients may be conducted in a more systematic and scientific manner." Franz did not merely champion the use of tests for detection of arbitrary mental functions, but actively sought interactions between neurological or psychiatric abnormalities and performance on psychometric tests. This methodology was summarized in his 1919 text, *Handbook of Mental Examination Methods,* where Franz presented tests for tactile localization, fine and gross motor coordination, praxis, language and speech, attention, vigilance, memory span, verbal and nonverbal recall, visuospatial skills, abstract reasoning, and general intelligence. Franz emphasized normative assessment and the importance of correlating the "facts" with the "environment." He expected test results to be interpreted in the context of extensive medical and psychological anamnesis. The debt modern neuropsychology owes to Franz is summarized in his obituary, which concluded that Franz "bridged the gap between experimental psychology on the one hand, and neurology and psychiatry on the other" (Fernberger, 1933).

Behavioral Neurology

Behavioral neurology and the aims of 19th century clinical neurology also contributed to the development of neuropsychology. Studies by Adolf Meyer, Karl Kleist, Walther Poppelreuter, and others contributed to a growing case literature on behavioral impairments and their relationships to specific types of head injuries (Levin, 1991).

Modern neuropsychology awaited the synthesis of localized brain impairment case studies with a theory of holistic cortical function. The English neurologist J. Hughlings Jackson developed such a comprehensive theory in the late 19th century, although his works were not published in the United States until 1958 (Golden, 1978). Jackson proposed that higher mental functions were not indivisible abilities, but were themselves composed of more basic skills. The ability to write, for example, depends not on a "writing center" but rather a combination of interrelated functions, including fine motor coordination, visual perception, spatial localization, kinesthetic-proprioceptive feedback, and so forth.

Jackson's theory combined the essential features of localization and equipotentiality. Brain function was considered to be a collection of basic abilities, and therefore localizable. However, since higher forms of behavior require these simple functions to interrelate, any higher cognitive ability must be considered as the product of the whole brain.

Alexander Luria has been the most recent and well-known proponent of a "dynamic localization of function model," where "any human mental activity is a complex functional system effected through a combination of concertedly working brain structures, each of which makes its own contribution to the functional system as a whole" (Luria, 1973, p. 38). Luria's functional systems are ever-changing interaction patterns of the basic brain functions necessary to complete a given behavior. With his extensive experience assessing the cognition, affect, and behavior of brain-injured patients, Luria extended Jackson's ideas by positing the existence of "primary, secondary and tertiary" areas of interaction. Primary areas work as analyzers of rudimentary sensory and perceptual input; secondary areas perform integration and elaboration of the input from primary areas; and tertiary areas, in turn, integrate the output of secondary areas, resulting in the most complex forms of cognition (e.g., reading, abstract reasoning).

Assessment of damage or recovery of these processing systems is accomplished by investigating patients' performance on standardized neuropsychological tests. While most modern neuropsychologists subscribe to a theory of brain–behavior relationships similar to Luria's, different strategies of assessment have been developed to elucidate these relationships.

Thus, clinical neuropsychology, cross-fertilized by the overlapping medical and psychological disciplines, developed as a clinical science to study the behavior of the individual as influenced by both neurological and psychological phenomena.

Neuropsychology and Neurotoxicity: Early Start and Long Interruption

Given the complex, multidisciplinary roots of neuropsychology, it might be expected that clinical toxicologic investigations would have been an important and continuing topic of investigation. In fact, promising investigations that might have developed into neuropsychological toxicology were performed as early as the late 1800s. Cognitive functions were assessed under various states of artificially induced neurotoxicological impairments, produced by alcohol, paraldehyde, trional, bromide of sodium, and caffeine (Hoch, 1904). Surprisingly, then, a proposal for functional testing of neurotoxicity did not reappear until 1963 with Ruffin's groundbreaking article entitled "Functional Testing for Behavioral Toxicity: A Missing Dimension in Experimental Toxicology," which advocated the use of psychological methodology to assess effects of neurotoxic exposure. Psychologists did not act on Ruffin's recommendation until Helena Hänninen's pioneering studies of carbon disulfide exposure in the 1970s—a 90-year gap between Hoch's 1904 review and the present.

There are several reasons for such a long interregnum, the most important being that the focus of mid-20th century neuropsychology turned away from analysis of diffuse impairments typically produced by neurotoxic exposure and toward the use of neuropsychological tests as "noninvasive methods to provide information regarding the presence or absence of brain damage, lateralization, localization, etiology of the cerebral dysfunction, and prognosis" (Horton & Wedding, 1984, p. 10). Neuropsychologists Ward Halstead and Ralph Reitan strongly influenced the growth and character of U.S. neuropsychology by emphasizing not only the detection of brain damage, but also the development of tests to localize focal higher cortical lesion. Matarazzo (1972) noted that "apparently so sure was Halstead that the psychological assessment techniques he and others were developing could help in effectively differentiating the brain-damaged individual from the patient who was not brain-damaged that he soon began to concentrate his research efforts, instead, on attempting to localize more precisely such lesions within the brain" (p. 384).

Halstead's belief that his tests "enabled him to predict lesions in the brain with great accuracy" (Parsons, 1986, p. 156) essentially defined the domain of neuropsychological inquiry for several decades, as lateralization, localization, and lesion detection became central to the discipline. "Diffuse" disorders, with the exception of government-funded alcohol research, were less frequently studied. More to the point, since neuropsychological abnormality was typically characterized as focal versus diffuse, diffuse disorders were not considered to be conceptually distinct from one another, and were therefore less obvious topics of investigation. By their failure to produce obvious focal effects on higher cortical functioning, neurotoxic disorders did not fit into the paradigm of the time and may have been unfairly slighted for research and clinical attention. The few early descriptive studies that were performed (i.e., investigations of carbon disulfide) did not provide sufficient momentum for further research (e.g., Raneletti, 1931; Pennsylvania Department of Labor and Industry, 1938).

MODERN NEUROBEHAVIORAL TOXICOLOGY

While neuropsychological toxicology has had a low profile and a late start in the U.S. scientific community, it has slowly grown in visibility among scientists, the general public, legislators, and union representatives. In the occupational medicine community, Joseph Ruffin, an industrial physician at Kaiser Steel Corporation, was one of the first to explicitly call for linking neuropsychological methods with industrial concerns in his article "Functional Testing for Behavioral Toxicity: A Missing Dimension in Experimental Toxicology" (1963). While much groundbreaking research accumulated during the next 20 years, a 1984 report sponsored by the National Toxicology Program reported that it was remarkable how few chemicals had been subjected to systematic behavioral analysis (Buckholtz & Panem, 1986). The same report identified the great need for neurobehavioral toxicology assessment of pesticides, cosmetics, food additives, and other commercial chemicals.

The capacity of industry to produce these materials has far outstripped basic research and clinical knowledge. The more than 1000 new compounds developed each year are added to the approximately 40,000 chemicals and 2,000,000 mixtures, formulations, and blends already in industrial use. There are more than 65,000 toxic chemicals in the Environmental Protection Agency's registry, and EPA receives about 1500 notices by manufacturers each year of intent to produce new products (*Neurotoxicity*, 1990). Three to five percent of these industrial chemicals, excluding pesticides, are potentially neurotoxic (O'Donoghue, 1986), which translates to 1800 to 3000 substances "most of which have never been tested for this [neurotoxicity] endpoint" (Claudio, 1992, p. 203). Approximately 1925 million pounds of substances already listed as neurotoxic are released annually into the air (Spencer, 1990c).

There are an estimated 32,000 hazardous waste sites, with an average of 6000 people living nearby each and whose health may be affected by the solvents and heavy metals they contain. More than 2,000,000 children, women of childbearing age, and elderly persons live near waste sites designated as national priorities for cleanup (Amler & Lybarger, 1993).

At least 600 pesticide ingredients are catalogued by EPA, a large percentage of which are neurotoxic. Very frequently, the neurotoxic or other toxic properties of new compounds have not been recognized before their introduction to the market (Landrigan, Kreiss, Xintaras, Feldman, & Heath, 1980). Moreover, with inadequate data on the toxic effects of more than 75% of chemicals produced in excess of 1 million pounds per year (Tilson, 1990), the need for neurotoxicity information is present and compelling.

In 1987, there were between 1023 and 1707 deaths that could be definitively linked to workplace neurotoxic exposure in the United States (Hessl & Frumkin, 1990). Neurotoxicity is considered by NIOSH (National Institute for Occupational Safety and Health) to be one of the ten leading causes of work-related disease and injury (*Neurotoxicity*, 1990). The number of disabilities or nonfatal impairments have not been tallied and would almost surely be underestimated, especially if diseases with unknown etiology are linked to neurotoxic exposure. It has been suggested, for example, that amyotrophic lateral sclerosis, parkinsonism, and some Alzheimer-type dementias may be linked to chronic, insidious effects of neurotoxic agents (Wechsler, Checkoway, & Franklin, 1991; Tilson, 1990). These forms of neurotoxic damage may not become manifest until exposure is long past (Spencer, 1990a), making epidemiological studies a necessity.

The problems of neurotoxic exposure are even more apparent in developing countries where economic concerns outweigh health concerns. About 100,000 workers in Korea are exposed to solvents (Lee & Lee, 1993). Workers in Mexico do not have the right to know what chemicals they use in the workplace (Ramirez et al., 1990), and in many Latin American countries, the materials themselves are unlabeled (Gimenez, Rodriguez, & Capone, 1990; Trape, 1990). In Mexico City, there are about 500,000 solvent addicts, many of them children (Cabrera, 1990). The World Health Organization (WHO, 1990) has estimated that approximately 3,000,000 cases of acute severe (hospitalizable) pesticide poisonings occur annually on a global level, with possibly even greater numbers of unreported mild intoxications and acute conditions. Jeyaratnam (1993) calculates that there are 220,000 pesticide-related deaths each year (including suicides) worldwide. Developing countries use only 20% of the agricultural chemicals produced but account for almost 99% of acute pesticide poisoning deaths (Jeyaratnam, 1993).

Other countries are somewhat ahead of the United States in recognizing the importance of neurotoxicity analyses. European and Scandinavian countries lead the United States in basic research, clinical application, and legislative support of neuropsychological toxicity investigations. In 1982, there were only about 700 occupational health researchers *including* neuropsychologists active in the United States, a country with over 60 million workers. These figures should be contrasted with those of Finland, a country with only 2 million people but which employs 500 occupational health researchers (Andersen, 1982). There are probably less than 50 full-time research and clinical practitioners of human neuropsychological toxicology in the United States at this time.

It is not surprising, given these statistics, to note that most of the basic neuropsychological research performed in the past two decades has been performed in European and Scandinavian countries and published there as well. In addition, Scandinavian countries have recognized the neuropsychological effects of one class of neurotoxic poisoning, "organic solvent syndrome," as a disability claim for over 30 years. Full compensation is awarded for symptoms of memory loss, lowered concentration, and loss of initiative as a consequence of chronic solvent exposure. "In [the United States], we're still debating whether [solvent syndrome] exists" (Wood, in Fisher, 1985, p. 14). Japan, Britain, and the European Economic Community, but not the United States, each have prescribed behavioral screenings to assess developmental neurotoxicity of new drugs (Tilson, 1990). Ongoing, multisite studies in North America and Europe may eventually produce similar standards in this country. Undoubtedly, differences in legal systems, insurance, and employee health policies have contributed to such differences.

Fortunately, an increasing amount of attention is being paid in the United States to these issues, and neurotoxicity testing guidelines at EPA are currently being revised (Tilson, 1990). Several journals now routinely publish neuropsychological investigations of toxic substance exposure. These include: *Environmental Research, Scandinavian Journal of Work Environment and Health, American Journal of Industrial Medicine, British Journal of Industrial Medicine, International Archives of Occupational and Environmental Health, Neurotoxicology and Teratology, NeuroToxicology,* and *Journal of Occupational Medicine.* Occasional articles can be found in related journals, including the *International Journal of Clinical Neuropsychology, Acta Neurologica Scandinavica, Acta Psychiatrica Scandinavica, Clinical Toxicology, Journal of Neurology, Neurosurgery and Psychiatry,* and others. U.S. neuropsychology journals including the *Journal of Clinical and Experimental Neuropsychology, The Clinical Neuropsychologist,* and *Archives of Clinical Neuropsychology* are also beginning to become more regular sources of research studies and case reports.

NEUROPSYCHOLOGICAL TESTING STRATEGIES

During the past four decades several different neuropsychological testing strategies have evolved:

Standardized (Fixed) Batteries

Single tests or groups of tests for which norms and impairment indices are available are used to compare a patient against ipsative or population norms. These tests (e.g., the Halstead–Reitan, the Luria–Nebraska) typically attempt global assessments of neuropsychological functions, although there are also standardized batteries for more specialized functions (e.g., motor steadiness batteries).

Flexible Batteries

The most widely used type of clinical neuropsychological investigations in the United States (Sweet & Moberg, 1990) consist of standardized neuropsychological tests developed separately by various researchers but collected and presented together as part of a larger examination called a "flexible battery." Some flexible batteries provide global assessment of neuropsychological functions (e.g., Wysocki & Sweet, 1985). Others could be compiled to assess more specific functions or disorders. For example, a "dementia battery" might load most heavily on memory tests, while a battery designed for left hemisphere cerebrovascular disorders would investigate language and speech more extensively. Flexible batteries are the most frequently used approach in human neurotoxicity investigations.

"Process" Approaches

Somewhat less methodologically rigorous than practitioners of the foregoing approaches, advocates of the process approach emphasize that the patient's approach to a neuropsychological task is at least as important as whether the task is passed or failed. Process approach advocates rely on the expertise of the clinician to intersperse standardized tests and impromptu procedures in the service of identifying the qualitative data leading to a correct or incorrect answer. The examiner subjectively interprets and integrates the continuum of patient behavior to propose a diagnostic formulation. More subjective than battery approaches, the process approach is a paradigm of the expert; the clinician's experience and sharp eye is at least as important as the result of the normed procedures. Thus, while less formally rigorous than the preceding evaluations, the process approach is valuable for insisting that variables including the patient's approach to testing, motivation, medical history, and other ipsative information are as important in their own ways as the statistical properties of the exam (e.g., Walsh, 1985).

Performance Testing

The history of performance testing overlaps with flexible battery approaches in that there is an attempt to develop a set of flexible, valid measures to assay behavior under different neurological or physical conditions. Rather than focusing on relatively static neurobehavioral correlates of medical or neurological disorders, performance testing has sought to analyze changes in behavior as a function of acute stressors. It follows that the methodologies developed by each set of practitioners would be quite different. For example, conventional neuropsychological measures usually strive to take a one-time "snapshot" of behavior, to compare against norms based on large numbers of naive subjects. Performance testing, by contrast, often utilizes small numbers of subjects with extensive training on a given test apparatus, and deviations from optimal performance under acute changes in test conditions are measured against the subject's own asymptote. The need for repeated presentations and collection of large numbers of data points has led to the adoption of computerized methods by performance testing researchers. Consider and contrast traditional, single-administration, paper-and-pencil, cognitive and neuropsychological measures with the task developed by Perez-Reyes, Hicks, Bumberry, Jeffcoat, and Cook (1988) for testing drug effects. A computer "displayed two-digit numbers in the central field of vision that changed approximately every half-second.... If these numbers exceeded a critical value (57), the subject was instructed to press button 3.... If these numbers fell below a critical value (53), then the subject was to press button 2.... Simultaneously, the peripheral displays ... were programmed to display one of four digits.... If either the left or right peripheral display changed from the steady value of 4 to a 5, the subject was instructed not to respond.... If the value changed to a 3, the subject was to press a corresponding left number

1 or right number 4 response button. . . . If the value changed to a 7 in either of the peripheral displays, the subject was to press [a] foot pedal" (p. 269).

Defense Department researchers have utilized performance testing methodology to determine behavioral effects of heat stress, low-oxygen atmospheres, and other conditions capable of influencing battlefield performance. The effects of subtle environmental influences on performance may have great relevance to neuropsychological toxicology, since these methods encourage repeat testing and require the same sort of complex whole brain activity that is often found to be impaired under neurotoxic insult.

Conclusion

These clinical approaches, combined with ongoing theoretical advances in understanding neuroanatomical–behavioral relationships, have already enabled neuropsychologists to make clinical inferences as to the nature of damage and recovery of brain injuries, including open- and closed-head injury, cerebrovascular disorders, dementias, and a variety of other impairments of brain function. The extension of these methods to trace the effects of nervous system toxicants is a logical and reasonable extension of clinical neuropsychology.

HISTORICAL VIEWS OF NEUROTOXICITY

The recency of neuropsychology's entry into toxicological research parallels perhaps the very late development of the science of industrial medicine. In fact, the history of neurotoxic exposure, prevention, and research has until recently been the province of informal literary anecdote coupled with formal medical neglect.

Anecdotal accounts of neurotoxins and their effects have been chronicled for centuries. Lead, perhaps the oldest human neurotoxin, was a subtle and ubiquitous poison for the Romans, who employed lead oxide to sweeten and preserve wine, cider, and fruit juices (Fein, Schwartz, Jacobson, & Jacobson, 1983, p. 1189). In 1535, Anglicus Bartholomaeus wrote of mercury, "The smoke thereof is most grevous to men that ben therby. For it bredeth the palsey, and quaking, shakynge, neshynge [softening] of the synewes" (Bartholomaeus, cited in Goldwater, 1972, p. 91). Carbon disulfide, a solvent used in the cold vulcanization of rubber, has citations in the medical literature since 1856 (Wood, 1981).

Unfortunately, this long anecdotal chronology did not have the effect of increasing medical knowledge or curiosity, and even as late as the beginning of the 20th century industrial medicine did not exist in the United States (Hamilton, 1985). Myths were allowed to stand without question, and body cleanliness was emphasized above environmental management of fumes and dust. "I remember the head surgeon of a great Colorado smelting company saying in a public meeting: 'It is not the lead a man absorbs during his work that poisons him but what he carries home on his skin'" (Hamilton, 1985, pp. 3–4).

It took detailed studies of lead's respiratory absorption (Aub, Fairhall, Minot, & Reznikoff, 1926) to finally convince U.S. physicians that workers required better advice than to "scrub their nails carefully for protection against metallic poisoning" (Hamilton, 1985, p. 4).

Other misconceptions have remained until relatively recently. For example, medical disease models of symptomatology carried over from the previous century led researchers to assume that neurotoxins exerted their effects in an "all or none" fashion. That there was such a thing as "subclinical" toxicity did not become apparent until the Minamata methylmercury poisoning incident in 1956 (Fein *et al.*, 1983) in which asymptomatic individuals contaminated by the output of a vinyl acetate plant continued to die from mercury poisoning long after the discharge was halted. Thus, these individuals were only *apparently* asymptomatic. Subtle signs of poisoning and continuing neuronal changes, which were initially invisible, continued to progress until they reached clinical threshold, without additional exposure (Fein *et al.*, 1983). Tragic events like Minamata redefined the domain of inquiry in industrial medicine, highlighting the need for research into, and tests of, subclinical neurotoxicity.

These tragedies notwithstanding, neurotoxicity continues to be underemphasized relative to other health concerns. One source proposed that a combination of ineffective legislation, weak regulatory agencies, and the low priority of occupational medicine within the medical community may continue to account for the underemphasis given neurotoxic syndromes (Hessl & Frumkin, 1990).

PRINCIPLES OF NEUROTOXIC DAMAGE

The influence of neurotoxic substances on behavior is the end product of biochemical, structural, and functional interactions on the human organism (Spencer, 1990b). Accordingly, understanding human neurotoxic poisoning requires some knowledge of the toxicological–neurological–general medical axis that forms the set of intervening variables interposed between the initial exposure and the final behavioral impairment. It is important to note, however, that "there is no unitary hypothesis for the mechanism of action of neurotoxicants. The nervous system consists of many different cell types; its functional state and sensitivity to exogenous influences are highly dependent upon developmental stage, and the fundamental structural components of such complex behaviors as learning are still largely unknown" (Silbergeld, 1990, p. 136).

Human neurotoxic damage can occur *directly* from toxic injury to the neuron, and/or *indirectly* insofar as injury to other body systems (e.g., pulmonary, renal) produces secondary neuronal damage with consequent neuropsychological dysfunction (Tarter, Edwards, & Van Thiel, 1988). Second, loci of neurotoxic disorder may occur principally in the (a) peripheral nervous system (PNS), where effects include segmental demyelination and axonal degeneration (e.g., hexane, *n*-hexane, methyl butyl ketone), (b) central nervous system (CNS) either by direct toxic effects on CNS neurons or by disruption of neurotransmitter metabolism (e.g., mercury, manganese, organophosphate pesticides, lead), or (c) combined CNS and PNS effects with degenerative neuropathology in both nervous systems (Baker, 1983a,b).

It has recently been proposed that, analogous to cancer development, neurotoxic exposure produces a syndrome characterized by "silent" subclinical neurological injury. Injury to nervous system components accumulates for months, or even years, but may not be apparent, except in the context of specialized neuropsychological or neurological tests, until it blossoms into a clinical syndrome many months or years later (Calne, 1991; Reuhl, 1991). Most of the evidence for neurotoxin-caused human neurodegenerative diseases is inferential, rather than direct causal, with inferences made from the observed similarity of certain neurotoxic effects to well-known degenerative diseases (e.g., MPTP or manganese-induced parkinsonism), and the fact that lesions produced in earlier developmental states (in animals) may not become apparent until the subject ages and age-related cell reductions deplete neuronal reserves below needed capacity. This latter finding has been termed the *event threshold* concept, and it may be particularly important in neurotoxin-induced parkinsonism, where "it has been estimated that loss of approximately 80% of dopaminergic neurons in the substantia nigra is required before a patient becomes clinically parkinsonian" (Reuhl, 1991, p. 343). Nevertheless, Reuhl states that "it is probable that silent toxicity does occur but is detected only relatively infrequently" (p. 342) and that "the possibility of silent damage, neural injury simply awaiting age or unmasking to be expressed, must be seriously considered in any neurotoxicology evaluation" (p. 345).

A variety of mechanisms may be responsible for neurotoxic injury, including neuropharmacological or neurodegenerative processes. For example, Spencer (1990a) suggests that three types of alterations in neural function are responsible for neurotoxic damage: (1) alterations of the excitable membrane, (2) interference with neurotransmitter systems, and (3) structural breakdown of the dendrite, perikaryon, or axon. For example, lead, thallium, and triethyltin may produce neurotoxic damage by interfering with or destroying the myelin sheath (Williams & Burson, 1985).

Anoxia is a frequent mechanism of neurotoxic interference with normal cell operation, since neurons are especially sensitive to oxygen deprivation and need as much as ten times the oxygen of nearby glial cells (Ruscak, Ruscakova, & Hager, 1968). Cell death can occur within minutes, either by decreasing the amount of oxygen carried in the blood, slowing the blood flow, or by blocking the utilization of oxygen. Norton (1986) described three types of anoxia resulting from neurotoxic exposure:

- *Anoxic anoxia:* from inadequate oxygen supply in the presence of adequate blood flow. Carbon monoxide poisoning is a common example, since oxygen is preferentially replaced in the hemoglobin molecule by unusable CO.
- *Ischemic anoxia:* any loss of oxygen caused by decrease in arterial blood

flow. Any substance capable of causing hypotension or interfering with cardiac function may indirectly cause nervous system damage via ischemic anoxia. Cyanide's capability of causing hypotension is an example.
- *Cytotoxic anoxia:* the result of direct interference with cell metabolism while oxygen and blood supply remain normal. Cyanide is also a culprit in cytotoxic anoxia, causing damage to both gray and white matter.

Each type of anoxia is capable of gradually increasing cellular damage, via loss of mitochondrial granules and edema. Cerebral edema, in turn, further worsens hypoxia and results in accumulation of lactate, ammonia, and inorganic phosphates.

The propensity of many neurotoxic materials to be *lipophilic* ("fat-loving" or tending to accumulate in fatty tissues) puts the brain at special risk since lipids comprise 50% of the dry weight of the brain, compared with 6–20% of other body organs (Cooper, Bloom, & Roth, 1982). In addition, the unique structural properties of the nervous system make it especially vulnerable to neurotoxic insult.

Selective Vulnerability of the Nervous System to Neurotoxic Damage (after Spencer, in *Neurotoxicity,* 1990):

1. Unlike other cells, neurons normally cannot regenerate when lost; neurotoxic damage to the brain or spinal cord, therefore, is usually permanent.
2. Nerve cell loss and other degenerative changes in the nervous system occur progressively in the second half of life and thus toxic damage may interact synergistically with aging effects.
3. Many neurotoxic materials cross the blood–brain barrier.
4. Certain regions of the brain and nerves are directly exposed to chemicals in the blood.
5. The architecture of nerve cells, with their long axon processes, exposes vast amounts of surface area to toxic interference or degradation.
6. The nervous system is highly dependent on a delicate electrochemical balance for proper communication. Any chemical capable of disrupting this balance can disrupt nervous system function.
7. Neurological, behavior, and other body functions may be profoundly disrupted by impairment or damage to even minor areas of the nervous system.

Neurotoxic damage varies with substance and exposure, with disruptions capable of occurring at any point along the biochemical and structural apparatus of the cell. Some substances interfere with intracellular biochemistry, causing changes in cell acidity, protein synthesis, or fluid dynamics. Others appear to selectively damage myelinated axons, interfering with neurotransmission. Damage may occur at specific sites in the cell or to the overall cell structure.

Synaptic Damage from Neurotoxicants

In the synapse, degradation of neurotransmission can occur at any level of *presynaptic* process, including synthesis, storage, release, and termination of neurotransmitters (Atchison, 1989). Neurotoxicants may also affect the nerve on the *receptor* end of the synapse, including transmitter binding to the receptor cell, "activation of the receptor associated ionic channel, and degradation of chemical transmitter" (Atchison, 1989, p. 393). Lead has a direct effect on synaptic action by presynaptic block of the end-plate potential and may also interfere with enzyme inhibition at several sites (Goetz, 1985).

Cellular Damage from Neurotoxicants

CNS and PNS structures appear to be differentially sensitive to toxic damage (O'Callaghan, 1989). In animal studies, it is possible to measure differential responses of nervous system cells to toxic damage by measuring the proteins unique to each type of cell. *Trimethyltin,* for example, causes extensive damage to certain areas of the rat hippocampus and cell protein degradations unique to those areas can be observed (O'Callaghan, 1989). One of the principal neurotoxic effects of *mercury* is related to that metal's inhibition of protein and RNA synthesis. *Toluene,* in rat brain studies, produces a 50% reduction of catecholaminergic neurons after 4 weeks of exposure to 250 or 1000 ppm toluene (Bjornaes & Naalsund, 1988). Sixteen weeks of exposure to toluene in rats produces "marked" changes in cerebellar and spinal cord proteins. Neuron-specific enolase (NSE), creatine kinase-B (CK-B), and β-S100 were elevated by 15, 20, and 55%, respectively. In spinal cord CK-B was reduced by 13%, while β-S100 increased by 35% (Huang, Kato, Shibata, Asaeda, & Takeuchi, 1993). Since CK-B and β-S100 are distributed primarily in the glial cells, their increase during toluene exposure may reflect the development of gliosis, an early sign of CNS (Huang et al., 1993).

Human studies, while not as detailed, have suggested the existence of similar cellular damage. For example, Rosenberg, Spitz, Filley, Davis, and Schaumberg (1988) performed magnetic resonance imaging (MRI) scans of 11 chronic toluene abusers. Three subjects had abnormal MRIs with "diffuse cerebral, cerebellar, and brainstem atrophy" that ranged in severity from mild to marked. Ventricular dilation was also seen, and other MRI findings suggested severe white matter damage. No improvement was observed on repeated MRI scans over an 18-month follow-up period, suggesting that toluene-induced physical injury to the brain may be irreversible.

Lead and *mercury* impair the functions of brain astrocytes, which control ionic and amino acid concentrations, brain energy metabolism, and cell volume (Rönnbäck & Hansson, 1992). Both lead and mercury have been shown to inhibit astroglial capacity to take up glutamate, causing secondary decrease in other neurotransmitters and corresponding increase in patient report of fatigue and loss of alertness (Rönnbäck & Hansson, 1992).

Subcellular components are also differentially sensitive to neurotoxic damage. Nissl substance, a structure with high ribosome content, has been shown to be destroyed by methylmercury. The small numbers of ribosomes in cerebellar and cerebrocortical neurons may explain why those areas are so easily affected by mercury poisoning. Chronic heroin intoxication has also been shown to affect Nissl substance in primate studies (Hirano & Llena, 1980). This suggests that Nissl substance-damaging neurotoxicants may affect the cell's ability to synthesize protein and thus regenerate or repair itself. Other structures of the cell known to be differentially affected by neurotoxicants include mitochondria (LSD-25, heroin), neurofibrils (aluminum, colchicine, vinca alkaloids) and the synapse (glutamate, organic mercury) (Hirano & Llena, 1980).

Damage to various components of white matter have also been tied to specific neurotoxins, including acrylamide, alcohol, triethyltin, *n*-hexane, methyl *n*-butyl ketone, 2,5-hexanedione, arsenic, carbon disulfide, and tri-ortho-cresyl phosphate (Hirano & Llena, 1980). White matter is vulnerable to a variety of neurotoxicity-related effects, including ischemia, with ensuing demyelination or direct toxic injury to the underlying axon.

Neurochemical Damage from Neurotoxicants

Neurochemical damage is perhaps the most interesting frontier in neuropsychology. A wide range of clinical syndromes are better understood by the damage they exert on neurochemical tracts than on structure in the way that neuropsychologists typically think of the term (e.g., lobes, sulci, hemisphere). Parkinson's disease is a case in point, both as an example of the interaction between neurochemistry, structural neuropathology, and neuropsychology, as well as for its possible role in chronic neurotoxic exposure.

Parkinson's disease is a clinical syndrome where dopaminergic neurons from the pigmented nuclei of the substantia nigra and locus coeruleus become selectively and severely damaged. These nuclei project to the striatum, the limbic system (nucleus accumbens, olfactory tubercle, and amygdala) as well as the frontal cortex (Cote & Crutcher, 1985). Among other symptoms, Parkinson's "profoundly disrupts" the motor act control mechanisms of the prefrontal cortex (Goldman-Rakic, 1987). Without understanding the cognitive neurochemistry of Parkinson's, the neuropsychologist is left to explain a diffuse and apparently unconnected set of deficits, including impaired voluntary movement, flattened affect, and eventual dementia. By linking Parkinson's to specific correlates of neurochemical lesion, however, the relationship of behavior to brain damage is far more consistent and interpretable.

Goldman-Rakic (1987) and others have argued that catecholamine impairments in dopamine and norepinephrine systems may explain a variety of behavioral disorders. In particular, the relationship between parkinsonism and neurotoxic exposure was made graphically and gruesomely clear with the discovery that an incorrectly distilled analogue of heroin (MPTP) produced "instant" and irreversible parkinsonism in drug abusers; in some individuals with a rapidity that left them "frozen" with the syringe still hanging from a vein. Manganese and other neurotoxic substances are also known to create parkinsonian states in exposed individuals, perhaps by the oxidation of dopamine that is in turn capable of producing injurious free radicals and quinones (Langston, 1988; Langston & Irwin, 1986). It may be that decline of striatal dopamine associated with aging interacts with further depredations on the dopaminergic system by neurotoxic exposure, resulting in parkinsonism in exposed individuals. It may also be the case that repeated toxic assaults on the nervous system take a chronic, insidious toll on system neurochemistry that young neurons are able to repair, but older ones fail to maintain adequate reconstruction (Langston, 1988).

Recent evidence suggests that many common neurotoxins act on brain neurochemistry to produce neurotoxic abnormalities that result in cognitive, affective, or behavioral impairments. For example, many common *solvents* "recognize dopamine as a selectively vulnerable target suggest[ing] that dopamine depletion . . . may have a role in solvent toxicity to the CNS" (Mutti & Franchini, 1987, p. 722) and may explain certain reversible impairments in human vigilance, psychomotor speed, and mood (Mutti, Falzio, Romanelli, Bocchi, Ferroni, & Franchini, 1988). Freed and Kandel (1988) reviewed animal and human data suggesting that selective attention is mediated by the locus coeruleus and its principal neurotransmitter, norepinephrine (NE). When dopamine β-hydroxylase (an enzyme involved in the production of NE) is inhibited, the result is a transient amnesia related to changes in NE levels. Inhibition of β-hydroxylase is also one consequence of exposure to carbon disulfide, a highly neurotoxic solvent (Costa, 1988). It is possible that further investigations of solvents may produce useful correlations between neurotransmitter dysfunction and neuropsychological impairment.

Many *pesticides* inhibit the normal degradation of neurotransmitters as a principal neurotoxic effect. For example, organophosphates and carbamates inhibit (possibly irreversibly) the enzyme acetylcholinesterase, whose function would ordinarily be to break down acetylcholine. The toxic effects of these pesticides, then, are the result of the overstimulation of the cholinergic system by an excess amount of acetylcholine. Disorders of other neurochemical pathways have also been implicated in *organophosphate* pesticide exposure, including noncholinergic systems involving biogenic amines, glutamic acid, γ-aminobutyric acid, cyclic nucleotides and others. These systems "may play important roles in the initiation, continuation and disappearance of organophosphorus cholinesterase inhibitor-induced neurotoxicity" (Ho, 1988, p. 151).

Much has been learned from the psychiatric literature about the behavioral consequences of cholinergic abnormalities. For example, *anticholinergic drugs* may cause delirium, agitation, and can cause a toxic psychosis that mimics dementia. Memory deficits are common consequences of neuroleptic medications with anticholinergic effects as well as the antiparkinsonian drugs with which they are often paired (e.g., Tune, Strauss, Lew, Breitlinger, & Coyle, 1982; Perlick, Stastny, Katz, Mayer, & Mattis, 1986). Both types of medications have anticholinergic properties (Hartman, 1988b). Since cognitive impairment over time cannot be tied to serum anticholinergic level, the use of neuropsychological tests may provide important information about the clinical extent of cholinergic impairment long after initial biochemical markers have ceased to be diagnostic.

Thus, rather than simply a "diffuse" influence on brain function, solvents and other neurotoxic substances may have a selective effect on neurotransmitter systems. Neuropsychologists, rather than searching for abnormal structural markers, may be better off reconceptualizing the systemic nervous effects of solvents and other neurotoxicants as *neurochemical* toxins. This paradigm shift of viewing neurobehavioral abnormalities as a function of changes in neurotransmitter systems rather than physical topographic damage, may also facilitate insight into nonindustrial neuropsychological abnormalities. Attention deficit disorder, a nebulous neuropsychological problem from a structural viewpoint, becomes a more interesting target of investigation when viewed as an abnormality of noradrenergic or dopaminergic metabolism (Oades, 1987; Freed & Kandel, 1988). Even classically "functional" conditions like posttraumatic stress disorder may prove to have neurochemical substrates (Kolb, 1987). Analysis of system neurochemistry in these disorders may generate useful neuropsychological hypotheses and help direct subsequent investigations of those systems.

When neurotoxic damage is indirect, it may influence other sites in the nervous system, including neuroimmune system function. Neurotoxic damage may also occur, not from the original substance, but from biotransformation into substances more toxic than the original. For example, recent evidence has suggested that neurotoxicity of MPTP depends not on the original compound, but on a breakdown product, MPP$^+$ (Langston, 1988). Neuropsychologists will be sensitive to additional factors that cause impaired performance on neuropsychological tests. For example, *indirect structural* damage to the nervous system may occur via other organ systems, e.g., hepatic and vascular abnormalities. The constellation of variables capable of affecting neuropsychological performance are further discussed in Chapter 2.

The capability of a substance to cause brain damage does not mean that its primary locus of effect lies inevitably in the brain. Tarter *et al.*, (1988) have cogently outlined how damage to various body systems, including pulmonary, renal, cardiac, and others, may indirectly cause neuropsychological impairment. An example of probable brain damage/neurotoxicity that appears to begin in the pulmonary system is the herbicide *paraquat,* a widely used commercial product that caused 1300 poisoning deaths in Japan alone (Crome, 1986, cited in Hughes, 1988). Cases of paraquat poisoning are often associated with suicide attempts or accidental ingestion of a 20% paraquat solution. Ingestion causes severe lung damage (pulmonary fibrosis) and the brain is severely affected by edema, possibly as a result of hypoxia and inflammation and hemorrhage of small cerebral blood vessels (Hughes, 1988).

Lead and mercury are two other well-known neurotoxins whose additional destructive preference for renal tissue may aggravate neurotoxic damage and serve to remind the reader that any neurotoxicant capable of damaging the kidneys, heart, or other system may also cause neuropsychological damage through a variety of mechanisms, including lowering the amount of brain blood, oxygen, or nutrients, or by allowing toxic waste products to circulate in the brain (Hartman, 1988).

Indirect reactive components of affect and personality may substantially influence neuropsychological performance. High levels of depression, fear of injury or death, posttraumatic stress, and other reactive personality components inevitably influence symptom composition.

Silbergeld (1990) has proposed that neurotoxic damage can be understood as a function of the particular mathematical model effects appear to follow. For example, a *nonmonotonic threshold* model appears to fit certain peripheral neurotoxicants (e.g., hexacarbons) that act on nervous system processes that degrade at certain thresholds. Another type of nonlinear model is that discussed earlier where toxicity depends on the elimination of functional reserves or redundancies within the system. Certain neurotoxins produce damage that is linear without threshold. *In utero* ionizing radiation, and childhood lead exposure are two neurotoxicants that fit this model. While models of the type described by Silbergeld do not describe functional deficits, they are templates to examine patterns of risk assessment that take into account chronic exposure and "silent" damage. Models of this sort may permit more accurate calculation of neurotoxic risk than simply tallying large-scale "body counts" of damaged human beings after the fact.

It is possible that neurotoxic exposure may be involved in the causation of many different neurodegenerative diseases. The amino acid contained in chickpeas, for example, has been implicated in the development of lathyrism (Spencer et al., 1986). High levels of aluminum have been found in patients diagnosed with amyotrophic lateral sclerosis (ALS) (Yasui, Yase, Ota, Mukoyama, & Adachi, 1991). Other exposures may induce degenerative processes.

Sites within the CNS may vary among neurotoxic substances, as Tables 1.1, 1.2, and 1.3 suggest.

TABLE 1.1. Mechanism of Action of Various Toxins[a]

Toxin	Mechanism of damage
Pyridines	MPTP parkinsonism; dopaminergic cell death
Glutamate, aspartate	Excitotoxins; excess excitation leads to cell death
Iminodipropionitrile	Neuronal degeneration similar to lathyrism
Hexacarbons	Exposure produces entrapment constriction of axon
Disulfides	Axonal swelling containing neurofilamentous material
Manganese	Depletion of striatal dopamine; parkinsonism
Iron	Extrapyramidal deficits; idiopathic parkinsonism
Aluminum	Brain accumulations found in Alzheimer's patients

[a]After Calne (1991).

TABLE 1.2. Selected Neurotoxins and Mechanisms/Sites of Damage

Substance	Damage site
MPTP	Substantia nigra
Tetrodotoxin (puffer fish)	Ion channel blocker
Saxitotoxin	Ion channel blocker
Scorpion toxin, DDT, pythethroids	sodium ion increase
Doxorubicin	Central and peripheral nerve axonal degeneration
Alcohol	Central–peripheral distal axonopathy
Clioquinol	Central–distal axonopathy
Vitamin B_6	Sensory neuropathy
Diphtheria toxin	Myelin cell bodies
Hexachlorophene	Glial mitochondria

TABLE 1.3. Nervous System Loci for Selected Neurotoxins[a]

Cortical gray matter	Subthalamus	Schwann cells
Azide	Azide	Acrylamide
Barbiturate	Carbon monoxide	Carbon disulfide
Carbon disulfide	Manganese	DDT
Cyanide	Internal capsule	Iminodipropionitrile
Lead	Azide	Isoniazide
Mercury	Carbon monoxide	Lead
Methyl bromide	Corpus callosum	Vinca alkaloids
Nitrogen trichloride	Azide	Sensory N. thalamus
Hippocampus	Carbon monoxide	Acetylpyridine
Acetylpyridine	Cyanide	Glutamate
Azide	Hexachlorophene	Lead
Barbiturate	Isoniazide	Mercury (organic)
Carbon monoxide	Lead	Anterior horn
Cyanide	Malononitrile	Carbon disulfide
Caudate/putamen	Triethyltin	Cyanide
Barbiturate	Optic chiasm	Iminodipropionitrile
Carbon disulfide	Barbiturate	Isoniazide
Cyanide	Carbon monoxide	Lead
Malononitrile	Cyanide	Vinca alkaloids
Manganese	Hexachlorophene	Hypothal. Vent. N.
Nitrogen trichloride	Isoniazid	Glutamate
Globus pallidus	Lead	Gold thioglucose
Carbon monoxide	Malononitrile	Mammillary bodies
Cyanide	Triethyltin	Nitrogen trichloride
Gold thioglucose		Tegmentum
Manganese		Azide
Methyl bromide		Cyanide
		Nitrogen trichloride

[a]After Northon (1986) and Hartman (1991).

LEGISLATIVE ISSUES AND NEUROTOXIC DAMAGE

Individual exposure to neurotoxic materials may intersect with legislation or ongoing legal debates (Morris & Sonderegger, 1984). One such controversy involves whether females of childbearing age should be allowed to work in fetotoxic (i.e., where they will be exposed to lead) environments. Another concerns whether a pregnant woman should be held liable for her drug abuse if it causes harm to the fetus.

Legislation also intersects with neurotoxicity as an issue of national regulatory policy, and national policy, like the agencies that regulate these substances, is fragmented and not always consistent. Table 1.4 lists the various agencies involved with regulating toxic substances.

There is general consensus that regulation of new and existing substances has lagged far behind need. Part of the problem is the sheer numbers of chemicals that require review. Critics have also suggested that the U.S. Environmental Protection Agency may be motivated by "a desire not to burden industry, rather than presumptive risk" (*Neurotoxicity*, 1990, p. 18). For example, out of the 60,000 chemicals in TOSCA's inventory, final rules were issued on only 25 chemicals or chemical groups, consent agreements were reached on 3 and nine proposed rules are pending in the period from 1977 to 1988 (*Neurotoxicity*, 1990). It is troubling that stringency of evaluation is tied to presumed risk, leading drugs to receive the highest level of scrutiny, but commercial chemicals to undergo

TABLE 1.4. Acts Governing Toxic Substances and Associated Agencies[a]

Act	Agency primarily responsible[b]
Toxic Substances Control Act	EPA
Federal Insecticide, Fungicide and Rodenticide Act	EPA
Federal Food, Drug, and Cosmetic Act	FDA
Occupational Safety and Health Act	OSHA
Comprehensive Environmental Response Compensation, and Liability Act	EPA
Clean Air Act	EPA
Federal Water Pollution Control Act and Clean Water Act	EPA
Safe Drinking Water Act	EPA
Resource Conservation and Recovery Act	EPA
Consumer Product Safety Act	CPSC
Federal Hazardous Substances Act	CPSC
Controlled Substances Act	FDA
Federal Mine Safety and Health Act	MSHA
Marine Protection, Research, and Sanctuaries Act	EPA
Lead-Based Paint Poisoning Prevention Act	CPSC
Lead Contamination Control Act	HHS
Poison Prevention Packaging Act	CPSC

[a]*Neurotoxicity* (1990); [b]CPSC, Consumer Product Safety Commission; EPA, Environmental Protection Agency; FDA, Food and Drug Administration; HHS, Department of Health and Human Services; MSHA, Mine Safety and Health Administration; OSHA, Occupational Safety and Health Administration.

the least (*Neurotoxicity*, 1990). Even more troubling is that neurotoxicity is not explicitly mentioned in most regulatory legislation, although it is implicitly considered under the scope of regulation since it is "toxic."

Another difficulty involves obtaining and sharing information regarding substances under scrutiny. Commercial concerns about confidentiality may limit transfer of information to government agencies, and may even hinder intraagency sharing of toxicological information (*Neurotoxicity*, 1990).

In this less than optimistic profile, there are some encouragements for the use of neuropsychological methods and neurotoxicity research. The Alcohol, Drug Abuse, and Mental Health Administration (ADAMHA) provides extensive funding for neurotoxicity research. The National Institute on Drug Abuse (NIDA) and the National Institute of Mental Health (NIMH) fund research grants related to neurotoxicity. Regulatory language in a number of acts implicitly encourage neurotoxicological research. For example, the 1970 Occupational Safety and Health Act attempts to "assure, so far as possible, every working man and woman in the nation safe and healthful working conditions" (Taft, 1974, p. 8). The act further mandated NIOSH to "conduct research into the motivational and behavioral factors relating to the field of occupational safety and health" (Fairchild, 1974, p. 3).

Environmental legislation has further stimulated the field of neuropsychological toxicology in the United States by calling for regulation of toxic materials and the evaluation of toxicity. These acts include the Clean Air Act of 1970; the Resource Conservation and Recovery Act, which regulates land disposal of toxic materials; the Marine Protection Research and Sanctuaries Act, which does the same for marine disposal of toxic wastes; the Clean Water Act and Safe Drinking Water Act; the Federal Insecticide, Fungicide and Rodenticide Act, and especially the Toxic Substances Control Act of 1976 (TOSCA); all implicitly or explicitly mandate research to support the development of safety regulations (Reiter, 1985).

There is general agreement, however, that the combined activity of government agencies in this regard is far behind the need for neurotoxicity review. In addition, such mandates do not customarily include neuropsychological assessment. In fact, with the exception of organophosphate pesticides, there are no requirements for routine neurotoxic assessments *of any kind* in the development of new environmental regulations (Buckholtz & Panem, 1986). Present-day statutes require special justifications to conduct such investigations, and there are no clear criteria for determining how behaviors or functions should be studied (Buckholtz & Panem, 1986).

While regulatory agencies have been somewhat slow to elicit neurotoxicity data, public health researchers are beginning to recognize neurotoxic syndromes as diagnostic entities, and to realize their importance to the quality of life. A background paper at a conference on long-term environmental research and development concludes:

> Even small degrees of central nervous system dysfunction are not for moral, ethical, and health reasons to be tolerated. The loss of five points in I.Q., fatigability, irritability,

or lethargy or slowed reaction times are significant losses. These are areas of human function that make life enjoyable and worth living. [Robins, Cullen, & Welsh, 1985, cited in Buckholtz & Panem, 1986].

Needleman (1990) characterizes a similar downward shift in I.Q. as a "social disaster," noting that moving the I.Q. distribution down 4–7 points will produce "a fourfold increase in the risk of severe dysfunction. . . . It also means that 5% of the distribution will be prevented from achieving superior function" (p. 334).

There are many potential places where neuropsychological research could interface with regulation. Neuropsychological data collection could occur at either the beginning or the end of the long chain of scientific and legislative actions that culminate in regulation. In the realm of industrial toxins, for example, individual neuropsychological studies could be used by the American Conference of Governmental Industrial Hygienists (ACGIH), the primary U.S. source of industrial exposure limit recommendations. A subcommittee of ACGIH representatives from government, academia, the National Institute for Occupational Safety and Health (NIOSH), and labor could incorporate the findings of neuropsychological research to influence their recommendations on toxic exposure effects (Anger, 1984).

ACGIH could then include neuropsychological data in its recommendations for what are called *threshold limit values* (TLVs). The latter are "airborne concentrations of substances that represent conditions under which it is believed that nearly all workers may be repeatedly exposed day after day without adverse effects" (Andrews & Snyder, 1986, p. 637). Originally proposed as unofficial standards for safe exposure to toxic substances in the workplace, TLVs have taken on the character of an "official" standard in the United States and many other countries. Anger (1984) described three types of TLV:

1. The TLVTWA (time-weighted average), which is the limit of exposure for an 8-h day/40-h workweek.
2. The TLVSTEL (short-term exposure limit), which is the highest level that should not be exceeded for a specified time limit (a 15-minute time-weighted average as of 1982).
3. The TLVC or "ceiling" concentration that must *never* be exceeded, even for an instant.

Recently, important questions have been raised about the adequacy and accuracy of TLVs as an index of safe exposure levels. Roach and Rappaport (1990) and Ziem and Castleman (1989) argue that TLVs are not true safety thresholds, but rather levels that were acceptable to industry. Supporting arguments include:

1. Many TLVs are reduced each year, sometimes to one-tenth of their original value, thus calling into question the adequacy of previous TLVs.
2. TLV protection of "nearly all workers" does not stand up to scrutiny when data for individual subjects are examined. In many cases, the incidence of adverse effects at or below TLV was substantially above zero, and in some substances, as high as 100%.

3. Studies used to assess TLVs are inadequate. The 1976 TLV for toluene, for example, was apparently set on the basis of results from only three subjects in a 100 ppm study, and two in a 50 ppm study.
4. Inferences from existing studies are inadequate. In the toluene studies above, two out of the three subjects exposed to 100 ppm toluene in air reported moderate fatigue and sleepiness, the third additionally experienced headache. One of the two subjects exposed 8 hours to 50 ppm toluene reported drowsiness and very mild headache (von Oettingen *et al.*, 1942, cited in Roach & Rappaport, 1990, p. 739).

Thus, rather than being a public health standard, TLVs may have been promulgated as being realistic for industry to achieve, leading Roach and Rappaport to conclude that some TLVs were blatantly "influenced by vested interests" and that the rest "are a compromise between health-based considerations and strictly practical industrial considerations with the balance seeming to strongly favor the latter" (p. 741).

While ACGIH's recommendations are voluntary, when OSHA (the Occupational Safety and Health Administration) adopts ACGIH's regulations they can be considered as a basis for federal regulation (Anger, 1984). Regulations, once in place, could serve as stimuli for further neuropsychological investigations.

TESTING SUBCLINICAL NEUROTOXIC EFFECTS

Diagnostic techniques that are currently available to monitor neurotoxic exposure can be divided into two basic categories: "internal dose indicators," which assay the chemical agents and their metabolites in the organism, and "indicators of effect," which "clarify the interactions between the external agent and the recipient organism" (Foa, 1982, p. 173). Since the "indicators of effect" provide greater information about the early (subclinical) effects of neurotoxic substances, than do internal dose levels (Lezak, 1984; Gamberale & Kjellberg, 1982; Foa, 1982), neuropsychological assessment may be considered a first-line diagnostic tool in the assessment of exposed patients. Neuropsychological methods have already been applied to a wide variety of neurotoxic materials to catalogue cognitive and affective alterations in function. Some of the effects are catalogued in Table 1.5.

ADVANTAGES OF NEUROPSYCHOLOGICAL TOXICOLOGY ASSESSMENT

Neuropsychological measures may be especially useful in providing the research validation required by TOSCA and other environmental regulations. The use of neuropsychological measures has several advantages in this regard:

TABLE 1.5. Neuropsychological Effects of Exposure to Common Neurotoxins[a]

Agent	Effects	Major uses/exposure sources
Arsenic	Memory impairment; CNS, PNS, and affect impairments; motor and visuomotor impairments; impairments secondary to cardiotoxicity	Pesticides, pigments Electroplating industry Seafood Smelters Semiconductor manufacturing
Carbon monoxide	Impaired arousal, vigilance, attention, motor speed, and coordination; visuospatial and visuomotor impairments	
Lead	Memory impairments; impairments in cognitive efficiency and flexibility; visuospatial defects; lowered IQ; dementia. Hyperactivity and retardation in children. Hallucinations and seizures from organolead. Depression; organic affective syndrome	Solder Leaded paint Moonshine Insecticides Auto body shops Battery manufacturing Foundries, smelters Drinking water from lead pipes
Mercury	Fine motor coordination, possible organic affective syndrome, e.g., "erethism," visuomotor impairments	Thermometers and scientific instruments Dental amalgam Electroplating industry Photography Feltmaking Textiles Pigments Taxidermy
Organic solvents (mixtures)	Memory, attention, visuospatial, and cognitive efficiency impairments; impaired fine motor coordination, organic affective syndrome	Dry cleaning fluid Semiconductor industry Degreasing Paint removing, gluing, lacquers, rubber solvents Adhesive in shoe and book industry Auto and aviation fuels
Organophosphates	Memory impairments; lowered arousal and attention. Organic affective syndrome; fine motor coordination and visuomotor impairments	Pesticides Agricultural industry

[a]After Hartman, Hessl, and Tarcher (1992).

1. *Concurrent and predictive validity.* It is commonly accepted that neurotoxic exposure symptoms may be validly assessed with psychometric techniques, and that neurotoxic symptoms may occur prior to the occurrence of other objective neurological sequelae (Gamberale & Kjellberg, 1982). Further, these disruptions may not be detectable *without* neuropsychological methods since subjects may develop alternative processing strategies to compensate for acquired neuropsychological deficits. Since neuropsy-

chological tests are more demanding, and can isolate elementary components of cognition and behavior, obfuscating effects of compensation can be minimized and specific impairments can be measured.
2. *Safety.* Neuropsychological tests are safe and noninvasive. There is no risk to the participant. Unlike certain medical tests, neuropsychological evaluation is neither painful (e.g., nerve conduction studies) nor potentially dangerous (e.g., CT with infusion). Tests may also be repeated to provide longitudinal data without harm to the patient. As Weiss (1983) succinctly states, "subjects' livers need not be fed into a blender to quantify damage" (p. 1174).
3. *Comprehensive and flexibility.* Neuropsychological evaluations can assess a wide variety of cortical and subcortical functions; tests may be employed to assess both global and local brain abnormalities. The neuropsychologist may choose among an array of assessment devices, from those that isolate specific nerve, muscle, and cortical connections (e.g., Finger Tapping) to those designed to measure overall intellectual, neuropsychological, or emotional abilities (e.g., Wechsler Intelligence Scales, Halstead–Reitan Neuropsychological Battery, Minnesota Multiphasic Personality Inventory-II).
4. *Objectivity and replicability.* Unlike more subjective forms of mental status examinations, neuropsychological tests can be made reliable; allowing for precise replication within or across individuals or patient populations. Standardization of patient variables, test administration, content, and scoring allows replication and validation attempts by other researchers unconnected with the original study.
5. *Cost.* Compared to medical and neurological workups, neuropsychological measures generally do not require elaborate or expensive facilities to provide assessment results. Professional time spent in the assessment rather than equipment is the main cost. There are usually no special facilities needed; examination is not limited to the laboratory, but requires only a single, relatively quiet room without distractions.
6. *Portability and convenience.* A valid and complete neuropsychological test battery for neurotoxic assessment can be carried in one or two small suitcases. Recent trends in computerized neuropsychological toxicology batteries (e.g., Baker, Letz, Fidler, Shalat, *et al.*, 1985) may eventually result in further convenience by automating test administration, data acquisition, and scoring procedures.
7. *Complementarity.* Neuropsychological evaluation can provide complementary information to conventional medical screens. For example, abnormal nerve conduction studies can be quantified in terms of behavioral dysfunction by employing neuropsychological tests of sensorimotor abilities. The meaning of CT-scanned cortical atrophy for the patient's real-world functioning can be quantified using neuropsychological measures that provide a comprehensive picture of cognitive, behavioral, and emotional status.

8. *Early warning.* Because neuropsychological methods have the sensitivity to detect very early neurotoxic dysfunction, early testing can prevent lasting brain damage. In addition, neuropsychological examination could serve to identify workers already impaired from toxic exposure and suggest transfer to less hazardous employment.

UNSOLVED CHALLENGES IN ASSESSMENT OF NEUROTOXIC SYNDROMES

The neuropsychological evaluation of neurotoxic exposure presents several unique difficulties and challenges; complications inherent to the neuropsychological evaluation of patients with possible neurotoxic syndromes. Several of these difficulties are related to diagnostic limitations in the field of neurotoxicology itself, and are thereby transferred onto the neuropsychologist whose service is requested.

Problems Inherited by the Neuropsychologist

1. *Lack of neurochemical markers.* While neuropsychological signs and symptoms are considered valid and sensitive measures of neurotoxic exposure, in many cases, test results cannot be buttressed by neurochemical markers of neurotoxic effects. In most cases, such markers do not exist and may not be discovered soon. The measurement of neurochemical markers of neurotoxic exposure are difficult to obtain, since the nervous system is generally isolated from easily sampled substances (e.g., urine, skin, blood), and we do not know enough about the relationship between most neurotoxins and biological markers to determine their clinical correlates (Silbergeld, 1993). While the neurotoxic effects of certain substances are fairly clear (e.g., n-hexane), these substances are the exception. Most toxicologic research provides little insight about expectable dose-related human functional impairments from particular toxicants. And since "virtually all toxicologic data are on the effects of exposure to a single agent" (Tilson, 1990, p. 298), there is almost no information about effects related to combinations or mixtures of toxicants. Since industrial or voluntary drug exposures often involve multiple neurotoxic agents, clinical prediction may require information as to whether these combinations are additive, synergistic, antagonistic, or completely novel in their neurotoxic effect (Tilson, 1990, p. 298). Concerning the neuropsychology of neurotoxic exposure, very little information is available to allow tracing neuropsychological toxicologic impairments to a specific type of nervous system lesion.

2. *Application and generalizability of animal models.* Current research with animal subjects provides information unobtainable from human subjects (e.g., lethal dose estimates, autopsy results). Several animal models have also been found useful to validate and examine in detail neurotoxic processes, i.e., n-hexane and MEK peripheral neuropathy, and MPTP parkinsonism (Winneke, 1992).

Nonetheless, animal models also suffer from limitations when extrapolated to human effects. First, animal models of neurotoxicity have not generally addressed behaviors of interest to human neuropsychological toxicologists, and have typically been limited to reinforcement paradigms. Functions often found to be impaired in humans—including vigilance, reaction time, complex psychomotor skills, memory, and emotion—do not yet have testable analogues in animal research (Cranmer & Golberg, 1986).

A second problem in the use of animal models is the difficulty generalizing from lower mammals to humans. Species-specific effects may mislead researchers who wish to extrapolate human neurotoxic sequelae from the results of animal tests. Primate experimentation might allow better extrapolation to human effects; however, such studies have rarely been performed.

3. *Exposure ambiguities.* In industrial settings at least, individual exposure data are often difficult to obtain. Often, measurements are simply not available. When single-exposure measurements are taken in the environment, these measures may not always reflect exactly where the individual has worked, or the variations of exposure intensity at any single site.

Thus, the incomplete state of the art in neurophysiological, neurochemical, and relevant animal models of neurotoxicity may come to haunt the neuropsychologist interested in human exposure effects by increasing diagnostic uncertainty. In addition, neuropsychological diagnosis of toxic exposure is complicated by the following factors.

Challenges to the Neuropsychologist

Acute versus chronic effects. Since an exposed individual may present for evaluation while remaining in contact with noxious substances at the workplace, acute and chronic effects may require partialing out. The task of neuropsychological evaluation is correspondingly more difficult than when the injury is discrete and localizable in time.

Difficulties assessing premorbid function. Assessing neurotoxic damage to the nervous system may be difficult in some neurotoxic exposure victims because of job selection factors. Patients who are hired for unskilled factory employment may show premorbid deficiencies in education or other areas. Test interpretation may suffer without adequate records or estimates of such premorbid difficulties. Since many factory employees perform well-learned repetitive tasks that do not require high levels of verbal ability, memory, or fine motor dexterity, low functioning on screening measures of neuropsychological toxicity may reflect premorbid abilities rather than recent, job-related impairment.

In our clinical experience it has been an unfortunate fact that the workers with the poorest premorbid educational or intellectual functioning are often the same individuals who suffer the most severe neurotoxic exposure. Job market pressures force unskilled or marginally skilled individuals into high-risk occupations; performing menial occupations in small, unregulated cottage industries (e.g., battery or metal reprocessing operations). When these individuals are finally evaluated for neurotoxicity, the neuropsychologist may find differential diagnosis particularly difficult.

Ruling out competing explanations. Research by Tarter et al. (1988) and many others have outlined the influence of various medical disease states on neuropsychological performance. Neuropsychologists attempting to uncover human neurotoxic effects have thus been likened not to searchers for the proverbial needle in a haystack, but rather for "the needle in a heap of needles" (Russell, Flattau, & Pope, 1990, p. 3). The importance of thorough rule-out investigation cannot be overemphasized.

Paucity of relevant test norms. Further complicating diagnosis is the lack of neuropsychological test norms for individuals most likely to have been exposed to neurotoxic substances. Few neuropsychological tests have been normed on the most common high-risk groups: factory employees and drug-abusing populations (i.e,. glue-sniffers). While there is some generalization possible for tests normed on individuals with comparable educational levels, more appropriate validation on relevant population groups would be preferable.

Very low premorbid educational levels or premorbid IQ results can invalidate interpretation of current neuropsychological tests since such tests may assume average educational or intellectual abilities.

Lack of corroborating medical evidence. The ability of neuropsychological evaluation to provide early warning of toxicity effects necessarily implies that such data will not always be supported by medical test results. The neuropsychologist cannot always depend on conventional medical diagnostic procedures (e.g., blood levels, CT, and EEG) to validate neuropsychologically observable but medically subclinical impairments. The counterargument is that sensitivity of neuropsychological testing and its superiority over conventional neurological and psychiatric methods is research proven in other areas (e.g., closed head injury, dementia) and should hold for neurotoxic exposure as well.

Differential diagnosis. The neuropsychological toxicologist will probably experience frequent diagnostic ambiguities in terms of classifying symptoms on the continuum of purely physical to mixed physical and psychological to purely psychological. For example, patients with acute industrial solvent exposure may suffer from cerebral atrophy and other structural brain damage (Hane & Ekberg, 1984, p. 8). They may also present with secondary impairments in the ability to cope with stress because of neurologically based fatigue, headache, decreased nerve conduction, etc. and so display maladaptive reactions to normal stressors (Fig. 1.1). Fear of brain damage itself may be a significant stressor, irrespective of actual toxic exposure. Psychogenic illness can occur without actual neurotoxic exposure if an individual or group simply believes it has been exposed to neurotoxins (Murphy & Colligan, 1979). Finally, exposure to human neurotoxins may provide predisposed individuals with opportunities to malinger or otherwise implement strategies of secondary gain, although newer neuropsychological methods have shown promise in detecting styles of malingerers (e.g., Rogers, 1988; Pankratz, 1988; Binder & Pankratz, 1987).

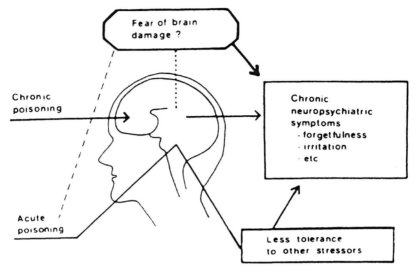

FIGURE 1.1. Differential diagnostic possibilities in neurotoxicity evaluations. (Reprinted, by permission of the publisher, from Hane, M., and Ekberg, K. (1984). Current research in behavioral toxicology. *Scandinavian Journal of Work Environment and Health,* Suppl. 1, 89.)

OTHER ISSUES

Economic Consequences

Neuropsychological methods of research and assessment have been implemented more slowly in the United States than in European and Scandinavian countries. The reasons for U.S. caution in entering the field may be as much sociopolitical as scientific. Neuropsychological toxicology cannot be considered a "value-free" scientific enterprise. Definitive research involving nervous system effects of workplace toxins could potentially generate profound economic consequences for industry and insurance companies. First, new and stricter regulations based on such research could mandate costly reengineering and renovation of the physical plant. Second, the expense of implementing preventive medical care may prove prohibitive. Costs could be substantial, in terms of work/management education as well as in monies allotted for testing and treatment. If not subsidized in some manner, the expense of such outlays might discourage employers from vigilant monitoring of chemical exposure. Smaller companies could be placed in the position of choosing between safety and solvency—a position that workers may find difficult to resolve.

The counterargument to this, of course, is that regulation and retooling will cost less in the long run than the payout in health insurance and loss of taxable worker income and productivity. However, government support of industries without the capital to retool may be necessary.

Legal Consequences

A third costly outcome of neuropsychological toxicology studies in our litigious society could be expensive legal judgments awarded for neurotoxic exposure. With 20 million U.S. employees exposed to neurotoxic substances in the workplace (Anderson, 1982), legally and scientifically valid neuropsychological methods could stimulate litigation that results in numerous and extremely large damage awards.

Political Impediments

Political as well as economic factors may influence the development of neuropsychological research. For example, recent executive branch decrees in the United States have increased the power of the Office of Management and Budget (OMB) to limit toxic chemical regulations to those that meet stringent "cost-benefit" criteria. There has been ongoing concern that the Environmental Protection Agency and OSHA under recent administrations have shown less motivation to implement more stringent exposure standards (Anderson, 1982). These factors, in turn, probably dampen the development of new regulations.

Limited Numbers of Researchers and Clinical Practitioners

Another limiting factor is simply the short supply of skilled professionals who could conduct research and administer clinical service. These professionals include physicians and neuropsychologists skilled in the biological and behavioral monitoring and treatment of neurotoxic illness, and health care workers able to work as a team with employees for prevention and detection of neurotoxic exposure. Unfortunately, previously cited limitations in the number of available occupational health researchers suggest that clinical demand far outstrips the supply. Increased support for training in occupational medicine and neuropsychological toxicology would be a prerequisite; however, graduate-level training in neuropsychological toxicology, at least, is very difficult to find.

Barriers to Company, Worker, and Patient Acceptance

Finally, even if the political, professional, and economic repercussions of neuropsychological toxicology research could be resolved, it is far from clear that basic research or neuropsychological monitoring will be particularly welcomed in the workplace. Companies may be unwilling to divulge proprietary formulas for competitive or liability reasons, even when failure to disclose such information hinders diagnosis or treatment (e.g., Rose, 1990). Worker confidentiality and privacy may not be easily maintained, and assurances may be viewed with skepticism in an adversarial labor–management climate. Many workers might perceive occupational health monitoring as a harmful intrusion into privacy, particularly since tests that screen for workplace toxins are also sensitive to effects of alcohol or drugs. An employee who is frequently monitored may be more quickly seen to demonstrate subclinical impairment, whether from neurotoxic exposure, alcohol abuse, psychological difficulties, aging, or unrelated illness. Without proper safeguards it is conceivable that positive findings on biological or behavioral tests could result in temporary or permanent loss of employment.

NEUROPSYCHOLOGY IN TOXIC SUBSTANCE RESEARCH

Neuropsychological investigation of toxic exposures occurs within a large, overlapping, multidisciplinary context. Neuropsychologists and other professionals who examine patients with toxic exposure histories will find themselves cooperating with specialists in medicine, psychology, labor, management, epidemiology, government, and law; each investigating relevant aspects of nervous system and behavior alterations induced by exposure to poisonous or toxic substances.

Neuropsychologists who become occupational health advocates may also find themselves embroiled in legal, political, medical, and scientific controversies. However, those clinicians and researchers who accept the multidisciplinary nature and impact of this somewhat controversial and exciting speciality may rightly consider themselves to be pioneers in a new form of clinical psychology. With special skills in psychometrics, research design, and the identification of brain–behavior relationships, the neuropsychologist is in a unique position relative to other mental health professionals to contribute to industry and the private sector.

Currently, few neuropsychologists are trained to provide such contributions, but even if every practicing neuropsychologist were familiar with toxicology research, their combined numbers would be far too small to handle the sheer magnitude of the investigatory task. Human beings are exposed to over 53,000 different substances, including pesticides, drugs, food additives, cosmetics, commercial and industrial substances (Fisher, 1985). OSHA sets standards for only 588 of these, and only 167 of these have been regulated for their neurotoxic effects. It is far from clear, however, that the remaining chemicals are free from risk, since a random sample of 100 chemicals revealed that almost none had actually been tested for neurotoxicity before being released on the market (Fisher, 1985).

The substances for which ACGIH has recommended threshold limit values in the workplace, because of their neurotoxic potential, are listed in Table 1.6.

TABLE 1.6. Chemicals for which ACGIH TLVs Were Set Because of Neurotoxicity[a]

Abate	Ethanolamine	Naphthalene
Acetonitrile (methyl cyanide)	Ethion	Nickel carbonyl
Acrylamide	Ethyl amyl ketone	Nitromethane
Aldrin	Ethyl bromide	2-Nitropropane
Allyl alcohol	Ethyl butyl ketone	Osmium tetroxide
Anisidine	Ethylene chlorohydrin	Parathion
Barium	Ethylene glycol dinitrate	Pentaborane
Baygon	Ethyl ether (diethyl ether)	Pentane
Benzyl chloride	Ethyl mercaptan	Perchloroethylene
Bromine pentafluoride	N-Ethylmorpholine	(tetrachloroethylene)

TABLE 1.6. (Continued)

n-Butyl alcohol	Fenamiphos	Phenyl ester
sec-Butyl alchol	Fensulfothion	Phenyl mercaptan
tert-Butyl alcohol	Fenthion	Phenylphosphine
p-t Butyltoluene	Halothane	Phorate
Camphor	n-Heptane	Phosdrin
Carbaryl	Hexachlorocyclopentadiene	Phosphorus oxychloride
Carbon disulfide	Hexachloroethane	1-Propanol (n-propyl alcohol)
Carbon tetrachloride	n-Hexane	Propargyl alcohol
Chlordane	Hydrogen cyanide	1,2-Propylene glycol dinitrate
Chlorinated camphene (60%) (Toxaphene)	Hydrogen selenide	Propylene glycol monoethyl ether
Chlorine trifluoride	Hydroquinone	Propylene oxide
Chlorobenzene	Iron pentacarbonyl	Pyridine
Chlorobromomethane	Isoamyl alcohol	Quinone
Chlorpyrifos (Dursban)	Isophorone	Ronnel
Cobalt hydrocarbonyl	N-Isopropylaniline	Selenium (and compounds)
Cumene	Isopropylether	Selenium hexafluoride
Cyanides	Lead, inorganic	Stoddard solvent (mineral spirits; white spirits)
Cyclohexylamine	Lindane	Strychnine
Cyclopentadiene	Manganese (and compounds)	Sulfuryl fluoride
Cyclopentane	Manganese cyclopentadienyl tricarbonyl (MCT)	Sulprofos
Decaborane	Manganese tetroxide	Tellurium
Demeton	Mercury, alkyl	1,1,2,2-Tetrachloroethane (acetylene tetrachloride)
Diazinon	Mercury, not alkyl	Tetraethyl dithionopyrophosphate (TEDP)
Diborane	Mesityl oxide	
Dibrom	Methomyl	
Dibutyl phosphate	4-Methoxyphenol	
Dichloroacetylene	Methyl acetate	Tetraethyl lead (TEL)
p-Dichlorobenzene	Methylacrylonitrile	Tetraethyl pyrophosphate (TEPP)
Dichlorodiphenyltrichloroethane	Methyl alcohol (methanol)	
2,4-Dichlorophenoxyacetic acid (2,4-D)	Methyl bromide (monobromomethane)	Tetrahydrofuran (THF)
Dichlorotetrafluoroethane (Freon 114)	Methyl n-butyl ketone (MBK)	Tetramethyl lead (TML)
	Methyl chloride (monochloromethane)	Tetramethyl succinonitrile (TMSN)
Dichlorvos (DDVP)	Methylchloroform	Tetranitromethane
Dicrotophos	Methylcyclohexane	Tetryl
Dieldrin	o-Methylcyclohexanone	Toluene
Diethanolamine	Methyl demeton	Tributyl phosphate
Difluorodibromomethane (Freon 12B2)	Methylene chloride	Trichloroacetic acid (TCA)
Diisopropylamine	Methyl ethyl ketone (MEK; 2-butanone)	1,1,2-Trichloroethane (vinyl trichloride)
Dimethylaniline	Methyl mercaptan	Trichloroethylene
1,1-Dimethylhydrazine	Methyl parathion	Tricyclohexyltin hydrochloride
Dioxathion	Methyl propyl ketone	Trimethyl benzene
Dipropylene glycol methyl ether (DPGME)	Methyl silicate	Trimethyl phosphite
Diquat	Metrizabin	Triorthocresyl phosphate
Disulfoton	Monocrotophos	Triphenyl phosphate (TPP)
Dyfonate	Morpholine	Xylene
EPN naphtha		

"Reprinted by permission of Pergamon Journals, Ltd., from Anger, W. K. (1984). Neurobehavioral testing of chemicals: Impact on recommended standards. *Neurobehavioral Toxicology and Teratology*, 6, 149–151.

For industrial agents, NIOSH has criteria documents listing the 36 agents with known nervous system effects at low concentrations (Anger, 1990). Since prescription drugs, many abused drugs, and substances whose neurotoxicity is presently being researched (e.g., styrene, other pesticides) are *not* included, this list must be considered an *under*estimate of the substances presenting immediate neurotoxic danger. Unfortunately, available research lags far behind the need to assess these substances. The effects of toxic substances on the human nervous system are still largely unknown. Those who survey the existing literature should therefore not be surprised by the number of questions that remain unanswered at each point on the continuum of neurotoxicological investigation, from the neuronal and neurochemical to the neuropsychological. Even the effects of lead, the most well-known metallic neurotoxin, have not been completely evaluated; only the PNS effects have been adequately documented (Krigman, Bouldin, & Mushak, 1980). For other metals, less commonly known to be neurotoxic (e.g., gold), *no* neuropsychological studies of their effects are available.

The situation is similar for other neurotoxins, including solvents and pesticides. Much basic research remains to be done to elucidate physiological and psychological nervous system function disturbances caused by these substances. Biological neurotoxins (e.g., animal venoms, plant poisons) are even less well researched, with no neuropsychological studies of their effects available. The task of providing such basic research is an enormous one.

Neurotoxic substances pervade our environment. The number of potentially neurotoxic chemicals in the workplace has been estimated to be as high as 850 (*Chemical Regulation Reporter,* 1986). Even this large estimate may not include neurotoxic prescription medications or abused drugs. Estimates of individuals exposed in their jobs to neurotoxic substances have ranged from a conservative 7.7 million (*Chemical Regulation Reporter,* 1986), to over 20 million (Anderson, 1982). Voluntary solvent inhalant abuse appears to be equally prevalent with estimates of up to 18.7% U.S. high school seniors having tried inhalants at least once (Giovacchini, 1985). Internationally, up to half a million children are said to be addicted to solvent inhalants (Montoya-Cabrera, 1990). In the United Kingdom, 1% of all deaths between 1971 and 1981 were attributed to solvent abuse (Ron, 1986).

The total number of individuals exposed to neurotoxic medications, and the total number of adults and children who voluntarily abuse other neurotoxic substances, have never been determined. There are some exposure data compiled for industrial workers. Table 1.7 has been compiled from various Centers for Disease Control documents and other sources (e.g., Anderson, 1982) and suggests the magnitude of neurotoxic exposure in the United States alone. Table 1.8 indicates the extent of worldwide exposure to neurotoxic substances.

WHO IS AT RISK?

The risk of adverse effects from neurotoxic exposure is shared by the employed and the unemployed, children, young adults, and elderly adults. While

TABLE 1.7. Subjects at Risk for Neurotoxic Exposure at the Workplace: United States

Neurotoxic substance	Estimated numbers at risk
Alcohols (industrial)	3,851,000
Aliphatic hydrocarbons	2,776,000
Aromatic hydrocarbons	3,611,000
Cadmium	1,400,000
Carbon disulfide	24,000
Carbon tetrachloride	1,379,000
Dichloromethane	2,175,000
Lead	
(In children—from inorganic paint exposure)	45,000
Lead oxides	1,300,000
Lead carbonate	183,000
Lead naphthenate	1,280,000
Lead acetate	103,000
Metallic lead	1,394,000
Manganese	41,000
Mercury	
Mercury sulfide	8,900
Mercuric nitrate	10,100
Mercuric chloride	51,000
Metallic mercury	24,000
Organic mercury	280,000
n-Hexane	764,000
Perchloroethylene (tetrachloroethylene)	1,596,000
Styrene	329,000
Thallium	853,000
Toluene	4,800,000
Trichloroethylene	3,600,000
Xylene	140,000
Pesticides (1979)	1,275,000
Rubber solvents (benzene and lacquer diluent)	600,000

TABLE 1.8. Worldwide Exposure to Selected Neurotoxic Substances[a]

Substance/country	Exposure extent
Pesticides	
Worldwide	At least 15,000 deaths, >1,000,000 poisonings
USA	4–5 million exposed workers; 300,000 cases/year
Honduras	32% of 1100 farmers revealed cholinesterase level decreases greater than 25%
Cuba	7000 exposed workers
Organic solvents	
Germany (former Fed. Rep.)	1–2 million exposed workers
USA	9.8 million exposed workers
Venezuela	7000 exposed workers
Lead	
USA	12.5 million children at risk
Venezuela	13,760 exposed workers
China	1.7% of 355,000 examined workers showed evidence of chronic poisoning

[a] Nakajimi (1993).

TABLE 1.9. Occupations at Risk

Occupational group	Substances
Agricultural	
Farm labor	Herbicides, insecticides, solvents, pesticides, mercury, pesticide manufacturing and distribution
Blue-collar	
Degreasers	Trichloroethylene
Steel	Lead, other metals, solvents
Textile (rayon)	Carbon disulfide, solvents
Painters	Lead, toluene, xylene, other solvents
Printers	Lead, methanol, methylene chloride, toluene, trichloroethylene, other solvents
Petroleum industry	Oil, gas, solvents
Plastics workers	Formaldehyde, styrene
Battery manufacturing	Lead, mercury
Lumber production	Pentachlorophenol, wood preservatives
Rubber, plastics	Solvents
Electrical	Polychlorinated biphenyls, solvents
Transportation workers	Lead (in gasoline), carbon monoxide, solvents
Trucking and distribution of industrial products	
Professional occupations	
Operating room technicians	Anesthetic gases
Pathologists	Solvents
Anesthesiologists	Anesthetic gases
Dentists and dental hygienists	Mercury, anesthetic gases
Hospital personnel	Alcohols, anesthetic gases, ethylene oxide cold sterilization
Service occupations	
Gas stations	Gasoline, solvents
Dry cleaners	Perchloroethylene, trichloroethylene
Technical	
Electronics workers	Lead, methyl ethyl ketone, methylene chloride, tin, trichloroethylene, glycol ether, xylene, chloroform, freon, arsine
Laboratory workers	Solvents, mercury, ethylene oxide
White-collar	
Office workers	Solvents, "sick building" effects
Other	
Hobbyists	Lead, toluene, glues, solvents

workers in many different blue-collar, white-collar, and professional occupations are at risk from exposure to neurotoxic substances, factory emissions and lead paint threaten the unemployed as well. Workers are an especially vulnerable risk group, with NIOSH concluding that neurotoxic disorders are one of the ten leading causes of work-related disease and injury.

Risk of exposure varies within and between occupations in that some worksites are better ventilated or use smaller amounts of neurotoxic material than others. Certain occupations are intrinsically more dangerous, especially those that utilize extremely neurotoxic materials, e.g., carbon disulfide and lead.

Membership in various ethnic or sociodemographic groups may increase risk for certain types of exposure (Vaughan, 1993). Epidemiological risk varies as a U-shaped curve with age, with the very young and the elderly in particular jeopardy. The National Academy of Sciences estimates that 7.5 million children under age 18 suffer from toxic exposure that may have caused or contributed to diagnosed mental disorder (National Academy of Sciences, 1989). Children have greater respiratory activity than adults, which increases inhalation of airborne toxins. Their rapid bone growth can lead to much higher accumulations of heavy metal (i.e., lead) per unit exposure, compared with adults. Zeigler, Edwards, Jensen, et al. (1978) estimate that children absorb 50% of ingested lead compared with only 5% for adults.

Developmental Neurotoxicity

The risk of neurotoxic abnormality to the developing organism had not been well recognized until 1960–1970 when many of the initial studies were performed and the discipline of *behavioral teratology* coalesced. Early behavioral teratology research investigated early malnutrition, administration of sex hormones to children, and the effects of radiation exposure on behavior. Many incidents of mass poisoning that also stimulated adult neurotoxicity research affected children across the life span, including Minamata syndrome, and cases of cerebral palsy possibly linked to maternal ingestion of mercury-treated seed grain. A similar, even more destructive outbreak in Iraq provoked the investigation of dose-dependent mercury exposure on the development of motor and cognitive impairments (Nelson, 1990).

Investigation of fetal alcohol effects has been a parallel and well-publicized offshoot of behavioral teratology research. Prenatal exposure to alcohol is "now recognized as a frequent cause of mental retardation, borderline intellectual development, and hyperactivity in children" (Nanson & Hiscock, 1990, p. 656). Worldwide estimates of alcohol embryopathy range from 0.4 to 3.1 per 1000; however, in communities with special risk, incidence as high as 190 per 1000 children has been recorded (Conry, 1990).

Current research indicates that developing fetuses are at risk from exposure to a variety of neurotoxic agents, including airborne environmental toxins, prescription or over-the-counter medications, and, most seriously, alcohol and abused drugs. The brain of the developing fetus and child is especially vulnerable to neurotoxic insult for a variety of reasons. First, the brain, with its 50% dry weight lipid content, is a preferred accumulation site for neurotoxic, lipophilic substances. Second, brain injury prior to or in the course of active cell division could cause especially severe injury by limiting the pool of dividing cells. Third, the developing brain is especially sensitive to the effects of hypoxia. Finally, neurotoxic injury may be comparatively more severe in developing brain tissue because the blood–brain barrier has not been fully developed (Claudio, 1992).

Some well-known exemplars of developmental neurotoxicants are listed in Table 1.10.

TABLE 1.10. Well-Recognized Human Developmental Neurotoxicants[a]

Ethanol	PBBs	Cadmium
Methylmercury	PCBs	Anesthetics
Other mercury	Pesticides	Cocaine
Lead	Cadmium	Methadone
Diphenylhydantoin (Dilantin)	Ionizing radiation	Heroin

The scope and cost to society of developmental neurotoxicity is as vast as it is troubling. Both industrial exposure and drug abuse contribute to the problem. The 1955 Minamata disaster in Japan first sensitized the world to the effects of congenital methylmercury poisoning, when at least 1600 adults and 26 infants developed mercury poisoning. Similar teratogenic effects of mercury were found in Iraqi infants exposed prenatally to methylmercury and later showed "gross impairment of motor and mental development, with cerebral palsy, microcephaly, deafness, and blindness," despite the fact that no clinical symptoms were observed in the mothers (Nelson, 1990, p. 302). Six million women of childbearing age use illicit drugs (Khalsa, in press), and up to 375,000 children are injured annually by substance abuse (National Association for Perinatal Addiction Research, 1988). Among 1987 high school seniors surveyed, almost 7% had used intoxicating inhalants (e.g., glue sniffing) and in certain Mexican communities these deadly neurotoxins are used *daily* by 22% of minors (Crider & Rouse, 1988).

Developmental damage from neurotoxic substance exposure is costly from a financial as well as moral perspective. A 1985 study on health costs from childhood lead exposure concluded that more than $500 million annually could have been saved if the neurotoxic effects of lead on children were reduced (*Neurotoxicity*, 1990). Drug abuse causes economic losses of at least $10 billion each year in the United States (Rufener, 1977). Fetal alcohol syndrome (FAS) is now considered the leading cause of mental retardation in the United States (Streissguth et al., 1991); its costs to society have been estimated from $3-21 million per year (from birth to age 21), to $1.4 million across the life span of a single FAS child (Streissguth et al., 1991). An estimated 2-3% of U.S. health care costs are thought to be related to neurotoxicity, at a cost of $1 billion annually (Nakajima, 1993).

Children are also at risk from the toxic materials brought home on the clothing and skin of their working parents. For example, there are positive associations between development of brain tumors in children or mothers exposed to various chemicals, and from fathers who work in aircraft, printing, chemical, or petroleum industries, or with solvents or paint (Roeleveld et al., 1990). Children of parents who work with lead and pesticides are also at risk (Garrettson, 1984).

Initial regulation of developmentally neurotoxic materials began in Great Britain and Japan in 1975, which incorporated developmental neurotoxicity testing requirements into their pharmaceutical test batteries (Rees et al., 1990). NIOSH, FDA, NIDA, and EPA are all currently empowered to promote testing and regulation of developmental neurotoxicants.

The term *developmental* is not limited to prenatal and neonatal effects but extends across the life span. The elderly are another group particularly susceptible to neurotoxic effects for a variety of reasons, including diminished capabilities for filtering toxins from the liver and kidney (Klaassen, Amdur, & Doull, 1986) and reduced effectiveness of the blood-brain barrier (Claudio, 1992). Reduced neurochemical reserves in the aging nervous system may potentiate effects of certain neurotoxins, as has been shown for MPTP, which in animal studies increases in neurotoxicity and lethality as a function of subject age (Langston, 1988). Since the brain loses approximately half of all cerebral cortical cells between ages 20 and 80, there is also far less cellular redundancy in the event of toxic injury (Claudio, 1992).

There is evidence to suggest that environmental neurotoxic insult in early or mid-life may combine with a 5-7% nigral cell loss per decade to induce eventual Parkinson's (Langston, 1988). Since 80% decrement in striatal dopamine is generally required for development of clinical parkinsonian symptoms (Riederer & St. Wuketich, 1976), neurotoxic substances could destroy a large percentage of dopamine tracts without clinical sequelae. Elderly individuals may therefore begin to show neurotoxic exposure effects accumulated over a lifetime of silent neurotoxic depletion of brain mass or neuronal reserves. Parkinsonian syndromes or other neurodegenerative dementing disease may be the end result of chronic subclinical exposure at a younger age (Spencer, 1990a). A pilot epidemiological study showed 26% of patients with Parkinson's disease to have been employed in farming versus 11% of male controls. Sixteen percent of patients, but no controls, were employed as welders (Wechsler *et al.*, 1991). Considering that 20% of the U.S. population is expected to be over age 65 by the year 2030, the problem of chronic, possibly neurotoxin-mediated neurodegenerative diseases may become a public health crisis, if left unexplored.

An additional risk for elderly individuals is increased susceptibility to prescription drug neurotoxic side effects. The Department of Health and Human Services reports that those over age 60 represent 17% of the U.S. population but account for almost 40% of drug-related hospitalizations and over half the deaths from drug reactions (*Neurotoxicity*, 1990).

Sex Differences

Gender may impact on risk in two ways. First, insofar as many occupations still employ predominantly either male or female workers, toxic exposures may be expected to impact on the sexes differently depending on the particular job's sex distribution. For example, the painting and plumbing trades remain predominantly male preserves, while the electronics industry employs many more women in occupations with high exposure to neurotoxic substances.

Second, there is growing evidence for biological differences in toxic susceptibility as a function of gender. Alcohol has been found to be less well digested in females because of relative lack of a certain stomach enzyme, leading to higher blood alcohol levels per unit ingested. A Soviet study exposed volunteers to minimally toxic levels of a cholinesterase-inhibiting pesticide. Females showed

20% greater decreases in blood cholinesterase than males, and experienced neurological and gastrointestinal symptoms of longer duration and higher intensity (Krasovskii et al., 1969, cited in Calabrese, 1985). Many pharmacologic agents also show sex-related differences, including lithium, lorazepam, nortriptyline, and oxazepam (Calabrese, 1985).

There is no corpus of neuropsychological studies on the effects of either industrially or pharmacologically neurotoxic substances as a function of gender. Ethanol, one of the few neurotoxins for which this type of information has been collected, has failed to yield definite gender effects (see Chapter 5). However, Calabrese (1985) has identified almost 200 toxic substances that (at least in animal models) exhibit experimental sex differences. It seems possible that human neurotoxins may eventually be shown to impact differently on the neuropsychological performance of males and females.

Occupational Risks

Table 1.11 is a partial listing of occupations along with neurotoxic substances potentially encountered in each.

Considering the variety of occupations, the large number of unresearched neurotoxic chemicals, and the even larger number of affected individuals, it is clear that the few psychologists currently engaged in this type of research can only begin to serve potential needs. While Weiss's (1983) call for "25,000 behavioral toxicologists" seems exaggerated, there is nevertheless a need for greatly expanded neuropsychological services in the industrial and private sector.

TABLE 1.11. Occupations and Activities at Risk for Neurotoxic Exposure

Occupations at risk	Neurotoxic substance
Agriculture and farm workers	Pesticides, herbicides, insecticides, solvents
Chemical and pharmaceutical workers	Industrial and pharmaceutical substances
Degreasers	Trichloroethylene
Dentists and dental hygienists	Mercury, anesthetic gases
Dry cleaners	Perchloroethylene, trichloroethylene, other solvents
Electronics workers	Lead, methyl ethyl ketone, methylene chloride, tin, trichloroethylene, glycol ether, xylene, chloroform, Freon, arsine
Hospital personnel	Alcohols, anesthetic gases, ethylene oxide (cold sterilization)
Laboratory workers	Solvents, mercury, ethylene oxide
Painters	Lead, toluene, xylene, other solvents
Plastics workers	Formaldehyde, styrene, PVCs
Printers	Lead, methanol, methylene chloride, toluene, trichloroethylene, other solvents
Rayon workers	Carbon disulfide
Steel workers	Lead, other metals, phenol
Transportation workers	Lead (in gasoline), carbon monoxide, solvents
Hobbyists	Lead, toluene, glues, solvents
Office workers	Solvents

Other Risks

The most obvious and ubiquitous risk of neurotoxic exposure is that of alcohol abuse and dependence. Within the past 12 months, from 7 to more than 10% of workers have experienced a problem with alcohol abuse (Roberts & Lee, 1993). Alcohol in combination with industrial exposure is a recent issue of concern since it may combine synergistically with solvents, pesticides, and metals to exacerbate neurotoxic effect. Alcohol also slows the clearance and metabolization of certain solvents, allowing them more time to exert deleterious effects in the body (Cherry, 1993).

Drug abusers make up a large and disparate group who voluntarily expose themselves to brain-damaging substances. Even among working individuals, it is estimated that 3.1 to 6% have experienced an episode of drug abuse within the past year (Roberts & Lee, 1993). Abused drugs form a class of neurotoxic materials capable of subtle and severe nervous system damage. Most recent concern has been focused on the effects of crack cocaine, sustained use of which may produce irreversible hypometabolic changes in chronic abusers, even with prolonged periods of abstinence (Strickland, Hartman, & Satz, 1991). However, more commonly available substances, e.g., inhalants, are known to induce neurotoxic effect with sustained use. Exposure to *pesticides* can be considered a fourth class of risk that is unique for both the acknowledged neurotoxicity of compounds in use and the extensive numbers of individuals potentially affected. "Workers exposed to pesticides are one of the largest occupational risk groups in the world [and] [t]he available evidence suggests there is a high probability for subtle adverse health effects" (Davies, 1990, p. 330).

SYMPTOMS OF CONCERN

Clinically, the types of subtle and gross damage exemplified by exposure in these circumstances, can produce a range of symptoms, from irreversible dementia to subtle psychiatric abnormalities without neuropsychological concomitants (e.g., sick building syndrome), emotional and behavioral dysfunction. Anger (1984) has compiled typical cognitive, emotional, and behavioral abnormalities from the reported neurotoxic effects from ACGIH documents through 1982 (Table 1.12). From a neuropsychological point of view, reported symptoms are somewhat unsystematically categorized; however, the list does suggest symptoms amenable to neuropsychological analysis.

A different way to categorize neuropsychological symptoms found in patients exposed to neurotoxins is shown in Table 1.13. The list is not meant to be exhaustive, but rather to suggest the major groupings of dysfunctions that can be testable with neuropsychological methods. Exact patterns of neuropsychological impairment will be dependent on the particular neurotoxic exposure, severity of exposure, individual susceptibility, and other factors. More information on these diagnostic factors is given in later chapters.

TABLE 1.12. Neurotoxic Symptoms Compiled From ACGIH Documents[a]

Motor		Cognitive		Affective/personality	
General motor		Alertness loss	1	Substance abuse	1
Activity changes	3	Impaired judgement	1	Anxiety	4
Incoordination	10	Memory loss	3	Asthenia/neurasthenia	2
Paralysis	9	Slurred speech	1	Belligerence	1
Performance changes	4			Delirium	2
Pupil constriction	2	*General changes*		Delusions	1
Rigidity	3	Analogy with other		Depression	4
Weakness	12	chemicals	25	Disorientation	1
		Anorexia	15	Excitability	3
Abnormal movement		Behavioral changes	3	Exhilaration	1
Ataxia	4	Cholinesterase inhibition	26	Giddiness	6
Chorea	1	CNS depression	10	Hallucinations	2
Convulsions/spasms	18	CNS edema	1	Inebriation	1
Gait, spastic	1	CNS stimulation	4	Insomnia	4
Movement disorders	1	Encephalopathy	1	Irritability	4
Nystagmus	2	Narcosis/stupor	28	Lassitude/lethargy	7
Tremor	19	Neuropathy/neuritis	6	Laughter	1
		Neurophysiological/		Malaise	2
Sensory		electrophysiological		Nervousness/nervous	
Auditory disorders	4	changes	10	disorders	11
Equilibrium disorders/vertigo/		Neurotoxicology	5	Psychological/mental	
dizziness	20	Operant behavior changes	1	disorders	4
Gustatory changes	5	Pathology, CNS/PNS	15	Psychosis	2
Olfactory changes	2	Psychic disturbances	2	Restlessness	1
Pain disorders		Unpleasant taste/smell	12	Sleepiness	5
(incl. anesthesia)	7	Weariness/fatigue/lethargy	8	Viciousness	1
Pain, feeling of	2				
Sensation deficits	1				
Tactile disorders	5				
Vision disorders	12				
Visual sense organ pathology	21				

[a]Reprinted by permission of Pergamon Journals, Ltd., from Anger, W. K.(1984). Neurobehavioral testing of chemicals: Impact on recommended standards. *Neurobehavioral Toxicology and Teratology, 6,* 152.)

SUMMARY

Neuropsychological toxicology is one of the newest branches of the field of neuropsychology, a clinical discipline concerned with understanding the relationship between brain damage and behavioral alteration. Neuropsychological toxicology, then, applies neuropsychological testing methods to assess the subtle but definite brain dysfunctions produced by neurotoxic substances.

The relatively recent development of neuropsychological toxicology reflects the youth of its founding fields, industrial medicine and neuropsychology. There remain many unanswered biochemical, anatomical, psychological, and behavioral questions that must be addressed before the field can deliver definitive pronouncements about the effects of toxic substances on the nervous system. In

TABLE 1.13. Common Neuropsychological Symptoms of Neurotoxicity

General intellectual impairments
  Intelligence (IQ) (with more severe exposures)
  Attention
  Concentration
  Abstract reasoning
  Cognitive efficiency and flexibility
  Global impairments (dementias)
Motor impairments
  Fine motor speed
  Fine motor coordination
  Gross motor coordination
  Gross motor strength
Sensory impairments
  Visual disturbances
  Auditory disturbances
  Paresthesias/anesthesias
  Tactile disturbances (PNS or CNS disorders)
Memory and learning impairments
  Short-term memory (verbal and nonverbal information)
  Learning (encoding of new information—verbal and nonverbal)
  Long-term memory (verbal and nonverbal)
Visuospatial impairments
  Constructional apraxias
Personality impairments
  Anxiety, depression, delirium, organic brain syndrome, organic affective disorder, other psychotic disorders, anger, tension, fatigue, irritability, posttraumatic stress disorder, autonomic arousal

addition, the legal, political, and economic complexities of this field have yet to be resolved. The number of multidisciplinary specialists in the field will likely have to increase considerably before such answers can be expected.

Despite current ambiguities in the field, the need for clinical service and research experimentation is immediate. Millions of workers, and an unknown number of other individuals, are exposed to neurotoxic substances as a by-product of their jobs, hobbies, prescription drug use, or voluntary substance abuse. The neurotoxic syndromes that result from this exposure can run the gamut from mild emotional disorder or other functional-seeming illness to complete and irreversible dementias.

U.S. neuropsychologists have lagged far behind their Scandinavian and European colleagues in expanding the frontiers of this multidisciplinary field. Current trends, however, are in the direction of greater worldwide participation in research and clinical service. The huge number of potential neurotoxins and exposed individuals suggests an immediate need for clinical and research programs in all phases of toxicology research. Hopefully, the entire field of toxicology in all of its clinical, legislative, and legal branches will see greater future growth. However, the need for neuropsychological research is particularly great. "Neuropsychological toxicologists" could provide immediately useful input to medicine, industry, and regulatory agencies; all those concerned with providing safe working and living conditions. Neuropsychological toxicology, as a conceptual entity, is the banner under which an entirely new area of research can be conducted and clinical services can be performed. The niche is there and the need is great.

· · · · ·

Many of the industrial toxins presented in the following chapters are preceded by general industrial hygiene information collected by NIOSH, from the computerized *Quick Guide*, Version 2.0 (1993), a version on disk of the NIOSH *Pocket Guide,* published by Industrial Hygiene Services, Inc., which summarizes toxicity information for many industrial substances. Unless specifically noted otherwise, all OSHA exposure limits are for 8-h time-weighted average (TWA) concentrations. *Ceilings* cited by OSHA "shall not be exceeded at any time, unless noted otherwise" (NIOSH *Pocket Guide*, 1985, p. 13).

Other abbreviations include: ppm, parts per million; TLV, threshold limit value; STEL, short-term exposure limit; IDLH level immediately dangerous to life or health, a "maximum concentration from which one could escape within 30 minutes without any escape-impairing symptoms or any irreversible health effects" (NIOSH *Pocket Guide*, 1985, p. 14).

2

Evaluation of Neurotoxic Syndromes

INTRODUCTION

Diagnosis and management of patients with neurotoxic syndromes can best be carried out by a specialized team of health care professionals. The respective contributions from physicians and neuropsychologists will obviously depend on the circumstances of toxic exposure. Individuals exposed to high concentrations of acutely toxic materials require immediate medical management to sustain life and, if possible, eliminate the toxicant from the body. In such a situation, neuropsychological evaluation can be a valuable follow-up procedure to assess recovery of cortical function. Alternatively, when the diagnostic issue is one of subclinical effects and/or low-level, chronic exposure, then neuropsychological evaluation can proceed in tandem with medical tests and become an important *initial* contribution to final diagnosis and subsequent management. In either case, the ideal strategy for either clinical or research evaluation of patients with claimed neurotoxic exposure is a multidisciplinary one (e.g., Becking, Boyes, Damstra, & MacPhail, 1993). A diagnostic decision tree of the typical sequence of clinical evaluation for neurotoxic syndromes at Cook County Hospital, Chicago, Illinois, is illustrated in Fig. 2.1. While the flowchart is specific to industrial neurotoxicity assessment, the flow of multidisciplinary, cooperative effort should be applicable to diagnosis of other types of neurotoxic syndromes.

Because more individuals are *chronically* exposed to neurotoxins, either at work or via prescription or "street" drug use, than are in danger from high-concentration *acute* effects, and because available neuropsychological literature has concentrated on initial diagnosis rather than follow-up, this chapter will chronicle the types of health questions and neuropsychological procedures that may typically be performed in such an initial evaluation.

INITIAL MEDICAL EVALUATION

Patients with verified toxic exposure should receive basic medical and neurological evaluation; toxicology screens of blood or urine and other tests may be

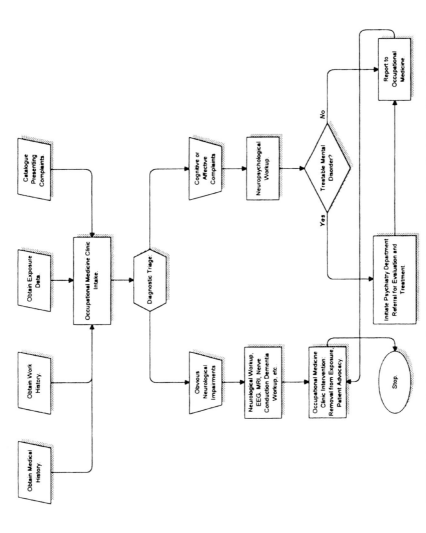

FIGURE 2.1. Patient referral sequence through the occupational medicine clinic at Cook County Hospital: multidisciplinary process. (The author thanks Stephen Hessl, MD, Director of Occupational Medicine, Cook County Hospital, for his collaboration in the development of this flowchart.)

initiated according to the types of substances that the patient has ingested or been exposed to. Clinical neurological examination of cranial nerves I through XII is recommended and "assessment of optic (II) and trigeminal (V) nerves is particularly useful in suspected neurotoxic conditions" (Spencer, Arezzo, & Schaumburg, 1985, p. 12). Spencer *et al.* (1985) also recommend examination of the ocular fundi, motor and sensory systems, reflex examination, and, in certain cases (e.g., anticholinergic drugs, pesticides), evaluation of the autonomic nervous system. While most patients receiving subclinical exposures of toxic substances will present with negative neurological findings, alternative medical or neurological illness occurring coincidentally with exposure should be ruled out.

HEALTH QUESTIONNAIRE CONSTRUCTION

While symptom questionnaires may not, by themselves, discriminate patients with neurotoxic exposure from controls or those with confounding diagnoses (Bukowski, Sargent, & Pena, 1992), clinical investigation can often be facilitated with administration of structured interviews or symptom checklists. In addition, the systematic collection of basic demographic, medical, and neuropsychological data allows for present or future research comparisons, an important consideration in any developing field. Questionnaires can be designed to be self-administered or given in structured interview by an examiner. Which type of questionnaire is chosen depends on many factors, among them the ability of the patient to work alone, the wish of the examiner to observe the patient's responses, and whether "in-depth" follow-up questions or elaborations are needed. Questionnaires may be constructed to acquire general information or they may be tailored to the specific substances under investigation. Hogstedt's questionnaire (Table 2.1) designed to assess solvent exposure symptoms is an example of the latter.

Beckmann and Mergler (1985) propose consideration of what might be considered "human factors" when designing and administering a questionnaire. These factors include:

1. Determining a "fatigue factor" for both examiner and patient in the administration of the questionnaire. Will exposed patients tire more easily than controls?
2. Will various forms of biases enter a patient's response set? Might there be differentially biased responding between exposed patients and controls? For example, will subjects be unwilling to accurately catalogue their alcohol intake for fear that such information will be made part of their personnel files?
3. How will impairment, if it exists, affect response quality?

In addition, the authors suggest that questions be as specific as possible. Questions that are tied to independently observable behaviors rather than purely subjective data are preferred (Beckmann & Mergler, 1985).

TABLE 2.1. Solvent Exposure Questionnaire

1. Do you have a short memory?
2. Have you ever been told that you have a short memory?
3. Do you often have to make notes about what you must remember?
4. Do you often have to go back and check things that you have done, such as turned off the stove, locked the door, etc.?
5. Do you generally find it hard to get the meaning from reading newspapers and books?
6. Do you often have problems with concentrating?
7. Do you often feel irritated without any particular reason?
8. Do you often feel depressed for any particular reason?
9. Are you abnormally tired?
10. Are you less interested in sex than what you think is normal?
11. Do you have palpitations of the heart even when you don't exert yourself?
12. Do you sometimes feel a pressure in your chest?
13. Do you perspire without any particular reason?
14. Do you have a headache at least once a week?
15. Do you often have painful tingling in some part of your body?

*Hogstedt, Hane, and Axelson (1980).

Caution must be urged in the interpretation of questionnaires from patients in litigation. A recent study suggested that U.S. patients with claims for personal injury related to emotional distress, but no history of head injury or toxic exposure, endorsed a variety of cognitive and emotional symptoms more frequently than either head injury/toxic exposure patients or controls (Dunn, Brown, Lees-Haley, & English, 1993). Subjective reactions to neurotoxic exposure depend not only on the exposure type and duration but also on the patient's general tendency to report medical problems and/or complain about discomfort (Seeber, Kiesswetter, & Blaskewicz, 1992).

Areas of Inquiry

At least three areas of inquiry are relevant to an evaluation of neurotoxic exposure: (1) medical history and current symptoms in major organ systems, (2) occupational history, and (3) psychological information. These data can be garnered from questionnaires specially designed for a particular patient or research population, or they may be collected via careful selection and use of available surveys.

Medical History

Existing questionnaires, like the Cornell Medical Index, may be an aid to the diagnostic assessment of neurotoxic syndromes. A physical symptom questionnaire also can be constructed for specific target symptoms with specialized questions addressing common neurotoxic complaints, e.g., dizziness, headache, incoordination, paresthesias. Any prior or current medical condition potentially capable of interfering with neuropsychological performance requires special inquiry. These conditions include, but are not limited to:

1. Central nervous system pathology, e.g., head injury, high fever, seizure disorder, periods of unconsciousness; diseases capable of affecting the nervous system, including lupus erythematosus, syphilis, AIDS or ARC, diabetes, etc., hereditary degenerative disease.
2. Peripheral nervous system (PNS) pathology, e.g., peripheral neuropathies, diabetes, carpal tunnel syndrome, PNS injury, chronic pain.
3. Nonneurological medical problems capable of affecting neuropsychological functioning: arthritis, liver disease, infectious or metabolic disorder, orthopedic/muscle injury in upper body producing motor impairment.
4. Drugs and medications: alcohol, drug abuse, caffeine intake, prescription drugs with neuropsychological effects (e.g., benzodiazepines).
5. Psychological/psychiatric problems (see Psychological History)
6. Other factors: recent sleep history, nutritional status (e.g., dieting, recent unusual weight loss or gain).

Occupational History

For workplace exposures, questions to be addressed include frequency, duration of and interval between exposures, and the relationship between these variables and symptoms. A detailed job description is essential for two reasons. First, job analysis will clarify the nature of the patient's contact with neurotoxic substances and will aid the examiner in assessing the severity of symptoms. Second, a detailed history of present and past job responsibilities can be compared with neuropsychological test results for consistency. For example, mild visuospatial difficulties in an "engineer" who draws and interprets blueprints might be seen differently from those same difficulties in an "engineer" whose job it is to sweep the factory floor. A list of occupational symptoms, illnesses, and injuries should also be included, either in the context of an occupational questionnaire or in another medical checklist. In addition, satisfaction with the work environment, relationships with fellow employees, and degree of work stress are important areas of inquiry to determine the influence of psychosocial stressors on symptom production.

Psychological History

Since many neurotoxic symptoms, by themselves, are nonspecific (e.g., headache, depression, forgetfulness), a psychological history will allow the clinician to consider alternative explanations for neurotoxicity complaints. There are at least six possible relationships of psychological symptoms to neurotoxic exposure:

1. Symptoms may be *primary* behavioral, cognitive, or affective concomitants of structural or neurochemical lesion. "Depression" as a result of lead poisoning, or slowed, confused thinking from pesticide exposure are examples.
2. Symptoms may be *reactive* to increased stress as a *secondary* consequence of real neurological disability. For example, if an architect were exposed to

toxicants that caused peripheral neuropathy, ensuing depression could be the psychological reaction to loss of fine motor control and subsequent inability to perform career duties. Similarly, an individual with carbon disulfide-induced cerebral vasculopathy may react catastrophically to perceived deficiencies in cognition. In both of these cases, real neurological disability may coexist with, or be exacerbated by, affective components.

3. Symptoms may be *reactive* to the *psychosocial* stressors of exposure, rather than to structural or neurochemical abnormality. For example, an individual might develop neuropsychological symptoms after working in a factory that has been found to be in violation of OSHA toxic exposure standards. Fears of possible exposure, job loss, nonneurological medical consequences of exposure (e.g., cancer), legal proceedings, and financial hardships may produce "learned helplessness" and other reactive effects. These reactions, in turn, may produce abnormal neuropsychological test behavior. The patient, family, friends, and coworkers may view toxic exposure as much more frightening and threatening than other illnesses, which reinforces the patient's use of medical services (Bolla, 1991). Knowledge of toxic exposure effects and fears of injury can occur in workers regardless of literacy or formal education (Vaughan, 1993).

4. Symptoms may be *unrelated,* linked to difficulties that existed prior to toxic exposure, or to stress that occur coincidentally with toxic exposure, e.g., continuation of premorbid alcoholism or personality disorder, death of a parent or spouse.

5. Symptoms may be *conditioned,* and obey the laws of Pavlovian conditioning in a subset of patients. Bolla (1991) describes a paradigm where the individual is exposed to a strong odor from a toxicant, which functions initially as an unconditioned stimulus to various symptoms. The odor alone serves as the *unconditioned stimulus,* and elicits similar symptoms. Generalization of response to similar odors may occur and symptoms may actually increase in severity after exposure ends as the result of generalization.

6. Symptoms may be *malingered.* Individuals in litigation, in particular, stand to gain significant compensation as a function of claimed damages (neuropsychological and otherwise). Ruling out the possibility of voluntary symptom production for such a purpose is or should be simply a part (albeit regrettable) of the clinical rule-out process for all neuropsychological evaluations.

Besides discrete symptom lists, psychological questionnaires may include the following items:

1. Social, marital, interpersonal adjustment (home and work adjustment).
2. Current symptoms and psychosocial stressors, e.g., the Symptom Checklist—90 Revised (SCL-90-R), "Personal Problems Checklist" (e.g., Schinka, 1984), and/or an event inventory (e.g., Holmes & Rahe, 1967).
3. Psychological history: prior or current history of mental health treatment, diagnoses; hospitalization and medication history; history of emotional functioning and psychological symptomatology.
4. Current behavioral and emotional adaptations/stressors. Representative questions might include queries pertaining to factors likely to interfere

with neuropsychological test performance (e.g., alcohol, substance abuse, smoking, sleep, mood state, interpersonal adjustment).
5. Personality and affect inventories: questionnaires that assay personality traits and current emotional state. Examples of the former include the Minnesota Multiphasic Personality Inventory (MMPI-2) or the Personality Assessment Inventory (PAI). Common inventories of current affect include the Beck Depression Inventory (BDI), the Profile of Mood States (POMS), and the Multiple Affect Adjective Checklist (MAACL). These are discussed in more detail later in the chapter.

Psychological questionnaire information may be obtained from family or collaterals if the patient is deemed to be an uncertain or inaccurate historian.

Other Factors

Complete personal history will sometimes produce relevant information unrelated to job, social adjustment, or psychosocial factors. Certain hobbies, for example, may include chronic exposure to neurotoxic substances, e.g., lead in stained-glass making, solder in electronic construction, toluene or *n*-hexane-based glues in plastic model construction. Certain recreational pursuits (e.g., boxing, soccer) have been linked to closed-head injury and diffuse neuropsychological impairment.

NEUROPSYCHOLOGICAL TOXICOLOGY TEST BATTERY CONSTRUCTION

Hänninen (1982), referring to the construction of neuropsychological toxicology batteries, stated that "not all neuropsychological assessment methods are good research tools" (p. 123). Similarly, neuropsychological instruments also vary in their clinical utility for toxicity diagnoses. Baker, Feldman, White, Harley, *et al.* (1983) suggest the following criteria for the selection of a neurobehavioral toxicology battery. The evaluation must be *comprehensive;* it must be *adaptable* to field conditions; it must be free of, or at least able to factor out, extraneous influences (e.g., age and personal habits); and results must be reproducible. Anger and Cassitto (1993) recommend test batteries take advantage of both the capacity of a human tester to elicit cooperation and detect subtle response factors, and also the accuracy of precision measurement with computerized methods. They propose that categorizing a battery as human versus computerized is artificial and limiting and that both high technology and human judgment are desirable features of neurotoxicity testing.

To this, several additional criteria can be added. First, the battery should be psychometrically sound and experimentally validated, if possible, among groups of patients with neurotoxin-related disorders. Second, while maintaining comprehensiveness, a battery of tests must also emphasize areas of neuropsychological functioning known to be correlated with subclinical changes in neuropsychological functioning. Some of the factors that typically show impairment include reaction time/motor response, fine motor coordination, cognitive efficiency, sustained attention, and affect.

Finally, it must be hoped that advances in the quantification and precision of neuropsychological toxicology batteries will proceed in tandem with similar advances in theoretical understanding and scientific vocabulary available to describe the syndromes. Letz and Singer (1985) believe the responsibilities of the specific community should include:

1. A commonly agreed-upon categorization of neuropsychological functions, an agreed-upon linking of specific tests to those functions, and a standard vocabulary for discussing those functions and tests should be established.
2. Rigid diagnostic criteria for clinical syndromes and specific definitions for subclinical effects associated with neurotoxicant exposure should be established.

Epidemiological versus Clinical Batteries

There is also a distinction to be made between neuropsychological tests developed for epidemiological research and those designed for individual clinical assessment. Initial epidemiological batteries were composed of the "best tests on hand at the time," usually traditional paper-and-pencil materials (Hänninen, 1991, p. 40). More recently, computerized batteries have become an "integral part of neurotoxicologic assessment" (Letz, 1993, p. 124), most notably the Neurobehavioral Evaluation System-2 (NES-2) (Baker & Letz, 1986).

There is no bright line separating epidemiological from clinical assessment batteries; there is substantial overlap in functional domains assessed. However, newer batteries developed for large-scale population assessment have had the luxury of what Helena Hänninen called a "progressive" approach; using newer, potentially more sensitive measures, or incorporating updated theories of the relationship between neurotoxic exposure and neuropathology. The risk, of course, is that loss of comparability with earlier tests of known clinical validity leave open the question as to whether an exposure "had no behavioral effect or whether the promising methods indeed were not sensitive enough to display them" (p. 47). Epidemiological batteries are not usually helpful in an individual clinical context where data from a single individual must be interpreted, since epidemiological batteries, by definition, assess group differences. The lack of more "clinical" norms to provide some index of impairment level, outcome, disability, or rehabilitation potential is a well-known and frustrating limitation of several promising epidemiological batteries.

The use of tests in group studies affords the epidemiological researcher some luxuries that clinicians cannot afford. Test reliabilities may be less stringent in epidemiological investigations, e.g., a reliability of 0.80 has been considered acceptable in group studies, but where clinical decisions are required, reliabilities of at least 0.90 are more common (Arcia & Otto, 1992). The development of group norms is less important for matched control designs, where group data comparisons take the place of population norm comparisons. When the issue is "statistical" difference rather than clinical impairment, there is less of an imperative to develop accurate clinical decision-making cutoff criteria.

Clinical Neuropsychological Assessment of Neurotoxic Exposure

Two general recommendations can be made about clinical neuropsychological toxicology testing. First, "blind" testing is inappropriate to clinical assessment of neurotoxicity. While well-designed research projects can justify isolating the neuropsychologist from information not collected during the actual test session, blind testing is almost never appropriate in a clinical context.

Patients with neurotoxic exposure should have their evaluations interpreted within the larger context of any available medical, demographic, educational, psychological, and sociological data that it is feasible to collect. Data from a period predating the patient's exposure to neurotoxic materials is a recommended context toward understanding the patient's current functioning. Developmental history, school records, premorbid personnel records detailing job performance and interpersonal adjustment are just a few of the data that may prove important to interpretation of the patient's current clinical behavior.

Second, neuropsychological toxicology batteries should, as much as possible, attempt to sample a *comprehensive* array of cognitive, affective, and neurobehavioral functions. An overly narrow approach that limits test selection to one or two "sensitive" measures does not do the patient justice from a diagnostic viewpoint; every patient should be approached with the possibility that alternative, "rule-out" diagnoses may be required. Failure to administer a comprehensive battery will limit the validity of neuropsychological examination by restricting the data from which the neuropsychologist may make a diagnostic judgment.

Test Structure and Content

Hänninen (1982), discussing the use of test batteries in neuropsychological toxicology, suggests that European batteries constructed for this purpose have a similar structure; all test *intellectual functions, memory, perceptual-motor speed and accuracy, and psychomotor abilities* (p. 124). She notes several tests from these categories that have consistently proven to be sensitive to neurotoxicity. These include: the Benton Visual Retention Test, several tests of perceptual speed and accuracy (e.g., WAIS-R Digit Symbol Test), and the Santa Ana Dexterity Test, a pegboard task similar to the Minnesota Rate of Manual Dexterity Test. Cassitto (1983) provided an expanded list of functions to be tested in neuropsychological toxicology studies. These include: vigilance, focused attention, visual and auditory acuity, time and space orientation, motor abilities, eye–hand coordination, memory, verbal skills, perceptual abilities, and abstract reasoning (p. 30). Summerfield (1983) additionally suggests that *reaction-time* tasks under time pressure or paced performance criteria may be more sensitive than unpaced tasks. He also advocates use of *parallel processing* or divided attention tasks that put the subject's neuropsychological functions under increased task demand, thereby increasing the likelihood that subclinical impairments will interfere with such complex processing and degrade performance.

Research by Morrow, Robin, Hodgson, and Kamis (1992) indicates that the use of a *continuous performance* test (e.g., scanning a computer screen and responding to a degraded image of the number "9") may be particularly sensitive to neurotoxic solvent-related effects. Using a signal detection paradigm, they found that d' (d *prime*) increased over successive blocks of trials for control subjects, but decreased for exposed patients. Thus, the ability to correctly discriminate targets from nontargets *decreased* for exposed subjects, while accuracy *increased* for normal controls. Such an extended CPT paradigm may be particularly useful in assessing neurotoxins that leave the brain with limited neurochemical reserves. While initial behavior is adequate, as reserves are depleted or not restored, behavior degrades—an effect opposite that of healthy individuals, who may actually exhibit facilitation with continuous performance.

Mergler and colleagues in several studies suggest the use of *color vision discrimination* tests to assess early effects of neurotoxic exposure (Mergler, Huel, Bowler, Frenette, & Cone, 1991; Mergler, Bowler, & Cone, 1990; Mergler, Belanger, DeGrosbois, & Vachon, 1988; Mergler & Blain, 1987). For example, they found impaired desaturated color vision on the Lanthony D-15 color panel in 17 out of 21 (81%) patients presenting with exposure to solvents, Freon, or pesticides at an occupational health clinic (Mergler *et al.*, 1990). Type II color vision loss (concurrent red-green and blue-yellow) which "most often results from damage to the optic nerve" was found in 52% of patients with diagnosed cognitive impairment (p. 671).

Certain tests do not appear to be sensitive to neurotoxic effects. Vocabulary tests, "achievement" tests of reading level or knowledge of school-learned facts are typical examples. These may be employed clinically as "hold" tests, i.e., tests that "hold up" or are resistant to subclinical neurotoxic effects. Both types of tests can be usefully included in a neuropsychological toxicology battery. So-called "hold" tests can also be used as a rough estimate of premorbid abilities to allow comparison with tests more sensitive to current function.

Clinical Testing Approaches

Neuropsychologists have tended to apply already familiar strategies of clinical neuropsychological evaluation to the investigation of neurotoxicity. These may or may not be appropriate to evaluating patients with neurotoxic exposure. For example, Christensen (1984) endorses the use of Luria's methods, although most research to date finds Lurian methods insensitive to neurotoxic exposure. Lezak (1984) advocates an approach tailored to the effects of specific substances, stating that "different kinds of conditions have different behavioral manifestations and therefore call for different assessment programs" (p. 28). This approach is probably an "ideal" that will not soon be realized; there have been few differentially diagnosable neurotoxic syndromes thus far uncovered that would require unique neuropsychological batteries. Bergman (1984) prefers the standardized Halstead–Reitan neuropsychological battery, an approach that requires long administration time, and has few direct assessments of memory.

Neuropsychological testing methodologies may be loosely grouped into "fixed" batteries versus "flexible" approaches. Each has its advantages and disadvantages. Fixed batteries with a rigidly specified set of tests are advantageous to researchers because they allow for easy replication of testing conditions and materials, and thereby encourage validation attempts. Alternatively, fixed batteries may not adequately sample the types of functions known to be impaired with certain neurotoxic substances. For example, the absence of verbal memory and learning tasks on the Halstead–Reitan would make that battery less than adequate for assessment of substances with documented effects on those functions, e.g., benzodiazepines or beta-blockers.

Flexible Approaches

Flexible or "process" approaches that tailor examination procedures and materials for individual substances can emphasize certain neuropsychological functions over others, and thus under certain circumstances may produce a more "in-depth" analysis of particular impairments. On the other hand, an overly flexible approach may emphasize "in-depth" analysis at the expense of a comprehensive survey of neuropsychological function. Failure to test for a complete array of neuropsychological functions may limit the investigation of effects to what is already known, and can result in an incomplete neuropsychological profile. In addition, when flexible approaches become too "flexible," i.e., when they use unvalidated procedures, immediate clinical conclusions and future research validation may be compromised.

An internationally agreed-upon neuropsychological battery for neurotoxic research could potentially allow vast amounts of data to be collected and compared from all parts of the world. On the other hand, flexible "core" batteries are subject to limitations of both fixed and flexible strategies; once designed, they may fail to show generalized application to the range of substances they may be subsequently called upon to test. A "required" battery could also stultify creative research that deviates from the agreed-upon approach. Finally, if a core battery is not designed for comprehensive assessment, it may share limitations of overly flexible approaches.

Examples of the neuropsychological test materials from fixed battery and flexible battery approaches include:

Standard Batteries

Two standard batteries have achieved international usage in neuropsychology: the Halstead–Reitan Battery (HRB) and the Luria–Nebraska Neuropsychological Battery (LNNB). The Halstead–Reitan may be the most commonly used standard neuropsychological battery (Bigler, 1984). It is composed of tests that were collected and introduced by Ward Halstead in 1947, and further developed by his graduate student, Ralph Reitan, who standardized the battery, eliminated some tests and added others. The present-day HRB includes the following subtests: Finger Oscillation Test (Finger Tapping), Tactual Performance Test, Category Test, Seashore Rhythm Test, Speech Sounds Perception Test, Lateral Dominance Test, Grip Strength (Dynamometer), Trail-Making Tests A & B, Aphasia Screening Test (Reitan–Indiana), and the Reitan Kløve Sensory-Perceptual Examination. One of the Wechsler Intelligence Scales is typically included as part of the HRB, originally the Wechsler–Bellevue, and currently the Wechsler Adult Intelligence Scale-Revised (WAIS-R). A version of the Minnesota Multiphasic Personality Inventory is also often included.

Comparatively few neurotoxicology studies have used the full HRB, probably because of the time involved in administration and scoring (over 6 h). Many researchers have utilized parts of the battery in neurotoxicology research, most frequently Finger Tapping and Trails, along with the Wechsler subtests of Block Design and Digit Symbol and Digit Span.

The LNNB was developed by Golden, Hammeke, and Purisch (1980) as a standardization of Russian psychologist Aleksandr Luria's methods. The authors used test items listed by Christensen (1975), included their own procedures, and constructed 11 clinical scales, initially called Motor, Rhythm, Tactile, Visual, Receptive language, Expressive language, Writing, Reading, Arithmetic, Memory, and Intelligence. The LNNB has had a somewhat controversial beginning; however, recent studies have demonstrated the LNNB to be statistically equivalent in diagnostic accuracy to the HRB (Golden *et al.*, 1981). While Luria's methods (and by inference the LNNB) have been suggested to be useful in neurotoxicology research (Christensen, 1975), there is little research with the LNNB with neurotoxic exposure patients.

Use of the LNNB for evaluation of neurotoxic syndromes may be problematic for several reasons. A major deficiency of the battery is the lack of complex, multicomponent tasks, the same types of tasks that are most often found to be impaired in neurotoxic exposure patients. This limitation is carried over from Luria's original testing philosophy, which emphasized the analysis of component abilities with systematic variation of simple tasks. Complex tasks were not given because their deficiencies could not be readily attributed to specific component abilities (Purisch & Sbordone, 1986). A second limitation in the LNNB is its relatively restricted investigation of memory, which is often impaired in neurotoxin-exposed patients. Given these limitations, it is not surprising that the LNNB is relatively insensitive to some types of alcohol disorders, the only major neurotoxin that has been investigated with this battery. Use of the LNNB in detecting early subclinical effects of other neurotoxins does not seem warranted at this time.

Flexible Batteries

Most investigations of neuropsychological toxicology have used "flexible" batteries. These batteries, like the WHO/NIOSH core battery, are typically compiled from a set of existing neuropsychological tests whose diagnostic utility and validity have already been demonstrated. Thus, the flexible battery follows Walsh's (1978) recommendation that "every effort should be made to reduce the variance produced by unstandardized and subjective methods of administration and scoring by adhering to standardized instructions, methods of presentation and presentation orders" (p. 301). Standardization benefits the research community by allowing easy replication of research findings. Use of existing tests saves research time otherwise spent on test construction and validation. It allows researchers to tailor the focus of the battery to investigate the specific neurotoxic effects of a substance.

There are a variety of neurotoxicity batteries that have been developed. Most generally, the batteries can be conceptually divided into those that are employed as epidemiological research tools, and those that allow for clinical decision-making on the individual level. Some attempt a comprehensive survey of functions, others are very minimal "screening" measures. The flexible batteries of neuropsychological tests discussed below are currently in use to detect neurotoxic symptomatology.

Microcomputer-Based Neuropsychological Batteries

Advantages. Automated neuropsychological batteries designed for personal computers have several advantages for neurotoxic assessment. They are easy to transport to a testing site, are unfazed by repetitive, standardized administration criteria, and can simultaneously administer, score, and statistically analyze test data. They surpass human administrators in efficiency and consistency of administration (Anger & Cassitto, 1993). These advantages make them particularly useful in large-scale epidemiological and laboratory studies. Computerized test batteries allow assessment of complex cognitive paradigms heretofore confined to the laboratory, including Sternberg memory tasks and signal detection paradigms. Other advantages that are just beginning to be addressed in newer batteries include improved comparability among tests, higher test-retest reliabilities as a result of better control of stimuli, innovative modes of testing (voice recognition, touch panels, eye-tracking systems) allowing testing of disabled patients, fewer data transfer errors, and utilization of tests with higher motivational value (Kennedy, Wilkes, Dunlap, & Kuntz, 1987).

Disadvantages. Computerized examinations also present a number of problems that are discussed in greater detail elsewhere (see Hartman, 1986a,b; Matazzo, 1985). A brief list of these difficulties follows.

1. Clinical norms may not be available for some computer batteries, making the use of these computer tests premature for individual clinical investigations of neurotoxic exposure.

2. Clinical evaluations require a broader range of neuropsychological functions than are currently available in computerized batteries (Letz, 1990, p. 198).

For example, motor tests, expressive language tests, assessments of 3-D praxis, and evaluations for astereognosis cannot effectively be performed on today's computers.

3. Use of computer batteries is deceptively easy, since many of the batteries essentially self-administer and thereby bypass professional administration, patient observation, or interpretation. This may encourage the inappropriate use or interpretation of computer-obtained data, e.g., use by untrained personnel, utilization by management to hire or fire workers, "drug screenings," or even mass testing for the sole purpose of generating income for the tester.

4. Computerized test results do not take the place of an individual professional evaluation of clinical behavior; mass testing paradigms may obscure differences in subject behavior that can influence testing, e.g., fatigue, computer anxiety, malingering. Test data are not meaningful without an observed context of test-taking behavior (Matarazzo, 1985). Subject cooperation, affect, and behavior must be independently observed by the clinician.

5. Most generally, computerized testing is not equivalent in data obtained or interpretations made within a clinical *assessment*. The latter requires integrated evaluation from a much wider domain of professionally obtained information and clinical data (Tallent, 1987).

6. A subset of the population is considered to be phobic or at least fearful of computerized methods. The possibility of "computer-phobia" should be addressed and ruled out. In addition, poorly designed software may exacerbate anxiety and frustration even in computer-literate populations. Inefficient performance on such software may be the result of computer-related factors rather than neurotoxic impairments.

Nonetheless, with proper validation and adherence to relevant ethical and methodological concerns, microcomputer batteries may greatly advance the science of measuring cortical function. With adequate patient observation and test validation, batteries such as these may prove to be powerful additions to the neuropsychologist's armamentarium.

There are several batteries currently being developed or validated that have been designed specifically for neuropsychological assessment of environmental or occupational exposure to neurotoxic substances. Of these, only the Neurobehavioral Evaluation System (NES) appears to be in relatively wide use at this time with extensive international support. Other batteries are less widely supported and/or based on obsolete computer systems (e.g., Apple II), and as such, may not prove useful or continue to be produced. The diversity of these offerings does suggest the potential for creative application of computerized methods in neurotoxicology.

Intelligence Tests

The revised Wechsler Adult Intelligence Scale (WAIS-R) is currently the most frequently used measure of "IQ" in the United States. Full-scale IQ scores computed from the Wechsler scales or other measures have not been found to be particularly sensitive to neurotoxic exposure effects. However, use of the WAIS-R or a valid screening equivalent (e.g., K-BIT) may be important for neurotoxicology examinations for at least two reasons. First, IQ correlates with many neuropsychological functions; the necessity to control for IQ has become increasingly apparent to researchers wishing to make valid statements about neurotoxic effects irrespective of premorbid abilities (e.g., Gade, Mortensen, Udesen, & Bruhn, 1985). Second, while overall IQ estimates from the WAIS-R are not diagnostic of neurotoxicity, several subtests appear to be sensitive to, and have been used quite frequently to test for, subclinical toxic exposure effects. These tests are Digit Symbol, Block Design, Similarities, and Digit Span.

NEUROPSYCHOLOGICAL TOXICOLOGY TEST BATTERIES

WHO Neurobehavioral Core Test Battery (NCTB)

Beginning in 1983, the World Health Organization (WHO) and the National Institute of Occupational Safety and Health (NIOSH) attempted to provide a compromise between fixed and flexible batteries with their advocacy of a so-called "core" battery approach to epidemiological investigations. The concept of a core battery is almost identical to the approach endorsed by neuropsychologist Arthur Benton, who recommended that a small group of tests be selected as a general screening measure and a possible prelude to more detailed testing if the screening is positive.

The WHO core battery was additionally intended to provide a rapidly administered culture-fair set of tests for international data collection. In crafting a battery with those goals, the developers of the WHO elected for the most part to stay with simple, easy-to-administer paper-and-pencil tests. Two of the tests (reaction time and Santa Ana pegboard), however, are expensive exceptions to that rule.

The NCTB's goal of international distribution and utilization has been quite successful, with studies conducted in the People's Republic of China and norms collected for populations in Hungary, Poland, Austria, Italy, France, and South Africa (Liang, Chen, Sun, Fang, & Yu, 1990; Cassitto, Camerino, Hänninen, & Anger, 1990; Dudek & Bazleywicz-Walczak, 1993; Nell, Myers, Colvin, & Rees, 1993). Data from over 2300 unexposed control subjects have come in from ten European countries, North and Central America, and Asia, in five age ranges between 16 and 65 (Anger *et al.*, 1993).

The battery includes the tests listed in Table 2.2 (Baker & Letz, 1986).

Factor analysis conducted by Hooisma, Emmen, Kulig, Muijser, and Poortvlient (1990) showed a five-factor solution to the NCTB, including perceptual speed, immediate memory, reaction time, motor speed and coordination, and learning and memory. Thus far, international performance has been most consistent across populations on Simple Reaction Time and the Benton Visual Reten-

TABLE 2.2. The WHO Neurobehavioral Core Test Battery

Test	Functional domain
1. Aiming (Pursuit Aiming II)	Motor steadiness
2. Simple Reaction Time	Response speed
3. Santa Ana Dexterity Test	Fine motor coordination
4. Digit Symbol Test	Perceptual speed
5. Benton Visual Retention Test (recognition form)	Visual perception/memory
6. Digit Span	Auditory attention/memory
7. Profile of Mood States	Currently experienced affect

tion Test (with data from Nicaragua being the exception). Santa Ana, Digit Symbol, Digit Span, and Aiming have provided more variable results. Particular problems were found administering the battery to an uneducated, largely rural and illiterate population from Nicaragua who were unfamiliar with geometric figures or holding a pencil, and whose results were much poorer than those found in other countries. The results indicate there are limitations in the universality of the WHO battery and that different tests may need to be selected for countries with illiterate or poorly educated populations (Anger *et al.*, 1993).

Pittsburgh Occupational Exposures Test (POET) Battery
(Ryan, Morrow, Bromet, & Parkinson, 1987)

An example of a flexible battery approach using well-known U.S. testing materials, the POET battery has been used for both clinical and epidemiological investigations. It employs many tests that are commonly used for neurotoxicity evaluations. Ryan *et al.* (1987) have normed the POET battery on 182 males employed in blue-collar occupations. Stratified norms are published for four age ranges, and factor analysis of the battery supports a four-factor solution, with associative and delayed memory recall accounting for the largest portion of the variance. Blue-collar norms make this an especially valuable battery, as it is normed on the population most likely to be evaluated. However, only 7% of the normative population had less than 9 years of education, possibly compromising interpretation of results from low-education workers. Tests in the POET include:

- WAIS-R Intelligence Subtests: Information, Similarities, Digit Span, Digit Symbol, Picture Completion, Block Design
- Visual Reproduction and Visual Memory: Immediate reproduction, direct copy and 30-minute delayed recall of Wechsler Memory Scale Form I design cards.
- Verbal Associative Learning: Ten pairs of unrelated nouns simultaneously read and visually presented for 2 seconds. First word of each pair was retrieval cue. Delayed verbal recall of ten pairs assessed 30 minutes later using first word as retrieval cue.
- Symbol Digit Learning: Visual presentation of symbols on Kodak Audio-

viewer for 3 seconds. Symbols presented at end of list as retrieval cue for digits. Response followed by presentation of symbol digit pair for 3 seconds. Four test trials. Thirty-minute delayed recall after all four trials completed.
- Incidental Memory: Subject asked to recall nine symbols from the WAIS-R Digit Symbol Substitution Test immediately after test.
- Recurring words: Continuous recognition paradigm. Fifty four-letter words individually presented either once, twice, or three times. Subjects determined whether they had seen each word before.
- Other tests: Boston Embedded Figures, Mental Rotation Test, Trail Making Test, Grooved Pegboard Test.

California Neuropsychological Screening–Revised (CNS-R)
(Bowler, Mergler, Huel, Harrison, & Cone, 1991)

Similar to the POET, the CNS-R has been used in various clinical and epidemiological investigations, including microelectronics workers exposed to solvents, and workers exposed to ethylene oxide. Also like the POET, the CNS-R is composed of neuropsychological tests in general use in the United States. Tests included are:

Verbal: WAIS-R Vocabulary
Attention and concentration: WAIS-R Digit Span, WMS-R Visual Memory Span, WMS-R Mental Control, Cancel H
Visuospatial and visuomotor speed: WAIS-R Digit Symbol, Trail Making Test A, Cancellation H Test
Cognitive flexibility: Digit Span Backwards, Stroop Color/Word Test, Trail Making Test B
Memory: WMS-R Verbal Memory Index, WMS-R Visual Memory Index, WMS-R General Memory Index, WMS-R Attention/Concentration Index, WMS-R Delayed Recall Index, WMS-R Information and Orientation, WMS-R Figural Memory, WAIS-R Digit Symbol Recall
Learning: WMS-R Verbal Paired Associates I, WMS-R Verbal Paired Associates II, WMS-R Visual Paired Associates I, WMS-R Visual Paired Associates II, WMS-R Logical Memory II, WMS-R Visual Reproduction II
Psychomotor and reaction time speed and grip strength: Lafayette Reaction Time Test (simple and choice), Fingertapping, Purdue Pegboard, Dynamometer

Baker, Feldman, White, Harley, Dinse, and Berkey (1983) Battery

Wechsler Memory Scale
  WAIS subtests:
    Vocabulary
    Similarities
    Block Design

Digit Symbol (modified—4th row from memory)
Continuous Performance Test
Santa Ana Dexterity Test
Tapping Tests (unspecified)
Profile of Mood States

London School of Hygiene Battery (Cherry, Venables, & Waldron, 1984)

According to Anger and Cassitto (1993), this battery has fallen into disuse as the developers have moved on to other activities. There has been no research using this battery in several years.

Trail-Making Tests
  WAIS subtests:
    Digit Symbol and Block Design
  Grooved Pegboard Test
  Dotting Test
  Visual Search Test
  Buschke Selective Reminding Memory Test
  Simple Reaction Time
  NART (Nelson Adult Reading Test) index of premorbid functioning

Institute of Occupational Health Battery (Hänninen & Lindstrom, 1979)

WAIS subtests: Similarities, Picture Completion, Block Design: including time score for first seven times and number of incorrect reproductions after the time limit, Digit Symbol
Wechsler Memory Scale subtests:
  Digit Span (forward, backward, and total scores)
  Logical Memory
  Visual Reproduction
  Associate Learning
  Modified Finger Tapping Test (Thumb Taps)
  Santa Ana Dexterity Test
  Mira Test
  Symmetry Drawing Test
  Rorschach Inkblot Test
  Eysenck Personality Inventory

Tuttle, Wood, and Grether (1976) Battery (Epidemiological)

Reaction Time: Simple, Choice
Santa Ana Dexterity Test
Critical Flicker Fusion
Neisser Letter Search

WAIS subtests: Digit Span, Digit Symbol
Feeling Tone Checklist

Valcuikas and Lilis (1980) Battery (Epidemiological)

This battery has not been used in several years (Anger & Cassitto, 1993).

Block Design
Digit Symbol
Embedded Figures
Santa Ana

Putz-Anderson et al. (1983) Battery (Epidemiological)

Choice Reaction Time
Michigan Eye-Hand Coordination
Neisser Letter Search
Short-Term Memory Span
Digit Span
Profile of Mood States
MMPI-Mania Scale
Orthorater

Smith and Langolf (1981) Battery (Epidemiological)

Memory Scanning
Short-Term Memory Span
Continuous Recognition Memory
Stroop Test
Figure Rotation

Williamson, Teo, and Sanderson (1982; Williamson, 1990) Battery (Epidemiological)

This set of tests proposed by Williamson and colleagues is grounded in information processing theory and experimental psychology, rather than clinical neuropsychology, as are most other batteries. Test-retest reliability is high (> 0.8 in most tests) and potential confounders have been investigated. The battery has been used effectively in populations exposed to mercury and lead (Anger & Cassitto, 1993).

Critical Flicker Fusion
Paired Associates
Simple Reaction Time
Memory Scanning (Sternberg)
Vigilance

Iconic Memory
Long-term recall of paired associates
Hand Steadiness (stylus in hole)

TUFF Battery (Hogstedt, Hane, & Axelson, 1980) (Epidemiological)

A Swedish acronym for a test battery for investigating functional disorders, the TUFF battery has been superseded in Sweden by the Gamberale, Iregren, and Kjellberg (1990) SPES (Swedish Performance Evaluation System) (Anger & Cassitto, 1993).

Solvent Screening Questionnaire
Vocabulary Synonyms and Antonyms
Figure Classification
Block Construction
Unfolding: operations on drawings of unfolded geometric figures and comparison with folded figures
Visual Gestalt Test: identifying a picture from successively more complete fragments
Digit Symbol Test
Dot Cancellation Test
Number Underlining Test
Bolt Test and Pin Test (tests of fine motor coordination and dexterity)
Cylinders (tests of fine motor coordination and dexterity)
Benton Visual Retention Test
Auditory Perception and Retention
Neurobehavioral tests used in NIOSH-supported worksite studies (from Anger, 1985)

Neuropsychological Screening Test Battery (Hänninen, 1990)

Test	Neuropsychological domain
Finger Tapping	Fine motor speed
Flanagan Coordination	Fine motor coordination
Mira Test Block Design (omit last 2 items)	Visuospatial/constructional
Memory for Designs (10/15 items new scoring rules)	Visual memory
Digit Span	Verbal memory/attention
WMS Associate Learning	Verbal learning

Based on a larger test battery (Hänninen, Eskelinen, Husman, & Nurminen, 1976; Hänninen, Hernberg, Mantere, Vesanto, & Jalkanen, 1978; Hänninen, Nurminen, Tolonen, & Martelin, 1978; Lindstrom, Harkonen, & Hernberg, 1976), Hänninen (1990) selected these tests to monitor the performance of Finnish factory workers exposed to organic solvents. The battery was designed for worker acceptability, ease of administration, and brief administration time. In addition, the use of standardized clinical tests allows the battery to be utilized for either research or clinical purposes.

Preliminary validation was conducted on 181 workers in paint production, rubber manufacture, and metal packaging of cosmetics and aerosols. Problems of unstable test–retest correlations suggest either that the battery is sensitive to acute effects or that behavioral deficit may wax and wane in specific neuropsychological domains. The author correctly indicates that the battery is presently suitable for health surveillance (i.e., screening of workers at risk) but more thorough examination is needed to confirm a neuropsychological deficit.

IMT (Instituto de Medicina del Trabajo) Battery (Almirall-Hernàndez, Mayor-Rios, del Castillo-Martin, Rodrigues-Notario, & Romàn-Hernàndez, 1987)

This battery has the unique distinction of being developed in Cuba and utilizing normed variations of common paper-and-pencil neuropsychological tests. Since the battery was designed for use in Latin American populations and uses inexpensive paper-and-pencil materials, it is not surprising that, according to Anger and Cassitto (1993), these countries have shown interest in adopting this Spanish language battery for neurotoxicity evaluations.

SPES (Swedish Performance Evaluation System) Battery

The SPES is a computer-administered battery for an IBM PC/AT equipped with an external clock card. It includes the following tests:

Simple Reaction Time
Choice Reaction Time
Color Word Vigilance
Color Word Stress
Search and Memory
Symbol Digit
Digit Span
Additions
Digit Classification
Digit Addition
Verbal Reasoning
Vocabulary
Finger tapping speed
Finger tapping endurance
Self-rating scales in mood, performance, acute symptoms, and long-term symptoms

NES (Neurobehavioral Evaluation System) Battery
(Baker, Letz, Fidler, Shalat, et al., 1985)

The NES battery is a computerized testing battery implemented for MS-DOS-based computers, to allow for on-site field testing of neuropsychological functions in working populations as a function of neurotoxic exposure. Most studies employing the NES have been in large epidemiological designs. The NES is the most extensively distributed of all microcomputer batteries used in neuropsychological toxicology studies, with 40 members of the NES user's group in North America and western Europe. It has been translated into eight languages, with 5400 subjects tested as of 1990 (Letz, 1990).

Tests employed by the NES system are, for the most part, fairly straightforward extensions of conventional neuropsychological tests. The authors elected to stay close to the series of "core" tests recommended by the World Health Organization and NIOSH for use in the epidemiological investigation of neurotoxic substances. Five of the NES tests are modified versions of WHO/NIOSH core tests and two others were recommended by WHO as supplemental tests. The NES battery consists of the following tests:

Finger Tapping (similar to Halstead–Reitan Finger Oscillation Test)
Hand–Eye Coordination Test: Subject traces the cursor over a sine wave drawn on the monitor, using a joystick
Simple Reaction Time
Continuous Performance Test: Subjects press a button when a tachistoscopically presented "S" is projected on the monitor
Symbol-Digit Substitution: Similar to WAIS-R Digit Symbol Test
Visual Retention Test (similar to Benton VRT)
Pattern Comparison
Pattern Memory
Switching Attention
Digit Span
Serial Digit Learning
Associate Learning
Associate Recall
Mood Scales (similar to Profile of Mood States)

NES performance appears to be minimally influenced by common covariates. Age and vocabulary/education accounted for most of the covariate influence. Surprisingly, chronic alcohol use added almost nothing to the variance, suggesting either the lack of reliability for the self-report measure, or the insensitivity of the NES to this common neurotoxin. Although the NES was not designed as a clinical tool, Arcia and Otto (1992) have compiled some preliminary normative and reliability data. Only the CPT mean latency, the switching condition of the Switching Attention Test, Digit Symbol Substitution, and Digit Span achieved reliabilities sufficient for clinical decision-making, but the authors justifiably caution that these tests cannot, by themselves, function as the criteria for diagnosis.

Many preliminary validation studies exist for the NES. Baker, Letz, Eisen, Pothier, Plantamura, & Larson (1988) found Symbol-Digit and Visual Retention impairments in a group of mixed solvent-exposed patients, and Baker, Letz, and Fidler (1985) found effects of age, education, and solvent exposure for several of the tests in the battery. Another study (Greenberg, Moore, Letz, & Baker, 1985) examined NES performance under acute exposure to nitrous oxide. Significant decrements were found on the Continuous Performance Test, Symbol-Digit Substitution, and Finger Tapping. The authors did not correct for the use of multiple t tests, but claim that the similarity in functions tested by these three tasks and known effects of nitrous oxide mitigate this methodological insufficiency. Unlike many of the batteries described, the NES continues to generate research. Spurgeon *et al.* (1992) used the NES to assess two separate groups of government employees exposed to solvents, finding significant associations with solvent exposure and the NES symbol digit subtest. The NES has been used in epidemiologic investigations in different countries and translations. The validity of the battery appears to remain robust across different languages. For example, Liang Sun, Chen, and Li (1993) found various NES-C (the Chinese version of the NES) subtests to be sensitive to low level mercury exposure in a chronically exposed population working at a fluorescent lamp factory. (For a review of international validation studies see Letz, 1990.)

MANS (Milan Automated Neurobehavioral System) Battery

This battery, like the NES, is an automated extension of the WHO Neurobehavioral Core Test Battery. The battery has two published validation studies (Agnew *et al.*, 1991; Bleecker, Bolla, *et al.*, 1992) as well as a set of clinical norms for adults by age, sex, and education (Fittro, Bolla, Heller, & Meyd, 1992). The battery consists of four subtests: Simple Visual Reaction Time, Serial Digit (recall), Symbol Digit Substitution, and the Benton Visual Retention Test. Advantages of the battery for large-scale testing are its brevity, and its ability to be run on a very inexpensive computer (e.g., 4.77 MHz 8086 CPU).

Psychometric Assessment System/Dementia Screening Battery (Branconnier, 1985)

This system is designed for the Apple II+ microcomputer with monochrome monitor. Versions of the Nelson Adult Reading Test (NART) are presented, along with WAIS subtests of Information, Similarities, and Vocabulary to obtain an estimate of verbal IQ.

Memory is assessed with a selective reminding-type task, until five trials or two consecutive perfect trials have occurred. Subsequent recall is tested in the presence of distractor items and a signal detection paradigm is used to compute response criterion ("beta") and criterion-free estimate of sensitivity (d'). What the authors call "amnestic aphasia" is tested with a word fluency task, requiring the subject to name as many items as come to mind beginning with certain letters.

Spatial disorientation is tested by the Money, Alexander, and Walker (1965) Standardized Road Map.

The principal aim of this battery is the identification of dementias. Since its sole reported validation has been on patients with Alzheimer's disease, it is unclear whether the battery will prove differentially diagnostic for less severely impaired patients in the early stages of neurotoxic exposure. The battery uses several novel approaches to neuropsychological function which may prove useful with further validation.

MTS (Microcomputer-Based Testing System) Battery (Eckerman et al., 1985)

Another system using the Apple II series of microcomputers, the battery attempts to test a set of cognitive factors derived by Carroll (1980) using factor analysis and hypothesized to be basic elements of neuropsychological function. The current program is written in PASCAL and tests the following functions (the reader is referred to the article for more elaborate descriptions of individual tasks):

1. Perceptual Apprehension: tachistoscope-like presentation of letters and objects.
2. Reaction Time and Movement: simple and choice reaction time using auditory modality.
3. Evaluation/Decision: comparison of similar pictures.
4. Stimulus Matching: comparing length of lines, running recognition of words.
5. Naming, Reading, Association: Stroop task.
6. Episodic Memory: a "Simon-says" task, digit span, supraspan digit learning, "which word came last" task, a "keeping-track" task. The authors correctly note that the current set of tests lack a longer-term 30-min recall task, and recommend inclusion of such a task in the neuropsychological battery.

Initial validation with parts of the MTS battery, using experimental exposures to alcohol and carbon monoxide, suggests that MTS tasks are sensitive to acute neurotoxic effects of these agents. The battery does, however, require specialized computer equipment not generally a part of the Apple series, including a "touch screen" and a special reaction-time clock. This may limit its acceptance somewhat.

Naval Biodynamics Laboratory Battery (Irons & Rose, 1985)

This Apple II system is designed for individuals working in "unusual environments" (Irons & Rose, 1985, p. 395). The following tasks are included:

1. Sternberg Memory Scanning Task
2. "Stroop-like" Task
3. Pattern Comparison Task

4. Math Test
5. Code Substitution Test (like Digit Symbol using letters and numbers)
6. Choice Reaction Time Task
7. Visual and Auditory Recognition Memory
8. "Spoke" Task (using a light pen to "tap" from the center outward to a series of lighted circles around the screen)
9. "Manikin" Task (a "mental rotation" task that asks the subject to identify the hand in which a computer-generated figure holds a box; the figure may be rotated in a number of orientations toward the subject)
10. Visual or Auditory Serial Addition (similar to the PASAT)
11. Auditory Digit Span
12. Maze Task (uses joystick)
13. "Logic Test" (requires subject to learn the correct numbered sequence of a set of 16 boxes drawn on the screen)

This battery is just beginning to be validated, and has several interesting tasks not found in other computerized batteries. It is hoped that future validating studies will include populations at risk for neurotoxic exposure.

Walter Reed Performance Assessment Battery
(Thorne, Genser, Sing, & Hegge, 1985)

The unique distinction of this battery is the authors' foresight to program equivalent versions for both the Apple and IBM-compatible disk operating systems. The programs are written in BASIC and are modifiable by the user. The following tasks are included in the battery:

1. Two-Letter Search—identification of letters in a string of characters
2. Six-Letter Search—as above with six target letters
3. Encoding/Decoding—letters to be translated to map coordinates
4. Two-Column Addition—mental arithmetic task
5. Serial add and subtract—computer-paced—requires sustained attention
6. Logical Reasoning—true–false decision after comparing a sentence describing letter order (B is followed by A) with a letter pair
7. Digit Recall—short-term supraspan memory test
8, 9. Pattern Recognition I and II—spatial memory tasks of increasing complexity
10. Visual Scanning—visual search task with minimal memory load
11. Mood-Activation Scale—POMS-like adjective/Likert scale
12. Mood Scale II—36-adjective, 3-point scale of affective states
13. Four-Choice Serial Reaction Time
14. Time Estimation I—identifies when a moving object will reappear from behind a barrier

The Walter Reed battery is available free of charge to agencies and professional users, on the conditions that users will exchange information about the battery, that it is not distributed to third parties without permission, and that published research will cite its source.

Other Computer Batteries

1. Cognitive Function Scanner (Laursen, 1990)
2. Automated Performance Test System (Kennedy et al., 1987)

3. ANAM

Automated Neuropsychological Assessment Metrics (ANAM) was not specifically designed as a test of human behavioral neurotoxicity, but rather as a measure of continuous human performance efficiency for the United States Defense Department. Because subclinical effects of neurotoxic substances almost invariably include impaired continuous processing efficiency, batteries like the ANAM, if clinically normed, may prove quite useful in neuropsychological toxicology. The authors of the ANAM (D. Reeves, K. Winter, S. LaCour, K. Raynsford, G. Kay, T. Elsmore, and F. W. Hegge) had as their explicit goal in developing the battery to produce "a library of automated tests ... to meet the need for precise measures of processing efficiency in a variety of cognitive domains. Many of the component tests in ANAM were derived from the UTCPAB/STRES Battery (Reeves et al., 1991) and Walter Reed Performance Assessment Battery (Thorne et al., 1985) and have been adapted for neuropsychological assessment."

The ANAM battery includes four standardized batteries designed for assessment of individuals ranging from superior (ANAUT) to a moderately impaired range (MILD) of intellectual functioning. An additional battery has been provided for Continuous Performance Testing. Tests include the following, with the origin of the automated version of each test listed in parentheses:

1. Memory Search-Sternberg (UTCPAB/STRES)
2. Mathematical Processing (UTCPABA/STRES)
3. Spatial Processing Successive (UTCPAB/STRES)
4. Spatial Processing Simultaneous (ANAM)
5. Procedural Memory (ANAM)
6. Running Memory (ANAM)
7. Pursuit Tracking (ANAM)
8. Moodscale 2 (WRPAB)
9. Stanford Sleepiness Scale (WRPAB)

ANAM has been designed for use on IBM compatible systems with an 80286 or faster microprocessor. Either keyboard or mouse response input is supported. An additional advantage is the modularity of the ANAM software, which allows the end user the option to run tests individually or as part of a standardized battery (Reeves, Winter, LaCour, Raynsford, Kay, Elsmore, and Hegge, 1992).

INDIVIDUAL TESTS

- *Sensory tests:* Orthorater, Vernier Acuity, Central Visual Acuity/Glare, Central Visual Acuity/Moving Target, Binocular Depth Perception, Peripheral Field Sensitivity, Eye Movement Fixation, Accommodation, Ainmark Perimeter, Farnesworth Color Vision Test, Audiometer, Tone Decay, Rail Balancing.
- *Motor tests:* Toe Pointing, Toe Tapping, Finger Tremor, Arm Tremor, Finger Tapping, Michigan Eye–Hand Coordination, Pencil Flipping Speed, Santa Ana Test, Dynamometer, Drop Reaction Time, Simple Reaction Time.
- *Cognitive tests:* Choice Reaction Time, Time Estimation, Arithmetic, Letter Recognition, Critical Flicker Fusion, Dual Task, Divided Attention, Light Flash Monitoring, Neisser Task, Pattern Comparison, Digit Span, Digit Symbol, Block Design, Raven's Progressive Matrices.
- *Affect/Personality* Multiple Affect Adjective Checklist (MAACL)
  Edwards Personal Preference Scale
  Feeling Tone Checklist
  Clinical Analysis Questionnaire (CAQ)
  Marlowe-Crowne Lie Scale
  Profile of Mood States (POMS)
  Mania Scale/MMPI

CHILDREN'S TESTS

Behavioral toxicology batteries for children have received comparatively little attention, which is unfortunate, since children are considered to be more at risk than adults for neurotoxic exposure to many substances (Winneke & Collet, 1985). Test selection strategies are similar to those of adults; a standardized battery may be used (e.g., the Young Children's and Older Children's versions of the Halstead–Reitan Batteries), or a flexible battery approach may be substituted. Winneke and Collet (1985) recommend that a children's neurotoxicology battery include tests of intelligence, language, memory, perception, attention, and motor functions. They suggest the following tests to be useful for the evaluation of childhood lead neurotoxicity: the WISC-R or WPPSI, the Bender-Gestalt, children's versions of the Trail-Making Tests, a delayed reaction time test, and a complex reaction time test (i.e., the Wiener Test) (p. 47). Other useful tests might include the Lincoln–Oseretsky or the Bruninks–Oseretsky tests of fine and gross motor control, cancellation tasks, a continuous performance test (e.g.,

Gordon Diagnostic System), and the Benton Visual Retention Test, which has child norms.

INTERPRETATION OF NEUROPSYCHOLOGICAL TEST RESULTS

The relationship between test results and clinical interpretation is necessarily mediated by an implicit model of neurotoxic action for a given exposure. Without such a model, meaningful clinical interpretation of test results is not possible. For example, Williamson (1990) gives the example of a study that finds lead-based impairments in tests of motor skills, learning, and memory. However, if each test also requires visual acuity, and lead impairs visual acuity, then "it is just as likely that lead is affecting only the sensory function" (p. 61).

SUMMARY

Reviewing neuropsychologists' testing recommendations, it is possible to see some commonalities, whether the authors favor standard or flexible battery approaches. All recommend measures of reaction time or motor speed. Fine motor coordination and finger dexterity testing is likewise required. Memory

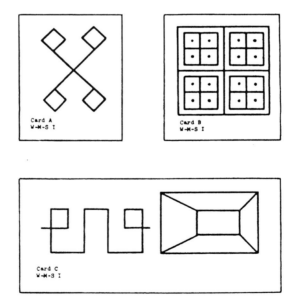

FIGURE 2.2. Wechsler Memory Scale, Figural Memory Cards, Form I. Copyright 1955, renewed 1974 by The Psychological Corporation. Reproduced by permission. All rights reserved.

assessment is considered to be a mandatory part of these neurobehavioral batteries (e.g., Memory Assessment Scales; see also Figure 2.2). Tests of cognitive efficiency and flexibility (e.g., Trails, Digit Symbol) are included on most batteries, as are tests that can suggest premorbid function (e.g., the Vocabulary Subtest of the WAIS). Some batteries, but not all, include personality testing as part of the examination.

From this it is possible to make some recommendations concerning the construction of a neuropsychological battery to test for neurotoxic sequelae. Because nonverbal abilities tend to be differentially more affected by common neurotoxins, a neuropsychological toxicology battery should load more heavily on these functions.

Table 2.3 can be considered a representative collection of commonly available tests to evaluate neurotoxic syndromes. Additionally, neuropsychologists might consider dusting off old psychophysics equipment, including rotary pursuit, graded weights, tachistoscopes, and the like. This equipment, well-normed in experimental psychology for use with intact subjects, might be of renewed utility to neuropsychologists investigating peripheral and central nervous system impairment.

TABLE 2.3. Representative Neuropsychological Testing Battery for Neurotoxicity Evaluations

Test	Functions tested
Finger Tapping Test	PNS and CNS motor strip
Grooved Pegboard Test	Visual–motor–spatial coordination/dexterity
Minnesota Rate of Manipulation Test	Visual–motor–spatial coordination/dexterity
Grip Strength (Dynamometer)	Gross motor strength
Digit Symbol Test	Cognitive efficiency and flexibility
Trails A and B	Cognitive efficiency and flexibility
Stroop Test	Reading, color naming, sustained attention under continuous distraction
Gordon Diagnostic System	Sustained attention, distractability
Paced Auditory Serial Addition Test	Auditory comprehension, processing of simple arithmetic problems under conditions of sustained attention
Category Test	Problem-solving/executive functions
Block Design	Constructional abilities
Memory Assessment Scales	Survey of memory functions
WAIS-R subtests	
Vocabulary	As estimates of general intellectual function
Information	
Kaufman Brief Intelligence Test (K-BIT)	IQ estimate
Medical	Complete clinical history, exposure data, drug and alcohol history, occupational and social functioning
Psychological (e.g., MMPI-2)	Used to identify variables relevant to neurotoxic exposure and rule out alternative diagnoses
Occupational (e.g., OSI)	

TABLE 2.4. Neuropsychological Measures to Assess Neurotoxic Exposure Effects

Test	Function
Halsted-Reitan Neuropsychological Battery Subtests	
Finger Tapping Test	Motor speed
Trail Making Tests: A&B	Attention, cognitive flexibility
Booklet Category Test	Integration of all cognitive functions, sustained attention, frontal lobe functions
Fingertip Number Writing Test	Peripheral sensation, parietal lobe function
Grip Strength Test (Dynamometer)	Gross motor function
Memory Assessment Scales	Comprehensive memory testing
Gordon Diagnostic System Model III (GDS)	
Adult Vigilance Test	Visual attention
Adult Distractibility Test	Visual attention with distraction
Other Tests	
Kaufman Brief Intelligence Test	IQ estimate
Wisconsin Card Sorting Test	Reasoning/frontal
Stroop Color-Word Test	Attention under distraction
MLAE Controlled Oral Word Association Test (COWAT)	Frontal lobe
Continuous Visual Memory Test (CVMT)	Visual memory
Block Design	Visual spatial
Hooper Visual Organization Test	Visual spatial
WAIS-R Digit Symbol Test	Psychomotor speed
Grooved Pegboard Test	Fine motor coordination
Minnesota Multiphasic Personality Inventory-2 (MMPI-2) and/or Personality Assessment Inventory (PAI)	Personality
Profile of Mood States	Present affect
Personal Problems Checklist	
Health Problems Checklist	
Neuropsychological History Questionnaire	
Harvard Occupational Health Questionnaire	
Rey Memorization of 15 Items Test	
Forced Choice Recognition Memory Testing	

The tests listed in Table 2.4 form one example of a flexible battery of commonly available neuropsychological measures that may be useful in assessing neurotoxic exposure effects. The collection began with a screening battery suggested by Wysocki and Sweet (1985) but has gradually "mutated" over the years as better-normed and more sensitive tests became available.

TESTING CAUTIONS

The diagnosis of subclinical neurotoxic syndromes can be a complex process that often involves significant tolerance of ambiguity. Neurotoxic syndromes presentations may interact with a variety of preexisting medical and psychological difficulties that could mimic neurotoxic presentations. For example, patients may have unadmitted drug or alcohol problems that are capable of causing neurotoxic syndromes similar in appearance but unrelated to a primary toxic exposure substance. Stress can cause decrements in neuropsychological performance, and to further complicate diagnosis, such stress may be functional and unrelated to neurotoxic exposure, *reactive* to physical or emotional limitations caused by exposure, or *primary* as a direct result of neurotoxicity. Many seemingly unequivocal test results may be mitigated by factors completely extrinsic to toxic exposure. In some cases, current theory and test applications simply do not allow more definitive diagnosis. Neuropsychologists unaccustomed to ambiguity or complex multifactorial diagnostic questions may find it difficult to adjust to the specialized diagnostic problems presented by this new field.

COMPLEMENTARY APPROACHES

The field of neuropsychology exists in a complementary and mutually beneficial relationship with medicine. Medical examination provides basic physical data and quantified measures of neurological integrity. Medical examination also allows the health team to rule out other illnesses that may mimic neurotoxic effects. Alternatively, neuropsychological data benefit the physician by providing a fine-grained and comprehensive measure of neuropsychological function, information that may support medical diagnosis and clarify the severity of the patient's neuropsychological impairment. Medical tests that have suggested utility for neurotoxicologic investigations include:

1. *Conventional neurological and cerebellar examination* (Juntunen, 1982b).
2. *Evoked potentials.* Sensory evoked potentials provide an index of the integrity of sensory pathways from the peripheral nervous system to the cortex. They are recorded from a sensory nerve in the upper or lower limb, or from the trigeminal nerve. *Short-latency evoked potentials* (less than 50 ms after stimulation) assess the conduction of the nerve pathway from the stimulation site to the somatosensory cortex. *Long-latency evoked potentials* (> 50 ms) are understood to include cortical processing and to be mediated by dosal column and lemniscal pathways (Araki & Murata, 1993). Damage from neurotoxins associated with peripheral nervous system dysfunction (e.g., *n*-hexane, methyl butyl ketone) can be quantified with evoked potential studies. The N20 component of short-latency evoked potentials is significantly associated with lead exposure indices, suggesting the use of this test when analyzing neurotoxicity to the somatosensory ascending pathway.

Visual evoked potentials are measured from the scalp, and reflect the integrity of the pathway between retina and visual cortex. Typical visual stimuli are either flashing lights, which can be used for animals, infants, or severely impaired patients, or pattern-reversing stimuli, which are considered to be more sensitive to visual system abnormalities. Promising results obtained from animal studies using visual (flash) evoked potentials suggest that VEPs and especially pattern reversal evoked potentials (PREPS) will be of future use in human studies (Wennberg & Otto, 1985). VEPs are prolonged by exposure to various solvents, including toluene, *n*-hexane, tetrachloroethylene, 2,5-hexanedione, and xylene (Araki & Murata, 1993), but are also affected by diving, and visual or mental fatigue. Otto and Hudnell (1993) list the following substances as able to alter VEPs in animals or humans.

- *Solvents:* Carbon disulfide, ethanol, hexane, sulfolane, tetrachloroethylene, toluene, xylene
- *Metals:* Cadmium, lead, manganese, methylmercury, organic mercury, triethyltin, trimethyltin
- *Pesticides:* Amitraz, carbamates, Chlordimefon, DDT, Deltamethrin, DFP, Dieldrin, Paraoxon, parathion, Permethrin, Triadimefon
- *Other:* Carbon dioxide, carbon monoxide, formaldehyde, methylpyridines, ozone, pyridine, styrene

Auditory evoked potentials (AEPs) are also recorded from EEG response, this time to a high-intensity click. Seven components are commonly recorded, ranging from very early (<10 ms latency) from the cochlear nerve to the brain stem (BAEP), through late (>1.5 s) components that include P300 measures and reflect changes in cognitive processing and cognitively mediated discriminations. Wennberg and Otto (1985) enthusiastically recommend the use of brain stem auditory evoked potentials (BAEPs) as the most promising sensory evoked potential test for neurotoxicity testing. BAEPs have been found to increase in a linear manner as a function of lead exposure (Otto *et al.,* 1985), especially above 25 μg/dl (Otto & Hudnell, 1990) and also have produced evidence to suggest lead-related impairment of the acoustic nerve and the brain stem (Hirata & Kosaka, 1993). BAEPs are also influenced by exposure to organic solvents, including *n*-hexane, 2,5-hexanedione, and toluene. Hearing loss is a possible confounder (Araki & Murata, 1993).

P300 event-related potentials occur when a subject attends to a stimulus and subjects it to some form of selective or discriminative processing that involves a cognitive component. Typically, a subject is required to discriminate one stimulus from another: the so-called "oddball" paradigm is often employed, where the patient is sequentially presented with binaural auditory stimuli, one of which occurs frequently, and one rarely (the oddball). Subjects are required to keep count of the oddballs, but not the usual tones. The P300 component is the EEG first wave peak that occurs between 250 and 500 ms after the oddball presentation and represents the latency required to cognitively categorize the stimulus (Araki & Murata, 1993). Alcohol and lead have been shown to affect P300 latencies. Covariates include education, intelligence, age, height, and skin temperature (Araki & Murata, 1993).

3. *Nerve conduction studies.* Slowed nerve conduction is characteristic of a number of neurotoxic substances. For example, Hirata and Kosaka (1993) found a significant association between radial nerve conduction and blood lead. Nerve

conduction studies, by themselves, may not be sufficiently sensitive to act as a screening measure, since "more than 50% of nerve fibers must be blocked or lost before the amplitude of evoked action potential is outside normal limits" (Bleecker, 1984, p. 216). Present knowledge concerning exposure effects on nerve conduction is also too uncertain to promote the use of nerve conduction studies for neurotoxicity screenings (Muijser, Juntunen, Matikainen, & Seppäläinen, 1985).

4. *Electroencephalography* (EEG). While the EEG has proven to have utility in the identification of neurological disorders—including epilepsy, tumors, and other diseases—its utility for the detection of subclinical toxicity has not been demonstrated. EEG abnormalities have been detected in exposure to certain toxic substances, including pesticides and organotins, but other substances (e.g., solvents) have not been shown to consistently produce EEG changes. At this time, EEG findings may be more appropriate for ruling out alternative diagnoses than for accurate detection of subclinical toxicity (Wennberg & Otto, 1985).

Quantitative EEG (QEEG), also known as brain electrical activity mapping (BEAM), is likely to prove more sensitive than conventional EEG, although it is still considered to be an experimental tool by the American Academy of Neurology, and not diagnostic by itself. Matikainen, Forsman-Grönholm, Pfäffli, and Juntunen (1993) studied the use of QEEG on 99 styrene workers and found a trend toward significance as a function of exposure but results at low exposure levels also showed a very high level of false positives.

5. *SPECT/PET*. Functional imaging techniques are highly promising, both as methods of detecting brain abnormalities in neurotoxic exposure and as "hard data" that validate neuropsychological test results. SPECT (single photon computed tomography) was able to detect functional brain abnormalities in patients presenting to an occupational medicine clinic with toxic encephalopathy (Callender, Morrow, Subramanian, Duhon, & Ristovy, 1993). Ninety-four percent of the 33 workers showed abnormal SPECT scans; the most frequently damaged areas in descending order were temporal lobes (67.7%), frontal lobes (61.3%), basal ganglia (45.2%), thalamus (29.0%), parietal lobes (12.9%), motor strip (9.68%), a cerebral hemisphere (6.45%), occipital lobes (3.23%), and caudate nucleus (3.23%). Approximately 79% of the patients' neuropsychological evaluations were also abnormal. Similar sensitivity of PET scan has been demonstrated in a case of tetrabromomethane encephalopathy (Morrow, Callender, Lottenberg, Buchsbaum, Hodgson, & Robin, 1990).

6. *Other tests*. Bleecker (1984) suggests that other neurological tests are useful in neurotoxicology diagnosis, including tremometry and quantitative sensory testing (e.g., the Optacon Tactile Tester). Electromyography (EMG) has also been used to investigate potential cases of toxic neuropathy (Seppäläinen, 1982a). Regional cerebral blood flow (rCBF) has been shown to provide useful correlates of cerebral metabolism in studies of abused drugs but is not in general clinical use (Mathew & Wilson, 1991b), Aratani, Suzuki, and Hashimoto (1993)

and Halonen, Holonen, Lang, and Karskela (1986) found some utility in assessing vibration threshold in organic solvent workers, as did Maurissen and Weiss (1980) for acrylamide exposure. Other tests that show promise but that have not yet been validated for neurotoxicology research include tests of warm/cold perception autonomic nervous system function.

3

Metals

HISTORY

Human contact with metallic substances is virtually coextensive with the development of civilization. It is likely, for example, that metallurgy accompanied the earliest cultures of the Tigris and Euphrates rivers around 6000 B.C. Metallic ornaments discovered at Ur in Babylonia suggest that "the casting of copper, silver and gold . . . began around 3500 B.C." (Marks & Beatty, 1975, p. 5). The Egyptians were working with lead as early as 3400 B.C.

Metals were also among the earliest neurotoxins produced by humans; their toxic effects have been chronicled for centuries. In 1473, Ulrich Ellenbog wrote a treatise on what today would be called industrial toxicology. It detailed some effects of occupational metal poisoning:

> This vapour of quicksilver, silver and lead is a cold poison, / for it maketh heaviness and tightness of the chest, burdeneth the limbs and oftimes lameth them / as often one seeth in foundries where men do work with large masses and the vital inward members become burdened therefrom. [Ellenbog, cited in Bingham, 1974, p. 199]

Somewhat more recently, Weiss (1983) corroborates that "excessive exposure to metals . . . has been implicated in a remarkable range of adverse signs and symptoms involving the central nervous system (CNS) and behavior" (p. 1175). Exposure to almost any metal can cause CNS dysfunction.

In addition to causing CNS damage, metals may be toxic to other parts of the nervous system. Peripheral nervous system (PNS) toxicity has been noted for a number of common metals (e.g., lead, arsenic). Several metals produce neurotoxic effects on the extrapyramidal system, damaging the caudate nucleus, putamen, globus pallidus and their connections to the thalamus, substantia nigra, and cerebellum (e.g., iron) (Feldman, 1982a). Finally, a number of metals are capable of damaging both the CNS and PNS.

Metallic exposures are not restricted to industrial sites. Metal products are ubiquitous in the environment, and metallic by-products of various industries have pervaded the atmosphere, been discharged into the water, and have entered world food supplies. Metals have been used as medicined, added to gasoline, and employed in insecticides and pesticides. While levels of certain metals have been declining in the environments, they may be replaced by others. For example, mercury and lead have been better regulated in recent years, and the number of individuals exposed to these toxins have been decreasing (though they still total in the millions). However, as environmental levels of these metals decline, the newer "high technology" metals may take their place in the environment. The future may see increased poisonings from less well-known metallic toxins including nickel, cadmium, and selenium compounds (Gossel & Bricker, 1984).

VARIABLES THAT AFFECT NEUROTOXICITY

Substance

Some metals are more toxic to the nervous system than others. For example, mercury and lead are far more neurotoxic than iron or inorganic tin. Different forms of the same metal may be differentially toxic to the nervous system. In general, organic metal compounds both penetrate the brain more effectively are absorbed more completely than inorganic metals. Thus, organometals may potentially cause greater neurotoxic damage than elemental forms of the same metals. Methylated mercury, for example, is far more damaging to the CNS than inorganic mercury. Elemental tin is another metal that poorly penetrates the nervous system in its inorganic state, but as an organotin is responsible for a wide range of neurotoxic and neuropsychological effects in humans.

Developmental Influences

The degree of neurotoxic damage from metal exposure is also influenced by other factors. The age of the exposed individual may affect toxicity, with the highest risk to fetal and perinatal development. Young children are also at greater risk than adults for the development of certain toxic effects (e.g., lead-induced hyperactivity).

Weiss (1978) makes the useful suggestion that developmental effects should not be considered "a shorthand for youth" (p. 24). Thus, neuropsychological effects of toxic metals may also occur in older individuals whose nervous systems may lose plasticity and the ability to compensate for neurotoxic damage. Unfortunately, the question of human developmental neurotoxicity is largely unexplored at this time. One important question for this subfield might include whether pre- or perinatal neurotoxic effects produce neuropsychological impairments in older adults when compensation mechanisms can no longer hide subclinical effects. Another question is whether chronic exposure to metal neurotoxins in older adults produces different neurotoxic symptoms in kind or degree, relative to younger adult exposure.

Acute versus Chronic Exposure

Certain metals may show one set of effects from an acute dose (e.g., arsenic), whereas chronic exposure produces other forms of neurotoxic damage. In industry there is increasing concern as to whether chronic exposure to low doses of metals can cause neurotoxic damage.

Individual Variation

As is the case for all neurotoxins, there appears to be wide interindividual variation in susceptibility to neuropsychological deficit. One cause of such variation may be the ability of the individual to eliminate toxins from the body. Aluminum compounds, for example, are not thought to be neurotoxic in healthy individuals. However, in persons with impaired renal functioning, aluminum has been cited as a cause of dementing syndromes.

The reasons for other interindividual variations remain obscure. Certain individuals with chronic exposure to lead and with documented blood levels above 70 µg/dl may show only mild neuropsychological abnormalities, while other individuals with lower blood levels and shorter exposures can present with more severe forms of neuropsychological impairment. The causes of such interindividual variability are presently unknown, but may be related to differential ability to eliminate toxic substances, metabolic differences, or perhaps variations in brain susceptibility or permeability of the blood–brain barrier to these toxins.

NEUROPSYCHOLOGICAL TESTING OF METAL EXPOSURE

The value of neuropsychological testing to describe effects of metal exposure has been described by Grandjean (1983):

> Studies of [metal] exposures have been greatly advanced by the use of modern neuropsychological methods ... while such psychological tests may be influenced by other factors not related to neurotoxicity, they provide a useful and relatively accurate assessment of the degree and character of neurotoxic effects in humans. [p. 331]

Despite clear endorsements and demonstrated value for individual cases, neuropsychological investigations of metal poisoning are difficult to locate in the research literature, perhaps because of the factors described in Chapter 1.

Lead and mercury are the metals whose neuropsychological effects are the most completely explored, although their exact pathophysiology is far from completely understood. Other neurotoxic metals have been described in the neurological literature, but have little or no accompanying neuropsychological data. Iron, cadmium, and tellurium are examples from this group. Because of the variety of neurotoxic effects capable of being produced, there is no general "metal encephalopathy" syndrome, even though some symptoms of poisoning are common to several metals. It is more informative to describe the unique neurological and neuropsychological features of exposure to individual metals. Accordingly, several of the more well-known effects of metallic neurotoxins are described in the following pages.

Where available, NIOSH and OSHA industrial safety levels precede each discussion.

ALUMINUM

History

Aluminum's use in medicinal compounds predates by thousands of years its isolation as a pure metallic ore. Salts of aluminum were used during the Roman empire for water purification. During the Middle Ages these aluminum salts were mixed with honey and employed as a treatment for ulcers (Crapper & De Boni, 1980, p. 326), a treatment not altogether different from current over-the-counter remedies.

Aluminum was discovered to be a CNS neurotoxin in the late 1800s and is of scientific value in this capacity; its oxide or hydroxide applied to the cortex of animal subjects produces a useful research model for studies of focal epilepsy (Ward, 1972).

Individuals at Risk

Common routes of aluminum accumulation include industrial exposure as well as ingestion of processed food (e.g., pickles), with the average diet containing about 22 mg/day of aluminum (Shore & Wyatt, 1983). Despite such continuous environmental exposure, most individuals with normal renal function appear able to clear aluminum effectively. An exception may be patients with Alzheimer's disease (AD).

The relation between aluminum accumulation and AD is unproven, and AD patients do not show differential levels of aluminum relative to controls in serum, CSF, or hair. In addition, the neurofibrillary tangles experimentally produced by aluminum differ from those that occur in AD (Shore & Wyatt, 1983). While both AD and aluminum intoxication have produced 10-nm neurofibrillary degeneration filaments, Alzheimer filaments are bundled pairs of filaments twisted in a helical array, whereas aluminum-induced neurofibrillary tangles are single 10-nm filaments (McLachlan & De Boni, 1980). It may be the case, however, that while aluminum does not cause AD, the damaged brain tissues of AD patients may be more vulnerable to this metal. It is also possible that the blood–brain barriers of AD patients may be damaged in the course of the AD process, rendering brain tissues increasingly vulnerable to aluminum accumulation from foods, antacids, and buffered aspirin (McLachlan & De Boni, 1980; Shore & Wyatt, 1983).

Patients with impaired kidney function or chronic renal failure may be especially vulnerable to aluminum neurotoxicity. Body and brain accumulation of aluminum in kidney-compromised patients has been linked to several factors. Dialysis using water that has not been deionized and/or that has been stored in aluminum tanks has been posited to cause aluminum accumulation (Crapper & De Boni, 1980). High aluminum concentrations can also be found in the tissues of some dialysis patients as a consequence of ingesting large amounts of antacids and phosphate binding gels (Russo, Beale, Sandroni, & Ballinger, 1992; Sedman, Wilkening, Warady, Lum, & Alfrey, 1984), although it had been previously thought that oral aluminum was poorly absorbed and therefore not neurotoxic. Sedman *et al.* (1984) report progressive neuropsychological deterioration and dementia from brain aluminum accumulation in a child whose single kidney functioned poorly and was unable to adequately excrete prescribed doses of antacids. Citrate, used to treat acidosis, increases gastrointestinal absorption of aluminum and may cause organic brain syndrome secondary to increased dietary aluminum absorption (Russo *et al.*, 1992).

Aluminum may also be neurotoxic to normal children when combined with exposure to other metals. A recent study suggests that combined exposure to aluminum and lead may synergistically affect visual-motor performance in children (Marlowe, Stellern, Errera, & Moon, 1985).

Dialysis Dementia

Clinically, some end-stage renal disease (ESRD) patients develop what appears to be a form of aluminum encephalopathy. Exposure to aluminum occurs via aluminum-containing dialysate and, probably somewhat less often, the chronic ingestion of aluminum-containing antacids, the latter being a usual concomitant of ESRD therapy. The disorder has been termed "dialysis dementia" and shows a rapid progression from personality changes to global intellectual impairment, accompanied by seizures, gait problems, asterixis, dysarthria, apraxia, and myoclonus. Abnormal EEG or CT may be found (Sedman *et al.*, 1984). Since not all dialysis patients who take large quantities of antacids develop dialysis dementia, unspecified individual differences may interact with aluminum ingestion to produce the disease.

Onset of dialysis dementia is apparently subtle, and initial symptoms occur after a mean of 37 months from initial dialysis. Severity is correlated with brain aluminum concentration. Uncorrected aluminum encephalopathy is progressive and potentially fatal. Brain aluminum concentrations of "15 to 20 times above control populations" have been found in patients with this disorder (Crapper & De Boni, 1980, p. 326). While overall percentages of patients who develop dialysis dementia have not been tallied, it is thought to be a leading cause of mortality among patients undergoing chronic hemodialysis (Rosati, De Bastiani, Gilli, & Paolino, 1980). It is not, however, an inevitable consequence of ESRD. Rosati *et al.* (1980) failed to find any significant neuropsychological deficits in a group of nine dialysis patients who *did* have significantly increased serum aluminum level. However, serum aluminum levels may have been too low to produce the syndrome. A more complete study of dialysis-related neuropsychological deficits did indicate that aluminum levels in dialysis patients were potentially neurotoxic. Bolla *et al.* (1992) examined 35 dialysis patients and found that alcoholic aluminum levels showed a significant negative association with visual memory. An interaction was found between vocabulary and aluminum effects, with low vocabulary patients showing a steeper decline in tests of frontal lobe functioning and tests of attention and concentration. Low vocabulary patients were also more depressed than high vocabulary patients. The authors suggest that patients with low vocabulary may "possess less well-developed compensatory strategies to overcome the neurocognitive effects associated with Al" (p. 1021). It may also be the case that lower vocabulary patients are less neurologically robust than patients with higher abilities and so may suffer earlier, more precipitous decline in mental abilities as a function of neurotoxic damage.

Other Diseases with Possible Links to Aluminum

Other diseases have been linked to elevations in brain aluminum. These include Down syndrome, Parkinson dementia complex of Guam, and striatonigral syndrome (Feldman, 1982a). Aluminum has also been linked to cerebellar Purkinje cell degeneration secondary to alcohol abuse (McLachlan & De Boni, 1980). It is not clear at this time whether genetic variations exist in the ability to handle this metal, causing certain individuals to be at higher risk than others.

Occupational Exposure

Welders exposed for about 20 years have 100-fold higher urinary concentrations of aluminum than nonexposed controls, with urinary aluminum half-life of 6 months or longer (Sjogren, Lidums, Hakansson, & Hedstrom, 1985; Sjogren, Elinder, Lidums, & Chang, 1988; Sjogren, Gustavsson, & Hogstedt, 1990). Welders who are exposed to aluminum fumes for more than 20,000 h (over a period of about 13 years) were twice as likely to report three or more positive responses on a neuropsychiatric questionnaire (Sjogren et al., 1990). Encephalopathic syndromes have also been reported as a consequence of industrial aluminum exposure. The first recorded industrial aluminum poisoning occurred in 1921 in a metal worker who developed memory loss, tremor, cerebellar signs, and loss of coordination (Wills & Savory, 1983). A later case study verified a syndrome of rapidly progressive encephalopathy in an aluminum powder factory worker who had been heavily exposed to aluminum dust (McGlaughlin et al., 1962). The worker developed left-sided focal epilepsy, became increasingly demented over the course of several months, and finally died of pneumonia. Brain aluminum concentration at autopsy was found to be 17 times greater than normal.

More recently, three patients who had worked for 12 years in the potroom of an aluminum smelter (which also included exposure to lithium) were discovered to have neurological and neuropsychological abnormalities (Longstreth, Rosenstock, & Heyer, 1985). Neurological symptoms common to all three included joint pains, severe lack of energy and strength, and tremor. Neurological examination revealed bilateral hearing loss, incoordination, intention tremor, and ataxic gait. Electrophysiologic studies were normal except in one individual

TABLE 3.1. Neuropsychological Performance of Three Aluminum Workers[a]

Test	Subject No. 1	Subject No. 2	Subject No. 3
WAIS-R			
Full Scale IQ	93	71	87
Verbal IQ	97	67	91
Performance IQ	87	78	84
Wechsler Memory Scale			
Memory Quotient	86	62	97
Problem Solving (unspecified)	Poor	Poor	Poor
Halstead-Reitan Impairment Index	1.0	1.0	0.6

[a]Reprinted, by permission of the American Medical Association, from Longstreth et al. (copyright 1985). Potroom palsy? Neurologic disorder in three aluminum smelter workers. *Archives of Internal Medicine, 145*, 1972-1975.

who showed bilateral delays in visual evoked potentials and delays at the medullar level with somatosensory evoked potentials. Neuropsychological performance showed clear deficit (see Table 3.1).

The exact etiology of these abnormalities cannot be verified. While results are consistent with known neurotoxic effects of aluminum, the patients' lithium exposure may have also contributed to the symptoms shown here. Furthermore, because blood or bone aluminum burden may not accurately reflect brain aluminum content, these patients' normal bone aluminum cannot rule out neurotoxic brain concentrations of the metal. The results are worth noting as an alert to possible human neurotoxic effects of prolonged aluminum exposure.

ANTIMONY

IDLH	80 mg/m^3
OSHA	0.5 mg/m^3
NIOSH	0.5 mg/m^3
(*Quick Guide*, 1993)	

Antimony is a silvery-white metal which is usually combined as part of an alloy with lead or zinc, for use in lead batteries, solder, sheet and pipe metal, bearings, castings, type metal, ammunition, and pewter. Little or no antimony is mined in the United States (ATSDR, 1991).

There are no adequate studies of antimony neurotoxicity in humans. Dogs show muscle weakness and hindlimb movement difficulty after ingestion of up to 6644 mg of antimony trioxide for 32 days. Neurological abnormalities have also been shown in rats and mice, leading the ATSDR (1991) to conclude that "because neurological effects have been observed in three species of animals . . . , these effects are also likely to occur in human exposure to high levels of antimony. Antimony may be fatal to humans at high doses" (p. 41).

ARSENIC

OSHA exposure limit (inorganic)	0.010 mg/m^3, 8 h TWA
NIOSH recommended exposure limit	0.002 mg/m^3, 15 min ceiling
As arsine (AsH$_3$)	
OSHA	0.05 ppm (0.2 g/m^3), 8 h TWA
NIOSH	0.002 mg/m^3, 15 min ceiling
IDLH	6 ppm
Odor threshold	< 1.0 ppm
(*Quick Guide*, 1993)	

History

Arsenic has the dubious reputation among metals as "historically the most important of the poisons used for criminal purposes" (Hamilton & Hardy, 1974, p. 31). It was employed in the pharmacopoeia of the Assyrians where it is mentioned in a papyrus dated from about 1552 B.C. (Marks & Beatty, 1975). Like many other metals, arsenic had been tried as a cure for many illnesses, including cancer, fever, herpes, ringworm, eczema, and ulcers (Marks & Beatty, 1975). A common treatment for syphilis over the past two centuries, arsenic was the cause of more injuries than cures, producing optic neuritis and encephalopathy as a consequence of its use (Windebank, McCall, & Dyck, 1984). With even less reputability, arsenic has been used in poison gases during wartime and as a method of homicide or suicide. The latter two applications account for most of the recently reported cases of arsenic poisoning.

Several mass poisonings related to accidental arsenic exposure have been reported, the first involving 40,000 individuals who became poisoned when arsenious acid was accidentally mixed with wine and bread. In another epidemic, sugar used in beer manufacture became contaminated with arsenic, poisoning 6000 people and resulting in 70 fatalities (Katz, 1985).

Current Uses and Individuals at Risk

Currently, the main legitimate uses of arsenic are in the pharmaceutical and agricultural industries. Arsenic is a principal element in many insecticides, weedkillers, and fungicides. Arsenic is found in wood preservatives, animal feed additives, and riot control gas (Morton & Caron, 1989). Calcium arsenate, an insecticide used in fruit orchards, is the most toxic of the arsenic compounds.

Other industries where workers may be exposed to arsenic include any industry that adds acid to iron and steel. Both of these metals contain small amounts of arsenic, and their immersion in acid frees this available arsenic. Metal "pickling" (acid treatment) plants may be dangerous in this regard (Hamilton &

Hardy, 1974). NIOSH (National Institute for Occupational Safety and Health, 1975a) estimates that approximately 1.5 million workers are exposed to inorganic arsenic. Industrial arsenic poisoning has become less frequent as a result of improvements in workplace safety, but suicide, homicide, and accidental pesticide intoxications continue to produce poisoning victims (Kelafant, Kasarskis, Horstman, Cohen, & Frank, 1993).

General symptoms of arsenic poisoning include rash, GI symptoms, weak pulse, cardiac arrhythmia, weakness, paresthesia of hands and feet, hair loss, brittle nails, respiratory irritations, and hoarseness (Morton & Caron, 1989).

Neurological Effects

PNS Effects

Peripheral neuropathy is a principal finding associated with inorganic arsenic intoxication. Schaumburg, Spencer, and Thomas (1983) report different profiles of neurotoxic damage, depending on whether the exposure was high level and acute (e.g., a suicide attempt) or prolonged and low level. Subacute, 310-day delayed neuropathy may develop in survivors of single high-level doses. Numbness, weakness in extremities, and "intense" paresthesias develop with the eventual outcome ranging from mild sensory neuropathy to "severe distal sensorimotor polyneuropathy" (Schaumburg, Spencer, & Thomas, 1983, p. 134).

CNS Effects

Bleecker and Bolla-Wilson (1985) speculate that CNS effects of arsenic may be similar to those found in thiamine deficiency, since arsenic prevents the transformation of thiamine into acetyl-CoA and succinyl-CoA. Validation for this hypothesis could potentially come from autopsy studies investigating neuropathological similarities between thiamine deficiency states produced by arsenic and those reported in the common alcohol-induced thiamine deficiency disorder, Wernicke–Korsakoff syndrome.

Neuropsychological Effects

Assessing neuropsychological dysfunction in arsenic exposure cases has been difficult because of potential confounding effects of alcohol intake. Feldman *et al.* (1979) report that 92% of arsenic-exposed smelter workers are also moderate drinkers. Bleecker and Bolla-Wilson (1985) report the case of a 50-year-old engineer exposed to chronic low-dose inorganic arsenic who consequently developed severe and chronic verbal memory and learning impairment on the Rey Auditory Verbal Learning Test and the Logical Memory Subtest of the Wechsler Memory Scale. However, interpretation was complicated by a family history of rapid-onset dementia. A larger study of a community that had been exposed to arsenic contamination from a wood processing plant was reported by Burns, Cantor and Holder, (1994). After excluding subjects for medical or substance abuse-related factors unrelated to arsenic exposure, they found that individuals who displayed arsenic related hyperkeratosis and or polyneuropathy showed psychomotor slowing, moderately impaired latency of recall, long term memory impairment, emotional abnormalities, slowed recognition of perceptual figures, and greater variability in auditory P300. The suggestion that arsenic causes "unequivocal" damage to the CNS and PNS (Windebank et al., 1984) implies that neurobehavioral evaluation could be employed as a useful index of damage and recovery. Frank arsenic encephalopathy can be accompanied by moderate neuropsychological impairments in memory, concentration, confusion, and anxiety, which may improve over time (Morton & Caron, 1989).

Emotional Effects

Organic arsenic-containing compounds may cause rapidly developing CNS symptoms, with progression from drowsiness to confusion and stupor. Organic psychosis resembling paranoid schizophrenia is a common result of organic arsenic intoxication and delirium may follow (Windebank et al., 1984). Fluctuating mental state, agitation, and emotional lability have been noted (Beckett, Moore, Keogh, & Bleecker, 1986).

Case Study: Long-Term Consequences of Short-Term Exposure

The following patient was accidentally exposed to arsenic compounds during his employment. Test results and case discussion clearly support the use of neuropsychological evaluation in arsenic exposure cases.

The patient is a 31-year-old tractor driver with an 11th grade education. In 1980 he was driving in the field when a crop duster sprayed the field and doused the patient with an arsenic-based insecticide. The patient reported no immediate effects, suggesting that he did not inhale the poison. However, the patient began to vomit after an unspecified time, which implies slightly delayed effects of skin absorption. He became unable to retain food and water and began to lose weight. The patient reported losing approximately 40 pounds in the month following his exposure.

The patient initially went to a physician who diagnosed his symptoms as "heat stroke." However, since symptoms did not remit, the patient came to the Occupational Medicine Clinic where he was diagnosed as having arsenic poisoning via an insecticide that contained arsenic acid.

The patient denied history of alcohol or drug abuse, head injury, high fever, or seizures. Hypertension was controlled with hydrochlorothiazide. Caffeine intake was about five cans of cola per day. He reported periodic (about once a month) attacks of vertigo that began immediately after his arsenic exposure.

Other symptoms reported included continuous paresthesias in the hands, lower arms, and the legs, beginning at the knees. The patient described his paresthesias as "tingling" rather than painful.

The patient was cooperative and appeared to try hard during testing; however, he had great difficulty performing tasks that required fine motor coordination and speed. When holding a pencil it was necessary for the patient to grasp it firmly with all of his fingers, in a balled fist instead of a more conventional grip.

In the neuropsychological examination the patient showed sensory changes consistent with peripheral neuropathy, including complete loss of sensation in the palms of his hands and poor ability to perceive numbers written on the fingertips.

The patient's verbal memory and learning was also impaired relative to memory for designs. The discrepancy between recall of the two Wechsler stories in immediate recall was probably related to educational limitations; however, poor recall of the first story after 30 mins, coupled with slower than usual verbal learning in the Associate Learning Test, suggest encoding difficulties irrespective of premorbid education. The patient's ability to maintain attention and concentration seemed likewise impaired, especially on the Trails Tests, where the patient was not hampered by sensorimotor dysfunction, but rather by the ability to correctly scan for numerical and alternating number–letter sequences.

On measures of current affect, the patient admits to severe depression that appeared to be reactive to real losses of function and ability to make a living. He was highly focused on his physical symptoms, exhibited early morning awakening, and mild loss of appetite, in addition to other "cognitive" forms of depression. In addition, the patient described himself as forgetful, grouchy, uncertain, discouraged, exhausted, and "extremely bitter."

The patient's current neuropsychological profile suggested a combination of structural and reactive emotional components. PNS damage, loss of attention and concentration, and probable verbal memory deficits may be direct effects of arsenic damage, but also certainly contribute to the patient's sense of a futile future as evidenced by severe depression noted in the Beck Depression Inventory. Recommendations for such a patient would include evaluation for supportive psychotherapy, antidepressant medication and reentry into the labor force in whatever capacity he is able.

BARIUM

IDLH	150 mg/m^3
OSHA	0.5 mg/m^3, 8 h TWA (as chloride or other soluble barium compound)
	5–10 mg/m^3, 8 h TWA (as dust)
NIOSH	0.5 mg/m^3
(*Quick Guide*, 1993)	

Barium is a silvery-white metal principally used by the oil and gas industry for drilling muds. Barium compounds also find use in paint, bricks, tiles, glass, rubber, and medical testing. Barium hydroxide is employed as a component of ceramics, insect and rat poisons, oil and fuel additives.

There is relatively little human exposure literature regarding barium. The most well-known form of barium exposure occurs when barium compounds are employed as radiopaque contrast media to study the GI tract. While barium is normally considered nontoxic and of limited solubility, potential toxicity may occur under conditions of colon cancer or perforations of the GI tract that allow barium to enter the bloodstream (Princenthal, Lowman, Zeman, & Burrell, 1983).

A single case study notes loss of deep tendon reflexes in a 22-year-old male who inhaled barium carbonate powder (Shankle & Keane, 1988). Symptoms of tingling around the mouth are considered to be an early indication of barium toxicity (Lewi & Bar-Khayim, 1964; Morton, 1945). Neurological symptoms may progress to partial or complete paralysis (Phelan, Hagley, & Guerin, 1984; Wetherill, Guarino, & Cox, 1981). An early autopsy study of barium poisoning revealed brain congestion and edema (McNally, 1925). There do not appear to be any neuropsychological studies of barium exposure.

BISMUTH

Bismuth has been used in large quantities by individuals in France and other countries to control chronic colon disorders. In the United States, bismuth drugs are prescribed to decrease odor and bulk of bowel movements, and are also used in colostomy patients after colon carcinoma surgery (Goetz, 1985). The phenomenon of bismuth encephalopathy has only been recognized since the early 1970s; however, by 1981 over 1050 cases and 72 deaths were attributed to this neurotoxic disorder (Buge, Supino-Viterbo, Rancurel, & Pontes, 1981; Le Quesne, 1982). The syndrome is characterized by a prodrome of several weeks or months of "depression, anxiety, irritability and tremulousness" (Le Quesne, 1982, p. 537). Further deterioration may occur without warning, sometimes after several years of exposure (Kruger, Weinhardt, & Hoyer, 1979). Symptoms include confusion, myoclonic jerks, dysarthria, and gait apraxia (Goetz, 1985; Nordberg, 1979). Goetz (1985) reports other neurotoxic effects of bismuth when injected intramuscularly as a systemic treatment of syphilis. Sequelae of multiple injections include acute myelopathy, flaccid paraplegia, and loss of bladder control (p. 50).

Little is known about the underlying neuropathology of bismuth intoxication although CT scans show hyperdensities in basal ganglia, cerebellum, and cerebral cortex. The pattern of gray matter cortical hyperdensities found in bismuth intoxication has not been seen in any other type of heavy metal intoxication (Buge *et al.*, 1981).

CADMIUM

IDLH 50 mg/m³ (as dust)
 9 mg/m³ (as fume)
OSHA 0.005 mg/m³ (dust or fume)
(*Quick Guide*, 1993)

Cadmium's uses include pigments, alloys, and electroplating. Principal effects of cadmium intoxication are on the kidney and lungs rather than the brain, although rat studies have shown CNS changes in early development and demyelinating PNS changes in adult rats. Rats also appeared to become irritable and showed lowered motor activity (Katz, 1985). Cadmium exposure in humans is associated with olfactory impairment. In a sample of 55 workers with chronic exposure to cadmium fumes, 44% were mildly hyposmic and 13% were moderately or severely hyposmic (Rose, Heywood, & Costanzo, 1992). No human neuropsychological studies of cadmium-poisoned individuals are available.

CHROMIUM

Cr (metal and insoluble salts) 1.0 mg/m³
 OSHA
Soluble 0.5 mg/m³
 OSHA
(*Quick Guide*, 1985)

Not heretofore thought to be neurotoxic, one human and several animal studies suggest that chromium may be neurotoxic. Duckett (1986) described three cases with unrelated encephalopathic diseases who showed evidence of CNS chromium neurotoxicity at autopsy. Sources of chromium were radiological contrast media, KCl solution, and Mylanta, an antacid. Chromium was thought to enter the brain through pathological pallidal blood vessels that showed vascular siderosis and possible breakdown of the blood–brain barrier. This appears to be the only report of pure chromium neurotoxicity in the human literature, and thus, the finding must be considered tentative. However, results suggest caution in the use of chromium-containing substances in older patients, or those with damaged vasculature.

Animal research has also revealed nervous system changes from chromium exposure. These include hypoactivity in rats provided with chromium-containing drinking water (Diaz-Mayans, Laborda, & Nunez, 1986) and cortical degenerative changes with chromolysis in rabbits given daily intraperitoneal doses of chromium compounds for 3 or 6 weeks (Mathur, Chandra, & Tandon, 1977). Thus far, there does not appear to be a relationship between exposure to chromium and expressed neuropsychiatric symptoms (Sjogren *et al.*, 1990).

COPPER

Dust and mist
    OSHA   1.0 mg/m^3
    NIOSH   1.0 mg/m^3
Fumes
    OSHA   0.1 mg/m^3
    NIOSH   0.1 mg/m^3
(*Quick Guide*, 1993)

Neurological and neuropsychological symptoms of copper toxicity occur during the development of Wilson's disease, a genetic disorder of abnormal copper metabolism. Copper is found in brain autopsies of patients with Wilson's disease, and it is likely that many neurological manifestations of the disease are caused by toxic accumulation of brain copper. Structural damage from copper toxicity includes "neuronal degeneration, spongy focal change in cortex, corpus striatum and central myelin" (Feldman, 1982a, p. 150). The basal ganglia and cerebellum are said to be particularly affected (Bornstein, McLean, & Ho, 1985). Generalized brain atrophy may or may not be present. Cirrhosis of the liver and deposits of excess copper around the cornea (Kayser–Fleischer rings) are also present in some degree.

Neurological symptoms of Wilson's disease include choreic movement, parkinsonian-like rigidity, and masked facies. Dementia and death within 4–5 years can result if copper is not chelated from the body. Wilson's disease is rare, and its exact pathophysiology remains open to debate, i.e., whether neurological changes are indirectly produced by copper accumulation in the liver or directly within the brain. It is also unclear whether an individual exposed to heavy amounts of copper in the environment can develop copper neurotoxicity without having a selective metabolic vulnerability to copper (Feldman, 1982a). Three of eight diagnosed Wilson's disease patients without overt neurological abnormalities showed abnormal increases in P100 wave latency, but SEPs and BAEPs were abnormal in only one patient (Aiello *et al.*, 1992). In patients with overt neurological hard signs, most studies find evoked potential abnormalities in 66 to 100% of patients (Aiello *et al.*, 1992).

Individuals with copper toxicity from Wilson's disease exhibit a variety of neuropsychological and emotional symptoms that appear to be a function of disease progression. Initial testing reveals abnormal motor functioning with intact IQ, while more advanced cases exhibit intellectual dysfunction, and the possibility of dementia in the late stage of the illness. Psychological abnormalities may also accompany Wilson's disease, and a 1985 review of case studies concluded that affective and behavioral or personality changes are the most common sequelae, with two less common profiles showing schizophreniclike or cognitive deterioration (Dening, 1985).

While the exact etiology of patients' psychological and intellectual deterioration is not known, patients with other types of extrapyramidal disorders (e.g., Parkinson's) have shown similar neuropsychological deficits. In addition, the generalized losses found in some Wilson's disease patients are consistent with a more global form of encephalopathy (Bornstein et al., 1985).

Reitan and Wolfson (1985) report neuropsychological effects in a 48-year-old high school-educated male with "a long history" of Wilson's disease. IQ was still intact, although Digit Symbol and Block Design were abnormally low. Spatial abilities were poor, with the patient showing constructional dyspraxia and markedly deficient left-hand performance on the Tactual Performance Test (TPT). Fine motor speed was also impaired as was cognitive flexibility (Trails B) and abstract reasoning (Category Test). The patient's impairment index on the Halstead–Reitan Battery was 1.0. Emotional complaints included high perceived stress, with a very low threshold of annoyance or irritation, and occasional outbursts of violent rage.

Bornstein et al. (1985) report an even more severe case in a 25-year-old college-educated male who, after experiencing what was thought to be mild depression for several months, presented with unsteady gait, dysphagia, slurred speech, and intermittent urinary incontinence. A neuropsychological examination was performed 4 months later, using the WAIS-R, Wechsler Memory Scale, Halstead–Reitan Battery, the Wisconsin Card Sorting Test, and several other measures. Verbal IQ at that time was 85 and Performance IQ was 70, scores quite inconsistent with the patient's previous position as a budget analyst. The patient's performance on all tests was impaired; the only tasks with scores less than one standard deviation below the mean were immediate verbal recall and a test of verbal concept formation. This patient's course became chronic despite chelation therapy, and at the time of the report he required total supportive care.

Finally, a somewhat more optimistic outcome for Wilson's disease was found in the case examined by Rosselli, Lorenzana, Rosselli, and Vergara (1987). The authors describe a case of Wilson's disease in a 30-year-old female born of consanguineous parents. The patient was neuropsychologically tested before and after penicillamine chelation treatment.

Initial presentation revealed a Performance IQ of 63 and a Verbal IQ of 73. There were marked impairments in memory, constructional abilities, motor speed, and slowing of verbal responses. Speech was dysarthric without aphasia. The patient showed tremor, rigidity, and bradykinesia.

The authors characterized the patient's improvement after chelation as "dramatic," although some deficits remained. Her Performance IQ increased to 91 while Verbal IQ remained constant at 72. Improvements were seen in both verbal and nonverbal memory for immediate and delayed recall. The patient's learning curve also showed significant improvement. The authors suggest the course of recovery is consistent with the reversibility of Wilson's disease dementia, speculating that remaining deficits in their patient were related to a 13-year prodrome. They hypothesize that earlier treatment would have arrested or eliminated expression of the disease in this patient.

GOLD

History

It has been suggested that the first recorded "medicinal" use of gold is to be found in the Old Testament, where Moses "took the calf which they had made and burnt it in the fire and ground it to powder, and strawed it upon the water, and made the children of Israel drink of it" (*Exodus*, cited by Marks & Beatty, 1975, p. 15).

Gold may have been used as a medicinal preparation by the Chinese as early as 2000 years B.C. The Taoist philosopher Pao Pu Tzu (253–333?) believed that gold, when taken internally, was second only to cinnabar in producing immortality. The 16th century Swiss physician Paracelsus used gold as a blood purifier, poison antidote, miscarriage preventative, and cardiac medicine (Marks & Beatty, 1975).

Current Uses and Neurotoxic Effects

Organic gold compounds are currently used to treat rheumatoid arthritis, lupus erythematosus, and bronchial asthma (Goetz, 1985; Hanakago, 1979), although "potentially severe toxic reactions limit their use" (Schaumburg *et al.*, 1983, p. 121). Gold neurotoxicity is rare and no neuropsychological studies are available in the gold toxicity literature. Neurological signs of ataxia, blurred vision, and tremor have been reported (Hanakago, 1979), although psychiatric symptoms of depression and hallucinations are possible without other CNS signs (Sterman & Schaumburg, 1980). Subcortical infarction and meningeal irritation have been cited in clinical case reports, and ventromedial hypothalamic damage has been produced in animals exposed to gold thioglucose (Sterman & Schaumburg, 1980). Peripheral neuropathy has also been reported in reaction to organic gold ingestion; however, it is unclear whether PNS effects are related to actual gold toxicity or a secondary allergic reaction (Schaumburg *et al.*, 1983). Polyneuritis from gold ingestion has also been associated with gold therapy (Hanakago, 1979), as has Guillain–Barre syndrome, radiculopathy, and myokymia (Goetz, 1985).

IRON

Iron oxide fume
  OSHA   10.0 mg/m^3
  NIOSH   5.0 mg/m^3
(*Quick Guide*, 1993)

Abnormal brain iron accumulation may cause neurotoxic effects in two medical disorders, namely hemachromatosis and Hallervorden–Spatz disease. In the former, iron absorption from a normal diet is abnormally high, with neurological and neuropsychological changes noted in adulthood. Accumulation of iron in the globus pallidus and the reticular zone of the substantia nigra, as well as severe degeneration and demyelination of neurons, are believed to induce dementia, ataxia, and neuropathy (Feldman, 1982a). Hallervorden–Spatz disease is an often familial, probably metabolic disorder that appears in childhood or adolescence. The disorder causes iron accumulation in the globus pallidus and the reticular zone of the substantia nigra, although there is no systemic disorder of iron metabolism. Demyelination may occur in the globus pallidus, and axonal swelling has been noted in the pallidonigral system and the cortex (Feldman, 1982a). The disorder produces choreoathetosis, rigidity, indistinct speech, and progressive impairment in intellectual functions. Hallervorden–Spatz disease is eventually fatal, with death occurring approximately 10 years after onset (Richardson & Adams, 1977).

LEAD

IDLH	700 mg/m^3
Inorganic, fumes and dust (Pb)	
OSHA	0.050 mg/m^3
NIOSH	0.100 mg/m^3
Tetraethyl lead [Pb(C$_2$H$_5$)$_4$]	
Conversion factor	1 ppm = 13.45 mg/m^3
IDLH	40 mg/m^3
OSHA	0.075 mg/m^3
NIOSH	0.075 mg/m^3
Tetramethyl lead [Pb(CH$_3$)$_4$]	
Conversion factor	1 ppm = 11.11 mg/m^3
IDLH	40 mg/m^3
OSHA	0.075 mg/m^3 (skin)
NIOSH	0.075 mg/m^3 (skin)

History

Lead is a ubiquitous metal, forming 0.002% of the Earth's crust. Easily smeltable from galena ore, it was one of the earliest neurotoxins produced by humankind. Lead beads have been found in Asia Minor dating back to 6500 B.C. (Wedeen, 1984). Lead was employed in ancient Egypt in ornaments, glass, and eye paint. The ancient Romans were extensive users of lead, producing an estimated 80,000 tons of lead annually at the peak of the Roman empire, using it in paint pigments, makeup, and elaborate piping and aqueducts (Windebank *et al.*, 1984). Wine was undoubtedly a major cause of lead poisoning for the

Romans, since their prominent philosophers advocated boiling wine in lead containers to increase sweetness and halt fermentation (Wedeen, 1984). The number of birth defects and developmental abnormalities produced by lead exposure in the Roman era can only be guessed, but may have been substantial, particularly since Roman women reportedly coated their cervixes with lead compounds to prevent conception (Wedeen, 1984). A high rate of sterility in the Roman aristocracy may have been induced by an elevated concentration of lead in wine and cooking utensils as well as by the use of lead compounds as sweetening agents.

Lead has no known biological benefit and, like all nonessential heavy metals, it is hazardous to living matter (Niklowitz, 1980, p. 27). The toxic effects of lead have been known almost as long as it has been mined. Two thousand years ago, Dioscorides wrote that lead "makes the mind give way" (Major, 1931). In 200 B.C., the physician Nikander of Colophon wrote of lead's poisonous effects:

> But may the hateful painful litharge [lead oxide]
> be unknown to you,
> If oppression rages in your belly,
> and all around your middle,
> Pent up in your howling bowels,
> winds roar,
> The sort which cause deadly disturbances
> of the bowels, which,
> Attacking with pains unexpected,
> overpowering mankind.
> The flood of urine does not stop it,
> but all around
> The limbs are swollen and inflamed,
> And, no less, the skin's color
> is that of lead.
> [Nikander, cited in Weeden, 1984, p. 14]

Both Benjamin Franklin and Charles Dickens wrote about the deleterious effects of lead (Needleman, 1993). The United States, however, was particularly slow in adopting lead regulations, and even in 1921, when the White Lead Convention proposed control of leaded paint and was ratified by 13 countries, the United States declined to join the signatories. The country was a large consumer and producer of leaded paint and the lead industry controlled dissemination of research funding and information related to lead through the late 1960s, when federally sponsored funding for lead research became available (Mushak, 1992).

Current Uses

Lead in its pure form and its various compounds continues to be utilized in diverse ways, with annual worldwide production estimated at about 5.8 million tons (Mushak, 1992). Although leaded gasoline and paint have declined in use, U.S. consumption of lead has increased since the 1980s, with 1.3 million tons of lead consumed in 1989 alone (Mushak, 1992). Lead continues to be employed in traditional heavy industry, and has found additional uses in electronic components, home radon barriers, television and CRT screens, medical imaging devices, nuclear shielding, and aviation soundproofing. Lead continues to be found in ceramic glazes, batteries, and paint pigments.

OSHA has identified five industries with high risk for significant lead exposure: primary and secondary lead smelting, battery manufacture, brass/bronze and copper foundries, and pigment manufacture (Froines, Baron, Wegman, & O'Rourke, 1990). Workers who employ lead soldering techniques are at risk; one study showed that 67% of automobile radiator repair workers had blood lead levels in excess of 25 μg/dl (Nunez, Klitzman, & Goodman, 1993). One of the most common sources of lead had been as an octane-boosting gasoline additive. When 12,000 neonates were examined, their umbilical cord blood lead level correlated 0.75 with sales of leaded gasoline (Rabinowitz & Needleman, 1982). However, growing knowledge of lead's deleterious health effects and ensuring legal regulation have reduced this form of lead pollution, with 90% of lead in gasoline having been removed in the past decade (Needleman, 1990).

Unfortunately, other source of lead pollution continue undiminished. Current world output of lead exceeds 3.5 million tons per year, an amount greater than for any other toxic metal. More than 2 million metric tons of lead are emitted into the environment annually (Mushak, 1992). More than 800,000 U.S. workers are exposed to lead in their jobs and up to 20% have elevated blood lead levels (Schottenfeld & Cullen, 1984). Further, "there is virtually no evidence that there has been a decrease in the prevalence and severity of the air lead concentrations since 1980 for 4 important industries, battery manufacture . . . brass/bronze foundries, and secondary lead smelting" (Froines *et al.*, 1990, p. 14).

Thus, in terms of quantity produced and number of workplace exposure victims, lead's potential effect on human health remains greater than for any other neurotoxin except alcohol. Perhaps because of this ubiquity and serious health risk, there is more information and research about lead than for any other neurotoxic metal.

Intake

In humans, lead is accumulated via inhalation and ingestion. It is transported throughout the body by the erythrocytes and is deposited in soft tissue and bone. Precise measurement of body lead is complicated by the variability in storage capacities and biological half-times among tissues; however, bone lead is considered to be the best estimate of cumulative lead exposure (Wedeen, 1992). Current estimates indicate that 95% or more of lead body burden is contained in bone and the biological half-life of bone lead is approximately two decades (Wedeen, 1992; Barry, 1975). Half-life varies somewhat according to the bone assayed, with estimated biological half-times of 16 years for the calcaneus, but 27 years for the tibia (Gerhardsson *et al.*, 1993). Determination of actual lead levels in various body compartments requires a much more detailed model of toxicokinetics of the sort worked out by Leggett (1993). Thus, blood lead levels may provide only a very preliminary approximation of body and brain lead burden and use of blood lead levels alone "might seriously underestimate lead body burden" (Erkkila et al., 1992).

EDTA chelation challenge has been termed the "gold standard" in accurate assessment of lead body burden in adults who are not currently exposed to excessive lead, although this is disputed by Tell et al. (1992) who find that chelated lead is not a good indicator of total body burden, but only of more readily mobilized lead in blood, soft tissue, and active exchangeable bone fractions (e.g., vertebra, rib, and surface cortical bones). In vivo tibial X-ray-induced X-ray fluorescence (XRF) is considered a promising alternate procedure (Wedeen, 1992) and X-ray fluorescence of the tibia is more precise than calcaneus XRF (Gerhardsson et al., 1993). Using a 20 µg/dl blood level limit, EDTA urinary chelation output of over 600 µg/dl and wet weight tibia bone levels above 20 ppm are considered elevated (Wedeen, 1992).

This implies that when lead workers retire, their dominant body burden of lead may reside in bone. Bone lead is highly correlated with urinary excretion of lead for 24 h after i.v. infusion with 1 g of calcium disodium edetate, a chelating agent (Tell et al., 1992). Bone lead was not significantly affected by this one-time procedure, suggesting to the authors that chelation mobilizes lead from the blood and soft tissue but not bone, the largest repository of lead body burden.

Zinc protoporphyrin (ZPP) level, a substance accumulated in the red blood cell when lead inhibits the cell's ability to contain iron, is thought to be a more sensitive indicant of biologically active lead on the nervous system than blood lead levels (Lilis et al., 1977). Since ZPP levels persist for the lifetime of red blood cells (approximately 120 days), ZPP screening may be useful for individuals who had been exposed to lead over this time frame, but whose blood or urinary lead levels have returned to normal (Zhang, 1993).

The body burden of lead thought to be neurotoxic has decreased rapidly as a function of research detecting neurotoxic lead effects at increasingly lower levels. Thus, before the 1960s lead levels under 60 µg/dl were considered benign, in contrast to EPA's scientific advisory committee, which has recommended 10 µg/dl as the maximum allowable body burden. Even this level has been challenged as too high for safety; it has been suggested that lead has *no* known safe level of exposure, and that toxic effects of lead may begin at the *nanogram,* rather than the *microgram* range of exposure (Davis, Elias, & Grant, 1993).

Neuropathology

Several mechanisms of lead-related neurotoxic damage have been proposed and the subject remains under investigation. Lead may compete with calcium, sodium, and/or magnesium in neurotransmission. Direct and indirect effects on nerve cell mitochondria, inhibiting phosphorylation, have been suggested (Silbergeld, 1982). In addition, "non-neural effects of lead on such processes as membrane transport, oxidative phosphorylation, and heme synthesis may affect

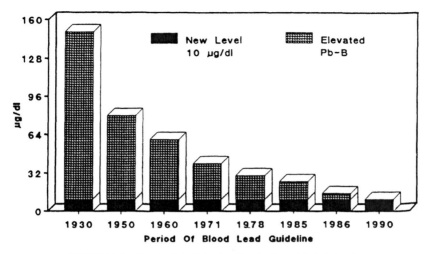

FIGURE 3.1. Graph of declining lead levels. From Mushak (1992).

neuronal function by depleting supplies of precursors, reducing energy sources or producing neurotoxic intermediates" (Silbergeld, 1982, p. 18). Rönnbäck and Hansson (1992) propose that low-dose lead and mercury are neurotoxic to astroglial structures and damage glutamate transmission, leading to secondary decreases in other neurotransmitter systems (and to patient complaints of fatigue and diminished concentration). Astrogliosis may be particularly evident in the hippocampus and cerebellum, which are particularly vulnerable to lead toxicity (Selvin-Testa, Loidl, Lopez-Costa, Lopez, and Pecci-Saavedra, 1994). Secondary neurological effects of lead-induced hypertension may also occur (Khalil-Manesh et al., 1994).

Evidence from primate and other mammalian studies indicates that "the brain has low uptake but tenacious retention of lead" (Leggett, 1993, p. 606). Leggett's model estimates a brain lead removal half-time of 2 years, but this only describes the time that lead remains in contact with brain tissue, not the possible permanency of damage during the time lead resides in the brain. Distribution of low levels of lead in the brain is significantly correlated with the brain's potassium concentration (Leggett, 1993), indicating that lead concentrates in areas of high cell density such as the hippocampus. Lead is taken up by mitochondria in brain cells and accumulates in areas of calcium localization. However, at high levels of lead exposure, the blood–brain barrier breaks down and lead gains direct access to neural tissue (Leggett, 1993).

There is evidence from animal research and some corroboration in human case studies that lead is a demyelinating agent (Fullerton, 1966; Behse & Carlson, 1978; Niklowitz & Mandybur, 1975). A possible relationship of chronic lead exposure and Alzheimer-like symptoms has been posited by Niklowitz (1979) who observed neurofibrillary tangles in animals and in the autopsy of a 42-

year-old survivor of childhood lead encephalopathy. Two studies suggest that cerebellar capillaries may be particularly susceptible to calcification as a result of lead exposure (Benson & Price, 1985; Tonge, Burry, & Saal, 1977). The latter study found 84% of 44 patients who were diagnosed with lead neuropathy at autopsy to show cerebellar calcification.

Chronic exposure to lead is associated with the development of peripheral neuropathy, in which the upper extremities are affected more than the lower, and the dominant hand shows primary involvement. "Wrist drop" involving the wrists and fingers is also prototypic. Overt cases that include these symptoms are rarely found today because of better control of lead in the working environment. Kajiyama et al. (1993) report one such case in a 25-year-old individual with a blood lead concentration of 100 μg/dl. The patient had mild decreases in median nerve motor conduction velocity and ulnar nerve sensory conduction velocity. Compound action potentials showed marked decreases in all examined nerves. Hand muscle atrophy was observed and there was a conduction block at the elbow, suggesting possible cubital tunnel syndrome, leading the authors to conclude that entrapped nerves may be particularly vulnerable to lead neuropathy. Data were considered consistent with peripheral neuropathy with predominant axonal degeneration.

Subclinical lead neuropathy is also detectable by tests of nerve conduction velocity (e.g., Singer, Valciukas, & Lilis, 1983). Hirata and Kosaka (1993) found a significant negative correlation between lead exposure and radial nerve conduction velocity, suggesting lead-related impairment of radial nerve conduction (Hirata & Kosaka, 1993).

CNS pathology has also been validated with neurophysiological testing. For example, the N145 component of visual evoked potential latency, which originates in the cerebral cortex, is correlated with duration of lead exposure (Hirata & Kosaka, 1993). The N20 component of short-latency evoked potentials is also significantly associated with lead exposure duration, indicating delay of conduction times from the wrist to the cerebral cortex. Auditory brain stem evoked response studies on lead by the same authors suggest that chronic lead exposure also impairs the conduction of the acoustic nerve and the brain stem. Murata, Araki, Yokoyama, Uchida, and Fujimura (1993) studied 22 workers in a gun metal foundry with blood lead concentrations of 12–64 μg/dl. Zinc (66–148 μg/dl) and copper (46–136 μg/dl) were also found in plasma. N75 and N145 latencies of the workers' visual evoked potentials were prolonged, as were the N9–N13 interpeak SSEP latencies. P300 was also significantly prolonged. Brain stem auditory evoked potential correlated significantly with lead absorption. Results, combined with significantly slowed motor and sensory nerve conduction velocities, indicated that peripheral, central, and autonomic functions were impaired by lead exposure.

Many questions remain unanswered, including the relative amounts of accumulation in different areas of the brain, and how metabolism or other individual differences mediate lead toxicity mechanisms. A significant advance in lead toxicology is Leggett's (1993) age-specific kinetic model of lead metabolism in humans that calculates the transfer of circulating lead into various organs, fluids, and structures.

Exposure

Occupational Risk

Adult exposure to lead usually occurs in the workplace. Common avenues of exposure include lead smelters, battery manufacturing or recycling plants, auto repair shops, and via contaminated bootleg alcohol. Lead shot manufacture and use of lead-containing industrial paint are also suggested to put the worker at risk for toxic lead exposure (Windebank et al., 1984).

Old paint is a particular hazard with respect to lead. In 1988, the United States Department of Health and Human Services estimated that 52% of all residences (42 million homes) contain paint with lead levels at or above the danger level as set by the Centers for Disease Control (United States Department of Health and Human Services, 1985, 1988). It has also been our experience that workers or home remodelers who remove old paint are at risk; we have seen high blood lead levels in transit workers assigned to burn old paint from elevated railroad track installations. Art conservators who restore antiquities to their original specifications may be at great risk. Fischbein et al. (1992) chronicled the case of an art conservator who attempted to restore a Peruvian tapestry with cinnabar, which contained 8.9 parts per thousand of lead. The fabric of the tapestry had a lead concentration of 380 parts per million. Blood lead levels in this patient reached a high of 127 µg/dl. Perhaps the most unusual form of occupational lead poisoning has been documented in Jewish scribes who prepare and work with a special ink that contains significant amounts of lead (Cohen et al., 1986).

Other Exposure Sources

Significant quantities of lead have been found to leach out of leaded crystal decanters, wine glasses, and baby bottles. Port wine stored in one such decanter showed 89 µg of lead per liter initially, but rose to 5331 µg per liter after 4 months. Brandy stored for 5 years in a similar decanter leached 21500 µg per liter of lead into solution (Altman, 1991). Apple juice and infant formula appear equally capable of leaching lead from crystal, contraindicating their use for those purposes.

Lead exposure may occur from hobbies that utilize lead, including stained glass-making, collecting or making lead figures (e.g., toy soldiers), oil paints with lead base, pottery-making with leaded glazes, and use of firearms. Blood lead levels in runners who train in populated areas with heavy automobile traffic may show almost three times the blood lead level of runners who exercise in rural areas (Orlando, Perdelli, Gallelli, Reggiani, Cristina and Oberto, 1994).

Prenatal and Childhood Exposure

Incidence

The first reports of childhood lead poisoning came from a pediatric house officer (Turner, 1897) and an ophthalmologist (Gibson, 1904). Lead poisoning in these affected children was traced to the porches and railings of houses in Brisbane, Australia. Perhaps as a consequence, a lead paint prevention act was passed in Australia 50 years before similar legislation was introduced in the United States (Needleman, 1993).

Until the 1940s the concept of subclinical lead toxicity was not accepted. Similar to the situation with mercury, it was thought that nonfatal lead exposure was without significant harm (McKhann, 1926). It remained for Byers and Lord (1943) to propose the idea of nonfatal lead toxicity, after they discovered that of 20 children who had "recovered" from lead intoxication, 19 had significant learning disorders or behavioral abnormalities. This paper has been cited as beginning the modern era of lead investigations (Needleman, 1993). Nevertheless, the severity of the problem was only slowly recognized by standard-setting agencies. For example, in the United States, initial standards for childhood lead toxicity were set at 60 μg/dl, now recognized as grossly inadequate to prevent lead toxicity.

Present-day extent of lead exposure in children appears to be almost as great as that in adults. One survey suggests that 3.9% or almost 700,000 U.S. children have blood levels at or above 30 μg/dl (Mahaffey, Annest, Roberts, & Murphy, 1982), which is 20 μg/dl above the current EPA cutoff for significant lead exposure. Racial and socioeconomic factors strongly influence the prevalence of lead elevation, with only 2% of white children showing elevated blood lead, as contrasted with up to 18% of poor, inner-city, black children. City living itself may be an additional risk. The U.S. Agency for Toxic Substances and Disease Registry found elevated blood lead levels in excess of 15 μg/dl in 17% of metropolitan-dwelling U.S. children (Agency for Toxic Substances and Disease Registry, 1988). Landrigan and Curren (1993) estimate that an appalling 68% of inner-city children in the United States have elevated blood lead levels.

The majority of these cases have been linked to *pica,* the ingesting of nonfood substances (Baloh, Sturm, Green, & Gleser, 1975). Dust, dirt, and especially old paint chips (some of which have been found to contain up to 50% lead) are common vehicles of exposure (Otto, Benignus, Muller, Barton, Seiple, *et al.,* 1982, p. 733). Houses built before the mid-1950s were commonly painted with interior and exterior lead-based paint. It has been estimated that flaking and peeling paint from such housing puts about 2,500,000 children under the age of 5 at risk (Baloh *et al.,* 1975). Approximately 12.6 million children under age 7 live in housing with lead-based paint (Agency for Toxic Substances and Disease Registry, 1988). A more recent report by the Department of Housing and Urban Development suggests that 57 million private homes inhabited by 9.9 million

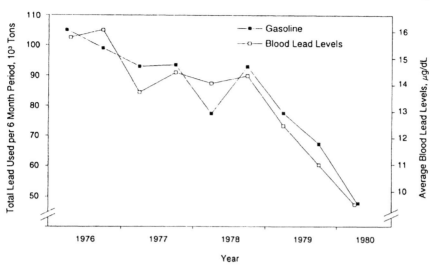

FIGURE 3.2. Relationship between blood lead and gasoline lead levels. From August (1983).

children under age 7 contain lead paint and 3.8 million of those homes are thought to contain peeling lead paint or excessive lead dust (cited in Borher, 1992).

Blood lead levels are also higher in children living within 100 feet of a major roadway, and can be positively correlated with the volume of automotive traffic in the vicinity (Caprio, Margulis, & Joselow, 1974), although federally mandated phaseout of leaded gasoline has produced a reduction in the blood lead levels of U.S. children (Centers for Disease Control, 1985).

Living in the vicinity of lead-utilizing heavy industry may also be a risk factor (Winneke et al., 1983). Landrigan et al. (1975) found that children living near a lead-discharging smelter had significantly decreased IQ and finger tapping scores compared with socioeconomic status (SES)- and race-matched controls. A parent working in a lead-utilizing industry may transport lead from the workplace via skin, hair, or clothing, thereby contaminating the home and its occupants. One study found that almost half of such children had blood lead elevations at or above the then-current clinical threshold of 30 μg/dl (Baker et al., 1977). Another study of 5000 women showed significantly higher cord blood lead levels when either parent worked in a lead-utilizing industry (Wang, Shy, Chen, Yang, & Hwang, 1989). Children of radiator repairers who use leaded solder are at similar risk for elevated lead levels (Nunez et al., 1993).

Where lead is not introduced by industry, paint, or auto, it may enter the household through the water supply, principally from plumbing contaminants rather than source water. The U.S. Environmental Protection Agency has estimated that between 20 and 40% of U.S. children's blood lead levels are derived from drinking water lead and that the percentage of children with elevated blood lead levels (>10 μg/dl) would decline 1.4% if all lead in drinking water could be eliminated (Davis *et al.*, 1993). Before lead piping and lead solder were banned for potable water supplies in 1986, the EPA estimated that 20% of the U.S. population was exposed to initial faucet output (after more than 6 h of nonuse) containing more than 20 μg/liter of lead. Many older communities may retain such pipe. The EPA's 1991 national primary drinking water regulation has set a goal of 0 μg/liter for drinkable water, and an action level of < 15 μg/liter at the 90th percentile of sampled taps (Davis *et al.*, 1993). The EPA is also considering restrictions on the sale of leaded plumbing materials and a ban on leaded solder for plumbers (Davis *et al.*, 1993).

Lead in soil is another potential source of exposure. While estimates of affected children are difficult to make, Davis *et al.* (1993) use a rough figure of 12 million children exposed to significant amounts of leaded soil and dust.

Finally, lead has been shown to cross the placenta, with fetal umbilical cord lead level correlating 0.75 with cord blood levels (Needleman, 1990). Lead in the blood at birth is assumed to be 85% that of the mother's blood level (Leggett, 1993). The risk of an infant being labeled "small for gestational age," "intrauterine growth retarded," or "low birth weight" was 1.5–2.5 times greater for infants exposed to cord blood levels greater than 15 μg/dl than for those with cord lead levels under 5 μg/dl (Bellinger *et al.*, 1991).

Relative Risk

Risk for lead intoxication is inversely correlated with age of the child (Niklowitz, 1979). Because children absorb more lead per unit of body weight than adults and have higher mineral turnover in bone, allowing more lead to be stored and released, they are considered to be at greater risk for lead poisoning than adults (Centers for Disease Control, 1985). High levels of lead in children have been found to increase concentrations of homovanillic acid, a dopamine metabolite, and vanillylmandelic acid, a norepinephrine metabolite, in 24-h urine collection (Silbergeld, 1993), suggesting lead's interaction with these neurochemical systems. Children's brain lead concentration "becomes extremely high and may approach or even exceed that of the liver and kidney" (Leggett, 1993, p. 606). The brains of infants and young children contain a much higher proportion of total body lead because the brain constitutes a greater percentage of their total body mass. As a result, lead deposition in the brains of infants up to 1 year old is three times higher than that of adults, and they may absorb 40% or more of ingested lead (Leggett, 1993). For low-level exposures, higher brain depositions of lead in children continue, but decline in a linear manner until age 5, when adult lead deposition values are achieved (Leggett, 1993). Individual childhood susceptibility to lead exposure also may increase when there are simultaneous neonatal illnesses or other neurotoxic exposure, e.g., neonatal jaundice, pesticide exposure, nutritional deficiency, or chronic disease (Grandjean, 1993).

Because of such factors, the Centers for Disease Control has lowered the definition of "elevated" blood lead levels from 25 to 10 μg/dl (CDC, 1985).

TABLE 3.2. 1991 CDC Child Lead Exposure Guidelines[a]

Blood lead level (μg/dl)	Category
<9	Not considered lead-poisoned
10–14	Consider more frequent screenings
15–19	Nutritional and educational interventions; more frequent screening; environmental investigation and intervention if level persists
20–44	Medical/environmental evaluation and investigation; consider pharmacologic treatment
45–69	Medical/environmental intervention; chelation
>70	Medical emergency

[a]Centers for Disease Control (1991).

However, even these lowered limits may not be adequate to prevent behavioral dysfunction as results from a recent metaanalysis of childhood lead effects not only failed to exhibit a threshold at 10 μg/dl, but actually showed a steeper rate of IQ decline at blood lead levels below 10 μg/dl (Schwartz, 1994).

Longitudinal and Chronic Prenatal Exposure Effects

While there is little evidence for a distinctive behavioral or cognitive "signature" from childhood lead exposure, studies consistently find deleterious effects of lead on longitudinal cognitive development. Long-term follow-up of children with lead exposure points to chronic neuropsychological impairments as a consequence of lead exposure. For example, Bellinger et al. (1984) classified several thousand 6-month-old infants by umbilical cord lead content into three groups: "low" (mean = 1.8 μg/dl), "mid" (mean = 6.5 μg/dl), and "high" (mean = 14.6 μg/dl). There was a difference of approximately 6 points on the Bayley Mental Development Index (BMDI) between the low cord lead group (< 3 μg/dl) and high cord lead subjects ($\geq$ 10 μg/dl). Results suggested that a cord blood lead level of only 10 μg/dl or greater "was associated with early developmental disadvantage" measured by the BMDI (Bellinger et al., 1984, p. 396). Further, when Bellinger, Leviton, Needleman, Waternaux, and Rabinowitz (1986) reassessed their subjects 6 months later, they found that high cord lead subjects continued to maintain about a 7-point deficit on the BMDI compared with low cord lead children. Since blood levels at 6 and 12 months in these two groups no longer differed from each other, prenatal lead levels were more predictive of BMDI impairment than later lead assays. Skills most strongly affected by lead at this age were "fine motor and interactional/linguistic skills" (Bellinger et al., 1986, p. 159).

Huel, Tubert, Frery, Moreau, and Dreyfus (1992) correlated maternal hair lead levels during pregnancy with performance on the McCarthy scales. General cognitive index, and quantitative and qualitative memory subscales were negatively correlated with maternal hair lead levels. Verbal but not perceptual or motor subscales were also significantly related to maternal hair lead content. Gestational age showed independent effects that were unrelated to lead exposure.

Longer-term outcome of childhood lead accumulation was addressed by Needleman, Schell, Bellinger, Leviton, and Allred (1990) who reexamined lead-exposed children after an average of 11 years. Neuropsychological impairment on the Neurobehavioral Evaluation System (NES) and other tests was found to be related to lead content of teeth shed at age 6 and 7. Children with dentin lead level greater than 20 ppm had "a markedly higher risk of dropping out of high school . . . having a reading disability . . . lower class standing in high school, increased absenteeism, lower vocabulary and grammatical reasoning scores, slower finger tapping" (p. 83).

Smaller but significant associations were found in a similar study of 1265 New Zealand children where dentine lead levels were assessed in children aged 6–8 and correlated with test scores, teacher ratings, and maternal/teacher ratings of attentional capacity at ages 12 and 13. Children with mildly elevated lead levels before age 8 did less well on objective tests, were seen by teachers as having inferior classroom performance, and "tended to be more often described as being restless, inattentive and lacking in concentration by parents and teachers" (Fergusson, Horwood, & Lynskey, 1993).

Lifelong residence near a lead smelter produced even more serious neuropsychological impairment. Benetou-Marantidou, Nakou, and Micheloyannis (1988) examined 30 children who lived near a smelter in Greece. Children were age 6–11, with blood levels ranging from 35 to 60 μg/dl and erythrocyte protoporphyrin levels greater than 100 mg/dl. Significantly more children elicited an abnormal neurological examination, and exhibited poorer performance on the Oseretsky test of motor development. School performance was also significantly worse in the lead-exposed group compared with controls.

At the other extreme, a 10-year follow-up of 148 middle-class U.S. 10-year-olds addressed the issue of exposure to relatively low lead levels (mean blood levels ≤ 8 μg/dl) (Stiles & Bellinger, 1993). Results showed few associations between lead and neuropsychological tests in this "best case" scenario. Blood lead at 6 months of age was negatively correlated with WISC-R Vocabulary but positively associated with Coding. Blood lead at 24 months was consistently associated with WISC-R IQ, with higher lead levels at that age associated with significantly lower Full Scale and Verbal IQ scores. Lead level at 57 months was positively associated with several error scores from the Wisconsin Card Sorting Test. The authors did not include tests of vigilance, arousal, or cognitive efficiency in the battery, functions that are often impaired as a result of lead exposure, so it is likely that their results underestimate lead-related cognitive impairments.

Chronic effects of lead in tooth dentin and bone were found to negatively impact on various attentional parameters in adolescents, by Bellinger *et al.* (1994). Dentin lead, even at the relatively low mean level of 14 μg/g negatively impacted on the Stroop Test, a cancellation task and Trail-Making, along with Wisconsin

Card Sorting Test scores. Tibia lead was significantly associated with poorer cancellation test performance and increases in time to complete to Digit Symbol test.

Neurophysiological Effects

High-level childhood lead poisoning more often presents with CNS neuropsychological symptoms than peripheral neuropathy. Childhood peripheral neuropathies, when they do occur, involve the legs more than the arms, whereas in adults the reverse is true (Windebank et al., 1984). Hearing loss and impairment of postural sway may also be found.

Neurophysiological measures have demonstrated the neurotoxicity of lead in children. Otto et al. (1982a) measured slow-wave potentials evoked by sensory conditioning in children with blood levels of 7 to 52 μg/dl and found data suggesting that even the lowest levels of lead were able to alter conditioning. Postural sway was significantly increased as a function of blood lead level in low-lead-exposed children (mean blood Pb = 11.9 μg/dl), implying lead-based impairment of vestibular and/or proprioceptive systems (Bhattacharya et al., 1993).

Hearing loss is also associated with childhood lead exposure. Increasing childhood exposure, from 6 to 18 μg/dl, was associated with 15% more children exhibiting impaired 2000-Hz hearing threshold (Schwartz & Otto, 1991).

Otto and Fox (1993) review the effects of lead on the visual system, the evidence for which comes primarily from rat and primate studies. Moderate-level developmental exposure to lead caused damage to rods in a rat study, while decreased neuronal volume and dendritic arborization in the primary and projection area of the visual cortex were found in chronically lead-exposed monkeys. The authors' review of the animal literature concludes that "a wide range of functional and neurochemical effects on retinal function occur ... at blood lead levels below 20 μg/dl." Investigatory focus on whether similar responses occur in lead exposed child is a clear and immediate priority. Effects on the infant auditory system are suggested by the work of Rothenberg, Poblano, and Garza-Morales (1994), who found that maternal blood lead level during pregnancy effected brainstem auditory evoked potentials during the first week of infant life. They suggest that intrauterine lead exposure may compromise brain structures involved in spatial localization of sound.

Teratogenicity

Lead passes through the placenta and penetrates the blood–brain barrier, accumulating in both breast milk and brain tissue. Neurological deficits may be produced both prenatally, and after birth, through breast-feeding. There are numerous studies linking prenatal lead exposure with "minor malformations, mental retardation, and impaired cognitive development" (Roeleveld et al., 1990,

p. 582). Maternal hair lead level during pregnancy is negatively correlated with cognitive, verbal, quantitative, and memory subscales on the McCarthy Scales of Children's Abilities (Huel et al., 1992).

Treatment of Childhood Lead Exposure

Calcium disodium ethylenediaminetetraacetic acid (EDTA) has been described as "the mainstay of chelation therapy for lead poisoning for the past 40 years" (Chisolm, 1992). Twice as much lead is excreted in the first 24 h of treatment using a combination of British Anti-Lewisite (BAL) and EDTA (Chisolm, 1992). Chisolm also suggests that a new chelation drug, dimercaptosuccinic acid (DMSA), has considerable promise in the rapid removal of lead, and in animal studies removes lead from brain tissue (whereas EDTA does not). No comparative human or primate studies testing this effect are available.

Neuropsychological Effects of Childhood Lead Exposure

There is some controversy concerning confounding influences in studies of lead-exposed children. Several researchers have questioned whether neuropsychological dysfunctions found in lead-exposed children are related to the effects of lead, or the nonspecific conditions highly correlated with lead exposure, including poverty, poor nutrition, and an understimulating environment (e.g., Baloh et al., 1975). Winneke and Kraemer (1984), in particular, argued that these confounding influences do not receive correct methodological treatment in most childhood lead exposure studies. Alternatively, Needleman (1993) argues against the "overcontrol" of so-called confounders, noting that father's employment, school placement, and other factors may themselves be related to lead exposure, and should not be arbitrarily removed from the statistical model, i.e., "to control for such outcome measures as school placement . . . hyperactive behavior . . . or developmental delay . . . may be to subtract out variance which properly belongs to the main effect, lead" (p. 164). It is possible, of course, to undercontrol factors that obscure the relationship between lead exposure and neuropsychological variables. The study by Ernhart, Morrow-Tlucak, Marler, and Wolf (1987) purporting to find no association between lead and IQ, examined a cohort where half of the subjects were born to women admitting excessive alcohol intake during pregnancy, an obvious confound.

Another complicating factor in the design of childhood lead studies is the lack of knowledge concerning childhood lead-ingestion patterns. This might cause inappropriate subjects to be grouped together. For example, if only acute blood lead levels are obtained, children with single acute exposures to lead and no body burden may be compared with others whose blood lead levels are equivalent but which reflect the slow release of lead from a large body burden accumulated over several years. In addition, children who sustained significant lead exposure as toddlers may not be examined until they are of school age, at which point their lead levels have decreased to "normal" levels and they are mistakenly grouped with an unexposed cohort (Bellinger & Stiles, 1993).

A third difficulty has been the correct selection of testing procedures that are most sensitive to lead effects. Some authors have criticized the use of insensitive testing procedures in the examination of childhood lead exposure (Needleman *et al.*, 1979). Overall, the problem of detecting the neuropsychological effects of environmental lead exposure has been characterized as detecting "weak signals embedded in a noisy background" (Winneke, Brockhaus, Ewers, Kramer, & Nef, 1990, p. 557).

Despite these methodological and practical complications, evidence has steadily accrued that ingestion of lead in childhood produces neurotoxic sequelae at both clinical and subclinical levels. Several studies show neuropsychological impairment to be associated with subclinical blood lead levels in children (e.g., Rutter, 1980). Hunter, Urbanowicz, Yule, and Lansdown (1985) found a small but significant slowing in reaction time in children with lead levels of 5–26 μg/dl (mean = 11.85 μg/dl). David, Grad, McGann, and Kolton (1982) found a significant negative correlation between supposedly "nontoxic" mean blood levels of 25 μg/dl and mental retardation that could not otherwise be explained by other factors. Yule, Lansdown, Millar, and Urbanowicz (1981) studied 141 children with blood levels from 7 to 33 μg/dl and found significant impairments in general intelligence and several achievement test scores (reading and spelling) that remained after social class had been partialed out.

Lead assays of shed teeth have also correlated with neuropsychological dysfunction. For example, Needleman *et al.* (1979) found WISC-R Full Scale IQs to be significantly lower by about 45 points between high- and low-lead groups. High-tooth-lead children also performed significantly more poorly on all subtests of the Seashore Rhythm Test, suggesting auditory processing dysfunction, and on a reaction time measure, suggesting impaired vigilance or attention. Winneke, Hrdina, and Brockhaus (1982) found impaired problem-solving and perceptual-motor integration in children with high lead dentine levels, although they also warn against confounding variables of "socio-hereditary background" in these types of studies (Winneke *et al.*, 1983). Rabinowitz, Wang, and Soong (1991) examined dentine from shed teeth in 493 children in grades 1–3 in Taiwan, and found significant negative correlations with Raven's Colored Progressive Matrices IQ. The effect was significant for all three grades, although by grade 3, the association became marginally nonsignificant. Adding family education and other demographic factors eliminated the significant association in boys, but not in girls. Bellinger, Hu, Titlebaum, and Needleman (1994) show that dentin lead levels continue to correlate with neuropsychological function through the end of adolescence. Dentin lead in a sample of 79 19- and 20-year-olds averaged 13.7 μg/g and was significantly associated with what the authors called a "focus-execute" factor that included performance on the Stroop Test, a cancellation test, and the Trail-Making Tests, particularly Trails B. Dentin lead level was also inversely related to the number of categories achieved on the Wisconsin Card Sorting Test and positively associated with WCST perseverative responses.

Personality and conduct-related variables may also be correlated with lead levels in shed teeth. Bellinger, Leviton, Allred, and Raboniwitz (1994) found that tooth lead level was significantly associated with total problem behavior ratings of eight year old children on a teacher rating form of the Child Behavior Profile.

Association of hyperactivity with lead has been noted (David, Hoffman, Clark, Grad, & Swerd, 1983) and replicated with a nondisadvantaged group of children (Gittelman & Eskenazi, 1983). Another study examined a total of about 1000 children and found that high lead levels were associated with impaired Verbal and Full Scale IQ on the WISC, especially in the Information, Comprehension, and Vocabulary subtests in a group of children matched for parent social status, sex, school, and neighborhood. The high-lead group also made significantly more errors on the Bender Visual Motor Gestalt and had poorer ratings in a behavioral rating scale (Hansen, Trillingsgaard, Beese, Lyngbye, & Grandjean, 1985). Similar deficits in other visuospatial tasks have also been found (Maracek, Shapiro, Burke, Katz, & Hediger, 1983). Such dysfunctions may be cumulative, and with repeated exposure the risk of permanent brain damage approaches certainty (Niklowitz, 1979).

A collaborative European effort attempted to partially standardize assessment techniques and general statistical design. The multisite study tested children from Bulgaria, Denmark, Greece, Hungary, Italy, Rumania, West Germany, and Yugoslavia on four subtests of the WISC, a background interference version of the Bender Gestalt, Trail-Making A, the "Vienna Reaction Device" (serial choice reaction time to auditory signals and colored light combinations), a delayed reaction time test, and Needleman's behavior rating scales. Total subjects were 1879, with a range of blood lead from < 5 to 60 µg/dl.

Significant results were obtained on the background interference Bender Gestalt, and the Vienna Reaction Device, suggesting that childhood visual-motor and attentional integration is affected by lead exposure. The set of WISC subtests employed (Vocabulary, Comprehension, Block Design, and Picture Completion) achieved borderline ($p < 0.1$) significance. Winneke et al. (1990) conclude that "a significant linear relationship was found between blood-lead concentration and visual-motor integration as well as serial-choice reaction performance" (p. 558).

Evidence for lead-related central auditory processing dysfunction can be found in Dietrich, Succop, Berger, and Keith (1992), who found that prenatal, neonatal, and postnatal blood lead levels were associated with increased difficulty perceiving speech in the presence of background noise, irrespective of IQ or hearing acuity. Primate experiments suggest that lead exposure either delays speech structure development, or alternatively causes compensation and reorganization of neural mechanisms in response to lead-mediated nerve damage (Otto & Fox, 1993).

A common criticism of lead research in children is that not all studies have found effects of low lead exposure. Needleman (1990) addressed this issue noting that some of the "insignificant" studies used small sample sizes and/or many covariates (e.g., Dietrich, Succop, Berger, Hammond, & Bornschein, 1991) reducing their effective power to make a no-effect inference. Further, since prenatal and some postnatal exposures are shared by parent and child, "correcting" for parental intelligence or caregiving quality may underestimate the variance related to lead and lead to the creation of a Type II error (Davis et al., 1993).

To answer the question whether childhood lead exposure studies taken together show significant effects, two important metaanalysis studies have been conducted. In the first, Needleman (1990) performed a metaanalysis on 13 lead studies (excluding his own) from 1974 through 1987. From such a subject pool of 2187 children, the joint p value was "3 in a trillion... a highly unlikely outcome" (p. 335), and one that strongly suggests an effect of lead on child development.

A second metaanalysis on the relationship between blood lead and IQ was performed by Schwartz (1994), which affirmed and extended Needleman's conclusions. An IQ decrease of 2.6 points was found with an increase in blood lead from 10 $\mu g/dl$ to 20 $\mu g/dl$, an effect that was not limited to disadvantaged children. In fact, lead-related estimated IQ loss was greater in nondisadvantaged children (2.89 ± 0.50 IQ points) than disadvantaged children (1.85 ± 0.92). Even more troubling was the fact that there was no evidence of a low-end threshold, even when blood lead levels reached 1 $\mu g/dl$. The slope of IQ loss was steeper at lower lead levels suggesting proportionally larger IQ losses in the 0–6.5 $\mu g/dl$ (an approximate 7-point drop in IQ) than in the 10–20 $\mu g/dl$ range. The Schwartz (1994) metaanalysis suggests that there is no threshold effect of childhood lead exposure at 10 $\mu g/dl$; in fact, the opposite may be the case; i.e., that lead induces proportionally greater damage at lower blood levels.

From a statistical viewpoint, however, metaanalyses may introduce their own artifacts, including (1) reliance on published significant findings, since insignificant results may not be approved for publication, (2) collapsing the results of studies heterogeneous in design, lead exposure markers, confounders, and statistics, and (3) the analysis is often based on "selective information provided by the authors rather than on independent analysis of the original data" (Winneke et al., 1990, p. 557).

Despite these potential problems, the overall weight of the literature strongly supports the continued finding that childhood lead exposure is neurotoxic to the developing child. Disputes may center on the long-term effects of very-low-level lead intoxication and on the methodology used to control for social and demographic factors that may be potential confounds and that may bias the state of the literature toward Type I error (Ernhart, 1992). Methodological criticisms of such studies must be taken seriously and may be used to guide future research; however, at this time there remains a preponderance of evidence on the side of a positive association between subclinical lead exposure and neuropsychological dysfunction. With both primate studies (reviewed in Schwartz, 1994) and a growing group of better-controlled human studies entering the literature, it is unlikely that the continuing finding of low lead neurobehavioral toxicity is artifactual. Schwartz's metaanalysis suggests that there may be no "safe" level of childhood lead exposure; other authors have argued that the neurotoxic effects of lead may occur as low as the *nano*gram rather than the *micro*gram level (e.g., Davis *et al.,* 1993). Even if data supporting low-level lead neurotoxicity were less convincing, it must also be pointed out that the human consequences of making a Type I error (incorrectly rejecting the null hypothesis and viewing this known neurotoxin as a subclinical poison) would be far less damaging in their clinical and social effects than would be the commission of a Type II error (that of incorrectly *accepting* the null hypothesis and assuming no effect of lead). Conversely, the societal benefits of reducing lead exposure would seem to far outweigh the drawbacks in any cost-benefit analysis that compares lead abatement costs to "the costs of medical care, compensatory education and school dropouts." Schwarts, 1994 p. 105). For example, Schwartz (1994) calculates that lowering blood lead levels of affected children just *one* microgram per deciliter would "lower the number of children with blood lead concentrations above 25 μg/dl by 145,000 and avoid 189 million in medical care costs" (p. 109).

Case Study: Childhood Lead Exposure

The patient (B.) was a 7-year-old black male referred for neuropsychological evaluation.

Tests Administered

Wechsler Intelligence Scale for Children-Revised (WISC-R): Full Scale IQ 55

Verbal tests	Sc. score	Performance tests	Sc. score
Information	02	Picture Completion	05
Similarities	02	Picture Arrangement	01
Arithmetic	04	Block Design	01
Vocabulary	04	Object Assembly	05
Comprehension	02	Coding	10
(Digit Span)	03	(Mazes)	08
Verbal IQ	55	Performance IQ	64

Other tests:
    Beery Buktenica Visual Motor Integration Test
    Bender Visual-Motor Gestalt Test
    Lateral Dominance Examination
    Trail-Making Tests (A&B): Children's Version
    Wepman–Morency Auditory Memory Span Test
    Finger Tapping Test
    Grooved Pegboard Test
    Purdue Pegboard Test
    Children's Hand Dynamometer
    Draw-A-Person Test
    Personality Inventory for Children (PIC)
    Vineland Adaptive Behavior Scales
    Children's Background Information Form
    Clinical Interview with Parents

Background Information. The patient is a 7-year-old African-American male. Mother is a high school graduate who attends evening classes in secretarial skills, father is an 11th grade graduate employed as an assembler of consumer products. Neither parent uses alcohol or drugs, either prenatally or presently. Mother had no unusual illnesses during pregnancy and has no chronic medical impairments.

Medical records report normal, uncomplicated pregnancy and a normal birth for B. Labor was relatively brief, concluding with an unremarkable vaginal delivery. Apgar score was good and B. weighed 9 pounds, 3 ounces at birth. He was able to leave the hospital at 2 days of age. The patient has two sibs, both of whom perform normally in school.

Family history was negative for asthma, allergies, anemia, cancer, diabetes, epilepsy, cardiac problems, hepatitis, hypertension, renal problems, or tuberculosis.

At 15 months, B.'s development was considered to be unremarkable, with age-appropriate skills, including drinking from a cup, using a spoon, and walking. Prior to age 2, B. was rated as "average" in the following behaviors: crying, weight gain, food tolerance, amount of sleep, amount of activity, and cuddling response.

As he grew out of infancy, however, B's mother noticed he had become "a very active child, always getting into mischief." Father recalled that B. "couldn't go to sleep like a normal child.... He'd stay up so late ... and he wouldn't want to sleep without us and it was the hardest thing to get him to sleep without us.... If I put him in the other room, he'd just cry and cry and we couldn't get any rest.... He would sleep in our bed with us."

X-ray reports noted B. to have densities suggestive of lead paint chips. Initial blood lead levels, drawn in December, 1983, were 38 µg/dl, with pica etiology, consistent with mother's statement on deposition and clinical interview that the walls of B.'s room were covered with chipping, peeling paint. Prior to that time, neither parent was aware of pica or the risks of lead poisoning from eating old paint chips.

Levels drawn 1 month later were 26 µg/dl of Pb. By August, 1984, the B.'s physician noted that "mother upset because B. *very* active 'driving me crazy' " and that "mom found B. recently eating lead last week." Levels taken on or about 8/84 showed blood lead at 50 µg/dl and ZPP of 288, and B. was classified having Class III intoxication.

At that time, B., his sibs, and his parents were living at B.'s birth residence. B. was the only child born while the family was living at the lead-painted house.

At 4 years of age, B. received psychological evaluation that showed him to be mentally retarded with the mental age equivalent of a 2-year-old child. B. showed poor ability to follow examiner instructions and execute test directions. The psychologist noted "major cognitive weaknesses" in verbal and numerical skills, perceptual-motor development, and memory skills. Gross motor coordination was inconsistent, with arm coordination at age level, but leg coordination reported to be at the 2.5-year level. "Organic involvement" was a suggested etiology for these impairments.

In November, 1985, the patient was seen four times at a local hospital for mothball ingestion. No sequelae were noted.

B. was placed on Ritalin when he was about 5 years old. Mother reports that Ritalin "kept him up even later" and she discontinued this medication about 6–8 months later after observing no improvement in B.'s behavior.

B.'s behavior disorder also manifested itself in school. In kindergarten, he disturbed other children, and mother reported that "he just can't sit still for 5 minutes, the other kids weren't used to that ... he was punching, talking and distracting them. ... Once he walked out of the classroom."

In 1986, the parents moved into a new home and there was no further evidence of pica.

In 1987, B.'s X-rays were found to show evidence for two past episodes of lead ingestion.

Complete blood levels for this patient are shown in Fig. 3.3.

Despite multiple chelation therapies, the patient continues to exhibit dangerous and impulsive behavior. Recently, B. threw his dog out the window, causing its death, stuck his hands inside an exercise bicycle's spokes, breaking two fingers, and sustained minor injuries when he impulsively ran into the street and was struck by a car.

Mother's contradictory account of her son's level of functioning continued when she was formally interviewed about B.'s interpersonal and cognitive capabilities using the Vineland Adaptive Behavior Scales. On many items, mother was unrealistically optimistic about B.'s level of function. For example, she endorsed

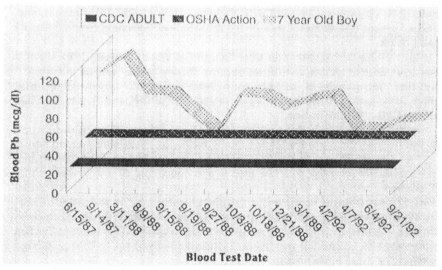

FIGURE 3.3. Graph showing 7-year-old males' blood Pb levels, 1987–1992.

several items on the Vineland Adaptive Behavior Scales indicating that B. knew certain skills, e.g., stating his complete home address. However, when B. was questioned directly, he was unable to demonstrate indicated knowledge. In addition, mother did not appear to view her son as hyperactive on a child personality inventory (PIC) although his behavior in the testing session was obviously hyperactive.

On the Vineland, mother rated B.'s abilities as 2 to 4 years lower than his current age. B.'s lowest scores were in play and leisure time skills, and in receptive language understanding. Mother evaluated both skills at a 3-year-old level. B.'s only behavioral strength, according to his mother, was in the area of personal hygiene and dress, where he scored in the 8 year, 9 month range. All other social skills were rated from 2 to 4 years lower than age expectation.

With the exception of B., mother stated that her other children are healthy and have no evidence of hyperactivity, mental retardation, or other educational or health problems. The other children attend regular school and are not considered problems by either parent.

Behavioral Observations. The patient is a slightly built African-American male child who arrived on time for evaluation, accompanied by both parents. B. was casually but neatly dressed in jeans, a blue denim shirt, and sneakers. His hygiene and grooming were good and he appeared to be well cared for.

Speech was normal in volume, rate, and timbre, although the patient was not spontaneously verbal. B. smiled and willingly accompanied me into the testing room. Initially quiet and cooperative, B. quickly became frustrated with testing, saying "Oh, come on!" "No way Jose!" in reaction to tasks he did not wish to perform.

Over the course of several hours, B. had increasing difficulty remaining in his seat. He became continually active, restless, and would constantly manipulate objects on the testing table. He was easily distractible and difficult to focus on task. These behaviors could not be controlled with verbal command, but slightly and temporarily improved with positive reinforcers (M&M's). There was no generalization to later behaviors, since when B.'s parents were being interviewed at the end of the examination, B. ran over and shut off my laptop computer during notetaking.

Neuropsychological Test Results. Intellectual, language, and general cognitive functioning. The patient currently performs at the juncture between Mild and Moderate Mental Retardation for Verbal and Full Scale IQ scores. Nonverbal skills were performed in the Mild range of Mental Retardation. B. scored almost uniformly at the bottom of measurable norms on tests of world knowledge, verbal abstract reasoning, social and interpersonal knowledge, and auditory attention span. Subtest scores comprising the patient's verbal IQ were not significantly different from one another.

The patient's performance on nonverbal intellectual tasks showed strongly significant differences, depending on the type of test. Tests that required repetitive, visual-motor matching skills were performed in a range expectable for a normal, unimpaired 7-year-old. In addition, B.'s ability to negotiate simple mazes was near average.

In contrast, two more complex tests that required planning, spatial information processing, or integration of linguistic and spatial information, were performed at the bottom of measurable norms.

Attention and concentration were impaired throughout the test battery and degraded the performance of several tests. B. was easily distractible and continually attempted to reach for nearby tests and papers on the testing table. He did not respond to direction and could not be left alone at the table because he would immediately begin manipulating test materials. B.'s behavior improved somewhat when he was reinforced for appropriate behavior with M&M's, but he was never able to maintain concentration for more than 5 minutes.

Specific neuropsychological tests of attention and concentration also showed deficit. B.'s performance on a test of simple visual attention was at the 5 year, 9 month level. His ability to focus attention on a series of photos (in order to arrange them in a meaningful sequence) was so impaired that he failed every item. Attentional limitations on this task contributed to a score below that of the average 5-year-old.

School readiness skills were almost uniformly several years below normal age expectation. Only B.'s ability to perform simple actions from verbal command was performed at normal age level expectation. In contrast, the patient's word knowledge, visual recognition of well-known storybook characters, and arithmetic skills were equivalent to those of a 4-year-old.

Sensorimotor functions. B. was right hand and right foot dominant for simple motor actions. He did not exhibit a preference for sighting through the right or left eye.

Motor skills showed reversed dominance for fine motor speed, fine motor coordination, and gross motor strength. Fine motor speed was average to mildly impaired for the dominant hand, and unimpaired for the nondominant hand. Fine motor coordination was severely impaired on a pegboard task with the patient showing performance equivalent to that of a 4-year-old.

Gross motor strength was average for the dominant hand and above average for the nondominant hand.

Memory and learning. B. showed poor performance on tests of digit span and verbal memory span. He was only able to remember three words or digits for immediate recall (a level equivalent to that of a 4.5-year-old), and could not recall any digits in reverse order of presentation. Visual immediate memory for hand movements was also impaired at three items, a score equivalent to that of a 5-year-old. Short-term memory for spatial locations was also performed at the 5-year-old level, as was B.'s ability to hear words and link them to appropriate pictures, a short-term memory task requiring both visual and auditory integration.

Visuospatial functions. B. showed a variety of deficits in spatial integration and construction. His ability to assemble cubes or triangles to match a template was significantly below expectation, with the latter task performed at the level of a 5-year-old. Similarly, copying of shapes and drawing of human figures were significantly impaired, compared with age norms. B.'s level of visual-motor integration was at the 5 year, 4 month level. The patient's performance was most severely impaired when visual information was presented sequentially, requiring both sustained attention and integration. B. was completely unable to perform this type of task.

Because of B.'s lack of ability to sit still, visual fields were not tested. However, there were no indications of visual field cut or loss of acuity.

Personality functioning. B. did not display any evidence of visual/auditory hallucinations, delusions, or other psychotic processes. There was no evidence of depressive affect. The patient's predominant behavior was hyperactive. For example, B. had difficulty remaining in his seat, showed continual activity, restlessness, and constantly manipulated objects on the testing table. He was easily distractible and difficult to focus on task. These behaviors could not be controlled with verbal command, but could be mildly and temporarily improved with positive reinforcers (M&M's).

Summary and Conclusions. This is a 7-year-old African-American male who was referred for neuropsychological evaluation after documented lead poisoning. Current neuropsychological test results reveal a global range of cognitive and behavioral deficits that place B. from 2 to 4 years below age-expected skills. Intellectual function was at the juncture between Mild and Moderate Mental Retardation, and B.'s only strengths were gross motor hand strength and performing simple, repetitive actions. He cannot integrate information, and cannot attend to materials long enough to make plans or remember them. He showed global impairment of intellectual functions and school readiness skills. Areas of neuropsychological impairment include concentration, fine motor coordination, visuospatial skills, abstract reasoning, planning, memory, and learning. In addition, hyperactivity and attention deficit add further severe interference to B.'s capacity for acquiring new learning.

Prior to his exposure to lead, B.'s birth and development were considered normal. There were no indications of significant injury, developmental delay, genetic defect, birth trauma, or familial impairment that would account for B.'s grossly impaired functioning. In contrast, blood lead levels on more than a dozen occasions during a 5-year period were in excess of the CDC's limits for pediatric lead poisoning. Several blood test results were greater than two to three times CDC limits, and at least one was in the Class III range of poisoning, a grossly neurotoxic pediatric lead level.

In B.'s case, lead was apparently repeatedly ingested and deposited in the bone, while he lived at his birth residence. This body burden of lead then gradually leached out toxic levels of lead into the bloodstream and the nervous system over a period of years. B.'s symptoms are easily explainable in relation to this chronic lead exposure and lead levels sustained.

Without lead poisoning, B. should have developed into a normal, healthy child capable of education and eventual employment. However, exposure to lead levels as high as three times CDC cutoffs have left B. cognitively and behaviorally retarded to the point of requiring continuous supervision. Despite special education, he has remained far below the capabilities of normal children and does not appear to be capable of cognitive remediation. Further, B. cannot control potentially self-damaging behavior and is therefore at risk for further self-harm or even serious injury or death resulting from hyperactivity-based impulsive behavior. B.'s running into the path of a car this year is an example that could easily be repeated.

In reviewing relevant factors in this case, including B.'s health records, neuropsychological test results, and known neurotoxic effects of lead, it is my professional opinion that B. suffers from permanent and disabling sequelae of childhood lead poisoning. B.'s ability to develop normal cognitive skills and eventually compete in the work force have been permanently destroyed. Lead paint ingestion while living at their former residence is the most likely explanation for his mental retardation and hyperactivity. Because of leaded paint ingestion, B. will now require lifelong structured living arrangements, medication evaluation and management, respite care for his mother, and lifelong social service intervention.

Diagnosis (DSM-III-R). I. 317.00 Mild Mental Retardation; I. 314.01 Attention-Deficit Hyperactivity Disorder—Severe; II. No diagnosis on Axis II; III. Lead poisoning (by history).

Recommendations

1. Consider psychiatric medication reevaluation. B.'s impulsive self-damaging behavior puts him at risk for injury and medication may lessen this risk.
2. Consider arranging for respite care for the mother.
3. Consider arranging for specialized behavior modification training for mother to assist her in controlling B.

Adult Lead Exposure

Organic Lead

In adults, lead is associated with two spectra of neuropsychological dysfunctions, depending on its composition. Organic (covalently bonded to carbon) lead, contained in leaded gasoline, solvents, or cleaning fluids, produces neuropsychological symptoms that include memory dysfunction and loss of ability to concentrate (Grandjean & Nielson, 1979). Emotional symptoms produced by organic lead exposure include psychosis with hallucinations, restlessness, nightmares, and impotence. In high concentrations, organic lead can produce delirium, convulsion, and coma.

Organic lead exposure exerts these effects by interfering with energy metabolism and possibly damaging the hippocampus, amygdala, and pyriform cortex (see Grandjean, 1983, p. 336, for a detailed discussion of these effects). One form of organic lead (tetraethyl lead) is metabolized into triethyl lead, which can cross the blood–brain barrier and disrupt cholinergic and adrenergic central pathways (Bolter, Stanczik, & Long, 1983).

To investigate the effects of low-level organic lead on neuropsychological function, Seeber et al. (1990) examined 38 employees involved in the production of tetraalkyl lead, a gasoline antiknock additive. Concentration of organic lead in the blood of these workers averaged 21 µg/dl, with an average exposure length of 14 years. Mean age of the workers was 41 ± 9 years. While neuropsychological test results were within the normative range, a digit symbol test showed significant negative correlation with total urine lead and with trimethyllead (urine). Simple and choice reaction times were also significantly correlated with total urinary lead. Significant correlations are consistent with a relatively higher neurotoxicity for organic lead compounds.

Most cases of organic lead intoxication are the result of exposure to leaded gasoline. Leaded fuels contain a variety of other neurotoxic substances, including benzene, xylene, triorthocresyl, and ethylene dichloride (Poklis & Burkett, 1977). Therefore, the neuropsychological results of organic lead exposure may actually be caused by combinations of triethyl lead and solvents, as well as from hypoxia in cases where intoxication has produced unconsciousness.

Case studies of organic lead intoxication are usually reported either as a by-product of accidental industrial exposure or through voluntary gasoline "sniffing." Such case reports are rare, and even fewer neuropsychological studies of organic lead intoxication have been performed.

Two cases of high-level gasoline exposure with hypoxia in a gasoline storage tank were reported by Bolter et al. (1983). One subject, unconscious for 20 min, experienced numerous cerebellar and cortical symptoms, including paresthesias, nausea, anxiety, and dysarthria. Neuropsychological testing showed intellectual and short-term memory functions to be in the borderline range. Other neuropsychological tests showed impaired motor functioning of the right hand, decreased bilateral tactile sensitivity, and impairment on phoneme and rhythm discrimination, expressive language, sensorimotor functions, and abstract reasoning. Two years later the patient complained of daily headache, numbness and weakness of the right hand, and continued to show lowered language and verbal memory abilities as well as impaired emotional functioning.

A second subject, unconscious for 2 min, also showed left hemisphere deficits which remained 3 years postaccident. Possible reactive emotional symptomatology was also seen in the form of phobic reaction to gasoline fumes, and symptoms of suspicion, distrust, and withdrawal. Lowered verbal functioning in both subjects was speculatively attributed by the authors to greater relative impairment of left hemisphere functions which show greater specialization for and potential for disruption of cholinergic activity.

Gasoline Sniffing. Neuropsychological and neurological effects of gasoline inhalation are also produced when vapors are deliberately inhaled to induce intoxication. Gasoline vapors are rapidly absorbed by the lungs and symptoms begin within 35 min. Voluntary gasoline abusers report that 15–20 breaths of gasoline vapor will cause a 5–6-h period of euphoric intoxication. Psychotomimetic concomitants can include visual, auditory, and tactile hallucinations, sensations of lightness or spinning, and alterations in shapes or colors (Poklis & Burkett, 1977). Rapid or more prolonged inhalation can result in "violent excitement followed by unconsciousness and coma" (Poklis & Burkett, 1977, p. 36).

Seshia, Rajani, Boeckx, and Chow (1978) examined 50 native American Indian patients in Winnipeg, Mannitoba, Canada, who had been voluntary abusers of leaded gasoline. The patients had histories of sniffing leaded gasoline from 6 months to 5 years. Forty-six (92%) had abnormal neurological examinations which resolved in all but one case after 8 weeks. All but one had blood lead levels above 40 $\mu g/dl$ and chelation was instituted in 39 patients. Twenty-seven of the forty-six patients tested had abnormal EEGs with very low voltage and an excess of diffuse slow activity. Ten of fifteen EEGs given 8–12 weeks later were normal. The most common neurological symptoms noted on initial intake included abnormally brisk deep reflexes (34 patients), intention tremor (29 patients), and abnormally brisk jaw jerk (28 patients). Various cerebellar signs were also observed and many of the abnormal findings correlated with mean blood lead or ALAD levels (Seshia, Rajani, Boeckx, & Chow, 1978). The authors also note that 17 (34%) of the patients resumed sniffing gasoline on discharge, a finding that could suggest effects of peer pressure or the addictive potential of gasoline inhalation.

An even more severe form of encephalopathy from organic lead intoxication has been reported by Goldings and Stewart (1982). The subject was a 15-year-old male with a several year history of "sniffing." Serum lead level on admission was 168 $\mu g/dl$. The patient developed a toxic delirium state with buccal and lingual dyskinesias, ataxia, myoclonus, and hallucinations of "bugs crawling on my skin." Snout reflexes, bilateral foot grasps, urinary incontinence, and a palmomental sign were noted. The patient's premorbid intellectual capacity was probably in the retarded range, limiting speculation about specific intellectual deficits from organic lead. However, symptoms of delirium and encephalopathy were similar to, if more severe than, other reported cases of organic lead intoxication (Goldings & Stewart, 1982). More recently, Prejean and Gouvier (1991) found lowered Verbal IQ and verbal memory in two young adult subjects who sniffed gasoline recreationally. The subject who sniffed leaded gas also showed bilateral fine tactile sensory loss that was slightly worse on the right side and slowed fine motor coordination in the right (dominant) hand. Serious impairment was noted on the Category Test.

Inorganic Lead

Exposure to pure or "inorganic" lead has been more extensively studied than has organic lead intoxication, probably because more individuals are exposed to inorganic lead. Inorganic lead appears to induce different neuropathological alterations, especially in cerebellar and hippocampal sites. The constellation of neuropsychological symptoms produced by inorganic lead also differs from that of organic lead in both cognitive and emotional spheres.

General Symptoms of Inorganic Lead Intoxication. Neurotoxic reactions to inorganic lead usually occur in the context of a systemic illness with classic symptoms. These include "abdominal pain, constipation, anemia and neuropathy." Gout may also be present (Windebank et al., 1984, pp. 213–245).

Exposed subjects do not typically complain of these symptoms when blood lead levels are less than 40 μg/dl. However, neurological investigations have found lead-related abnormalities below the level of subjective complaint. Both PNS and CNS effects have been reported. For example, Araki and Honma (1976) found that nerve conduction velocities of median and posterior tibial nerves were significantly correlated with blood lead levels in a group of 39 lead workers whose mean blood lead concentration was 29 μg/dl. Another study found a "highly significant decrease in visual sensitivity" (and presumed damage to the optic nerve) in 35 workers with subclinical blood lead levels (Cavalleri, Trimarchi, Minoia, & Gallo, 1982, p. 263). This latter result is consistent with Niklowitz's (1979) observation that "Pb is taken up by the brain even when blood-Pb levels are quite low" (p. 28), and suggests that vulnerable brain tissue has no protection against lead damage, even at subclinical exposures.

Lead exposure is associated with hearing impairment. Schwartz and Otto (1991) found increased risk of elevated hearing thresholds at all four reference frequencies (500, 1000, 2000, and 4000 Hz). Elevated hearing thresholds are also observed at frequencies above 4 kHz and in one study were found in the range of 0.5–8 kHz (Otto & Fox, 1993; Repko, Morgan, & Nicholson, 1975). Workers with lead levels above 70 μg/dl are particularly likely to show lead-related hearing impairments (Otto & Fox, 1993). Increases in blood lead from 6 μg/dl to 18 μg/dl, were associated with 2-dB hearing loss. In children, an increase from 6 to 18 μg/dl was associated with 15% more children with impaired 2000-Hz hearing threshold. Etiology is unclear, although there is animal evidence that lead reduces glucose utilization in highly metabolically active auditory centers (Bertoni & Sprenkle, 1988).

Balance on measures of postural sway may also be impaired, and has been seen to improve with chelation treatment (Linz, Barrett, Pflaumer, & Keith, 1992).

At higher blood lead levels, "wrist drop," a motor peripheral neuropathy with little sensory involvement, is probably the most frequently cited neurological symptom. However, clinical observations of wrist drop are less common today than in previous years, possibly because exposure levels are more closely monitored in many industries.

Neuropsychological Effects

Subjective Complaint. As is the case with neurological effects, lead-induced changes in neuropsychological function are not well correlated with subjective complaint. Subclinical blood levels of 40–50 μg/100 ml are the lowest levels where overt neuropsychological changes have been detected; however, few patient complaints are noted in this range. For example, Hänninen et al. (1979) studied a group of 20 workers whose blood lead levels never exceeded 50 µg/dl, and compared them with both a group of 25 workers with blood lead from 50 to 69 µg/dl as well as a group of controls. Length of exposure in the groups ranged from 2 to 9 years. Subjective complaints were measured using a questionnaire and a Finnish adaptation of the Eysenck Personality Inventory. Relatively short (2 year) exposure to lead in welders produced three times the risk of endorsing three or more neuropsychiatric symptoms on questionnaire (Sjogren et al., 1990).

Significantly more low-lead-exposure subjects than controls admitted to fatigue after work, sleepiness, depression, and apathy. Cognitive symptoms were reported only in the higher-exposure group. The group of subjects with slightly higher lead exposure added forgetfulness and sensorimotor complaints to their subjective list, which also included fatigue, restlessness, apathy, and gastrointestinal complaint (Hänninen et al., 1979).

In patients with lead levels higher than 69 µg/dl, subjective complaints may again be somatic or emotional, rather than cognitive. This could reflect the relative saliency of common accompanying symptoms of gastritis and joint pain compared with more subtle changes in cognition. Alternatively, attention and abstraction deficits characteristic of lead intoxication may so seriously impair introspective abilities as to militate against the patient's realizing the extent of his or her cognitive losses.

In either event, there is evidence that neuropsychological decrements may precede entry of those complaints into subjective awareness. Neuropsychological evaluation appears to provide an early warning of lead neurotoxicity, and is a more sensitive indicant than subjective complaint alone.

Cognitive Effects. Neuropsychological functions shown to be deleteriously affected by lead include visual intelligence and visuomotor functions (Hänninen, Hernberg, et al., 1978), general intelligence (Grandjean, Arnvig, & Beckmann, 1978), choice reaction time (Stollery, Broadbent, Banks, & Lee, 1991), and memory, "particularly in learning new material" (Feldman, Ricks, & Baker, 1980). Williamson and Teo (1986), using an information processing model of memory, found significant lead-related decrements in sensory storage memory (brief tachistoscopic presentation of letter pairs), Sternberg-type short-term memory scanning, and a paired associate learning task. In combination with lowered critical flicker fusion thresholds, the authors interpreted their results as consistent with arousal deficit and/or degradation of retinal or visual pathway input.

Reductions in psychomotor speed and dexterity (e.g., Williamson & Teo, 1986; Repko, Corum, Jones, & Garcia, 1978) have been cited in many studies, perhaps related to either reduced nerve conduction velocities or damage to the motor areas of the cortex (Hänninen, 1982). Matsumoto et al. (1993) found that lead workers' tapping speed decreased with higher blood lead levels, that tapping recovery after a 30-s rest was worse in workers with lead levels above 30 µg/dl, and that workers with 30 µg/dl could not sustain initial tapping levels.

When lead levels exceed the subclinical threshold of approximately 30 μg/dl, subjects may develop neuropsychological deficits relatively rapidly. Bleecker, Agnew, Keogh, and Stetson (1982) examined 13 workers who had been exposed a median of 2.5 months (range of 2 weeks to 8 months). Workers displayed significant deficits on Block Design, Digit Symbol and a decreased rate of learning on the Rey Auditory Verbal Learning Test. Marginally significant impairment of visual memory on the Wechsler Memory Scale was also noted.

The clinical appearance of more severely lead-intoxicated patients suggests generalized losses in memory, attention, concentration, and abstract thought processing. In severe cases these symptoms can present as globally dementing impairments.

Emotional Effects. Lead-induced emotional alterations include depression, confusion, anger, fatigue, and tension (Baker, Feldman, White, & Harley, 1983). These symptoms have been observed both in chronically exposed patients, as well as in individuals with lead exposure histories of 2 weeks to 8 months (Bleeker et al., 1982, p. 255). These symptoms may go unrecognized by mental health personnel untrained in neuropsychological effects of neurotoxins (Schottenfeld & Cullen, 1984). Because the emotional concomitants of lead intoxication can resemble major affective disorder, misdiagnosis could potentially occur and delay appropriate administration of chelation therapy (Schottenfeld & Cullen, 1984).

The causes of lead-induced emotional changes are uncertain. Direct organic brain dysfunction may produce depression via cortical and subcortical tissue damage. Hypothalamic alterations and changes in catecholamine metabolism have also been suggested (Schottenfeld & Cullen, 1984). Secondary emotional reactions to diminished cognitive functioning may also contribute to affective changes.

Neuropsychological Recovery from Lead Neurotoxicity. Baker, White, et al. (1985) examined 160 foundry workers during the course of improving hygienic conditions at the plant. Individuals with an initial blood level over 50 μg/dl showed a 20% reduction in tension, 18% reduction in anger, 26% reduction in depression, 27% reduction in fatigue, and 13% reduction in confusion on the Profile of Mood States (POMS). Significant associations between exposure indices and neuropsychological performance were also found on the Santa Ana Dexterity Test (both hands) and a paired associate learning test. Workers failed to show improvement in other lead-impaired tasks.

Case Report—Probable Lead Encephalopathy

The patient is a white male in his late 40s with a ninth grade education. He had worked for 20 years at a steel refinery where his job was to heat-treat steel and quench it in molten lead. He was not issued any protective mask or clothing until OSHA came to the plant in 1979, after the patient had already worked there for about 18 years. The patient reported that he started noticing medical problems around 1976 when he began to feel abnormally fatigued. By 1977–1978 he felt "extreme" fatigue in his arms. Other symptoms noted at that time included weight loss, loss of appetite, and poor memory. Although the patient has not worked with lead for approximately 3 years, a blood lead level taken a month before testing indicated 50 µg/dl. The patient's highest measured blood lead while working with lead was over 80 µg/dl.

Current symptoms include constant body aches "like the flu," and "terrible" hearing in addition to poor memory and fatigue. There is no prior psychiatric history, nor any reported work difficulties prior to the mid-1970s.

This patient showed neuropsychological deficits that appear to be fairly typical of moderate effects of chronic lead exposure. While his education may limit his abilities somewhat, it seems likely that memory and cognitive efficiency deficits shown here are related to chronic lead exposure and a continuing body burden of lead. Personality evaluation was interpreted as consistent with depression and concern over significant decrements in health and well-being. However, the patient's depression could also be a direct neurological effect of lead intoxication (Schottenfeld & Cullen, 1984). The patient's hearing loss could be the result of factory noise damage to the ear or of lead-related CNS damage.

Case Report—Severe Lead Encephalopathy

This patient is a 53-year-old black male who received longitudinal testing to monitor effects of chronic cognitive changes as a result of lead exposure. As can be seen from selected test results, the patient shows global and severe

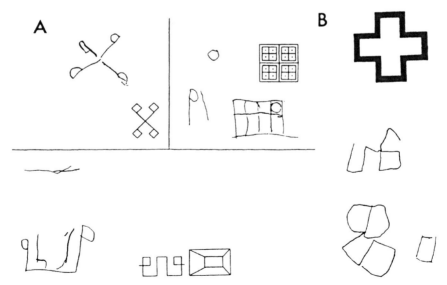

FIGURE 3.4. Severe effects of chronic lead intoxication. (A) WMS: Figural memory–immediate recall, initial testing. For reader comparison, card images were inset subsequent to the evaluation. (B) Spatial Relations Test, first testing (direct copy).

neuropsychological impairment, with marked verbal and nonverbal memory deficit, and poor ability to comprehend instructions and generate internal plans (Fig. 3.4). Complicating the inference of causality to these data is the fact that the patient was raised in rural surroundings and has had no schooling whatsoever. However, the patient did perform capably in a lead refinery for 20 years, was able to follow directions, and showed good work performance before he was finally fired for incompetence. It therefore seems likely that premorbid functioning was substantially less impaired than current data suggest. This patient's current cognitive status was so impaired that he essentially voiced no complaints about cognitive difficulties. In this he was typical of other patients with severe lead encephalopathy/dementia. Like such patients, his lack of complaint probably reflected an inability to observe and reflect on those difficulties. Further, joint and stomach pain tend to be very salient in patients with severe lead encephalopathy and such symptoms may tend to obscure less noticeable neuropsychological abnormalities.

LITHIUM

See Chapter 6.

MANGANESE

Fume and compounds (as Mn)
  OSHA   1 mg/m^3
    STEL   3 mg/m^3
    Ceiling   5 mg/m^3 (fume)
  NIOSH   1 mg/m^3
    STEL   3 mg/m^3
Note: different RELs and PELs exist for cyclopentadienyl manganese tricarbonyl, methyl cyclopentadienyl manganese tricarbonyl, and manganese tetroxide)
(*Quick Guide*, 1993)

The first reported medical use of manganese was as a purgative. Another early observer noted that manganese miners were "uniformly cured of scabies and other cutaneous affections during their stay at the works," which led to the use of manganese to treat these disorders (cited in Marks & Beatty, 1975, p. 128). As early as 1837, however, a Glasgow physician described five workers who inhaled manganese oxide and developed paraparesis, masked facies, drooling, weak voice and propulsion, in the absence of tremor (Cawte, 1985).

Although required as a nutritional trace element, manganese is neurotoxic in large amounts. Customary route of exposure is via inhalation, which can result in serious neurological disturbance. Thousands of workers are said to have developed manganese neurotoxicity (Politis, Schaumburg, & Spencer, 1980).

Manganese miners, welders, and workers in dry-cell battery plants are primary exposure victims (Grandjean, 1983, p. 335). Other occupations where manganese is used include steel alloy plants, and factories that produce ceramics, matches, glass, dyes, fertilizers, welding rods, oxidizing solutions, animal food additives, germicides and antiseptics (Katz, 1985). Poor ventilation systems in worksites where manganese is present increase the risk for neurotoxic exposure (Wang, Huang, et al., 1989).

In miners, initial signs of intoxication have been reported in as few as 6 months and as long as 24 years (Politis et al., 1980). Full-time welders exposed to manganese for more than 2 years had six times the risk of answering a neuropsychiatric questionnaire with three or more positive responses (Sjogren et al., 1990). Symptoms are not inevitably correlated with exposure or high tissue levels, but appear to depend on a combination of contributing factors including "genetic predisposition, age, nutrition, anemia or alcohol" (Cawte, 1985, p. 216).

Neurological and Neuropsychological Aspects

Manganese poisoning is thought to occur in three stages. The initial manifestations of manganese intoxication last approximately 13 months and include sleepiness, poor coordination, ataxia, and impaired speech (Baker, 1983a) as well as other more psychiatric-seeming symptoms, including asthenia, anorexia, insomnia, hallucinations, "mental excitement, aggressive behavior and incoherent talk" (Chandra, 1983; Politis et al., 1980; Cawte, 1985). This latter set of early symptoms has led to the term "manganese mania." Arousal, judgment, and memory deficits have also been reported (Rosenstock, Simons, & Meyer, 1971).

The second stage of manganese intoxication is characterized by greater numbers of neurological symptoms, including abnormal gait, expressionless facies, speech disorder, clumsiness, and sleepiness. In "third stage" severe cases, the syndrome is called "manganism" and presents with "parkinsonian-like" symptoms of "asthenia, staggering gait, muscular hypertonia and hypokinesia, and tremor" (Grandjean, 1983, p. 335). Frontal lobe dysfunction, dementia, and emotional lability have also been reported to occur at this final stage (Cawte, 1985).

Characteristic EEG parameters of manganese intoxication include decreased amplitude and weakened rhythms, described as an "encephalopathic" EEG type (Izmerov & Tarasova, 1993).

Few neuropsychological studies of manganese exposure have entered the literature; however, their results are similar. Roels et al. (1987) surveyed asymptomatic Belgian manganese workers and found effects on memory, reaction time, and motor coordination for exposure levels as low as 1 mg/m^3. More recently, Iregren (1990) compared 30 workers exposed to manganese levels from 0.02 to 1.4 mg/m^3 of air, with a mean exposure of 9.9 years. Exposed workers were compared with workers from two other industries without manganese exposure. Subjects were administered a partially computerized performance test battery, the Swedish Performance Evaluation System (SPES). Compared with an age- and verbal ability-matched control group, exposed subjects performed significantly worse on tests of simple reaction time and finger tapping. Memory was not tested. Adding their results to those of Roels *et al.* (1987), the author suggests that tests of reaction time, memory, and motor function may function as early indicators of low-level manganese neurotoxicity and that exposure standards as low as 2.5 mg/m^3 remain too high to protect workers from neuropsychological damage.

Wennberg, Hagman, and Johansson (1992) examined very-low-level manganese exposure in 30 men with a mean exposure duration of 9.4 years. Mean levels of manganese dust were 0.18 mg/m^3 at one factory and 0.41 mg/m^3 at another. Despite Mn values that were less than 10–20% of the Swedish occupational limit value, diadochokinesic movements of exposed workers were lower in frequency (turns of the wrist) than controls. Workers' P300 values were slightly longer in latency but results did not achieve significance. EEG did not differ between the two groups.

Mergler *et al.* (1994) compared manganese alloy workers exposed to a higher level of manganese dust (geometric mean 0.89 mg/m^3) with a matched pair control group. Whole blood manganese levels were 1.03 µg/dl (geometric mean). Exposed workers and controls reported more neuropsychological symptoms, and showed inferior motor speed and steadiness, along with poorer cognitive flexibility (e.g., Stroop) and odor perception.

Neurophysiological Aspects

While the neurochemical mechanism of manganese-induced aggression remains unknown, Chandra (1983) hypothesizes that increases in activity level characteristic of manganese exposure may be related to elevated striatal catecholamines observed in manganese-intoxicated patients. Cawte (1985) provides a more elaborate hypothesis that catecholamines are first displaced from the ATP complex in the adrenal medulla with consequent "flooding" of the brain by catecholamines, thereby producing the psychiatric phase of the disorder. Later, more parkinsonian stages are by-products of neurotransmitter depletion. Cawte notes that support for manganese interaction with the dopamine pathway is provided by the observation that L-dopa may relieve many neurological symptoms of chronic manganism (Cawte, 1985).

MERCURY

Inorganic (as Hg and compounds)
OSHA 0.1 mg/m^3 ceiling
NIOSH 0.05 mg/m^3, 10 h TWA
ACGIH 0.05 mg/m^3
IDLH 28 mg/m^3

Organic and alkyl compounds
                                                   0.01 mg/m^3
                                                   0.04 mg/m^3 ceiling (ACGIH)
IDLH 10 mg/m^3 (NIOSH, 1985)

History

    Mercury's unusual liquid state at room temperature, coupled with its silvery appearance, have garnered the attention of physicians, philosophers, scientists, and charlatans throughout history. In the second and third centuries A.D. the Chinese philosopher Pao Pu Tzu recommended mixing pills of three parts cinnabar (red mercuric sulfide) and one part honey to induce immortality: "Take ten pills the size of a hemp seed every morning. Inside of a year, white hair will turn black, decayed teeth will grow again. . . . The one who takes [cinnabar] constantly will enjoy eternal life and will not die" (cited in Marks & Beatty, 1975, p. 18). Hippocrates was said to have employed mercuric compounds in his pharmacopoeia around 400 B.C. (Chang, 1980) and Indian physicians of the Brahman period (800 B.C.–1000 A.D.) employed mercury to treat smallpox and syphilis (Marks & Beatty, 1975). Thomas Dover (1660–1742) declared that "to take an ounce of quicksilver every morning . . . [is] the most beneficial thing in the world" (Marks & Beatty, 1975, p. 54).
    Children were not spared the "curative" effects of this metal. For example, Antonio Musa Brassavola (1500–1555) gave children 220 grains of mercury to help them to expel worms. Samuel Bard (1742–1821), a professor of medicine at Kings College, New York, prescribed 30 or 40 grains of calomel (mercurous chloride) to children for "sore throat distemper" and proclaimed "the more frequently I have used [mercury compounds], the better effects I have seen from them" (Marks & Beatty, 1975, p. 55).
    Mercury was cited as a "cure" for syphilis as early as 1000 A.D.; one of many metals that have been tried on this disease, including antimony, arsenic, bismuth and copper (Goldwater, 1972). That mercury cured syphilis "may have been the most colossal hoax ever perpetrated in the history of [medicine]" (Goldwater, 1972, p. 226). As was most likely the case with other metallic cures for syphilis, the belief in mercury's potency may have had more to do with the fact that the early stages of syphilis disappear within a year. If mercury was applied during that time, an association with a remission of symptoms might erroneously be inferred, since visible lesions disappear as the virus enters a second stage. While organized medicine has abandoned the systematic use of p.o. mercury, the metal continues to be included in folk medicines. For example, various Chinese patent medicines have been found to be the source of several cases of severe mercury poisoning, including in a New York City 4-year-old whose mother systematically fed him a Chinese medicine called Tse Koo Choy, which contains mercurous chloride. The child developed progressive neurological deterioration, with drooling, dysphagia, and irregular movements. Recovery was

incomplete, with continuing movement disorder and school-related problems (Kang-Yum & Oransky, 1992).

Like lead, mercury's toxicity has been recognized for centuries; one of the earliest recorded accounts of occupational poisoning is from the supervisor of a mercury mine in 1616: "sooner or later even the strongest mitayos succumbed to mercury poisoning, / which entered into the very marrow of their bones / and made them tremble in every limb" (Pereira, cited in Bingham, 1974, p. 199).

Occupational mercury poisoning was noted by Bernardino Ramazzini (1700, cited in Goldwater, 1972) in miners, gilders, those who treat syphilis, chemists, mirror makers, and painters, among others. Hatters were especially at risk because of their exposure to *le secret* or *secretage*, the closely guarded proprietary compound (mercury nitrate) used in the 1600–1800s by members of the hatting guild to cure felt (Hamilton, 1985). Their use of *le secret* caused hatters to develop symptoms of "erethism," a syndrome of irritable and avoidant behavior, in addition to the tremors which became known as "hatter's shakes."

Some authors have understood Lewis Carroll's remarkable "Mad Hatter" to be a caricatured victim of mercury poisoning. Notwithstanding the very real behavioral abnormalities produced in hatters by mercury poisoning, it is more likely that Carroll was burlesquing a furniture dealer near Oxford, who was known for his eccentric behavior and inventions; among the latter was an "alarm clock bed" that tossed the sleeper on the floor at the predetermined hour (Gardener, cited in Goldwater, 1972, p. 274). Similarly, the phrase "mad as a hatter" may have been a variant on the cockney phrase "mad as an adder" (p. 274). Medical recognition of mercury symptoms among hatters appears to have first occurred in New Jersey, USA, with Freeman's account of "Mercurial Disease among Hatters," as published in the *Transactions of the Medical Society of New Jersey*. Because of competition and emphasis on employment over health, from 10 to 80% of hatters and felt-makers developed mercury-induced tremor (Wedeen, 1989).

It has been proposed that George Washington's need for false teeth was indirectly tied to his use of calomel. This mercury-containing medicine may have caused gum inflammation, ulceration, and tooth loss as a result of mercurialism (Goldwater, 1972).

The history of mercury exposure contains three recent and tragic mass poisonings from the highly toxic organic form of mercury. The first was at Minamata Bay, off the coast of Japan, where mercuric chloride discharge from an industrial plant was converted to the highly, neurotoxic methyl mercury either by microorganisms or from concentration of small amounts of organic mercury along the food chain (Schoenberg, 1982). Whatever the etiology, many individuals died or suffered severe neurological and neuropsychological damage from eating bay shellfish. Many symptoms were noted, including "ataxia, dysarthria and constriction of the visual fields" (Takeuchi, Eto, & Eto, 1979, p. 17). In child victims a "decortication" syndrome not found in adult cases was observed, with apraxia, aphasia, agnosia, disorders of consciousness, and "loss of all mental activities" (Takeuchi, Eto, & Eto, 1979, p. 17). By 1990, over 2000 individuals have been certified as having "Minamata disease" by the Japanese government (Kinjo, Higashi, Nakano, Sakamoto, & Sakai, 1993).

An even larger outbreak of methylmercury poisoning occurred in Iraq during 1971–1972 when grain treated with a mercuric fungicide was taken by peasants and made into bread or cereals rather than kept for seed grain as had been intended. A total of 6530 individuals were hospitalized and over 500 deaths were recorded (World Health Organization, 1976). Many individuals did not enter the hospital and more than 50,000 persons were thought to have been affected with over 5000 deaths (Weiss & Clarkson, 1986).

The third outbreak occurred in Ghana where maize seeds treated with methoxymethylmercury acetate were eaten. One hundred and forty-four individuals were poisoned and twenty died (Durban, 1974).

Modern Industrial Risks

The yearly industrial output of mercury is substantial. In the United States, more than 11 million pounds of mercury is utilized by various electrical, paint, pharmaceutical, and other industries (Chang, 1980). While mercury in its elemental form is poorly absorbed from the gastrointestinal system, it is also quite volatile and may be absorbed from the air in unventilated areas. Because elemental mercury becomes volatile at room temperatures, it may enter the indoor atmosphere from an unsealed container and produce chronic intoxication (Feldman, 1982b). The vapor pressure of pure mercury is also high enough to produce hazardous air concentrations at normal outdoor temperatures (World Health Organization, 1976).

Occupational Exposure

Mercury miners are at obvious risk for poisoning and intoxications. Kishi, Doi, *et al.* (1993) surveyed a cohort of 76 ex-mercury miners from Hokkaido, Japan, who had worked in a mine that opened in 1939 and closed in 1970. The survey was conducted approximately 18 years after exposure had ceased. Exposures were high at over 1.0 mg/m^3 and many of the miners had a history of overt intoxication, with exposure of more than 100 times the present Japanese TLV. While there was a decrease in symptoms 18 years after exposure, prevalence of hand tremor, headache, and slurred speech was higher in the poisoned group than controls. For example, 72% of miners with a history of intoxication reported tremors. Neuropsychological testing of workers showed that performance in tests of grip strength, eye–hand coordination, color reading, block design, and digit span were significantly impaired relative to controls in a matched pair analysis.

Occupations where the volatility of mercury causes risk to workers include those that employ or manufacture barometers or mercury vapor lamps. Also at risk are technicians who prepare dental amalgams, and possibly dentists themselves, since it is reported that 15–20% of U.S. dental offices exceed OSHA limits for ambient mercury exposure (Uzzell & Oler, 1986). When dentists in Singapore were compared with controls with no exposure to mercury, dentists had higher aggressive mood scores and scored significantly more poorly on a variety of neuropsychological measures, including Trails A&B, Symbol Digit, Digit Span, and WMS Verbal and Visual Memory (immediate and delayed recall). Neuropsychological impairments and emotional changes were related to cumulative exposure to mercury vapor (Ngim, Foo, Boley, & Jeyaratnam, 1992). In a study of Chinese dentists, duration of exposure to mercury was associated with declines in performance on Digit Span, Symbol Digit, and Grooved Pegboard, after controlling for age, sex, education, and consumption of Chinese medical products thought to contain mercury. Dentists' urinary mercury levels have declined from 1975–1983, when the mean urinary mercury level was 14.2 ng/ml, to 1985 and 1986, when levels declined to 5.8 and 7.9 ng/ml, respectively. However, 10% of the 1985–1986 surveyed group had mercury concentrations about 20 ng/ml (Fung & Molvar, 1992). Thirty nanograms is considered acceptable.

Other occupational groups at risk for mercury exposure include felt-makers, photoengravers, and photographers (Feldman, 1982b). Industrial workers who manufacture electrical switches and batteries, those using mercuric salts in plating operations, tanners and embalmers are also at risk. Employees who produce or apply organic mercury compounds as pesticides, fungicides, disinfectants, or wood preservatives may also be endangered (Feldman, 1982b; Hamilton, 1985).

Nonoccupational Exposure

Mercury vapor exposure is not limited to the workplace. The most common and controversial nonoccupational mercury exposure occurs via implants of dental amalgam to correct the effects of dental caries. Approximately 100 million amalgam fillings are implanted each year by dentists in the United States (Fung & Molvar, 1992). Amalgam is a metallic alloy, usually composed of 25–35% silver, 15–30% tin, 2–30% copper, and other metals which are then mixed with between 40 and 54% mercury. An additional concern is that mercury may be converted by common mouth bacteria (e.g., *Streptococcus sanguis, S. mutans,* or *S. mitiors*) to methylmercury in the mouth. Chewing with recently implanted amalgam fillings increases mercury vapor levels 4–15 times resting state, and 3 times OSHA safety levels (Gay, Cos, & Reinhardt, 1979; Svare *et al.,* 1981; Wolff, Osborne, & Hanson, 1983). Mercury levels from fillings continue to be detectable for as long as 1 year after amalgam implant, albeit below current safety levels (8.5–15 ng) (Fung, Molvar, Strom, Schneider, & Carlson, 1991). Individuals with many fillings (more than 36 restored surfaces) absorb more than 10–12 µg of mercury per day, double that of individuals without significant amalgam exposure. Brain and kidney concentrations of mercury are significantly higher in autopsied individuals with amalgam fillings, than in those with amalgam restoration (Nylander, Frieberg, & Lind, 1987). The neurotoxic effects of this long-term body burden of mercury are largely unknown, although relatively few cases of overt allergic response are known and no cases of mercury neurotoxicity from this route have yet entered the literature (Fung et al., 1991; Fung & Molvar, 1992). Removing old amalgam, or careless dental techniques, may also expose patients to elevated levels of mercury vapor.

Wolff et al. (1983) warn that "patients [with amalgam fillings] could receive a chronic low level exposure to elemental Hg for many years.... It is generally agreed that if amalgam was introduced today as a restorative material, they [sic] would never pass F.D.A. approval" (p. 203). Mercury from amalgam fillings may cross the placenta in pregnant women, and may be especially dangerous to the fetus under typical conditions of chronic, low dose exposure. "Experimental evidence suggests that it would be prudent to avoid placing dental 'silver' amalgam fillings in pregnant women" (Lorscheider & Vimy, 1990, p. 1579).

The possibility of amalgam toxicity has prompted the Dental Products Panel of the U.S. Food and Drug Administration to consider opening the issue for further regulatory study. While studies of amalgam exposure on neuropsychological function have not yet been conducted, known toxic effects of mercury and recent toxicological findings strongly indicate the need for this avenue of investigation.

It has been suggested that routine eating of mercury-containing fish constitutes a greater source of daily mercury exposure than amalgam. Thimerosal contact lens disinfection solutions also contain mercury and may likewise constitute another source of daily exposure to mercury. Neither has been investigated for their relationships to neurological disorders or neuropsychological impairments (Fung & Molvar, 1992).

Neurotoxic exposure to mercury has also occurred in home accidents. Spilled mercury sinks into floor cracks and carpeting and may be difficult to see or remove. Attempts to take up the metal via a vacuum cleaner may actually propel the metal through the vacuum and expel it as an aerosol, facilitating its spread.

A broken thermometer spilling mercury onto a carpet in a small, floor-heated room was the cause of poisoning in three children (Muhlendahl, 1990). Windebank et al. (1984) report the more unusual case of ataxia, anorexia, and lethargy in three members of a family after their 9-year-old some attempted to make "silver bullets" by heating lead shot and thermometer mercury in a frying pan. Air samples in the family's home showed four times the maximum industrial limit of mercury. In two similar and more tragic scenarios, adults attempted to smelt silver from a quantity of amalgam dental filling. The first case caused his own death and the death of three other family members. Displayed symptoms included severe diffuse alveolar damage, fibrosis, renal tubular necrosis, vacuolar hepatotoxicity and CNS symptoms of ischemic necrosis, gliosis, and vasculitis (Kanluen & Gottlieb, 1991). The second case caused the death of four individuals and required the demolition of the house (Taueg, Sanfilippo, Rowens, Szejda, & Hesse, 1992).

Mercury remains an ingredient of unregulated ethnic patent medicines that, taken internally, can cause severe neurological disorder (Kang-Yum & Oransky, 1992). A related and neglected source of mercury contamination occurs in the context of occult religious ritual. Afro-Caribbean and Latin American religions, especially Santeria and Voodoo, routinely dispense metallic mercury to be carried on the person, kept in the house, put in bath water, mixed in perfume soap, devotional candles, ammonia, or camphor (Wendroff, 1990). Mercury is legally available in herbal pharmacies called botanicas across the United States and Central America and proprietors recommend sprinkling mercury on the floor or mixing with soap and water to mop the floor. As Wendroff correctly points out, "any of these procedures would liberate mercury vapour directly into the room's atmosphere" (p. 623). While no cases of mercury intoxication from this source have yet been identified, the nonspecific psychosomatic symptoms of mercury poisoning would likely be misdiagnosed as psychiatric illness unless special inquiry and blood screens were conducted on symptomatic patients from high-risk populations.

Neuropathology—Methylmercury

The methylated form of mercury is a far more potent CNS poison than mercury in its inorganic state (Weiss, 1978). Methylmercury is almost completely absorbed from the gastrointestinal tract. By contrast, less than 15% of inorganic mercury is retained (World Health Organization 1976). Methylmercury intoxications cause symptoms of toxic encephalopathy, either early symptoms of poisoning that include weakness, unusual taste in the mouth, gum hemorrhage, sleep disturbances, and sometimes auditory and visual hallucinations (Koloyanova & El Batawi, 1991). Methylmercury intoxication produces neurotoxic effects that differ from those of inorganic mercury. Takeuchi, Eto, and Eto (1979), reporting on autopsied Minamata victims, found that methylmercury-induced lesions occurred predominantly in the "calcarine cortex, pre and post-central gyri, superior temporal gyrus and central portion of the cerebellum" (p. 18). The basal ganglia may also show severe destruction (Chaffin & Miller, 1974; Feldman, 1982b). In contrast, elemental mercury lesions have been thought to occur in the occipital cortex and the substantia nigra (Grandjean, 1983). For example, Takahata, Hayashi, Watanabe, and Anso (1970) autopsied the brains of two mercury miners with symptoms of ataxia and tremor. Mercury concentration in the two brains was highest in the occipital cortex (33.56 and 14.80 ppm) and the substantia nigra (23.05 and 18.00 ppm), and also the parietal cortex (6.21 and 13.80 ppm). Characteristic EEG is said to be of the "paroxysmal type" with high-amplitude, 2–3 wave/s α waves against a background of generalized cerebal changes (Izmerov & Tarasova, 1993).

In addition, chronic exposure to methylmercury produces systemic CNS damage, including constriction of visual fields, ataxia, dysarthria, partial deafness, tremor, and intellectual impairment (Windebank et al., 1984). Feldman (1982b) reports one case of methylmercury poisoning where members of a family ate pork from animals accidentally fed seed grain treated with methylmercury fungicide. Several individuals developed initial symptoms of ataxia, agitation, decreased visual acuity and stupor. Evaluation 10 years later showed blindness, ataxia, retardation, choreoathetosis, myoclonic jerks, and abnormal EEGs.

TABLE 3.3. Dose-Related Neurotoxic Effects of Mercury[a]

	Human	Nonhuman primate	Small mammals
High dose (12–20 ppm)	Decreased brain size Cortical, basal ganglia, and cerebellar damage Sparing of diencephalon Ventricular dilation Myelinated fibers Ectopic cells, gliosis Disorganized layers Misoriented cells Cell loss	Decreased brain size Cortical, basal ganglia damage Sparing of diencephalon Gliosis Cell loss Sparing of cerebellum Ventricular dilation Ectopic cells Disorganized layers Myelin loss Cell loss Cell misorientation	Decreased brain size Cortical, basal ganglia, hippocampal, and cerebellar damage Sparing of diencephalon Ventricular dilation
Moderate dose (3–11 ppm)	No information	No information	Decreased brain size Damage to cortex, cerebellum Myelin loss
Low dose (<3 ppm)	No information	No information	Decreased brain size Cell loss

[a]From Rees et al. (1990).

Developmental Neuropathology

Rees et al. (1990) have summarized the variety of neuropathological effects on brain development with methylmercury exposure to humans and animals in Table 3.3. Developmental screening using psychological tests has placed significant exposure risk at 6 ppm hair level or approximately 0.3 ppm fetal brain level (Kjellstrom, Kennedy, Wallis, & Mantell, 1986). There are few studies of *in utero* exposure of human fetuses; autopsy studies are reported from victims of the Iraqi grain incident, where pregnant women ate grain contaminated with methylmercury. One infant lived for 33 days and the other died 7 h after birth. Autopsy studies revealed "ectopic and disoriented cells in the cerebrum and cerebellum, including the white matter ... enlarged, reactive astrocytes were seen in all regions where damage was present" (Burbacher, Rodier, & Weiss, 1990, p. 192).

In patients who live through fetal methylmercury poisoning, neuropathology is permanent. Eight patients with fetal methylmercury poisoning were examined with CT scan, 25–28 years later. CTs of all eight patients "showed mild or moderate enlargement of the cerebral fissures or sulci ... indicating decreased cerebral cortex." Lateral ventricles were mildly enlarged in five patients and two patients showed mildly enlarged third ventricles (Hamada et al., 1993, p. 104). CT scans were described as "relatively mild" compared with the patients' clinical presentations, with five of the eight patients either "bedridden" or confined to a wheelchair. Patient presentation resembled a "mixed" type cerebral palsy, except for the degree of mental retardation.

When 1144 nonfetal exposure cases were surveyed 30 years after exposure, it was clear that mercury poisoning had lifelong consequences. The Minamata disease group had significantly higher rates of complaints than controls, and impairment in activities of daily living were made worse with increasing age (Kinjo et al., 1993).

Developmental effects of prenatal mercury exposure may be minimized by confounding toxicokinetic effects. Grandjean, Weihe, Jorgensen, Clarkson, Cernichiari, and Videro (1992) found that consumption of fish was associated with prolonged gestation period, which would protect high-mercury-exposed children from effects of low birth weight. In addition, alcohol consumption was associated with lower levels of mercury in cord blood, both because of an unexplained toxicokinetic effect and because many of the female drinkers did not eat fish. These data suggest that certain common confounders may obscure the measurement of prenatal mercury exposure and bias results toward the null hypothesis.

Neuropathology—Inorganic Mercury

Opinions differ as to the neuropathology of mercury vapor poisoning. Feldman (1982b) suggests that toxic effects are related to mercury-induced enzyme inhibition. Since mercury penetrates and damages the blood–brain barrier (Chang, 1982), and accumulates in the brain, it may also produce direct structural damage. Several authors have speculated about brain localization of inorganic mercury damage. Grandjean (1983) suggests that brain damage from inorganic mercury vapor may be localizable to the occipital cortex and the substantia nigra. Other authors have suggested that inorganic mercury appears to show uniform brain distribution. Conservative estimates of brain mercury able to induce neurotoxic symptoms correspond to a blood level of 150 μg or greater (Nordberg, 1976; Goetz, 1985). To prevent development of neurotoxic effects, urinary mercury should not exceed 50 μg (Osterloh & Tarcher, 1992). It has been proposed that mercury exposure inhibits astroglial ability to take up glutamate, leading to glutamate receptor hypersensitivity, decreased cortical glutamate, and secondary decreases in other neurotransmitters. Decreased activity in the locus coeruleus will lead to symptoms reported by patients with low-dose exposures, e.g., fatigue and decreased alertness (Rönnbäck & Hansson, 1992). At higher concentrations, extracellular glutamate may become toxic to neurons.

Acute and Neurological Effects

Acute exposure produces chest tightness, breathing difficulty, coughing, inflammation of mouth and gums, headache, and fever (Ehrenberg et al., 1991). Neurological sequelae from inorganic mercury exposure begin approximately 24 h following acute exposure. Fine tremor in the eyelids, lips, and fingers are early signs of toxicity. Acute exposure to mercury vapor may produce insomnia, excitation, vasomotor problems, gingivitis, and kidney dysfunction (Fung &

Molvar, 1992). Levels of mercury in urine greater than 0.5 mg/liter have been linked "to the development of mild neurotoxic effects" (Feldman, 1982a, p. 147). There are also reports of toxicity at lower levels (WHO, 1980). For example, EMG disturbances said to be suggestive of subclinical neuropathy were seen in workers who were exposed to mean mercury vapor concentrations of 0.027 mg/m^3 with urinary mercury excretion of 93.4 ± 30.4 (Zedda *et al.*, 1980, cited in Soleo, Urbano, Petrera, & Ambrosi, 1990).

A case of very high mercury exposure is reported by White, Feldman, Moss, and Proctor (1993) in a thermometer factory worker who presented with an initial urinary mercury level of 690 µg/liter. Neurological examination showed upward gaze nystagmus, peripheral neuropathy, tremor, nerve conduction abnormalities, and diminished reflexes and pain sensation. MRI revealed numerous small foci dispersed throughout white matter and bilaterally in the upper pons. Another MRI 6 months later additionally suggested multiple foci of high signal intensity, including periventricular white matter and possible multiple foci in deep cerebral white matter and pons. MRI 9 months later was slightly improved but continued to show multiple deep white matter foci.

Low-level mercury exposure may produce EEG abnormalities, especially when exposure is chronic. Forty-one chloralkali workers exposed for 15.6 years to low levels of mercury (average 25 µg/m^3 in air) showed significantly slower and more attenuated power density spectra on quantitative EEG than the referents (Piikivi & Tolounen, 1989). Abnormalities were most prominent in the occipital region, gradually decreasing in the parietal lobe, and were "almost absent" in the frontal lobe. A dose–effect relationship was not found.

Tremor is a common concomitant of urinary excretions above 250 µg/dl of mercury (Grandjean, 1983). With low-level exposure, there is no difference in average occurrence of tremor between exposed patients and controls, but tremor power spectrum is different between the two groups (Chapman *et al.*, 1990). For workers with mean urinary mercury concentration of 23.1 µg/liter, upwards shifts in spectral power and frequency distributions were noted (Chapman *et al.*, 1990). Hand and finger tremor, progressing to tremor of the face and eyelid, occurs at a frequency of about 5 Hz (Grandjean, 1983). Tremor eventually involves the head and trunk (Feldman, 1982a). These and other changes may be progressive, continuing to worsen after cessation of exposure. Moreover, toxic neurological abnormalities may be permanent. Mercury poisoning may be fatal, causing adult respiratory distress and respiratory failure (Taueg *et al.*, 1992).

Since the 1970s, research correlating mercury body burden with neuropsychological effects has suggested that relatively low levels of mercury are correlated with behavioral dysfunction. However, low levels do not always produce neuropsychological impairments. Symptom complaints and indices of affective changes (i.e., POMS, EPI) appear to be more sensitive markers of subtle CNS-related mercury impairments (Longworth, Almkvist, Soderman, & Wikstrom, 1992). Hair samples are significantly correlated with cerebral and cerebellar mercury levels, although correlations are higher for organic mercury than for the inorganic form of this metal (Suzuki *et al.*, 1993). (See Table 3.4.)

TABLE 3.4. Effects of Mercury Exposure on Chloralkali Workers

Hg level	Effect	Authors
>100 μg/m³ in air	Increased symptoms	Smith et al. (1970)
>50 μg/liter urine	Forearm tremor frequency, impaired psychomotor tests	Millet et al. (1975)
108 μg/195 μg urine	Impaired digit span	Smith et al. (1983)
96 μg/g urine creatinine	Eye–hand coordination, arm–hand steadiness	Roels et al. (1982)
29 μg/liter blood		
58 μg/liter urine–20 μg/liter blood	Coordination (pegboard)	Piikiv et al. (1984)

Neuropsychological Effects

Mercury intoxication may be produced by inhalation, ingestion, or, in the case of a 19-year-old who attempted suicide, injection (Zillmer, Lucci, Barth, Peake, & Spyker, 1986). Neuropsychological effects are present in all modalities of exposure. In general, disturbances in behavior and cognition are "the earliest signs of chronic mercury intoxication" (Feldman, 1982b, p. 203), and thus may be precede other neurological effects. Therefore, tests of neuropsychological function are considered "the only tool[s] with which to monitor ... workers exposed to low mercury concentrations and to detect any preclinical signs in the central nervous system" (Soleo et al., 1990, p. 108). Intellectual disturbance from mercury intoxication has been found to occur in tests of visuospatial abilities, visual memory, nonverbal abstraction, cognitive efficiency, and reaction time (Angotzi et al., 1982). Block Design, Digit Symbol, and Raven's Progressive Matrices appear to be sensitive indicators of toxic effects, as are Digit Span and tests of visual memory (Hänninen, 1982; Soleo et al., 1990). Lowered eye–hand coordination and decreased finger tapping scores are significantly correlated with urinary mercury levels (Langolf, Chaffin, Henderson, & Whittle, 1978), as are foot tapping scores (Chaffin & Miller, 1974). Mercury exposure increases test fatigue, and decreases initial learning of word associations.

The emotional dysfunction called "erethism" is also thought to be a very common emotional abnormality related to mercury exposure. This syndrome, derived from the French and Greek word "to irritate," represents an unusual and morbid level of overactivity of the "mental powers or passions" *(Oxford English Dictionary,* 1971, p. 890). In current use the term also connotes symptoms of avoidant, irritable, and overly sensitive interpersonal behavior, depression, lassitude, and fatigue (Hänninen, 1982, p. 167). Ross and Sholiton (1983) found some evidence of this syndrome when they interviewed nine laboratory technicians exposed to mercury vapor. These authors describe symptoms of irritability in six individuals and "shyness" in three. Two of the nine individuals displayed most of the classical symptoms of erethism; however, three others showed none of the classical symptoms. In another study, Uzzell and Oler (1986) found significant elevations on the SCL-90-R, a psychopathology inventory, in subclinically exposed individuals compared with controls. Similarly, Ehrenberg et al. (1991)

assessed workers at a mercury thermometer plant for the presence of 28 symptoms. Difficulty sleeping and a metallic taste in the mouth were the only symptoms that significantly differentiated mercury workers from controls. However, because of multiple comparisons, even these symptoms may be artifactual. Even so, mercury workers reported significantly more symptoms than controls and "prevalences of specific symptoms ... among thermometer workers exceeded those among control plant workers for 21 of the 28 symptoms queried" (p. 500). Langworth, Almkvist, Soderman, and Wikstrom (1992) examined a group of 89 chloralkali workers with median blood Hg concentrations of 55 nmol/liter, serum Hg 45 nmol/liter, urine Hg 14.3 nmol/mmol, and creatinine 25.4 µg/g. POMS scores for fatigue and confusion, and neuroticism on the Eysenck Personality Inventory were significantly higher in the exposed group than controls, as were complaints of disturbed memory and concentration. Soleo *et al.* (1990) examined a group of eight mercury lamp manufacturing workers using an 18-item personality profile (Gordon Personal Profile) and found overall significant increases in overall pathology compared with controls.

For the suicidal adolescent who injected himself with metallic mercury, neuropsychological effects were particularly severe (Zillmer *et al.*, 1986). In the context of a normal neurological examination at 6 weeks postinjection, the patient exhibited a Halstead–Reitan Impairment index of 1.0 and a Wechsler Memory Scale MQ (memory quotient) of 59, both indicating severe impairment of neuropsychological function. Overall WAIS-R IQ scores, while somewhat depressed, did not appear to be particularly sensitive to the patient's neuropsychological abnormalities.

A deliberate oral ingestion of mercury in another individual was reported to have produced evidence of "poor concentration and a defect of recent memory" on psychological testing. EEG showed diffuse, predominantly left hemisphere cortical dysfunction (Lin & Lim, 1993). However, oral ingestion was apparently less neurotoxic than exposure by injection described earlier.

Hänninen's (1982) review concluded that mercury-induced neuropsychological abnormalities fell into three major groups: (1) motor system abnormalities (e.g., fine motor tremor), (2) intellectual impairment (gradual and progressive deterioration of memory, concentration, and logical reasoning), and (3) emotional disability. Occupational studies of mercury exposure have found support for individual symptoms of impairment within these categories. For example, Williamson *et al.* (1982) examined 12 mercury-exposed workers who were exposed to the supposedly less toxic organic and inorganic forms of mercury, including mercury-based fungicide, and inorganic mercury amalgam used in gold refining. Exposure length in total time and hours per day was quite variable, from 1 to 8 h per day and total duration from 3 months to 8 years. Nonexposed controls were age-, sex-, race-, and education-matched. The authors administered an extensive neuropsychological battery to all subjects, including various tests of memory, a paired associate learning test, critical flicker fusion, a pursuit rotor to measure visual pursuit, a Sternberg task, and tests of hand steadiness, reaction time, and vigilance. No significant differences were found for flicker fusion or simple reaction time. Speeded tracking of a pursuit rotor was impaired at the higher speed but not at the lower (Williamson et al., 1982). Other significant impairments in the mercury group were in the areas of hand steadiness, short-term (but not iconic or long-term) memory, verbal learning, and the Sternberg paradigm. The only test that correlated significantly with total duration of mercury exposure was the visual pursuit rotor performance. Only hand steadiness correlated significantly with current urinary mercury levels. The authors explained their results as consistent with mercury-induced effects on "psychomotor coordination and short term memory" (Williamson et al., 1982, p. 285).

Triebig and Schaller (1982) also report a significant relationship between mercury exposure duration and short-term memory, but failed to find other significant neuropsychological deficits. Subjects were exposed a mean of 8 years, and had blood mercury averaging 68 μg/liter in one group and 30 μg/liter in another. Changes in nerve conduction velocity were not consistently found at this level of exposure.

Uzzell and Oler (1986) used X-ray fluorescence to detect low-level mercury exposure in dental technicians, and compared their performance to dental assistants without detectable mercury levels. Subjects were similar in age, education, and years employed in dental work (approximately 15). Tests employed included the WAIS, the Rey Auditory Verbal Learning Test, Recurrent Figures Test, the PASAT (Paced Auditory Serial Addition Test), Finger Tapping Test, the Bender–Visual Motor Gestalt Test, and the SCL-90-R, a symptom checklist. The subclinically exposed subjects showed deficient performance in the Recurrent Figures Test (an examination of visual recognition memory) as well as significantly elevated SCL-90-R profile compared with controls. The study's small number of subjects in each group (13) may have militated against achieving significance in other neuropsychological measures, since mean subject performance of the experimental group was also worse than the control group's in tests of attention, memory, and visual-motor abilities.

Since subjects remained employed and able to perform their work, the authors' neuropsychological results may be indicative of a very early "pretremor" phase of mercury intoxication. While elevated levels of psychopathology shown by subjects on the SCL-90-R are suggestive of "erethism," there is no information provided about whether subjects knew their group membership. If subjects had been told of their mercury body burden, the personality results shown could be equally well explained as a reactive consequence of that knowledge.

A fourth study examined subjects with longer exposures, higher average blood mercury levels, and a wider range of blood mercury. Piikivi et al. (1984) studied the neuropsychological effects of mercury vapor in 36 workers exposed for a mean of 16.9 years. They found significantly lowered Similarities, Digit Span, and Visual Reproduction (WMS Figural Memory) in workers with blood mercury levels greater than 75 nmol/liter compared with controls with lower blood levels. Individuals with higher blood mercury levels (280 nmol/liter) and those exposed to higher peak levels also showed significant deficits in motor coordination on the Santa Ana Dexterity Test. The authors' results are consistent with mercury-induced memory, abstract reasoning, and motor coordination deficit. Unfortunately, they did not also test for indications of personality or affective disorder. The authors recommended a maximum metallic mercury exposure limit of 25 µg/m^3 to avoid neurotoxic effects.

A more recent study (Langworth et al., 1992) found performance of chloralkali workers on tests of hand-eye coordination, alternating finger tapping, and Sternberg task to be negatively correlated with number of blood mercury peaks > 150 nmol/liter in the previous 5 years.

Chronic low-level effects of mercury vapor were examined in 158 Chinese workers at a fluorescent lamp factory by Liang et al. (1993). Ambient air level of mercury was 0.033 mg/m^3, higher than the Chinese occupational standard of 0.01 mg/m^3. Urinary mercury excretion was 25 µg/liter (range 0.005–0.190 mg/m^2). Workers were compared with age-matched, unexposed controls on a Chinese version of the Neurobehavioral Evaluation System (NES; see Chapter 2). Workers showed poorer mental arithmetic, two-digit search, switching attention, and slower finger tapping. Although age was controlled, subjects with longer job tenure performed more poorly, suggesting a possible cumulative effect of mercury exposure, or an accelerated aging effect. On a Chinese version of the Profile of Mood States, workers showed higher levels of fatigue and confusion than controls. The authors suggested that psychological test changes were found with an average air mercury level of 0.03 with a range of 0.005–0.19 mg/m^3 and a urinary mercury output as low as 0.025 mg/liter.

Teratogenicity

Methylmercury-induced teratogenicity was an obvious by-product of the tragic mass methylmercury poisoning at Minamata, Japan. Children born of mothers who ate contaminated shellfish showed severe neurological effects, including "cerebral palsy, ataxia, pathological reflexes, disturbed psychomotor development, and mental retardation often accompanied by microcephaly" (Roeleveld et al., 1990, p. 582). The literature addressing teratological effects of inorganic mercury is comparatively small, but Roeleveld et al. (1990) suggest that mercury may lodge in the placenta and impair fetal growth. Mercury vapor is fetotoxic and several cases of severe congenital brain damage from maternal occupational exposure to mercury vapor have been reported (Gelbier & Ingram, 1989). Mercury is also stored in breast milk, where it becomes a risk to postnatal development of breastfed infants.

NICKEL

OSHA　　　　　　　　　　　　0.1 mg/m^3 (soluble compounds)
　　　　　　　　　　　　　　　1 mg/m^3 (metal and insoluble compounds)

NIOSH 0.015 mg/m³ 10 h TWA
ACGIH 0.015 mg/m³ (soluble)
(*Quick Guide*, 1993)

Very little information is available concerning the neuropsychological effects of nickel. Feldman (1982a) describes two cases of toxic nickel encephalopathy after an accident occurred in an electroplating plant where workers were exposed to nickel chloride and nickel sulfate. Symptoms included severe frontal headache, nausea, and vomiting. One victim had showed restricted visual fields, and both men's CT scans showed small ventricles, suggesting cerebral edema. With penicillamine chelation complete recovery was reported within 1 month. No neuropsychological testing of nickel-exposed individuals appears to be available.

PLATINUM

Soluble salts as Pt
OSHA 0.002 mg/m³
NIOSH 0.002 mg/m³
(*Quick Guide*, 1993)

Platinum is commonly used in the electronics and chemical industries, and is also used in jewelry. Pure platinum has not been linked to neurological symptoms; however, platinum-containing compounds used in cancer treatment are neurotoxic. "Cis-platinum" or cisplatin is one of the most often used forms of chemotherapy to combat testicular, bladder, and head and neck cancer.

CNS and PNS damage have been reported as a result of cisplatin use (Walsh, Clark, Parhad, & Green, 1982). PNS damage takes the form of progressive symmetrical sensory neuropathy (Schaumburg & Spencer, 1983), although intraarterial injection may induce reversible unilateral cranial nerve palsy. CNS effects may include damage to the optic nerve and spinal cord (Pomes *et al.*, 1986). A postmortem study suggested that cisplatin induced degeneration of the optic disk, optic nerve, and long tracts of the spinal cord (Walsh *et al.*, 1982). There are no neuropsychological studies of cis-platinum intoxication. LoMonaco *et al.* (1992) examined the course of cisplatin neuropathy in 16 patients treated with three courses of cisplatin with a cumulative dose of 600 mg/m². Initial symptoms of polyneuropathy were detected in 4 of 9 patients after the second course of chemotherapy. All 16 patients showed clinical and/or neurophysiological signs of polyneuropathy 3 months after CDDP administration, and progressed in severity for up to 6 months after treatment. Clinical signs of polyneuropathy, i.e., diminished vibration sense and reduced tendon reflexes, appeared 1 month after the last course of CDDP and preceded detectable neurological impairments. Prognosis for recovery is guarded, as nearly 40% of patients in another study complained of paresthesias 4–9 years later (Roth, Griest, Jubilis, Williams, & Einhorn, 1988) and symptom relief has been ascribed to symptom adaptation rather than nerve regeneration (Hansen, Helweg-Larsen, & Trojaborg, 1989).

SELENIUM

Selenium and compounds (as Se)
OSHA 0.2 mg/m^3
NIOSH 0.2 mg/m^3
(*Quick Guide*, 1993)

Selenium is found in industrial settings as a result of mining or processing selenium-containing minerals including copper, lead, zinc pyrite, roasting lime, and cement. Selenium is used commercially in the production of glass and ceramics, and other substances including steel, rubber, brass, paint and ink pigment, and photoelectric materials (Katz, 1985).

In very small doses, selenium is an essential trace element in the human diet. In larger amounts from voluntary or industrial exposure, selenium has been associated with neurotoxic effects, including headache, vertigo, convulsions, and a motor neuron disease similar to amyotrophic lateral sclerosis, a disease reported in four ranchers who grazed their cattle in high-selenium soil and ate meat with high selenium content (Goetz, 1985). Neurotoxic symptoms may be expected as secondary features of systemic pulmonary or hepatic disease stemming from selenium exposure. Exposure to hydrogen selenide may cause pulmonary edema or obstructive pulmonary disease. The lowest level of daily selenium intake estimated to be dangerous is 500 μg (World Health Organization, 1977). There are no neuropsychological studies of selenium intoxication.

SILICON AND SILICONE

OSHA 6 mg/m^3 (silicon dioxide)
NIOSH 6 mg/m^3 (silicon dioxide)
(*Quick Guide*, 1993)

Silicon shares properties of both metal and glass, and therefore can be appropriately included in this section. It is widely used in the glass and electronics industries. Silicon has been controversially linked to dementia in several studies cited by Goetz (1985). However, it is unclear whether elevated silicon levels found in demented patients are causal or secondary to a neuropathological disease process.

Silicon forms about 28% of the Earth's crust by weight. It is found naturally in the human skeleton, and in the environment as an oxide or a silicate. Silicon is a principal component of glass, clay, and microelectronic devices. Silicone processes pure silicon through a variety of steps into a high-molecular-weight polymer. Of late, there has been considerable concern related to undesirable side effects caused by silicone breast augmentation implants. Case reports and anecdotal accounts have suggested the existence of silicone-mediated autoimmune diseases, including scleroderma, Reyaud's syndrome, systemic sclerosis, systemic lupus erythematosus (SLE), Sjogren's syndrome, but the incidence is thought to be rare (e.g., 39 to 115 cases per million for development of SLE) (Yoshida, Chang, Teuber, & Gershwin, 1993). Yoshida, German, Fletcher, and Gershwin (1994) have noted a substantial similarity between reactions to the "oily substance" released from silicone breast implants and autoimmune dysfunction from adulterated cooking oil (toxic oil syndrome), and adulterated l-tryptophan [eosinophilia myalgia syndrome (EMS)]. Several of these disorders have clear neuropsychological sequelae (e.g., EMS, SLE), but the connection between silicone exposure and immunologic or neuroimmunologic abnormalities is inconsistent and uncommon, and symptom etiology has not been verified as of this writing. No neuropsychological studies of silicone-mediated immune disorders have entered the literature.

TELLURIUM

OSHA	0.1 mg/m^3
NIOSH	0.1 mg/m^3
(*Quick Guide*, 1993)	

Tellurium exposure occurs in miners, and in the alloy, rubber, glass, stainless steel, copper, and lead industries (Katz, 1985). Tellurium exposure in humans has been associated with mild neurological complaints, including headache and sleepiness. Symptoms seem to be the same whether exposure is acute or chronic. Rat studies have suggested that tellurium exposure may produce muscle weakness and behavioral abnormalities. PNS damage is more common than CNS effects in experimentally induced toxicity studies. There are no neuropsychological studies of tellurium-exposed individuals.

THALLIUM

Soluble compounds (as Tl)	
IDLH	20 mg/m^3
OSHA	0.1 mg/m^3 (skin)
NIOSH	0.1 mg/m^3 (skin)
(*Quick Guide*, 1993)	

Thallium was initially used in the 19th century to treat "night sweat" symptoms of tuberculosis. Currently the main uses of this metal are for rodent and pest control, although thallium is also used in tungsten filaments, jewelry, alloys, fireworks, and high-refractive-index glass (Katz, 1985). Thallium poisoning is relatively rare and most case reports consist of homicide attempts or accidental ingestions. Hair loss (alopecia) is a unique identifying sign of thallium poisoning.

Damage to the PNS occurs within 24–48 h of thallium ingestion and may progress for several weeks after ingestion of a single dose (Windebank *et al.*, 1984). CNS effects may occur with larger doses, and organic psychosis may ensue. Chromolytic changes in the motor cortex, substantia nigra, and globus pallidus and third nerve nuclei have been reported (Bank, 1980). Optic neuropathy has been reported in children who have ingested this metal, and a single case report of an exposed worker who also developed optic atrophy has also been cited (Hamilton & Hardy, 1974).

TIN AND ORGANOTINS

Inorganic (compounds except oxides) (as Sn)	
IDLH	400 mg/m^3
OSHA	2 mg/m^3
NIOSH	2 mg/m^3
Organic (all organic compounds except cyhexatin)	0.1 mg/m^3
OSHA	
NIOSH	0.1 mg/m^3
(*Quick Guide*, 1993)	

Tin's first recorded medical use was during the Middle Ages as a vermifuge. It was also employed to purge fevers, induce sweat, and "cure" venereal disease and hysteria (Marks & Beatty, 1975). Tin is presently used to plate containers, but is also found in solder and as part of other metal alloys and dental amalgam (World Health Organization, 1980). Organotins are used as polyvinyl chloride (PVC) stabilizers and elsewhere in the plastics industry. Trisubstituted forms of organotin have biocidal properties useful in agriculture and other circumstances where fungicides, miticides, bactericides, or similar substances are required (World Health Organization, 1980).

Inorganic Tin

Inorganic (elemental) tin is poorly absorbed and rapidly turned over in tissue, hence its toxicity is very low compared with organotin compounds. Although animal studies show that inorganic tin can produce ataxia, muscle weakness, and CNS depression, no such effects have been cited thus far in human studies (World Health Organization, 1980).

Organotin

Organotins are used as pesticides, herbicides, rodenticides, disinfectants, and as a component in antifouling marine paints. In contrast to inorganic tin, organic tin compounds are very neurotoxic; their effects have been documented for over a century (Reuhl & Cranmer, 1984). Of these, diethyl, trimethyl, and triethyl tins are primarily toxic to the nervous system, in contrast to other organotins which target other organ systems. Exposure to diethyltin diiodide caused symptoms of cerebral edema, visual impairments, psychological disturbance, nausea, and vomiting. Fatalities were over 50% of the 210 individuals studied in one exposed group (Stoner, Barnes, & Duff, 1955).

Triethyltin

Triethyltin exerts its neurotoxic effects by causing "massive" myelin edema with consequent raised intracranial pressure (Reuhl & Cranmer, 1984, p. 187). The mechanism of edema is not known. In France, during 1954, 200 individuals became intoxicated, and half eventually died from the neurotoxic effects of a compound called "Stalinon," which contained triethyltin impurities and purportedly treated "osteomyelitis, skin infections ... anthrax" and acne (Hamilton & Hardy, 1974, p. 177; Watanabe, 1980). Severe diffuse white matter edema and changes in glial cells were found in four autopsied victims of Stalinon (Watanabe, 1980), and it is probable that most of the Stalinon deaths could be attributed to cerebral edema (World Health Organization, 1980). Clinical symptoms of exposed individuals included "nausea, vertigo, visual disturbances, papilledema and convulsions," typical symptoms of elevated intracranial pressure (Reuhl & Cranmer, 1984, p. 195).

Trimethyltin

Trimethyltin has different neuropathological properties than triethyltin. Trimethyltin specifically targets the cortex, limbic system, and brain stem. Experiments with primates have shown damage to the hippocampus, amygdala, pyriform cortex, and neuroretina (Reuhl & Cranmer, 1984). It is capable of also decreasing γ-aminobutyric acid in the hippocampus and dopamine levels in the striatum (Hanin, Krigman, & Mailman, 1984). One case study reports two chemists exposed to trimethyltin who subsequently experienced epilepsy, mental confusion, headache, and psychomotor dysfunction, but who did not show characteristic manifestations of intracranial pressure common to triethyltin victims (Fortemps, Amand, Bomboir, Lauwerys, & Laterre, 1978).

Reports of neurotoxic exposure to organotins in the workplace are uncommon. One study examined 22 chemical workers who had been exposed to spilled trimethyltin chloride over a period of a month (Ross, Emmett, Steiner, & Tureen, 1981). Significantly more high-exposure workers reported nonspecific symptoms of forgetfulness, fatigue, loss of libido and motivation, and periods of headache and sleep disturbance.

Neuropsychological Effects

Neuropsychological tests were also administered in the Ross *et al.* (1981) study, although the authors did not name the tests or report actual scores. Using nonparametric statistics they reported significant decrements in verbal memory, finger tapping, fine motor eye–hand coordination, as well as deficits in "visual-motor integration and learning" (Ross *et al.*, 1981, p. 1093; Ross & Sholiton, 1983).

Personality Effects

The authors suggested that a unique pattern of mood disorder was found significantly more often in organotin-exposed workers, which consisted of alternating bouts of rage and deep depression lasting from several hours to several days. Personality deterioration may be long-lasting; when 4 of the 12 individuals receiving the highest exposure were reevaluated between 9 and 34 months postexposure, all were said to suffer long-term deterioration of personality functioning.

As the authors noted, it is not known whether these changes are direct effects of organotin or reactive via posttraumatic stress disorder. It is also unclear whether premorbid personality factors or coping style influenced postexposure behavior.

TUNGSTEN CARBIDE

Tungsten carbide is one of the so-called "hard metals" that are alloys of metals which include tungsten carbide and cobalt. Alloys of this type are employed where conditions require exceptional strength, rigidity, or resistance to heat and corrosion.

The hard metals are associated with a variety of exposure-related disorders, most typically dermatological, cardiac, or pulmonary. So-called "hard metal disease" caused by hard metal dust produces asthma and progressive, sometimes fatal, pulmonary fibrosis.

Neuropsychological consequences of hard metal disease were recently investigated in workers who had developed pulmonary symptoms of hard metal exposure, complained of memory deficits, and had been removed from the workplace (Jordan, Whitman, Harbut, & Tanner, 1993). Compared with a control group, the exposed cohort showed impaired verbal memory on several tests. The attribution of this symptom to neurotoxic effects of hard metals is premature, however, since the present study was not able to rule out alcohol use, solvent exposure, or respiratory disease *per se* in the production of symptoms.

VANADIUM

As dust	
IDLH	70 mg/m^3
OSHA	0.05 mg/m^3 (resp)
NIOSH ceiling	0.05 mg/m^3 (15 min) (resp)
(*Quick Guide*, 1994)	

Vanadium is present in steel hardening, photographic processes, and insecticides. Trace amounts are present in milk, seafood, cereals, and vegetables. Normal serum levels of vanadium are 35–48 μg/100 ml (Goyer, 1986). Vanadium poisoning in animals "is characterized by marked effects on the nervous system..." and nervous depression has been seen with human industrial exposure (Goyer, 1986, p. 628).

ZINC

Zinc chloride (fume)	
IDLH	90 mg/m^3
OSHA	1 mg/m^3
NIOSH	1 mg/m^3
STEL	2 mg/m^3
Zinc oxide (fume/dust)	
IDLH	600 mg/m^3
OSHA	5 mg/m^3 (fume)
STEL	10 mg/m^3
NIOSH	5 mg/m^3 (fume/dust)
STEL	10 mg/m^3
	15 mg/m^3 15 min ceiling
(*Quick Guide*, 1993)	

Zinc is a common metal, found in sheet metal, batteries, and various alloys. Three neurological syndromes have been linked to toxic exposure. Oral ingestion of zinc chloride or of acidic foods stored in galvanized containers has been fatal, with survivors showing residual neurological symptoms of dyspnea, weakness, muscle spasm, and lethargy (Goetz, 1985). Inhalation of zinc vapor may produce neurological and psychiatric symptomatology, including irritability, "upper extremity coarse intention tremor, incoordination and ataxia" (Brazier's disease) (Goetz, 1985, p. 57). Neuropsychological tests of zinc toxicity have not been performed. Exposure to zinc phosphide has produced symptoms of irritability, psychomotor stimulation, drowsiness, and stupor. Egyptian workers exposed to zinc phosphide reported memory impairment, decreased attentional capacity,

TABLE 3.5. Other Metals with Neurobehavioral Effects

Metal	Effect
Barium	Severe neurotoxicity, flaccid paralysis, including respiration
Boron	CNS depression
Vanadium	Tremors, "nervous depression"

psychomotor hyperactivity, and rapid fatigue. EEG abnormality was present in 17.4% of the sample (Amr *et al.*, 1993).

OTHER METALS

Metals as a group tend to be neurotoxic. However, there is little or no neuropsychological research on many metals with common industrial applications, e.g., thallium or platinum. There is also no neuropsychological information available concerning the neurobehavioral effects of exposure to the substances listed in Table 3.5. Further neuropsychological, neurophysiological, and epidemiological studies are clearly needed.

4

Solvents

INTRODUCTION

Solvents must be included in any discussion of neuropsychological toxicology, since in the 100 years that organic solvents have been produced, evidence has slowly accrued to suggest their involvement in overt and subclinical neurotoxic syndromes. Approximately 9.8 million workers inhale or come into skin contact with solvents every day in the United States, suggesting a sizable population at risk for neurotoxic exposure effects (*Organic Solvent Neurotoxicity,* 1987).

The term "organic solvent" is a generic classification for a chemical compound or mixture used by industry to "extract, dissolve or suspend" non-water-soluble materials including fats, oils, resins, lipids, cellulose derivatives, waxes, plastics, and polymers (*Organic Solvents,* 1985, p. 3; Arlien-Søberg, 1985). Most solvents are liquids and their chemical composition may be simple, like carbon tetrachloride, or complex, like those solvents derived from petroleum (MacFarland, 1986).

Solvents are also grouped into major classes according to their chemical composition. One major class of solvents is that of the *hydrocarbons,* which, in turn, are divided into two subgroups:

1. *Aliphatic hydrocarbons* include hexanes, pentanes, and octanes (MacFarland, 1986). They are derived from petroleum and mineral spirits and are 90% of the aliphatic solvents used in industry. Common uses of aliphatic hydrocarbons are dry cleaning agents, and degreasers. They are common ingredients in paints, varnishes, and insecticides.

Members of this group are classified according to their boiling points, on a continuum that includes petroleum ethers, rubber solvent, varnish makers' and painters' naphtha, mineral spirits, and Stoddard solvent (also known as "white spirit").

2. *Aromatic hydrocarbons:* Made from coal tar, they have a characteristically pleasant odor that makes them frequent targets of inhalant abusers. Benzene, toluene, styrene, and xylene are common examples. Acute exposure to aromatic solvents has been shown to affect catecholamine neurons.

Other solvent classes include the following:
- *Halogenated compounds*—of which the chlorinated hydrocarbons are among the most toxic—include carbon tetrachloride, methylene chloride,

perchloroethylene, and 1,1,1-trichloroethane. Chlorinated hydrocarbons have been imputed to have special effects at low concentrations on the vestibulo-oculomotor system in trichloroethylene (Larsby et al., 1986) and other chlorinated solvents (Tham, Bunnfors, Eriksson, Lindren, & Odkvist, 1984). Jarkman, Skoog, and Nilsson (1985) found that electroretinograms are affected by exposure to low doses of several chlorinated solvents. Further, chlorinated solvents are "among the few agents that result in changes in brain fatty acid pattern" and, although the significance of this finding for neuropathology and function impairments is not yet known, "there are indications of the necessity of a normal brain fatty acid composition for proper intellectual functioning" (Kyrklund, 1992, pp. 19–20).
- *Alcohols:* including methanol (wood alcohol) and ethanol (grain alcohol), also known as *oxygenated solvents*
- *Glycols,* including ethylene and propylene glycol
- *Ketones,* including acetone methyl ethyl ketone and methyl *n*-butyl ketone
- *Complex solvents:* Other classes of materials have solvent properties, including esters, aldehydes, and substances like cyclohexane, dioxane, isophorone, pyridine, and turpentine (MacFarland, 1986).

Common characteristics. Curtis and Keller (1986) state that all solvents share common characteristics including *volatility* (significant vapor pressure) and *solvency* (pass through intact skin). In addition, solvents are *lipophilic* with an affinity for nerve tissue, and are *soluble* in blood and pass rapidly through lung tissue. Further, solvents have relatively *unobjectionable odors* which may not provoke irritation until high concentrations are achieved. Finally, exposure to solvents may cause *chemical dependency.*

EXPOSURE

Individuals are typically exposed to solvents accidentally and involuntarily in the workplace. Exposure can also be deliberate in the pursuit of an intoxicating solvent "high." Exposure is widespread in each category.

Industrial exposure to solvents began approximately 100 years ago in small cottage industries. Present-day solvent exposure sites include large-scale manufacturing applications involving millions of workers. Workers involved in the manufacture of paints, adhesives, glues, coatings, dyes, polymers, pharmaceuticals, and synthetic fabrics are commonly exposed. In addition, workers who directly employ solvents in their trades are also at risk, including painters, varnishers, and carpet layers (*Organic Solvents,* 1985).

Solvent abuse is also widespread, with one study estimating that 18.7% of U.S. high school seniors had tried solvent-based inhalants (Johnson, Bachman, & O'Malley, 1979).

Neurophysiological Effects

Despite long-standing industrial use and individual abuse, changes in neurophysiological and neuropsychological function as a result of solvent exposure did not enter the health literature until 1940, while large-scale investigations were not undertaken until the late 1960s and early 1970s. Experiments on animals have shown that neurotoxic effects span a continuum from molecular changes to cell death. Analysis of these effects can be accomplished to some degree by considering solvents as a class of neurotoxicants. Alternately, it is unlikely that a unitary mechanism will be found to the explain the actions of all solvents, as several have been found to show distinct morphological and biochemical effects (Arlien-Søberg et al., 1992). The exploration of these commonalities and differences remains an active and ongoing field of investigation.

Intake

One route of solvent exposure is via direct absorption into the skin. The speed of absorption into the skin is related to the solvent's solubility in water. Solvents as a group are quite volatile so they may just as easily enter the bloodstream via inhalation as from direct absorption through the skin. Solvents entering the bloodstream through the alveoli of the lungs diffuse quite rapidly into the blood. Solvents accidentally ingested (or purposely swallowed in suicide attempts) are easily absorbed from the gastrointestinal tract.

The amount of solvent retained is dependent on a combination of solvent-related and human-related factors. Some of the former include blood and tissue solubility, and *a priori* toxicity of the solvent (Astrand, 1975). Solvents may also show great intraindividual variations, and their toxicity may vary according to diurnal metabolic cycles, alcohol use, and other individual variations. Individual differences may also account for variation in solvent toxicity; for example, solvents are stated to have longer biological half lives in obese persons than in thin ones (Cohr, 1985). Solvents are excreted unchanged in expired air and their water-soluble metabolites are eliminated in urine. Physiological measurement of solvent uptake is more usefully obtained with urine sample than with blood concentrations or exhaled air. These latter measures "are usually only of value during the immediate exposure interval in view of the short half-life of these compounds in the blood" (Baker, Smith, & Landrigan, 1985, p. 214).

General Toxic Effects

As a class, solvents produce several types of exposure effects. Some common toxic effects do not involve the nervous system. For example, because of their defatting and corrosive qualities, solvents can irritate the mucosal membranes of the eyes and upper respiratory tract, and cause nausea, loss of appetite, vomiting, and diarrhea (Arlien-Søberg, 1985; James, 1985). In addition, solvents can be hepatotoxic, nephrotoxic, and can also induce cardiac arrhythmia by sensitizing the heart to catecholamines (James, 1985).

Neurotoxicity

All solvents depress CNS activity via anesthetic action. Solvent-exposed individuals are rendered progressively less sensitive to stimuli as a function of solvent concentration and/or exposure duration until unconsciousness, coma, or death occurs. Other acute effects can include light-headedness, feelings of drunkenness, and ataxia. Solvents appear to be the cause of sleep apnea in a significant proportion of occupationally exposed males exposed for longer than 10 years (Edling, Lindberg, & Ulfberg, 1993). Spencer and Schaumberg (1985), in reviewing solvent neurophysiological research, found lasting neurological sequelae for five individual solvents, including the hexacarbons n-hexane +/methyl ethyl ketone (MEK), methyl n-butyl ketone ± MEK, carbon disulfide, trichloroethylene, and voluntarily abused toluene. Vibration perception threshold may be impaired in workers with chronic exposure, suggesting peripheral nerve damage (Aratani *et al.*, 1993). Solvent exposure appears to be one of the factors involved in elevated risk of developing motor-neuron disease (Gunnarsson, Bodin, Söderfeldt, & Axelson, 1992).

Some organic solvents have also been suggested to interact with neurochemical mechanisms controlling pituitary actions, producing hormonal alterations. Svensson, Nise, Erfurth, and Olsson (1992) found a reversible, dose-dependent decrease in gonadotropin hormones (HL and FSH) in printers exposed to toluene. The authors suggest that solvent effects on the hypothalamic–pituitary axis is the most likely etiology.

Blood solvent levels may significantly underestimate the brain concentration of these lipid-soluble substances. Experimental exposures suggest that brain concentration may lie between one and three times blood solvent levels, depending on the solvent's degree of lipid solubility (Kyrklund, 1992).

Moen, Kyvik, Engelsen, and Riise (1992) analyzed the cerebrospinal fluid (CSF) of 16 patients diagnosed with chronic toxic encephalopathy from organic solvents. The control group comprised 16 patients with symptoms of backache and/or myalgias. Solvent encephalopathy patients had elevated levels of total protein and albumin, possibly resulting from protein leakage through the blood–brain barrier, or from reduced CSF drainage. Taurine concentrations were significantly negatively correlated with solvent exposure indices, which may be related to calcium flow or reductions in cell membrane protection. The authors note that reduced CSF taurine is also found in patients with "hereditary mental depression and parkinsonism and in patients with a reduced level of consciousness" (p. 280).

Teratogenicity

Maternal occupational exposure to solvents may be related to increased risk for birth defects. Risk for spontaneous abortion in wives of workers exposed to high levels of toluene and miscellaneous solvents is significantly increased (Lindbohm, *et al.*, 1992). Because most solvents are capable of passing through the placenta and being excreted in breast milk, exposure may occur both pre- and postnatally. The literature in this area is comparatively scarce. A recent review noted only four studies linking maternal laboratory work with increased risk of congenital malformations (Roeleveld et al., 1990). Solvent-induced spermatotoxicity may also produce fetal abnormalities, with increased risk in fathers exposed to solvents or employed as painters (Olsen, 1983).

Psychiatric Disorder

Evidence is mixed as to whether chronic exposure to organic solvents is associated with increased risk of psychiatric disorder. Several studies report increased risk of having a psychiatric diagnosis as a function of solvent exposure (Axelson et al., 1976; Olsen & Sabroe, 1980; Mikkelsen, 1980). There does not appear to be an increased risk of first-time nonpsychotic psychiatric hospitalization (ICD-9 Codes 300–316) associated with chronic solvent exposure (Labrèche, Cherry, & McDonald, 1992), although risk is higher for organic mental disorder-related psychiatric hospitalization (Labrèche et al., 1992).

Neuropsychological Symptoms

Neurological examination is usually normal, except in the most severe cases of solvent exposure; however, "subclinical" neuropsychological effects tend to be seen much earlier in the subject's history of exposure. Neuropsychological abnormalities include behavioral, cognitive, and emotional dysfunction. Commonly reported are complaints of headache, dizziness, fatigue, paresthesias, pain, and weakness. Subjective complaints of memory disturbances have been frequently claimed (Edling, 1985). Severe exposure is capable of causing dementia, defined by DSM-III as "a loss of intellectual abilities of sufficient severity to interfere with social or occupational function. The deficit . . . involves memory, judgment, abstract thought and a variety of other cortical functions. Changes in personality and behavior also occur "(American Psychiatric Association, 1980, p. 107). Neuropsychological symptoms of solvent-induced dementia appear to suggest diffuse cortical and subcortical damage.

Symptoms found in solvent-exposed workers tallied by questionnaire and interview by Mikkelsen, Browne, Jorgensen, and Gyldensted (1985) include impairments in memory, concentration, general intelligence and problem-solving, speed, and initiative. After excluding heavy alcohol users and those who took antihypertensive medication and/or antipsychotic drugs, Wang and Chen (1993) found that exposure severity self-report symptoms include acute headache and chest tightness, with chronic dizziness, rapid fatigability, depression, and palpitations. Self-reported solvent exposure is also significantly associated with increases in work-related slips, trip, and falls, especially when exposure varies over time (Hunting, Matanoski, Larson, & Wolford, 1991). Affective and personality abnormalities may accompany chronic solvent exposure. Inventory-assessed symptoms of fatigue, depression, anxiety, irritability, and emotional lability are reported.

Morrow, Ryan, Goldstein, and Hodgson (1989) evaluated MMPI results of chronic solvent exposure (mean duration 7.3 years) with no self-report history of alcohol abuse and no documented history of neurological or psychiatric disorder. MMPI results showed very high clinical elevations on scales 1, 2, 3, and 8, a pattern that resembled former World War II prisoners of war, "half of whom met full DSM-III criteria for PTSD" (Morrow *et al.*, 1989, p. 745).

Disturbance of olfactory perception and sensitivity to environmental odors have also been documented for solvent-exposed individuals. General decrements in olfactory sensitivity were found by Schwartz *et al.* (1990), who examined workers at two paint manufacturing facilities using the University of Pennsylvania Smell Identification Kit (UPSIT). Dose-related impairment in olfactory sensitivity was found, but only among workers who had never been smokers. Mean UPSIT score was below the 5th percentile for age-adjusted norms. Ryan, Morrow, and Hodgson (1988) studied a small ($n = 17$) group of subjects with a mean solvent-exposure duration of 7.4 years. Eight of the subjects described a heightened sensitivity to specific odors like gasoline or perfume, sufficient to produce aversion and nausea. This symptom was related to scores on verbal learning and immediate visual reproduction in a regression analysis. In a later study of 32 solvent-exposed workers, 60% reported cacosmia (Morrow, Ryan, Hodgson, & Robin, 1990).

DIAGNOSTIC PROCEDURES

As is true for metals, the constellation of cognitive and emotional alterations produced by solvent intoxication is "most reliably measured by psychological test methods" (Lindstrom, 1981, p. 48). However, neuropsychological analyses of solvent effects usually occur in the context of a general medico-diagnostic workup. While neurological examination and other medical procedures tend to be less sensitive to early subclinical solvent effects, they are nevertheless essential for several reasons. First, a complete evaluation is usually needed to rule out alternative diagnoses (e.g., unrelated CNS or PNS lesion). Second, neurological and physical evaluations provide tests for functions not examined by neuropsychologists (e.g., liver function tests, tests of cranial nerve function). Finally, conventional medical examination may complement and further validate neuropsychological findings.

Neuroradiological and Neurophysiological Techniques

Techniques in common use include electroencephalography (EEG) to measure CNS effects, and electroneuromyography (ENMG) to detect abnormalities in the electrical activity in nerves and muscles. Tests of regional cerebral blood flow have also been found to be useful. CT scan may be employed to rule out alternative diagnoses, or to expose more serious structural damage found in chronic exposure to certain solvents. However, it is not sensitive to early subclinical effects and Triebig and Lang (1993) found only a tendency toward more atrophy in spray painters chronically exposed to solvents. There were no other dose-effect relationships. Overall indices of atrophy were also not correlated with neuropsychological test results.

Magnetic resonance scans (MRI) are more sensitive to solvent-induced atrophy. There are several reports of an association between chronic solvent exposure and mild-to-moderate atrophy (Lorenz et al., 1990; Aaserud et al., 1990; Myint, 1990). Rosenberg et al. (1988) showed that T2-weighted chronic toluene exposure increased white matter signal intensity in periventricular regions and produced diffuse loss of differentiation between gray and white matter. Single case studies of mercury exposure and 2,6-dimethyl-4-heptanone patients have shown abnormal MRI following exposure (White et al., 1993).

A study that compared CT and MRI indicated that both are often able to image chronic solvent encephalopathy, but not early subclinical effects. Leira, Myhr, Nilsen, and Dale (1992) examined ten patients with diagnosed solvent encephalopathy on CT and MRI. MRI provided a more sensitive measure of brain injury than CT, picking up two additional cases of ventricular atrophy and one additional case of cortical atrophy in the cohort. In addition, MRI revealed greater severity of ventricular atrophy in one case and cortical atrophy in another. MRI also showed additional features not resolved by CT; six patients had grade 1 atrophy of the vermis cerebelli, with five also showing posterior fossa cistern enlargement. MRI additionally revealed a lacunar left thalamic infarction in one patient and two small brain stem lesions in another. Because of its increased sensitivity, the authors recommended MRI as the diagnostic imaging method of choice over CT in cases of solvent-induced encephalopathy.

Functional imaging techniques also appear to show promise in detecting solvent-related neurotoxic disorders. One recent study found SPECT (single photon computed tomography) to detect abnormalities in patients presenting to an occupational medicine clinic with toxic encephalopathy (Callender et al., (1993). Ninety-four percent of the 33 workers showed abnormal SPECT scans; the most frequently damaged areas in descending order were temporal lobes (67.7%), frontal lobes (61.3%), basal ganglia (45.2%), thalamus (29.0%), parietal lobes (12.9%), motor strip (9.68%), a cerebral hemisphere (6.45%), occipital lobes (3.23%), and caudate nucleus (3.23%). Approximately 79% of the patients' neuropsychological evaluations were also abnormal. Similar sensitivity of PET scan has been demonstrated in a case of tetrabromomethane encephalopathy (Morrow, Callender, et al., 1990).

One of the more promising diagnostic neurological techniques available is the use of P300 event-related potentials. Morrow, Steinhauer, and Hodgson (1992) compared 12 solvent-exposed individuals with 19 controls. Both N250 and P300 latencies were significantly longer in the exposed group with P300 latency positively correlated with length of exposure. P300 latencies of solvent-exposed workers were also longer than those found in a group of schizophrenic patients examined in the same study. P200 amplitude was significantly higher in solvent-exposed patients, possibly reflecting delayed sensory registration of stimuli, an effect that occurred for both simple counting and choice reaction time. Mean exposure duration of evaluated patients was 3 years, although the range was less than 1 day to 30 years. None of the subjects were currently exposed to solvents and their last exposure occurred 2 years previously (range 2 months to 10 years). Missing from the study is any indication of alcohol use or other potential confounds within the sample.

Visual evoked potentials (VEP) may also be sensitive to chronic solvent exposure. Urban and Lukas (1990) examined 54 rotogravure printers exposed to toluene from 1 to 41 years, and compared them to controls. Twenty-four percent showed one or more features of abnormal VEP and a variety of other differences were seen between exposed workers and controls, but workers also were heavier alcohol users than controls. Of the percentage of workers who drank more than 50 grams daily, 69% had abnormal VEP, and only 34% were normal. Percentage of abnormal VEP increased with ethanol intake. The authors warn that "frequent use of alcohol in toluene exposed workers ... must be taken into account when etiology of VEP changes is considered" (p. 822).

International Diagnostic Criteria for Solvent Syndromes

Just as there is no single test capable of diagnosing solvent intoxication, there is also no single conceptual system to categorize the effects of solvent exposure. Systems of diagnosis and requirements for impairment both vary somewhat from country to country. Denmark, for example, defines "chronic toxic encephalopathy" in terms of impairment of intellect, memory, affect, and personality. Neuropsychological test results are crucial to the diagnosis, and while neurological deficit may occasionally be found, the latter is not essential for diagnostic attribution. In fact, severe neurological deficits may argue for an alternative diagnosis (*Organic Solvents*, 1985, p. 8).

Other Scandinavian countries use criteria similar to Denmark's. Each seeks to verify qualitative and quantitative effects of solvent exposure by clinical history as well as by documenting a symptom picture that may include both objective and subjective data. All seek to rule out other causes of brain dysfunction and all subject the potential exposure victim to a variety of neurological, neuropsychological, and neurophysiological tests, including CT, MRI, EEG, electromyography (EMG), evoked potentials (EP), sensory nerve action potentials (SNAP), F-response, muscle biopsy, and nerve biopsy. Sweden specifies exposure parameters more than its sister countries, by suggesting that more than 10 years of long and/or intensive exposure is necessary to the diagnosis, which they term "psycho-organic syndrome."

The World Health Organization's (WHO) 1985 Copenhagen conference suggested that solvent intoxication effects should be termed "slight or mild" chronic toxic encephalopathy, depending on exposure severity. "Slight" encephalopathy was characterized by evidence of subjective complaint and personality impairment without neurological signs or symptoms. As exposure progressed, discernible deficits would be noted in "cognition, memory and neurological deficits, characterized by balance deficits and mild ataxia" (*Neurobehavioral Methods*, 1985b, p. 31).

There is currently no official medico-diagnostic category in the United States that corresponds with "psycho-organic syndrome" or its equivalents. Baker and Fine (1986) suggest that the DSM-III diagnostic criteria for organic affective syndrome are met for the early, reversible symptoms of fatigue, irritability, depression, and apathy encountered in solvent-exposed individuals. Chronic exposure to high concentrations of some solvents (e.g., toluene) may also produce global intellectual deterioration and thereby satisfy the criteria for dementia.

A research conference, held in the United States the same year as the WHO conference, attempted to more completely characterize the progressive nature of solvent neurotoxicity. The panel members at the conference partitioned solvent effects into four separate syndromes, the mildest of which was termed a "Type 1" disorder (Baker & Seppäläinen, 1986). Type 1 solvent intoxication is also equivalent to earlier terminology of "neurasthenic syndrome." This syndrome, consisting of "fatigue, irritability, depression and episodes of anxiety," is not accompanied by deficits on neuropsychological tests, but can be documented on instruments that record subjective complaints, e.g., symptom checklists. A syndrome of this type has been hypothesized to presage neurobehavioral dysfunction (Baker, Smith, & Landrigan, 1985, p. 213). Type 1 symptoms are also said to be reversible if exposure is discontinued (Baker & Seppäläinen, 1986).

When exposure continues, neuropsychological methods can identify the next stage, which has been variously termed "psycho-organic syndrome" (Flodin, Edling, & Axelson, 1984) in the Swedish literature, "mild dementia" in Danish writings, and "mild toxic encephalopathy" by WHO (*Neurobehavioral Methods*, 1985b).

The 1985 International Workshop on the Neurobehavioral Effects of Solvents held in North Carolina further divided this middle stage into its emotional and cognitive components. For research purposes, a Type 2A exposure syndrome was said to be characterized by disturbances of mood and personality, fatigue, impulse control, and motivation, while Type 2B syndromes would be recognized primarily as intellectual disturbances in memory, learning, and concentration, psychomotor difficulties and impairments in other neuropsychological functions (Baker & Seppäläinen, 1986). The relationship between these two intermediate-severity solvent syndromes and their reversibility has yet to be determined.

The time course for development of a psycho-organic or Type 2B syndrome may be dependent on intensity of solvent exposure. Olson (1982) found that tool cleaners who work with vats of solvent appear to show a more rapid onset of solvent symptoms than did painters. Cleaners who were exposed for a mean of 4.3 years performed significantly more poorly in tasks of reaction time, concentration, and memory. Incipient symptoms may begin as early as 3 years with chronic industrial exposure (e.g., Flodin *et al.*, 1984).

The most severe stage or "Type 3" solvent syndrome is characterized by progressive and global neuropsychological, intellectual, and emotional decline which fulfills the medical criteria of dementia and is equivalent to WHO's "severe

TABLE 4.1. Categories of Solvent-Induced CNS Disorders

Severity	Category of CNS disorder	
	Identified by WHO/Nordic Council of Ministers Working Group, Copenhagen, June 1985[a]	International Solvent Workshop, Raleigh, NC, October 1985[b]
Minimal	Organic affective syndrome	Type 1
Moderate	Mild chronic toxic encephalopathy	Types 2A or 2B
Pronounced	Severe chronic toxic encephalopathy	Type 3

In view of the difficulty of categorizing these disorders, correspondence between the two systems of nomenclature is not exact.
[a]Source: *Neurobehavioral Methods* (1985b).
[b]Source: Baker and Seppäläinen (1986).

toxic encephalopathy" (World Health Organization, 1985). This latter stage is rarely seen clinically, and usually only after many years of workplace exposure or severe recreational abuse (e.g., Fornazzari, Wilkinson, Kapur, & Carlen, 1983). We have also seen one patient who developed a similar syndrome within 1 year, working with solvents in a small, enclosed, and unventilated art studio.

It must be emphasized that the tripartite division of solvent effects proposed at the 1985 conference were *working hypotheses* of which not all researchers, even those at the same conference, agreed. For example, Mikkelsen (1986) suggested that solvent effects should instead be classified according to three dimensions: (1) particular combinations of neurotoxic signs and symptoms, (2) severity of effects, and (3) reversibility of effects.

Table 4.1 shows the approximate correspondence between proposed categorizations of solvent syndromes.

Diagnostic Process

The patient who is under evaluation for solvent intoxication should optimally be evaluated by a team specially trained in the diagnosis of solvent syndromes. The initial evaluation should be sufficiently broad to rule out alternative diagnoses, and sufficiently fine-grained to describe in detail the pattern of the individual's deficits. Thus, a chronic solvent exposure victim might receive neurological, neuropsychological, hematological, and neuroradiological evaluations as part of a standard workup. Baker and Seppäläinen (1986) have suggested that data collected from the evaluations can be used to diagnose, control exposure, and provide appropriate health-related referrals to patients claiming solvent intoxication.

Conventional medical/neurological techniques are usually employed to obtain objective physiological correlates of behavioral dysfunction. These techniques have included blood, urine and breath assay, EEG, CT, nerve conduction, and, more recently, MR scans, as well as several types of evoked potential studies (e.g., Otto, 1983b; Urban & Lukas, 1990; Morrow, Steinhauer, & Hodgson,

1992). Regional cerebral blood flow studies may also be of value (Risberg & Hagstadius, 1983).

SOLVENT MIXTURES

Mixtures of solvents are studied for their neurotoxic properties for several reasons. First, industrial personnel are more commonly exposed to mixtures or multiple solvents than to a single solvent (Lindstrom & Wickstrom, 1983). Painters, for example, have often been exposed to solvent mixtures, including "white spirit," a group of paint solvents with boiling points midway between gasoline and kerosene. Although techniques used to study mixtures versus single solvents are similar, one difference seems to be that solvent mixture research proceeds on the assumption that since all organic solvents depress the functions of the central nervous system and therefore their neurotoxic effects should be similar in most aspects (Gregersen, Angelso, Neilsen, Norgaard, & Uldal, 1984).

Acute Effects

Acute solvent mixture effects are prenarcotic, usually reversible and dependent on dose levels (Winneke, 1982). Symptoms often disappear following an exposure-free interval; workers who have experienced progressive discomfort during the week may experience renewed well-being by Monday (Arlien-Søberg, 1985). Cherry, Venables, and Waldron (1983) investigated acute behavioral and mood changes in workers exposed to solvents pre- and post-workshift. Workers rated themselves significantly higher in sleepiness, physical and mental tiredness, and believed themselves to be in poorer health than did controls. Neuropsychological findings were inconsistent, with paint solvent and styrene workers showing somewhat greater effects of exposure than methylene chloride workers. Impairments in attention and concentration have also been related to acute effects (Gregersen *et al.*, 1984).

When exposure levels are high enough, however, acute solvent exposure may precipitate toxic encephalopathy with chronic neuropsychological impairments and neurological concomitants of dementia. In one such case, a 38-year-old laborer spray-painted a truck on two occasions in an unventilated room, with total exposure time of approximately 24 h over four consecutive days with no loss of consciousness (Welch, Kirshner, Heath, Gilliland, & Broyles, 1991). Solvents were a mixture of toluene and MEK. Initial headaches were followed by memory and concentration difficulties, respiratory distress, dysarthria, and ataxia. A variety of neuropsychological impairments were noted and seen to persist over a period of 2 1/2 years. Fluid collection over the left parietal area was found on two consecutive MRIs but relationship to exposure is unclear. Neurological examination showed inability to perform tandem gait, intention tremor, and dysmetria in all four extremities. A test of malingering (Ray 15 Item) was normal.

Chronic Effects

Chronic effects are much more frequently researched than acute effects (Savolainen, 1982). Workers with chronic solvent exposure are significantly more likely to be hospitalized with organic brain damage than for other psychiatric diagnoses, particularly when there is also heavy use of alcohol (Cherry, Labrèche, & McDonald, 1992). Neurological and neuropsychological effects have been documented for many solvents and may occur separately from, or in combination with, acute effects.

Physiological

Gross clinical signs of CNS impairment are usually "minor, if any" (Seppäläinen, Husman, & Martenson, 1978) and conventional neurological tests "are frequently too imprecise and unchallenging" (Bleecker, 1984). However, chronic CNS and PNS deficits can be found by sensitive tests. Chronic neurological effects include vestibular dysfunction (Arlien-Søberg, Zilstorff, Grandjean, & Pedersen, 1981) and cerebral atrophy (Seppäläinen, Lindstrom, & Martelin, 1980; Arlien-Søberg, Bruhn, Gyldensted, & Melgaard, 1979). Juntunen, Hupli, Hernberg, and Luisto (1980) examined 37 patients with suspected chronic solvent poisoning. Pneumoencephalography (PEG) suggested patterns of "slight asymmetric central atrophy and/or localized cortical atrophy" in 62% of patients. ENMG results suggested peripheral neuropathy in 23 of 28 patients examined.

EEG abnormalities are frequently associated with chronic solvent exposure (Seppäläinen & Antti-Poika, 1983). Physiological changes have been correlated with development of psycho-organic syndrome. Seppäläinen (1985) reviewed EEG changes in chronic solvent-exposed workers and found abnormal EEGs in 17 to 67% of patients. She suggests that EEG results are indicative of an acute loss of neurons or neuronal functions as indicated by increased low-frequency irregular or rhythmic theta or delta activity. This phenomenon is most easily seen in acute solvent-exposed patients, while slowly developing increases in slow-wave activity are noted with chronic solvent exposure (Seppäläinen, 1985).

Significant reductions in cerebral blood flow have been observed in painters exposed to solvents for a mean of 22 years (Arlien-Søberg, Henriksen, Gade,

TABLE 4.2. Percentages of Workers with Abnormal EEGs[a]

	Percentage with abnormal EEG	Reference
72 painters	17%	Seppäläinen and Lindstrom (1982)
102 car painters	31%	Seppäläinen (1978)
233 workers	40–72%	Seppäläinen (1973)
107 solvent-poisoned patients	65%	Seppäläinen et al. (1980)
87 solvent-poisoned patients	67%	Seppäläinen and Antti-Poika (1983)

[a]Reprinted, by permission of the publisher, from Seppäläinen, A. M. (1985). Neurophysiological aspects of the toxicity of organic solvents. *Scandinavian Journal of Work Environment and Health, 11* (Suppl. 1), 61–64.

Gyldensted, & Paulson, 1982), while Risberg and Hagstadius (1983) observed small but significantly lowered regional cerebral blood flow in workers chronically exposed to solvent mixtures for a mean of 18 years. The latter study showed greater differences in bilateral frontotemporal and left parietal regions. Larger differences were also found in those individuals with the highest exposure. Since alcohol intake and other demographic variables were similar for patients and controls, the authors speculate that chronic solvent exposure may cause premature aging of the brain by a process similar to that of chronic alcoholism.

Polyneuropathy is also frequently associated with long-term exposure to solvents (Seppäläinen, 1982b; Husman & Karli, 1980). Loss of vibration sense and diminished sensitivity to pain and light touch have been reported (Husman & Karli, 1980), as have lowered sensory and motor nerve conduction velocities and abnormal EMGs (Seppäläinen, 1982b). Autonomic nervous system abnormalities and symptoms have also been found in a group of patients who had been chronically exposed to solvents and subsequently diagnosed as showing features of organic solvent intoxication (Matikainen & Juntunen, 1985; Ng, Ong, Lam, & Jones, 1990; Steinhauer, personal communication, 1993). Somatosensory evoked potentials in solvent workers showed slight decrease of peripheral conduction velocity and increase in central conduction time; an effect potentiated by simultaneous alcoholism (Massioui et al., 1990). Solvent-exposed workers also had minor impairments modulating attentional resources in N1, N2, and P3 components (Massioui et al., 1990).

Patients with diagnosed chronic (Type 2B) solvent encephalopathy showed elevated concentrations of protein, albumin, and IgG and reduced concentrations of phosphoethanolamine, taurine, homocarnosine, ethanolamine, α-aminobutyric acid, and leucine compared with controls (Moen et al., 1990). Protein and albumin levels may be associated with either protein leakage through the blood–brain barrier, or impaired drainage of CSF. The significance of amino acid changes is unknown at this time, although decreased CSF taurine concentration has been seen in patients with familial depression and parkinsonism (Moen et al., 1990).

Psychiatric/Emotional

Psychiatric symptoms of organic solvent exposure may range from mild depression or irritability to organic psychosis. While emotional impairments as a result of solvent exposure have been hypothesized to occur and coded in the international diagnostic schema for exposure effects (Type 2A), there has been less systematized research in this area. Acute and chronic effects were studied by Morrow, Kamis, and Hodgson (1993) in 30 men and women who were relatively young (mean age 38.6) and relatively well educated (mean education 13.2 years). Magnitude of psychiatric symptomatology for this group was two to three times that of controls on the SCL-90-R Global Severity Index, and the Beck Depression Inventory. Exposed subjects experienced significantly more somatic distress, concentration difficulties, depression, anxiety, fear, and interpersonal alienation, compared with controls. The effect is cross culturally robust, with similar findings reported in Chinese solvent workers (Ng et al., 1990).

Neuropsychological

There is general agreement that short-term conditions of very high exposure may result in solvent poisoning with long-lasting neuropsychological deficits, especially if levels are high enough to produce unconsciousness. Houck, Nebel, and Milham (1992) provide evidence that persistent CNS effects may also be produced by high levels of what might be called "intermediate-term" exposure. Workers cleaned a poorly ventilated basement with mixtures of either alkylated benzenes/monoethanolamine foam, or 1,1,1-trichloroethane and perchloroethylene spray. Five of the twenty-seven workers reported memory complaints after being employed on the project from 3 to 58 days. Neuropsychological testing was described as "grossly abnormal" in two of the five, with both workers unable to resume previous employment, and one unable to live independently.

We have seen a similar case, a college-educated artist without prior neurological illness who developed slurred speech, gait ataxia, cerebellar and neuropsychological impairments after a summer spent painting and cleaning art materials in an unventilated basement studio. Although solvent levels were not available, the artist reported that substantial amounts of solvents were in the air, and absorbed from spills into a carpet. A toxicologist consultant concurred that very high levels of ambient solvents would have been produced by these factors.

Case Study A: Chronic Solvent Exposure without Unconsciousness in an Artist

Reason for Referral. The patient was referred for evaluation after reporting neurological impairments coincident with heavy exposure to solvents while working in her home art studio.

Tests Administered

Halstead–Reitan Neuropsychological Battery Subtests
    Finger Tapping Test
    Trail-Making Tests: A&B
    Computer Category Test
    Fingertip Number Writing Test
    Grip-Strength Test (Dynamometer)
Wechsler Memory Scale (Russell Revision)
    Logical Memory—immediate and 30-minute delayed recall
    Figural Memory—immediate and 30-minute delayed recall
    WAIS-R subtests: Digit Symbol, Block Design
Gordon Diagnostic System Model III (GDS)
    Adult Vigilance Test
    Adult Distractibility Test

Other Tests
  Kaufman Brief Intelligence Test (K-BIT)
  Wisconsin Card Sorting Test (WCST)
  MLAE Controlled Oral Word Association Test (COWAT)
  Recognition Memory Test (RMT)
  Grooved Pegboard Test
  Minnesota Multiphasic Personality Inventory-2 (MMPI-2)
  IDS Adult Neuropsychological History

Clinical Interview
Medical history. The patient is an artist who uses oil paints as her primary media. The patient reports that she has been exposed to solvents since 1988. Exposure began two days per week at her art school, and one day each week at her home studio. The patient states that ventilation was poor in both places. At her art school she experienced daily headaches and noted that the "place was covered in paint and you knew you were breathing in turpentine, terpenoid and mineral spirits." At her art school, all painting took place in a large (50 × 50 feet) high-ceiling room with no windows and no fresh air. The patient states that she remained in the room from 9:00 a.m. through 4:00 p.m., excepting lunch. She wore a lab coat and latex gloves while painting. However, at the end of the day, she would take her gloves off and clean her brushes in mineral spirits or turpentine, and would also manually clean the bristles by hand, without gloves.

The patient also has an art studio in the basement of her home, where she painted one day each week while she went to her art school, and now paints there four days each week. The room is described as 15 × 15 feet with full-sized windows that remained permanently closed to limit allergen exposure. The patient reports that her symptoms became particularly severe during one summer when the ventilation and air-conditioning system were turned off because of unexpectedly cool weather. From approximately May through August, the patient reports that she worked with no ventilation in the studio and had at least one wet painting drying in the room at a time.

It was during this period of time that the patient began noticing other symptoms, i.e., "I got very, very slow. I started to get a very dazed feeling in my head, I thought I was having sinus trouble. I started to get very lethargic, like I was dazed. I'm not a laid-back person or slow moving. I called a psychiatrist, and asked 'do you think this is psychological?'" According to the patient, the psychiatrist did not believe that her symptoms were psychological in origin.

Other symptoms experienced during this time period included "a very dull feeling" in her head and chest tightness, with the perception that she was having difficulty getting enough oxygen. Additionally, the patient noticed that she would frequently stumble, trip, and fall over to one side, a problem that persists to this day. She began to lose strength in her extremities, and had particular loss of strength in her right hand and right ankle. In September or October, when the patient visited her family physician, she reported walking "like I was 80 years old and experienced slow and slurred speech. Her internist tentatively diagnosed "sinus trouble" and prescribed medication, which the patient believes relieved "about 50%" of her symptoms.

A visit to an ENT physician that same month ruled out inner ear problems and found no hearing loss. The patient went on vacation for one week and made an appointment with a neurologist, who informed her that "something wasn't right" and the problem was consistent with neurotoxic exposure. She subsequently became so weak that she could not leave her bed. When this was occurring, the patient had been painting in her studio through September 1st, and she reported that "solvents were in everything," including the carpets, and that "everything was being wiped down with solvents, including large plastic sheets under the desks."

In January, the patient underwent an MRI, which was reported to be normal. However, she reports that during this same time, her muscles became so weak that she had difficulty carrying groceries, and her coordination was so impaired that she had to manually place a pen in her hand with her opposite hand, and could not move her toes volitionally. She also developed stress incontinence, a problem she had never before experienced. The patient noted that a visiting friend said to her that she thought the patient had "a neurological disease."

Currently, some of the patient's symptoms have improved. While she feels better able to coordinate and get dressed, she continues to feel "shaky, slow, and weak."

The patient denied any personal or family history of multiple sclerosis, Parkinson's disease, or other degenerative neurological disease. Recent tests for blood lead, cadmium, plasma cholinesterase, and red cell cholinesterase were negative. Blood pressure is reported to be normal-to-low. There is no history of head injuries, seizures, unconsciousness, or arthritis. She is a nonsmoker and does not drink alcohol. There is no history of alcohol abuse. Previous medical history is positive for a hysterectomy and hemorrhoid operations. The patient drinks two cups of decaffeinated coffee per day and eats two pieces of chocolate. She denies any episodes of severe depression, significant weight loss, or loss of appetite. She has no significant psychiatric history, although she consulted counselors twice for problem-focused therapy, once during the course of a relationship and once to better her communication and parenting style with her child.

Current medications include Premarin, Entex, Per-Colace, and topical Cleocin and Sulfacet.

The patient admitted to another source of potential neurotoxic exposure from long-acting Dursban®, a chlorpyrifos insecticide that she reports was applied to her house every other month.

Social history. The patient is married and has a child from her present marriage and two grown sons from a previous marriage. She admits to a number of current and past psychological stressors, but none appear related to her symptoms, nor has there been any previous episode of physical symptom production under stress. Earlier stressors include interpersonal difficulties with a grown son and a recent conflict with several of her relatives.

Educational history. The patient states that she received good grades in high school and was a National Honor Society awardee. School grades were reportedly "good," in the top 10% in a class of 600. The patient's best subject was math, for which she received A's. Spelling was a relative problem.

Behavioral observations and mental status. The patient is a slim, well-dressed female. The patient was oriented to time, place, person, and situation. Thought content was coherent with no indication of circumstantiality, tangentiality, or psychotic process. There was no evidence of delusional or paranoid behavior. Speech was normal in volume, rate, and timber with no indication of aphasia or dysarthria. Motivation was good throughout interview and testing. The patient appeared to try hard throughout the evaluation with no indications of either exaggeration or minimization of symptoms or test performance.

Intellectual, language, and general cognitive functioning. The patient's current IQ was estimated to be in the Average range, based on a test of expressive

TABLE 4.3. Neuropsychological Test Results

Test	Score	Rating
Category Test (errors)	71	Heaton T=42
Trails A (seconds)	100	Heaton T=17
Trails B (seconds)	132	Heaton T=32
Finger Tapping DH	22.8	Heaton T=23
Finger Tapping NDH	29.8	Heaton T=35
Grooved Pegboard DH	140"	Heaton T=17
NDH	123"	Heaton T=21
Fingertip # Writing	DH=6 errors	
	NDH=4 errors	
Dynamometer	DH=16 kg	
	NDH=15 kg	
WAIS-R Block Design	SS=8	Heaton T=48
WAIS-R Digit Symbol	SS=6	Heaton T=39
WCST		
Trials	128	
Correct	86	
Perseverative responses	22	Heaton T=45
Nonperseverative errors	22	
Perseverative Errors	20	
Categories	4	
Learning to Learn	−3.35	
K-BIT	V=94/M=105	Comp IQ est.=99
COWAT	35 words	(average)
Recog. Memory Test	Words 49/50	SS=14
	Faces 44/50	SS=09
Gordon Dx System III	Vigilance=28/30	Distractibility 22/30
Stroop	Words	T36
	Colors	T47
	Colored Words	T41
Russell WMS	Logical immediate	Russell 1.5 (good)
	Logical delayed	Russell 2.0 (mild)
	Figural immediate	Russell 1.0 (good)
	Figural delayed	Russell 1.0 (good)
MMPI-2	F=T48, K=T70	No clinical scores > T65

vocabulary, word identification, and interpretation of nonverbal analogies. Current test results appear to underestimate the patient's premorbid high school and college grades. There were no indications of aphasia, alexia, or agraphia during testing or during clinical observation.

The patient was at the border of mild impairment on a test of sustained inductive problem-solving that is thought to be a measure of both general cortical tone and prefrontal cortical integrity (Category Test).

On a test of visual scanning, sequencing, visuomotor speed, and attention—which required her to sequentially connect with a pencil line 25 randomly spaced numbers on a page—performance was considered moderately impaired. The patient showed mild to moderately impaired performance when given the more complex version of this same test, which requires rapid drawing of lines between successive alternating numbers and letters randomly spaced on a page (e.g., A-1-B-2-C-3, etc.). Rapid perceptual matching of abstract line drawings to Arabic numerals according to a template (the subtest on the WAIS-R most highly correlated with overall brain damage or integrity) was performed at a level consistent with mild impairment.

The patient performed in the bottom 1% of a normative population on a "continuous performance" test, which required visual identification and motoric response to predefined sequences of numbers (i.e., 1-9) rapidly displayed on a computer panel. She was also severely impaired in the number of misplaced responses (commissions) made on this task. Performance improved in the more difficult variation of this test, which immediately followed the simpler version. On the Distractibility subtest, which measures visual attention under highly distracting conditions, i.e., requiring visual scanning of number sequences while distracting numbers flashed on either side (GDS), the patient performed in the low average range for correct items responded to, but again was worse than 95% of the population in her number of incorrect responses selected.

Visual attention under distraction that required verbal rather than motor response (Stroop) was low average.

Sensorimotor functions. The patient displayed several abnormalities on bedside neurological tests, including bilateral intention tremor on finger-to-nose testing, worse in the right hand. The patient displayed severe impairment in attempting to execute palm-dorsum hand slaps on her knee with her right hand. She failed a tandem Romberg with her shoes off, falling almost immediately to her left.

Fine motor speed on a tapping test was moderately improved bilaterally, with the dominant hand worse than the nondominant hand. Fine motor coordination on a pegboard task was severely impaired bilaterally, again with dominant hand performance worse. Grip strength on a dynamometer was moderately impaired, with the same pattern of greater impairment of the dominant hand.

Tactile sensitivity was mildly impaired bilaterally, worse on the right, for numbers traced on the fingertips. The patient was unable to correctly distinguish one versus two points touched to her fingertip at 5 mm distance on the right hand, but correctly performed the task with her left hand.

Memory and learning. Memory for stories and abstract designs was good and unremarkable for both immediate and delayed recall. Recognition memory for both words and faces was also good and showed no evidence of deliberate exaggeration.

Visuospatial functions. Visuospatial constructional skills assessed through the assembly of block design patterns from a template were average and unimpaired.

Personality functioning. On an objective test of personality functioning (MMPI-2), the patient showed a somewhat defensive profile that is often found in educated subjects. The profile displayed is more common among women and older persons than among men and younger persons. Psychiatric patients with this code usually receive somatoform disorder diagnoses, although the test profile is not distinguishable from patients with genuine medical illness. Severe anxiety and depression usually are absent, as are clear psychotic symptoms.

Individuals with similar profiles may repress resentment and hostility toward others, particularly those who are perceived as not fulfilling their needs for attention. Most of the time they are overcontrolled and likely to express their negative feelings directly and passively, but they occasionally lose their tempers and express themselves in angry, but not violent, ways. Behaving in a socially acceptable manner is important, as is the need to convince other people that they are logical and reasonable, conventional and conforming in their attitudes and values.

Summary and Conclusions. This is a female with a variety of neurological and cognitive complaints of acute origin corresponding in time to heavy use of solvents in an unventilated artist's studio. Current neuropsychological evaluation shows moderate to severe bilateral impairments in fine motor speed and coordination, and tactile sensation. Most indicators were worse in the dominant (right) hand. In combination with impairment on tests sensitive to attention and diffuse cortical processing, the patient's results suggest both central and peripheral components. Both cerebellar and cerebral functions appear to be at least mildly impaired, with cerebellar functions apparently worse. Psychological factors do not explain the nature and extent of current neuropsychological deficits.

The relationship of neuropsychological symptoms to solvent exposure is likely, although the relative severity of sensory-motor and cerebellar symptoms seems unusual for a relatively brief exposure. However, if the patient was indeed working in a completely unventilated studio with large amounts of solvents, it is possible that her exposure was extremely high over that time. The patient's symptoms appear most similar to those seen with perchloroethylene and mixed solvent exposure and apparently involved both inhalation and direct dermal contact. Before final diagnosis is made, neurological evaluation of other similar nervous system disorders should be explored, e.g., multiple sclerosis, unrelated cerebellar disease. It is unlikely Dursban® exposure contributed to the patient's current symptoms, as cholinesterase levels were within normal limits. Similarly, lead levels were normal and did not reflect significant exposure. Considering the recent history of cognitive and behavioral improvement reported coincident with removal from solvent exposure, it appears that extremely heavy solvent exposure is the most likely explanation for the patient's recently developed symptom constellation.

Diagnosis (DSM-III-R)

I. Organic Mental Disorder Not Otherwise Specified: Solvent Encephalopathy with PNS and CNS involvement
II. No diagnosis on Axis II

Recommendations

- Filter studio air so fan can be run continuously; open windows.
- Replace all studio carpets with ceramic tile or other nonporous flooring.
- Consider nerve conduction studies to determine extent of PNS involvement.
- Avoid further solvent exposure and switch to non-solvent-containing painting and cleaning media.

Long-Term Chronic Exposure

Because industrial painters or paint manufacturers are more likely to have chronic exposure to solvent mixtures than most other occupational groups, many recent neuropsychological studies have used workers in these occupations as subjects. Most of these studies have found neuropsychological decrements that seem related to chronic solvent exposure. For example, a recent South African study examined workers of a paint factory with greater than 5 years' solvent exposure. After age, education, and alcohol consumption were forced into the regression model, average lifetime solvent exposure significantly predicted performance on four NES-2 tests: continuous performance latency time, switching attention latency time, mean reaction time, and pattern memory.

Painters exposed for a mean of 6.7 years to approximately "several hundred" parts per million of hydrocarbon solvents showed deficits in visuospatial functions and psychomotor coordination (Hane *et al.*, 1977).

Lee and Lee (1993) studied a group of high- and low-exposure solvent workers who had worked an average of 7.8 years, and compared them with nonexposed controls. Solvent workers performed significantly more poorly than controls on the Benton Visual Retention (BVR) Test, Digit Symbol, and the Santa Ana Dexterity Test (left hand), but when age, alcohol consumption, and smoking were entered into the model, only the BVR remained significant, while Digit Symbol became marginally significant.

Lindstrom (1980) studied a population of painters, dry cleaners, degreasers, and others exposed for a mean of 9.1 years, and specifically referred for previously diagnosed occupational disease. Relative to controls, subjects performed significantly more poorly on Digit Span, Digit Symbol, Block Design, and several visuoconstructive tasks, a pattern characterized by the author as lowered visuomotor performance and decreased freedom from distractibility (Lindstrom, 1980).

Eskelinen, Luisto, Tenkanen, and Mattei (1986) investigated a group of 21 individuals already diagnosed as being in various stages of an organic solvent syndrome. All but one of these patients had been exposed to solvent mixtures. For the latter patient the mono-exposure solvent was styrene. Comparison groups included 16 patients with vertebrobasilar insufficiency, 16 patients with miscellaneous cerebral traumata, and 15 tension headache patients. None of the comparison groups had solvent exposure histories.

The groups were given a neuropsychological battery developed at the Institute for Occupational Health in Helsinki, which consisted of various items from the Wechsler Adult Intelligence Scale (WAIS), the Wechsler Memory Scale (WMS), tests of fine motor speed and coordination, and several spatial-constructional tasks (the Mira Test and the Symmetry Drawing Task). A symptom questionnaire was also included.

The results of the study suggest that workers with chronic solvent exposure show a specific pattern of neuropsychological impairment that differs from patients with other forms of PNS or CNS impairment, as well as from tension headache controls who might be expected to show similar emotional symptoms. A discriminant analysis performed by the authors suggested that the best combination of tests to distinguish solvent patients from other groups included Similarities, Block Design Speed and No. of Failures, Digit Span—forward, and the Visual Memory subtest of the WMS. Five of the nine subjective symptom scales were included in the model for best power. These included "memory difficulties, fatigue, sleep disturbances, neurovegetative symptoms, and headache" (Eskelinen et al., 1986, p. 249). The study suggests that chronic solvent exposure patients may have a unique diagnostic profile, and that their impairments in intellectual abilities and memory rather than spatial or sensorimotor dysfunction distinguish them from other neurological patients. However, since motor tasks were impaired relative to the only nonneurological group (headache patients), solvent-exposed patients may still be said to be impaired in these functions relative to normals.

The study can be criticized for its use of small numbers of patients, using solvent patients with a slightly lower premorbid educational level, and for its failure to include a matched control group of normals without medical or psychological symptoms. It is likely, therefore, that future replication studies may observe different discriminative patterns than those found here. Nonetheless, since results found by the authors are consistent with known solvent effects, it is also possible that basic neuropsychological impairments shown here may be real and not artifacts of the study design.

A mixed group of Chinese spray painters, printers, and paint manufacturing workers exposed to solvents for an average of 9.4 years were investigated by Ng et al. (1990). Significant differences between workers and age- and alcohol-matched controls showed that workers performed significantly more poorly on the WAIS-R Digit Symbol Test, a test of choice reaction time, Digit Span, and WMS Associate learning. WAIS-R Block Design and Logical Memory scores were marginally significant ($p < 0.09$). Subjects, however, had one year less education than controls, so it is unclear how much educational differences influence scores.

Kishi, Harabuchi, Katakura, Ikeda, and Miyake (1993) studied 81 painters from the Japan Railway system and two additional painting shops, where mean solvent exposure duration was 14.2 years (range 1–43 years). Half of the workers were exposed more than 5 h per day and more than 21 days per month. The majority were exposed to toluene, but more than 25% were also exposed to xylene, mineral spirits, methanol, lacquer, or gasoline. Twenty of the workers (mean age 39.5) were examined neuropsychologically. Multiple regression analysis that controlled for age, education, and alcohol intake, found a significant relationship between exposure duration and poor performance in Block Design and Digit Span tests. However, when test scores were additionally regressed against vocabulary, all other associations vanished except for Santa Ana and a high Confusion score on the Profile of Mood States.

Another study used only a questionnaire to query car painters who had been exposed for a mean of 14.8 years, longer than previously described patients. The workers described themselves as being excessively tired after work, and having difficulty with concentration and memory. They were also significantly more bothered by itching, and admitted to prenarcotic acute symptoms during work including nausea, vomiting, feeling drunk, and being absentminded (Husman, 1980).

Orbaek *et al.* (1985) investigated workers in a paint production plant exposed to solvents for an even longer period—a mean of 18 years. Compared with controls matched for age and education, solvent workers showed a significant 4% decrease in cerebral blood flow and EEG changes. Neuropsychological tests that measured sustained attention and concentration were impaired in the subgroup with the greatest solvent exposure. Solvent-exposed workers also admitted to significantly more neurological and neuropsychological symptoms on a 60-item rating scale, including inner tension, hostile feelings, and short-term memory problems.

Ekberg, Barregard, Hagberg, and Sallsten (1986) surveyed general health and neuropsychological performance of solvent-exposed floor layers with age-matched carpenters who had long (> 20 years) or short (5–10 years) relevant work experience. Floor layers experienced depressive illness and received antidepressant medication significantly more often than carpenters. The authors also noted small but significant associations between neuropsychological performance on several tests, and exposure to contact adhesives or alcohol-based glues after age effects were controlled. Unfortunately, there was no information regarding relative educational attainment between groups. It may also be argued that 5–10 years of exposure to relevant substances may not be an appropriate control, since subclinical solvent effects may have begun to appear by that time.

Painters exposed for a mean of 22 years to white spirit mixtures were compared with matched controls in the construction industry and showed significantly lower performance in reaction time, Block Design, Digit Symbol, Figural Memory (Wechsler Memory Scale), and visuo-constructive tests (Lindstrom &

Wickstrom, 1983). Arlien-Søberg et al. (1979) studied a group of painters with a 27-year mean exposure selectively referred for evaluation (like Lindstrom, 1980), because of suspected organic solvent intoxication or dementia. They found 39 out of 50 patients to suffer neuropsychological impairment, after ruling out alternative etiologies. CT or pneumoencephalogram (PEG) showed cerebral atrophy in 31 of these cases with significantly widened sulci in 38 of the 50. Deficits of many neuropsychological functions were found in these individuals, including Visual Memory, Digit Span, Paired Associate learning, and Sentence Repetition.

Several studies suggest that high-technology workers are not immune from neurotoxic exposure effects. One such study examined workers in a videotape manufacturing facility who were exposed to mixed solvents, including MEK, cyclohexanone, tetrahydrofuran, and toluene (Chia, Ong, Phoon, Tan, & Jeyaratnam, 1993). Workers were tested after a weekend to limit acute effects and mean exposure duration was only 40.2 months. Controls were matched for age and education, but IQ level was not assessed. Workers had a significantly higher prevalence of headache, nose and eye irritation compared with controls. Exposed workers also did significantly more poorly than controls on a complex coordination task (Santa Ana—Both Hands), Digit Span, and the Figural Memory cards from the Wechsler Memory Scale.

Workers in the microelectronics industry may also be at risk for mixed solvent-related neuropsychological deficit. Bowler et al. (1991) studied 67 workers who were formerly involved in electronic assembly and exposed to a variety of solvents, including chlorofluorocarbons, trichloroethane, trichloroethylene, and xylene, and compared them with a matched control group who had not worked in the electronics industry. Exposure had been for an average of 6.7 years. Both groups were administered a flexible battery (CNS-R) constructed by the authors and containing the Wechsler Memory Scale-Revised, subtests of the WAIS-R, Trails, a cancellation task ("Cancellation H"), Stroop, and several motor tasks. Bonferroni α corrected t tests showed significantly lower performance by the former workers on WMS-R indexes, Trails, cancellation, Digit Symbol, Stroop, and tests of motor speed and strength. Unfortunately, several methodological problems mar the interpretation of the study, most notably failure to adjust for significant vocabulary differences favoring the controls, and the fact that the experimental cohort had completed successful workers' compensation litigation. In addition, the use of workers who had previously been evaluated by an occupational health clinic for possible work-related illnesses, suggests the possibility of a *sick worker effect,* i.e., the cohort examined may have been especially vulnerable to neurotoxic insult. Nonetheless, the apparent uniformity of the Bowler et al. results merits follow-up with a population of currently employed, nonlitigating, microelectronics workers and better matched controls.

More subtle neuropsychological impairments were noted in solvent workers across several studies by Mergler and her colleagues (e.g., Mergler et al., 1990, 1991). In the 1991 study, for example, the authors used a test of desaturated color perception (Lanthony D-15) on microelectronics workers exposed to a variety of solvents and alcohols and found that 18 of 23 workers showed color perception deficit (dyschromatopsia), a pattern that was not found in matched controls. Dyschromatopsia may not be found, however, when solvent exposure remains below Western occupational exposure limits (Nakatsuka *et al.,* 1992).

The preceding studies, while using diverse procedures for patient selection and neuropsychological evaluation, nevertheless appear, in aggregate, to support a syndrome of deficits related to solvent exposure. Critics, however, have countered that the Scandinavian literature may not be replicable in U.S. and European settings for a variety of sociocultural reasons. For example, several Scandinavian countries view solvent syndromes as de facto grounds for disability pensions, while procedures to prove solvent damage favor the worker. In Denmark, for example, the state must prove that a worker does *not* suffer from solvent poisoning for a disability claim to be denied. In addition, widespread publicity concerning details of solvent intoxication syndromes, as well as the difficulty of *disproving* a syndrome, may serve to encourage claims (A. Gade, personal communication).

Increasingly, however, studies that support the existence of organic solvent psychosyndromes are becoming available outside of Scandinavia. Recent supportive research in the United States was done by Linz *et al.* (1986), who compared neuropsychological results on 15 industrial painters who worked extensively with solvents to 30 controls, at the Oregon Health Sciences University. Painters were impaired on a variety of neuropsychological measures, including Trails A, Digit Symbol, Seashore Rhythm, and Halstead–Reitan Speech Sounds Perception tests. Tests of memory, learning, construction, and abstract reasoning were also impaired relative to controls. Twelve out of fifteen painters had HRB impairment indexes > 0.5 with painters also reporting significantly increased numbers of neurological symptoms on a questionnaire, including decreased coordination, personality change, and decreased memory. Neurological evaluation showed 4 of the 15 painters to have mild distal neuropathy, 5 to have sensorimotor peripheral neuropathy measured by EMG and nerve conduction studies, and 3 to have abnormal EEGs.

Another U.S. study examined 74 male shipyard painters who were exposed to solvent compounds used in primers, shellac, varnishes, and enamels. History included exposure in confined areas to "methyl isobutyl ketone, xylene, perchloroethylene, ethylene glycol and mineral spirits" (Valciukas, Lilis, Singer, Glickman, & Nicholson, 1985, p. 48). Compared with a control group carefully matched by age, education, race, and sex, the painters performed significantly more poorly on the Block Design Test and an embedded figures test that required subjects to identify superimposed outline drawings of common objects. In addition, painters who admitted to chronic solvent-related symptoms in a questionnaire performed significantly more poorly than asymptomatic individuals on Block Design, Digit Symbol, and an overall index of performance (Valciukas *et al.,* 1985).

Morrow, Ryan, *et al.* (1990) reported the neuropsychological test results of 32 solvent-exposed workers who presented for the first time at the Occupational Medicine Clinic at the University of Pittsburgh over a 3-year period. Trichloroethylene and toluene were the most frequently reported solvents, although all workers had been exposed to multiple solvents. Mean exposure duration was 9.0 years (range 1 month–19 years). Almost all neuropsychological measures from their POET screening battery were significantly poorer in the exposed group compared with age- and education-matched controls and four of five cognitive factors were significantly more impaired in the exposed group. Impaired factors included: learning and memory, visuospatial, psychomotor speed and manual dexterity, and attention/mental flexibility. A factor labeled General Intelligence did not differ between the two groups.

One of the more recent U.S. studies of neuropsychological effects of solvent exposure was able to provide good estimates of individual solvent exposure while controlling for age, race, alcohol use, vocabulary, and smoking. Bleecker, Bolla, Agnew, Schwartz, and Ford (1991) examined 187 workers in two paint manufacturing plants, to determine possible effects of mixed solvent exposure. None of the psychiatric structured interviews (Zung, Present State Examination) correlated with exposure, but neuropsychological measures and vibration threshold correlated with increasing exposure. Exposure intensity correlated with digit symbol substitution, serial digit learning, a measure of reaction time, trail-making tests, and vibration sensitivity in the toe. More florid "psycho-organic" profiles were not found.

Studies of abnormalities in solvent effects have also been shown in non-Western countries, where less of an interaction between social values and symptoms may be expected. For example, Ng *et al.* (1990) compared controls with a group of printers and paint workers in Hong Kong, who were exposed to a mixed group of solvents for a mean of 9.4 years. Compared with an age- and education-matched group, the solvent-exposed workers were significantly more likely to complain of fatigue, depression, and irritability. Neuropsychological performance on WAIS-R Digit Symbol and a choice reaction time test were also significantly worse in the exposed group.

Studies using computerized methods have also had some success in detecting solvent-related neuropsychological impairments. Spurgeon *et al.* (1992) used the Neurobehavioral Evaluation System (NES) on two solvent-using populations: brush painters and a mixed group of solvent-exposed government employees. After controlling for premorbid ability, alcohol consumption, and acute solvent exposure, significant effects were found on the NES Symbol Digit subtest. One of the groups with greater than 10 years of solvent exposure also had poorer performance on a paired associate learning test, learning fewer word pairs than controls, although short-term memory for word pairs learned was similar for subjects versus controls. Morrow, Robin, *et al.* (1992) used a computerized continuous performance paradigm with a group of workers exposed to a mixed group of solvents. Exposed subjects were slower and showed decreased d' across blocks of trials, suggesting possible early fatigue or loss of neurochemical reserves.

The studies listed above are reasonably consistent in their attribution of neuropsychological sequelae from solvent exposure. Nonetheless, criticisms have been made of cross-sectional or noncontrolled designs, where exposure estimates are approximate or unknown, and premorbid abilities of subjects must also be estimated, often from minimal data. Study designs that eliminate objections common to cross-sectional research are presently being attempted. Williamson and Winder (1993) report an ongoing longitudinal study of 200 first-year apprentice vehicle spray painters and compared longitudinally against industrial workers who are not exposed to significant amounts of solvents. Workers are assessed on critical flicker fusion, hand steadiness, simple reaction time, visual pursuit, Sternberg digit identification, paired associates memory (immediate and delayed), and a brief neuropsychiatric questionnaire.

Perhaps the strongest design for a group study of neurotoxic effects would require the subjects and controls to be monozygotic twins, differing only in exposure to solvents. Hänninen, Antti-Poiki, Juntunen, and Koskenvuo (1991) found just such an unusual group of subjects where the experimental group were exposed through various painting or gluing occupations for a median duration of 13 years (range 5–30 years). All subjects were administered a 2-h flexible battery of neuropsychological tests.

Three of seven cognitive tests were performed significantly more poorly by the exposed twins, compared with their nonexposed sibs, although degree of impairment was termed "slight." Individuals presenting clinically would not have been identified as having solvent syndrome. Impairments were found on a modified WMS associate learning test (five easy, five difficult), Block Design, and Digit Span. Psychomotor tests were inconsistent, with exposed subjects performing more rapidly in the Digit Symbol Test, but more slowly for nondominant hand Santa Ana (pegboard). Small sample size and generally low exposure levels mitigated the findings. Similar conceptual abilities on the Similarities test would suggest generally equivalent verbal manipulation skills between groups. It is to be hoped that the Hänninen *et al.* (1991) study can be replicated in other twin groups, possibly where solvent exposure is more marked.

The following cases are presented as examples of chronic solvent exposure effects. The patients' histories have been changed, and portions of the original report have been omitted to protect their confidentiality.

Case Study B: Chronic Solvent Exposure and Neuropsychological Impairment

Reason for Referral. The patient is a white male in his 50s with a history of approximately 13 years' exposure to a mixed group of neurotoxic solvents and chemicals, including toluene, xylene, aromatic hydrocarbons, butyl alcohol, ethylene glycol, monobutyl ether, and methyl ethyl ketone (MEK). Neuropsychological testing was initiated to determine whether current cognitive, affective, and behavioral performance patterns can be related to the patient's exposure history.

Tests Administered

Finger Tapping Test
Wechsler Memory Scale with Russell Modification
Trail-Making Tests A&B

TABLE 4.4. Neuropsychological Test Scores in a Patient with Chronic Solvent Exposure

1. Finger Tapping: RH mean: 41.5	Russell 3
(Dom. hand: R) LH mean: 39.2	Russell 2
2. Wechsler Memory Scale:	
(a) Logical	
Immediate recall: A: 6.5 B: 4.5 Total: 11	Russell 4
Delayed recall: A 5.0 B: 1.5 Total: 6.5	Russell 4
(b) Figural	
Immediate recall: A: 3 B: 3 C: 5 Total: 11	Russell 1
Delayed recall: A: 0 B: 0 C: 2 Total: 2	Russell 4
3. Trail-Making Test: A: 39 (seconds),	Russell 2
B: 97 (seconds)	Russell 2
4. WAIS-R Digit Symbol Test: Raw score: 41	Scaled score 6
	Age corrected: 8
5. Stroop Test:	
No. of Words: 93 + age corr.: 8 Total: 102	T score 47
No. of Colors: 51 + age corr.: 4 Total: 55	T score 33
No. of Colored Words: 25 + age corr.: 5 Total: 30	T score 35
6. Luria–Nebraska Pathognomonic Scale:	
Critical level 66.27	
No. of errors: 23	T score 61
7. Spatial Relations: (Greek Cross) Best: 1 + Worst: 2–3	Russell 1
8. Pegboard: (Grooved) RH: 81" T 39 LH: 85" T 42	
9. Grip Strength: RH: (1) 55 (2) 50 LH: (1) 48 (2) 48, (kg) Mean	
RH: 52.5 Mean LH: 48	
10. Shipley IQ estimate = 85	
11. Block Design: Raw score: 28 Scaled score: 9 (age corr.: 11)	
12. Category Test: No. of Errors: 94	
13. Associate Learning: 5;0, 5;0, 6;1	
14. Digits Forward: 8 Digits Backward: 4	
15. Personality Test: MMPI, POMS	

MMPI Scores (T scores, K corrected where appropriate):

L 53	Hy 71	Pt 74	A 49
F 55	Pd 64	Sc 69	R 74
K 68	Mt 49	Ma 45	Es 51
Hs 88	Pa 50	Si 67	MAC 34
D 85			

Note: 0–5 Russell ratings are from Russell (1975) and Russell, Neuringer, and Goldstein (1970) with higher scores indicating greater impairment.

WAIS-R subtests:
  Block Design
  Digit Symbol
Stroop Color-Word Test
Luria–Nebraska Neuropsychological Battery:
    Pathognomonic Scale
Spatial Relations Test
Grip Strength Test
Grooved Pegboard Test

Shipley Institute of Living Test (SILS)
Harvard School of Public Health Occupational Health Questionnaire
Minnesota Multiphasic Personality Inventory (MMPI)
Personal Problems Checklist
Profile of Mood States (POMS)
Kent EGY Scale D
Category Test
Clinical Interview

Background Information. The patient is a right-handed white male who reports an exposure history to a mixed group of solvents and chemicals over a period of 13-14 years. The patient reports that during this time he was responsible for cleaning a spray paint booth with either MEK or toluene. MEK was applied by hand, using a brush. In addition, the patient used toluene to clean the paint from his hands. After each application the patient states that he felt intoxicated and that his skin itched. There was no ventilation in the paint booths during cleaning as the ventilating fans were turned off. He also states that he did not have protective gloves during the application of either MEK or toluene.

The patient's current symptoms include weakness, shaking, and pain in his legs accompanied by involuntary muscle contractions (myoclonus). He reports that the pain in his legs is so intense "that it feels like a toe is about to break off," a sensation that can last from 10 to 15 minutes at a time. The patient also reports shoulder and back pain with limitation of movement. Other symptoms include numbness in the arms and shoulders.

The patient has seen many health care professionals for diagnosis and relief from these symptoms, including staff at a nationally known clinic, a chiropractor, and a psychiatrist. At the time of testing, the psychiatrist was empirically assessing the effects of Haldol on the patient for "restless leg syndrome," but the psychiatrist reports that the patient's symptoms have not diminished thus far. The patient does report, however, that his sleep has improved somewhat after being given Haldol. Medical reports were not available as of this writing; however, the patient's psychiatrist reports that the patient does have documented denervation of the left gastrocnemius, as well as spinal defects of a possibly degenerative nature in L4, L5, and S1.

Neuropsychological symptoms noted by the patient include increased fatigue and memory loss. Appetite is reported to be good with no recent changes in weight. Sleep is poor, with awakening after 2 or 3 hours. The patient reports that he occasionally gets depressed but denies suicidal ideation. There is no history of mental health treatment. There have been occasional episodes of hypertension but blood pressure is currently controlled. Current blood pressure is reported by the patient to be 120/70.

With the exception of hypertension and previously noted symptoms, the patient has no other current medical problems. He denies history of diabetes, arthritis, epilepsy, chronic neurological illness, fracture, frostbite, or orthopedic injury. Though he is a pack-per-day smoker, he denies cardiac or pulmonary disease. The patient's medical history includes a car accident where his head struck the windshield and cracked the glass; however, there was no loss of consciousness and the patient reports that he recovered fully within a few hours. The patient's current medications include: Tenormin, 1/day, Haldol 1 mg/day, and Hydrochlorothiazide 50 mg. Alcohol use is currently low, approximately six beers during the last calendar year. Starting 13 years ago, for about 10 years, the patient had a social drinking history of approximately two six-packs of beer per weekend when he played cards with friends. Current coffee intake is 4–6 cups per day.

The patient graduated from the 9th grade and he reports that his favorite subject was arithmetic and his worst was spelling. He does not remember his grades. The patient is currently employed as a storeroom clerk and has not had any contact with solvents or other chemicals for approximately 2 years.

Behavioral Observations. The patient is a heavyset, white male of less than average height. Speech was clear and somewhat laconic, spoken with accentless midwestern pronunciation. Affect appeared to be blunted or depressed and the patient's "flat" expression appeared similar to that found in patients with Parkinson's disease. The patient's responses to history-taking questions tended to be short and somewhat difficult for this examiner to organize. Many questions had to be asked to determine the relationship between time, chemical exposure, and symptoms.

The patient's behavior was appropriate at all times during the testing session. He was cooperative and appeared to try hard during all tests. He was able to maintain attention throughout the examination. The only unusual behavior displayed during the testing session was a convulsive arm twitch, triggered by a light pencil touch to the palm.

The patient denied or minimized cognitive changes during the past decade. His principal cognitive complaint was of increasingly poor memory for day-to-day information. The complaints for which he has received multiple medical consults are somatic and related to limb pain.

General Cognitive Functioning. The patient was oriented to time, place, person, and situation for the testing session. Speech was unremarkable in volume and timber, although somewhat slow in rate. There were no indications of aphasia or alexia.

The patient's estimated premorbid intellectual functioning was in the High Average range or above for most abilities. In contrast, many of the patient's abilities appear to have deteriorated from this level. During testing, the patient displayed several types of general cognitive deficits, most notably in cognitive efficiency/flexibility, and abstract reasoning ability. Each of these was moderately impaired. For example, the patient showed reduced ability to separate essential from distracting stimuli (Stroop Test) and made many errors in a task that required inductive reasoning (Category Test). In the latter test, the patient made almost double the number of errors beyond a pathognomonic cutoff level.

Memory and Learning. The patient showed moderate to severe deficiencies in memory and learning. Both verbal and nonverbal encoding were involved. Verbal memory items were poorly encoded and poorly remembered later. Educational limitations may have affected scores somewhat on this particular task. However, nonverbal memory, which is less dependent on education, was also affected. In a task requiring the memorization of line drawings, the patient was able to perform adequately only when he was asked to recall the drawings immediately (Fig. 4.1). When recall was requested 30 minutes later, the patient scored in the range of severe impairment, recalling almost nothing of the previously presented figures (Fig. 4.2).

When presented with a task requiring the memorization and learning of easy and difficult word associations (e.g., north–south, versus crush–dark), the patient was almost completely unable to learn any of the more difficult word pairs, even after three trials.

Sensorimotor Functions. Sensory and motor abilities appear to be mild to moderately impaired. Fine motor speed and dexterity were mild to moderately impaired in the dominant hand. Left hand performance was closer to normal in fine motor speed, but also shows mild decrements in dexterity. Pencil drawings of the patient appeared to suffer from a very fine tremor in their execution (e.g., Fig. 4.3). Gross motor strength was normal bilaterally.

The patient did not show any basic sensory deficits in his hands; however, when his palm was lightly touched with a pencil during a test of body location, his entire arm twitched convulsively. In addition, the patient showed mild to moderate deficits in translating proprioceptive cues into verbal descriptions. For example, he could not correctly name letters that were drawn on the back of his wrist. The patient showed no evidence of hearing impairment.

Visuospatial Functions. The patient showed no evidence of constructional apraxia; he was able to correctly draw complex figures, and could correctly name objects placed in his hand. In addition, the patient performed well in a task that required him to assemble blocks in abstract patterns.

Personality Functioning. Personality testing suggests that the patient experiences significant depression and is very concerned with his abnormal physical symptoms. He appears to cope with anxiety and tension by repression and being concerned with details rather than the "big picture." In several inventories of emotional symptomatology, the patient reported extreme inability to concentrate, and forgetfulness, as well as fatigue, uneasiness, and exhaustion.

Summary and Recommendations. This is a white male in his 50s referred by his psychiatrist to determine if the patient's symptoms were consistent with a 13-year exposure history to various central and peripheral neurotoxic solvents. Current neuropsychological evaluation suggests that the patient functions signifi-

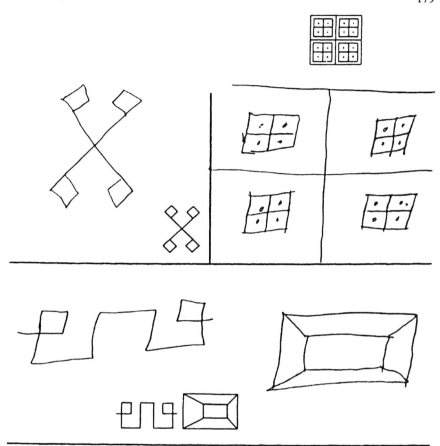

FIGURE 4.1. Chronic exposure to solvent mixtures. WMS: Figural memory—immediate recall. For reader comparison, card images were inset subsequent to the evaluation.

cantly below estimates of premorbid abilities in memory, cognitive efficiency and flexibility, and abstract reasoning. The pattern of impairment is consistent with known neurotoxic effects of some of the solvents to which the patient has been exposed. Chronic exposure to toluene, for example, has been linked to both CNS and PNS deficits including depression and flattened affect, memory impairment, and other cognitive abnormalities, including dementia. MEK is a peripheral neurotoxin, and ethylene glycol may cause sensorimotor neuropathy and CNS abnormalities.

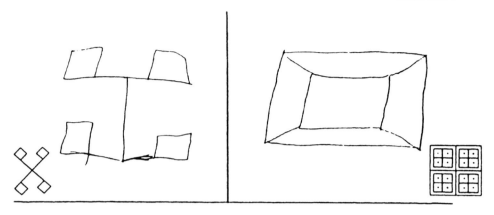

[no recall]

FIGURE 4.2. Chronic exposure to solvent mixtures. WMS: Figural memory—30-minute delayed recall. For reader comparison, card images were inset subsequent to the evaluation.

Overall, the pattern of impairment in this patient appears to be consistent with chronic exposure to neurotoxic solvents and chemicals. Prognosis for improvement after exposure to mixed neurotoxins is somewhat cautious, although if patients do not show abnormalities in CT or EEG, their prognosis is better than if such tests are positive. The prognosis for recovery of solvent-related peripheral neuropathy is controversial, and while some reviews cite eventual recovery, other authors report continuing symptoms 3 years postexposure.

Suggestions for the patient's further care include:

- Trial of antidepressant medication.
- Referral to an occupational medicine clinic where the patient can be further evaluated for solvent exposure effects.
- MR scan, PET scan, or regional cerebral blood flow study may yield further information about the level of activation in various brain areas.

SOLVENTS 181

FIGURE 4.3. Chronic exposure to solvent mixtures. Spatial Relations Test (direct copy).

Case Study C: Neuropsychological Consequences of Chronic Exposure to a Solvent Mixture of Toluene and MEK

Reason for Referral. The patient is a white male in his 30s with chronic exposure to a mixture of toluene and MEK.

Tests Administered

Finger Tapping Test
Grooved Pegboard Test
Grip Strength Test
Fingertip Number Writing Test
Two-Point Discrimination Test
Minnesota Rate of Manipulation Test
    Turning Test
    Displacing Test
    Placing Test
    One Hand Placing and Turning Test
Wechsler Memory Scale (Russell Revision):
    Logical and Figural Memory—immediate & 30-minute delayed recall
    Associate Learning Test
Wechsler Adult Intelligence Scale-Revised:
    Digit Symbol Test
    Block Design Test
    Information Test
    Vocabulary Test
Shipley Institute of Living Test
Stroop Color-Word Test
Luria–Nebraska Pathognomonic Scale
Spatial Relations Test
Minnesota Multiphasic Personality Inventory (MMPI)
Beck Depression Inventory
Personal Problems Checklist: Adult
Harvard Public Health Occupational Health Questionnaire
Psychological/Social History Form
Clinical Interview

Background Information. Prior to obtaining his current job with occasional exposure to solvents, the patient worked inside a warehouse for 7 years and was heavily exposed to solvents, primary toluene and MEK. Further details of his solvent exposure history are contained in his medical records.

During the last 2 or 3 years of exposure, the patient began experiencing concentration difficulties. He believes that a growing inability to pay continuous attention had adversely affected his simultaneous pursuit of a master's degree in public administration and caused him to take a year longer than was usual to complete his degree. Loss of concentration is a continuing problem, "I lose track of my thoughts, I can't seem to keep my mind on things." The patient also reports several "close calls" while driving.

The patient's first symptoms of muscle fatigue were partially relieved with prednisone, which he continued to receive intermittently until recently. Currently

TABLE 4.5. Neuropsychological Test Scores in a Patient with Toluene and MEK Exposure

1. Finger Tapping: RH mean: 34.2	Russell 3
(Dom. hand: R) LH mean: 34.4	Russell 3
2. Wechsler Memory Scale: Logical Memory:	
Immediate recall: A: 10 B: 6 Total: 16	Russell 3 %R=59
Delayed recall: A: 7.5 B: 2 Total: 9.5	Russell 3 %R=33
Figural Memory:	
Immediate recall: A: 3 B: 4 C: 5 Total: 12	Russell 0
Delayed recall: A: 1 B: 2 C: 1 Total: 4	Russell 3 %R=33
3. Trail-Making Test: A: 41 (seconds)	Russell 2
B: 70 (seconds)	Russell 1
4. WAIS R Digit Symbol Test: Raw score: 43	Scaled score 6
5. Stroop Test: No. of Words: 74 + age corr.: 0 Total: 74	T score 33
No. of Colors: 51 + age corr.: 0 Total: 51	T score 30
No. of Colored Words: 29 + age corr.: 0 Total 29	T score 34
6. Luria–Nebraska Pathognomonic Scale:	
68.8 + Age value: X − Education value X = Critical level 48.97	
No. of Errors: 13	T score 45
7. Spatial Relations: (Greek Cross) Best: 1 + Worst: 2-3	Russell 1
8. Associate Learning: 1)E5 H1 2)E6 H2 3)E6 H3	
9. Block Design: Raw score: 22	Scaled score 7
10. IQ Estimates	
SILS: V: Raw 30 T50, A: Raw 30 T57	
Pred. Abs. from, Age Ed. E, Vocab. 28 AQ Conversion 104	
Estimated WAIS-R Full Scale IQ from Total SILS by Age 102	
WAIS-R: Information Raw 21	Scaled score 10
Vocabulary Raw 52	Scaled score 11
11. Category Test: No. of Errors: 54	Russell 2
12. Grip Strength: RH: (1) 48 (2) 51.5 LH: (1) 46.5 (2) 46 (lb/kg)	
Mean RH: 49.75 Mean LH: 46.25	
13. Pegboard: (Grooved) RH: 89 T19 LH: 100 T12	
14. Fingertip Number Writing	

	4635	3546	6543	5463	6354
RH:	7634	2590	6543	5792	6354
LH:	9284	8548	6543	8928	6354

15. Personality Test
MMPI: (T scores)   L t62, F t66, K t48, Hs t77, D t94, Hy t74, Pd t48, Mf t57, Pa t62, Pt t73, Sc t76, Ma t50, Si t78, A t64, R t54, Es t33, Mac t41
16. Other: 1-2 point touch discriminations: pt. could not perform accurately, even at 20 mm, compass point distance

the patient states he would "rather work through muscle fatigue by forcing yourself to go on" than worry about side effects from continuing the medication. This year, he has again experienced a recurrence of fatigue symptoms. He reports "waking up tired as much as when I ended the day." The patient also states that his "forearms get real tired when I use a pipe wrench at work" and that he becomes winded after climbing two flights of stairs. He also reports that he has difficulty keeping his eyelids open and that they have a tendency to droop. Other symptoms include occasionally blurred vision, especially in the right eye, and pain in the left knee the cause of which was reported to the patient as polyneuropathy.

The patient has attempted to build up his endurance with a home exercise program that includes "push-ups, sit-ups, windmills, and 15 minutes on an exercise bike." Despite having exercised in this manner for 3 years, the patient reports little increase in capacity, stating that he "feels tired and dead" on the days that he exercises.

The patient denied any episodes of fatigue prior to working at his company. He reports that he spent 4 years in the service and never had a problem with muscle fatigue.

The patient currently receives Synthroid subsequent to diagnosis of hyperthyroidism in 1981 and thyroidectomy. He reported that his thyroid levels were normal at last check.

The patient denied any other ongoing medical problems, and in fact, stated that prior to his employment he "never went to a doctor except once as a kid and once in the service." The patient is a nondrinker with no history of drug abuse. There was no history of head injury, hand injury, psychiatric disorder, or any other problems capable of interfering with neuropsychological performance.

The patient is married and lives with his wife and their two children. While he reports that his marriage is "fine," the patient also admits that his sex life has deteriorated, not because he is impotent, "but because I'm so tired all the time."

Behavioral Observations. The patient is a white male of heavy build and shorter than average height. He was casually but neatly dressed for the testing session in slacks and a blue oxford shirt, giving him a somewhat collegiate appearance. The patient's grooming and hygiene were excellent. The patient was oriented to person, place, time, and situation and appeared to cooperate fully with all phases of testing. The patient's display of effort was particularly evident in a sustained performance task (Minnesota Rate of Manipulation Test) which appeared to become unusually tiring over time.

Affect was calm throughout the evaluation, with no evidence of depression or anxiety. In the interview, the patient presented his symptoms in a laconic, straightforward manner that did not suggest exaggeration. In fact, the patient appeared to underestimate the degree of fingertip sensory deficit later elicited in examination.

General Cognitive Functioning. The patient's estimated premorbid intellectual abilities were in the Average to High Average range. There were significant discrepancies between verbal school-learned abilities and nonverbal skills. The patient's vocabulary and recall of school-related information were in the average range. In contrast, his ability to perform tasks that require nonverbal, spatial or constructional skills was significantly impaired and well below estimates of premorbid function. Verbal and nonverbal abstract reasoning abilities were also significantly impaired considering the patient's level of education.

The patient also performed significantly below expectation on several tests of sustained attention and concentration.

Sensorimotor Functions. The patient showed moderate to very severe bilateral impairments in sensory and motor abilities of the hands. For example, fingertip sensation was severely reduced with the patient being unable to discriminate sharp versus dull stimulation. Nor was he able to reliably differentiate one- versus two-point touch on the middle finger, even when the points were 20 mm apart. Fingertip number writing also showed severe impairment. The only consistently correct number identifications were on the patient's fifth finger, bilaterally. Identification of numbers traced on all other fingers was severely impaired. Identification of larger objects in the hand was more successful, although the patient performed these identifications at a slower than average rate.

Motor functions were similarly decremented. Fine motor speed was in the range of moderate impairment, with the patient failing to show the expected preference in speed for the dominant hand. Fine motor dexterity was also quite limited; the patient's performance on a pegboard task placed him in the bottom 20% of the population. Motor tasks requiring extended use of the hands and arms showed even poorer performance. On a task requiring approximately 10 minutes of speeded and coordinated hand/arm movement, the patient scored in the lowest 1-2% of the normative population sample.

Gross motor strength was approximately normal.

Memory. The patient showed mild to moderate impairments in short-term memory. Most notable was the loss of nonverbal information over the course of 30 minutes when tested by surprise recall. While immediate recall was quite good, after 30 minutes, the patient was only able to recall about one-third of the visual information previously presented. Verbal recall was moderately impaired for both immediate and 30-minute delayed recall.

When verbal material was repeated presented, the patient was able to retain information adequately. His learning curve on a verbal paired-associate task was normal.

Visuospatial Functions. The patient showed no significant deficits in simple constructional tasks; however, when spatial tasks included an abstract reasoning component (i.e., assembling abstract block patterns to match a template) the patient's performance was below average.

Personality Functioning. While the patient admitted to several difficulties in social adjustment, most of his current difficulties appeared to be reactive to the stress of ongoing litigation, and concerns about his current and future health. He admitted to anxious and depressed feelings and was very worried about his physical health. Perhaps as a result of these stressors, the patient is introverted and suffers from low self-esteem. The patient appeared to be coping with his difficulties by concentrating on individual details and may not have either the preference or emotional resources to plan a more integrated life strategy at this time.

Summary and Conclusions. This is a white male who has been referred for evaluation of neurobehavioral status secondary to chronic solvent exposure. At the present evaluation, the patient experienced severe sensorimotor deficits that were most apparent when performance had to be sustained over several minutes. The patient also appeared to have significantly impaired sensory thresholds in his fingertips; he was unable to discriminate sharp versus dull, one versus two points, and numbers written on the fingertips. Verbal short-term memory showed moderate impairment for immediate and delayed recall, while nonverbal short-term memory is good initially but shows significant loss over time. Other significant deficits relative to the patient's estimated premorbid abilities included abstract reasoning and the ability to sustain attention and concentration.

The combination of central and peripheral deficits shown here is quite consistent with the effects of chronic exposure to central and peripheral neurotoxic solvents (i.e., toluene and MEK) and the patient's symptoms have been documented in the solvent literature. The patient's deficits do not seem easily explainable by any other mechanism, and they are not consistent with abnormalities related to thyroid hormone level.

The majority of the patient's emotional symptoms are consistent with reaction to sustained stresses of poor health and subsequent litigation, although exacerbation of preexisting personality traits cannot be ruled out. Supportive counseling may help improve coping and lessen the patient's psychological concomitants of his physical injury.

Diagnosis

- Sensorimotor, cognitive, and memory impairments probable solvent-related etiology.
- Adjustment disorder with mixed emotional features.

SOLVENT ABUSE

Intentional inhalation of conscious-altering substances may have begun with the Greeks at Delphi, where a seer is reported to have foretold the future while inhaling carbon dioxide. Columbus, landing in the Americas, was said to have observed West Indians inhaling various psychotropic substances. Solvents became the inhaled euphoriants of choice during the late 1800s and early 1900s at parties where alcohol, ether, nitrous oxide, or chloroform were inhaled (Giovacchini, 1985, p. 18). In the mid-1900s, tons of ether were consumed as a substitute for alcohol in Ireland (Barnes, 1979). Today, solvent abuse is a problem of epidemic and international scope. In Mexico City, there are about 500,000 children addicted to various substances, most commonly to organic solvents because of their low cost and euphoriant effect. Children as young as 4–7 years old have been reported with "marked addiction" to organic solvent inhalation (Montoya-Cabrera, 1990).

In the United States, the typical solvent abuser is young, male, and between the ages of 7 and 17. Solvents are inhaled out of boredom, or to reduce stimulation from an environment perceived as hostile. These abusers are also likely to show delinquency, low self-esteem, poor academic records, and social isolation. While many are from low-income minority groups, middle-class abusers have also been reported. Estimates of adolescent abuse rates range from 2.3 to 50% (Giovacchini, 1985). Solvent abuse is more prevalent in some populations than others, e.g., very high use of gasoline inhalation has been reported in native American populations, and Hispanics tend to be overrepresented among solvent sniffers (Barnes, 1979). Gasoline inhalation, whether leaded or unleaded, can cause neuropsychological impairments. Prejean and Gouvier (1994) report the case of a polysubstance abuser who inhaled unleaded gasoline exclusively on a daily basis for over 1 year. The patient was said to demonstrate deficits in tactile sensory perception and tactile memory, with unremarkable psychomotor performance. The patient refused to complete the Category Test because of a high error rate. Significantly poorer verbal IQ compared with performance IQ was noted although it is unclear whether this was related to premorbid deficiencies.

Solvents are typically placed in a plaster bag and inhaled; soaked on a piece of cloth which is then sniffed; or inhaled directly from the original container (Barnes, 1979). Unlike other forms of adolescent rebellion or experimentation, solvent inhalation can be fatal. Bass (1970) reports 110 deaths related to inhalation of trichloroethane and fluorinated hydrocarbons, possibly from cardiac toxicity. It has been estimated that solvent abuse currently causes at least 100 deaths per year of young abusers (Cohen, 1978). This has not, unfortunately, diminished its popularity as an intoxicant, and young abusers seek out and experiment with an astonishingly diverse variety of substances containing solvents. These include substances shown in Tables 4.6, 4.7, and 4.8.

It would be difficult in some cases, and impossible in others, to substitute nontoxic alternatives for the solvents in these ubiquitous products. Thus, the problem of solvent abuse requires more than careful labeling and judicious product development by industry. Professionals in medicine, clinical psychology, law, and education will need to work together on the problem of solvent-related chemical dependency. Neuropsychologists may contribute to this effort in both research and clinical capacities; by performing basic neurobehavioral solvent research, and by clinically detecting, monitoring, and suggesting rehabilitation strategies for solvent-abusing individuals.

Acute Effects

Solvent vapors are rapidly absorbed into the lungs and transferred to the brain and other organs with high lipid content. Symptom onset is within seconds.

TABLE 4.6. Substances Containing Abusable Solvents[a]

Aerosol cocktail chillers	Degreasers	Aerosol pain relievers
Nonstick frying pain aerosols	Wax strippers	Aerosol fly sprays
Cold-weather car starters	Lighter fluids	Aerosol antiperspirants
Contact cements	Nail polish removers	Hair sprays
Dry-cleaning fluids	Aerosol shoe polishes	Aerosol paints
Transmission fluids	Refrigerants	Paint lacquers and varnishes
Windshield washer fluids	Sanitizers	Typewriter correction fluids
Brake fluids	Polishes	Anticough aerosols
Fire extinguisher fluids	Disinfectants	Room deodorizers
Liquid waxes	Aerosol deodorants	Gasoline

[a] Reprinted, by permission of Academic Press, from Giovacchini, R. P. (1985). Abusing the volatile organic chemicals. *Regulatory Toxicology and Pharmacology*, 5, 18–37.

While peak blood levels occur between 15 and 30 min postinhalation, peak tissue levels occur somewhat later (Linden, 1990). Acute toxic reactions to solvent abuse occur through anoxia because solvent vapors decrease partial pressure of oxygen. Neurotoxicity appears related to alteration in the neuronal membrane, rather than degradation of neurotransmitter function. Aliphatic and aromatic hydrocarbons, alkyl halides, and ketones are the most closely associated with permanent neurotoxic damage (Linden, 1990). Direct damage to pulmonary structures may further worsen hypoxemia. Other mechanisms of injury via acute exposure include secondary effects of cardiovascular depression, respiratory depression, impaired liver and kidney function with respect to the alkyl halides (especially those containing chlorine), and release of carbon monoxide (in the

TABLE 4.7. Constituents of Common Solvent-Containing Products[a]

Product	Solvent
Aerosols	Dichlorodifluoromethane (propellant 12), trichlorofluoromethane, isobutane
Fingernail polish	Acetone, aliphatic acetates, benzene, alcohol
Gasoline	Petroleum hydrocarbons, paraffins, olefins, naphthenes, aromatics (benzene, toluene, xylene)
Household cements	Toluene, acetone, isopropanol, methyl ethyl ketone, methyl isobutyl ketone
Lacquer thinners	Toluene, aliphatic acetates, methyl, ethyl, or propyl alcohol
Lighter fluid/cleaning fluid	Naphtha, perchloroethylene, carbon tetrachloride, trichloroethane
Model cements	Acetone, toluene, naphtha
Paint strippers	Methylene chloride
Plastic cements	Toluene, acetone, aliphatic acetates (e.g., ethyl acetate, methyl-cellosolve acetate)
Typewriter correction fluid	Trichloroethane, trichloroethylene

[a] Reprinted, by permission of PMA Publishing Corp., from Goetz, C. G. (1985). *Neurotoxins in clinical practice*. New York: SP Medical and Scientific Books.

TABLE 4.8. Solvent Classes and Exemplars with Potential for Abuse[a]

Structural class/exemplars	Principle source/use
Aliphatic hydrocarbons: methane, ethane, propane, butane, isobutane, petroleum ether, benzine, hexane, naphtha, gasoline, mineral spirits, kerosene, jet fuel	Dry cleaning fluid, spot remover; solvents for grease, plastic, and paint
Alkyl halides: monochloromethane (methyl chloride), dichloromethane (methylene chloride), monochloroethane (ethyl chloride), trichloromethane (chloroform), 1,1,1-trichloroethane (methyl chloroform), 1,1,2-trichloroethane (TCE; vinyl trichloride), trichloroethene (trichloroethylene), tetrachloromethane (carbon tetrachloride), 1,1,2,2-tetrachloroethane (acetylene tetrachloride), tetrachloroethylene (perchloroethylene), dichlorodifluoromethane (Freon 12), dichlorodibromomethane (Freon 12-B2), dichloromonofluoromethane (Freon 21), trichlorofluoromethane (Freon 2 & 11), trifluorobromomethane (Freon 12-B1), trichlorotrifluoroethane (Freon 113)	Aerosol propellant (freons), dry cleaning fluid, spot remover, fire extinguisher propellant and refrigerant fluids (freons), skin anesthetic (ethyl chloride); solvents for: adhesives, cements, glues, cements, oils, waxes, solvent-based typewriter correction fluid (TCE)
Alkyl nitrites: propyl/isopropyl nitrite, butyl/isobutyl nitrite, amyl/isoamyl nitrite	Cyanide antidote kit (amyl nitrite), jet fuel, room deodorizer, vasodilator (amyl nitrite)
Aromatic hydrocarbons: benzene, methylbenzene (toluene), dimethylbenzene (xylene), trimethylbenzenes, tetramethylbenzene (durene), ethylbenzene, propylbenzene, isopropylbenzene (cumene), butylbenzene, vinylbenzene (styrene)	Solvents for: acrylic paints, adhesives, cements, glues, indelible ink, plastics
Ethers: dimethyl ether, diethyl ether (ether), diisopropyl ether, dibutyl ether, divinyl ether, methyl propyl ether	Gasoline engine primer, rocket fuel, solvents for: alkaloids, dyes, esters, gums, lacquers, oils, paints, plastics, resins
Ketones: dimethyl ketone (acetone), methyl ethyl ketone (MEK; butanone), methyl *n*-propyl ketone (MPK; 2-pentanone), methyl isopropyl ketone (MIK; 3-methyl-2-butanone), methyl *n*-butyl ketone (MBK; 2-hexanone), methyl *n*-amyl ketone (MAK; 2-heptanone), ethyl *n*-butyl ketone (3-heptanone)	Nail polish remover (acetone); solvents for: adhesives, cellulose derivatives, fats, gums, lacquers, oils, resins, varnishes

[a]After Linden (1990).

case of methylene chloride and similar compounds, e.g., chlorobromomethane). The global oxygen depriving effects of CO produced by these solvents make them systemic asphyxiants. Aromatic hydrocarbons, especially benzene, may induce brain–behavior abnormalities through secondary damage to bone marrow, kidneys, muscle, and liver (Linden, 1990).

Symptoms of solvent inhalation abuse have been partitioned into four stages (Wyse, 1973). Initial intoxication is characterized by euphoria, excitation, dizziness, visual and auditory hallucinations. By the second stage, CNS depression has set in and symptoms include "confusion, disorientation, dullness, loss of self-control, tinnitus, blurred vision, diplopia" (Barnes, 1979). By stage 3, increasing CNS depression may produce sleepiness, incoordination, ataxia, dysarthria, diminished reflexes, and nystagmus. In the fourth stage, seizures and EEG changes occur (Barnes, 1979). Paranoia, tinnitus, and bizarre behaviors may be noted. Symptoms of CO poisoning may be noted with methylene chloride intoxication.

Chronic Effects

Chronic solvent abusers present with a variety of symptoms that are not related to acute intoxication. Weakness, anorexia, fatigue, weight loss, paresthesias, and a variety of psychiatric symptoms may be evident. Neurological exam may indicate cerebral, cerebellar, muscular, or optic nerve atrophy, nerve deafness, as well as a "stocking-glove" sensorimotor peripheral neuropathy (Linden, 1990).

The two solvents most commonly noted to produce neuropsychological impairments are methyl butyl ketone and toluene. The former has been linked to the development of peripheral neuropathies while its effects on cortical functioning are unknown. Several studies of toluene abuse suggest that chronic exposure can cause structural brain damage to the cerebellum and cortex (e.g., Fornazzari *et al.*, 1983) and the thalamus (Grigsby, Rosenberg, Dreisbach, Busenbark, & Grisby, 1993). Continued toluene inhalation produces a variety of neurological and neuropsychological toxic effects, including ataxia and dementia.

Allison and Jerrom (1984) used neuropsychological measures to examine a group of ten subjects, average age 15, who had a 4.6-year mean history of solvent abuse. The type of inhalant used by this group of Scottish adolescents was "EVOSTIK," a combination of toluene and acetone. The neuropsychological battery consisted of the WISC Vocabulary and Block Design tests, the Wechsler Memory Scale (WMS) with immediate and 10-minute delayed recall of the Logical and Figural Memory subtests, and the Paced Auditory Serial Addition Task (PASAT). Compared with a control group matched on age, socioeconomic status, and education, the solvent abusers scored significantly more poorly on Block Design, the PASAT, and the WMS, the latter test showing impairment on all subtests except Information and Orientation. The PASAT score of the inhalers was 65%, significantly poorer than the 25% error score obtained by controls.

The study has some shortcomings in design, among them lack of verification as to whether subjects continued to use solvents at the time of testing and the fact that the psychologist testing the adolescents was not blind to the purpose of the study, or the group to which each subject belonged. The results are, however, consistent with other studies of solvent abuse.

Another, better controlled, investigation examined the neuropsychological performance of 37 chronic solvent abusers, and did control for acute effects by eliminating any individual with positive urine toxicology screen from the study.

Testing was conducted by a technician blind to the group membership of the subjects (Berry, Heaton, & Kirby, 1977). Mean exposure duration was 5.5 years, average age was 18.3, and nationality was mixed among Anglo, Hispanic, and American Indian. Controls were 11 subjects described as peers or siblings of the experimental subjects, matched on age, ethnicity, education, sex, and other sociodemographic variables.

An extensive battery of neuropsychological tests was administered, including the WAIS, the Halstead–Reitan, as well as various memory and motor tests (e.g., Grooved Pegboard). Approximately half of the neuropsychological battery was performed significantly more poorly by the solvent abusers, including Verbal and Full Scale IQ. The Halstead Impairment index was also significantly more impaired in the solvent group, although both groups were under pathognomonic cutoffs established by Reitan. Memory was found to be impaired on the Tactual Performance Test (TPT) and a Story Memory Task. Hand Dynamometer, Maze Coordination Errors, and total time per block on the TPT were also impaired.

While not guilty of the same inadequacies as the previous study, the study can be criticized for not specifying the types of solvents under study, and apparent failure to correct for multiple comparisons when using t tests. In addition, verbal, rather than performance, IQ was more impaired in the experimental group, a finding that could suggest lower premorbid functioning in the impaired group relative to controls. Lower premorbid functioning could also have predisposed these individuals to solvent abuse rather than solvents causing current impairment (see the section on Toluene later in this chapter).

Inhalation abuse of other solvents has not been the subject of neuropsychological studies at this time.

INDIVIDUAL SOLVENTS

While solvent mixture researchers often consider solvent effects as part of a unitary "solvent syndrome," other investigators have attempted to find unique patterns of neurological and neuropsychological damage from individual solvents. Researchers who work with individual solvents state that "exposure data indicate that greater differences exist in the neurotoxic potential of even the commonest industrial solvent" (Savolainen, 1982) as evidence that solvent exposure should not be considered a unitary concept. Experimental studies have validated that a wide variation of acute neurotoxicity exists among different solvents (Frantik, Hornychova, & Horvath, 1994). It is logical to conclude that human neurotoxic outcomes may vary considerably among different types of solvents.

Several other solvents have been documented to cause direct structural damage to the nervous system, including carbon disulfide, styrene, toluene, trichloroethylene, *n*-hexane, and methyl butyl ketone. Spencer and Schaumburg (1985) have suggested that several other compounds be considered as putative human neurotoxicants, including diethyl ether, ethylene chloride, nitrobenzene, pyridine, tetrachloroethane, perchloroethylene, and xylene, and that further research be performed to determine possible neurotoxicity of the chromogenic aromatic hydrocarbons (e.g., benzene, indane, tetraline, diphenyl) and methylene chloride. The effects of selected solvents entering the neuropsychological and neurobehavioral literature are reviewed below.

Acetone

Acetone has been considered one of the less toxic solvents in industrial use. Neuropsychological studies are inconsistent, with some showing performance effects below 500 ppm and others showing no effect at higher exposure levels (Matsushita, Goshima, Miyakaki, Maeda, & Inoué, 1969; Dick, Setzer, Taylor, & Shukla, 1989; Naaki, K. 1969; Seeber, Kiesswetter, Vangala, Blaszkewicz, & Golka, 1992). Acute exposure parameters correlated with worker "annoyance" and complaints of fatigue (Kiesswetter, Blaszkewicz, Vangala, & Seeber, 1994). No neuropsychological studies of long-term exposure to acetone vapor appear to have entered the literature.

Carbon Disulfide

Odor threshold	0.016–0.42 ppm
IDLH	500 ppm
OSHA	4 ppm (12 mg/m^3)
STEL	12 ppm (36 mg/m^3) (skin)
NIOSH	1 ppm (3 mg/m^3)
STEL	10 ppm (30 mg/m^3) (skin)
(*Quick Guide*, 1993)	

CS_2 is a sweetish-smelling, clear liquid discovered accidentally by the German chemist Lampodius in 1797. In the 1800s, CS_2 was subsequently tried out as a narcotic and analgesic agent (Spyker, Gallanosa, & Suratt, 1982) and saw its first industrial application in that century when it was used to cure natural rubber. CS_2 was the first solvent to be recognized as neurotoxic (Delpech, 1856) with psychological symptoms following the use of CS_2 noted immediately after its industrial debut. The book *Dangerous Trades* (Oliver, 1901) chronicled "the extremely violent maniacal condition" of carbon disulfide workers of that time. "Some of them have become the victims of acute insanity, and in their frenzy have precipitated themselves from the top rooms of the factory to the ground" (Oliver, cited in Wood, 1981, p. 397). Unfortunately, the factories' solution at the time was not better ventilation or otherwise safe working conditions, but rather the placing of bars on the windows to prevent poisoned workers from "precipitating" themselves. Despite the striking observations contained in *Dangerous Trades,* Hamilton (cited in Wood, 1981) noted that no major investigation of carbon disulfide occurred in the United States prior to 1938.

Current uses of carbon disulfide include soil fumigation, perfume production, as a component of lacquers, varnishes, insecticides, and especially as a solvent in the rubber and rayon industries (Spyker et al., 1982). WHO suggests that only these latter workers in the rayon or cellophane industries are routinely exposed to CS_2 concentrations high enough to adversely affect health (World Health Organization, 1979a). Alcoholics taking the prescription drug disulfiram (Antabuse) are also exposed to CS_2.

Neurotoxicity

CS_2 is possibly the most globally neurotoxic of all solvents, and it appears to be the case that virtually all nervous system functions are affected by CS_2. Most clinical cases of CS_2 toxicity are the result of inhalation. Inhalation of "4800 ppm for 30 minutes will usually result in acute poisoning, including narcosis, vomiting, injury to the central nervous system and death" (Spyker et al., 1982, p. 90). Symptoms experienced during chronic workplace exposure commonly include nausea, dyspnea, headache, vertigo/dizziness, fatigue, feelings of inebriation, and decreased memory (Aaserud et al., 1990). Exposure exceeding 60 mg/m^3 has been associated with signs of mild, diffuse encephalopathy of vasculopathic origin in medium- to low-level exposures, or a full "psychoorganic" syndrome with long-term, low-level exposures (Cassitto et al., 1993). Doses as low as 20 ppm, the current U.S. safe exposure limit, have been shown to produce neurological damage, and workers exposed to a time-weighted average of only 11 ppm have reported headaches (Spyker et al., 1982). Ruijten, Salle, Verberk, and Muijser (1990) estimate that lifetime exposure below 4 ppm would be required to prevent "small observed effects."

Chronic exposure to CS_2 produces a variety of neurological and neuropsychological impairments. Ruijten et al. (1990) found that workers exposed to CS_2 in the range of 1–30 ppm had slowed nerve conduction velocity of slow fibers and increased refractory period in the peroneal nerve. In a study of 16 Norwegian viscose rayon workers with a mean exposure of 20 years, 15 had abnormal neurological examination, 11 had abnormal neurography, and 14 showed abnormal neuropsychological performance on an extended Halstead–Reitan battery (Aaserud et al., 1990). Clinical assessment of patient history and CS_2 exposure suggested that 8 of the 16 patients had encephalopathy primarily attributed to CS_2, while 6 others showed encephalopathy at least partially caused by CS_2. Minor, mixed sensory and motor neuropathy was present in 6 patients, motor neuropathy in 4, and pure sensory neuropathy in 1, most visible in the lower extremities. CT showed cerebral or cerebellar atrophy in 13 of the workers; 9 were termed "slight," 6 were "moderate," and 1 was characterized as severe. It is unclear whether current workplace CS_2 will produce such significant pathology, as many subjects had been exposed to high levels of both CS_2 and hydrogen sulfide. In the first 10 years of exposure, concentrations reached 600–1000 mg/m^3 for one group of workers, and several workers suffered "an acute loss of consciousness at high gas exposures" (Aaserud et al., 1990, p. 27). Characteristic EEG of CS_2 intoxication is described as having frequent sinusoidal β waves at 18–20 per second, described by the authors as an "irritative" EEG (Izmerov & Tarasova, 1993).

The most extensive neuropsychological study of carbon disulfide workers has been that of Cassitto et al. (1993), who examined a total of 493 subjects in the investigation of a viscose rayon factory from 1974 to 1990. In the first study during 1974–1975, Digit Symbol had the best discrimination among workers with different classes of risk. Timed visuomotor test performance was negatively correlated with exposure indices. Scores improved in the second survey during 1988–1989, and by 1989–1990 there was relatively little evidence of neurotoxic morbidity. The trend toward improved neuropsychological performance was associated with lower exposure levels, but there was no improvement in memory function. The authors suggest that symptom complaints and neuropsychological difficulties may occur even when exposure levels do not exceed 8 mg/m^3.

Neuropathology

Very little is known concerning the CNS neuropathology of CS_2, though central and peripheral axon destruction have been observed (Wood, 1981). Toxicological effects are thought to be produced through alkylation of neuronal protein by semi-stable epoxide intermediates (Savolainen, 1982). Other possible, albeit hypothetical, mechanisms suggested to account for CS_2 neuropathological damage include: chelation of essential trace metal metabolites, enzyme inhibition, disturbance of vitamin, catecholamine, or lipid metabolism, or liver interactions, interference with protein synthesis, heme synthesis, and hormone releasing factors (Cassitto et al., 1993; World Health Organization, 1979a). Animal studies have identified severe destruction of the putamen, caudate nucleus, global pallidus, and substantia nigra (Wood, 1981). Peripheral nerves show neuropathy that may be irreversible even at subclinical exposures (Seppäläinen, Tolonen, Karli, Hänninen, & Hernberg, 1972).

CS_2 has been found to cause general and cerebral atherosclerotic changes in exposed workers (Tolonen, 1975; Key et al., 1977). Workers with 10 years' or greater exposure to CS_2 have from two to five times the mortality from ischemic heart disease relative to nonexposed controls (Tiller, Schilling, & Morris, 1968; Tolonen, Nurminen, & Hernberg, 1975). Low-density lipoprotein cholesterol concentrations and diastolic blood pressure show significant linear relationships with measures of CS_2 exposure (Egeland et al., 1992). Thus, secondary neurological and neuropsychological effects of CS_2 could occur as a result of CS_2-induced vasculopathy. Workers exposed to levels as low as 30 mg/m^3 for 10 years or more have been suggested to have increased death rates from heart disease (WHO, 1979a). It is not unheard of for rayon workers in their 30s to present for neuropsychological evaluation having already received single and occasionally multiple cardiac bypasses. In a comprehensive review of the vascular effects of CS_2, Tolonen (1975) concluded that "disturbed ocular microcirculation" is the first subclinical vascular sign of CS_2 "excessive morbidity" (p. 73) as well as being one of the most common symptoms of CS_2 intoxication (Wood, 1981).

Neuropsychological Effects

Neuropsychological studies of CS_2 suggest that symptom questionnaires will be sensitive to very low levels of CS_2 exposure (i.e., under 20 ppm). However, few neuropsychological correlates of exposure have been detected at this level. For example, Putz-Anderson *et al.* (1983) examined 131 workers in a rayon plant who had at 1 year or more of CS_2 exposure "generally below 20 ppm." These workers, who were also screened for illness with neurological sequelae and substance abuse, were compared with controls having "inconsequential" CS_2 exposure.

Exposed subjects reported significantly more symptoms than did controls on a health inventory of CS_2 effects. However, when subjects' neuropsychological test results were examined and corrected for multiple comparisons, only an eye–hand coordination task significantly differentiated the two groups.

Another study that examined higher and longer exposures to CS_2 in somewhat older subjects, suggested that CS_2 is capable of creating more serious forms of neuropsychological impairment. Hänninen, Nurminen, *et al.* (1978) studied a group of workers exposed for a mean of 17 years to between 10 and 30 ppm of CS_2 [subjects in the Putz-Anderson *et al.* (1983) group were exposed for a mean of 12 years with only 28% of the workers exposed to levels in excess of 10 ppm]. The impairments of the Hänninen *et al.* subjects were categorized by the authors as occurring in the areas of motor speed, emotionality, energy level, and psychomotor performance (Hänninen, Nurminen, *et al.,* 1978).

Findings similar to those of Hänninen, Nurminen, *et al.* (1978) have been generated by a U.S. study conducted by NIOSH (Tuttle, Wood, & Grether, 1977). Various reaction time measures, vigilance, visual-motor functions, and constructional abilities showed significant correlations with duration of exposure. Partial correlations, with age removed, revealed several neuropsychological tests to be sensitive to duration of CS_2 exposure, including simple, choice, and drop reaction time, Santa Ana (right hand), a Neisser vigilance task, Digit Symbol, and, to a lesser degree, Block Design.

Neuropsychological measures have also been used to compare the cognitive function of CS_2 workers with and without toxic profiles generated by "clinical neurological, otoneurological, and neuro-ophthalmological examination..." (Hänninen, 1971). When Hänninen (1971) compared these two groups to a population of nonexposed controls, she discovered that both the "exposed" and the "poisoned" group showed neuropsychological abnormalities with differing discriminant functions for each group. Poisoned workers were characterized by the author as showing "retarded speech, low vigilance, clumsiness... inaccuracy of motor functions, diminished intellectual capacity, low level of spontaneous behavior and impoverished capacity of visualization" (p. 376). Workers exposed to CS_2 but not overtly poisoned also displayed neuropsychological deficits compared with controls, albeit with somewhat fewer personality and psychomotor deficits than the clearly poisoned group. Hänninen (1971) characterized these subclinically exposed workers as depressed with mild motor and intellectual impairments. They showed diminished functioning relative to controls in Digit Span, Picture Completion, Block Design, and Digit Symbol tests, as well as on tests of coordination and visual memory.

CS_2-related deterioration of emotional status is a consistent concomitant of other forms of neuropsychological deterioration. Acute manic–depressive psychotic symptoms chronicled in early CS_2 literature are uncommon today because workplace exposures have been regulated to much lower levels than had been the case. However, even these regulated levels of CS_2 appear to be capable of producing a personality syndrome of gradual onset that includes stereotyped behavior, neurasthenic syndrome, depression, irritability, and insomnia (Wood, 1981; Hänninen, 1971).

Results from these and other studies, in the absence of consistent neurological test detection of subclinical CS_2 abnormalities, have led WHO to conclude that "psychological test results are almost the only way of assessing involvement of the central nervous system" (World Health Organization, 1979a, p. 56).

Testing for CS_2 Effects

A battery approach to neuropsychological testing is recommended to quantify CS_2 damage since there is no one pattern of deficit displayed on neuropsychological tests (Wood, 1981, p. 398). Based on the results of the preceding studies, neuropsychological tests sensitive to subclinical CS_2 poisoning include reaction time measures, tests of vigilance and cognitive efficiency like the WAIS-R Digit Symbol, constructional tests (e.g., Block Design), and tests requiring coordination and manual dexterity (i.e., the Santa Ana Dexterity Test).

Carbon Tetrachloride (Tetrachloromethane, Halon 104, Freon 10)

Conversion factor	1 ppm = 6.39 mg/m^3
Odor threshold	140–584 ppm
IDLH	300 ppm
OSHA	2 ppm (12.6 mg/m^3)
NIOSH	2 ppm (12.6 mg/m^3), 60-min ceiling

(*Quick Guide*, 1994)

Once in wide use as a dry-cleaning solvent, carbon tetrachloride has been replaced by less toxic substitutes (e.g., perchloroethylene). Carbon tetrachloride shows initial preference for nervous system toxicity, causing "headache, lethargy, weakness, vertigo, tremor, ataxia and blurred vision" (O'Donoghue, 1985a, p. 128). These symptoms can be followed by confusional state, coma, and convulsions (O'Donoghue, 1985a). Central and peripheral neuropathy have been reported including cerebellar signs, blurred vision, constriction of visual fields, optic atrophy, and sensorimotor neuropathy. Neuropathological studies suggest pontine, cerebellar, and, to a lesser extent, cortical hemorrhage (O'Donoghue, 1985a). There do not appear to be any neuropsychological investigations of carbon tetrachloride exposure in the literature.

Freon (Fluoroalkanes and Nitromethane)

Freon is the generic name for a variety of fluorinated solvents, e.g., fluorotrichloromethane, 1,1,2,2-tetrachloro-1,2-difluoroethane, each of which has specific NIOSH and OSHA Time-Weighted Average levels and IDLH values. All freons cause narcosis and may produce nervous system damage, but there are little occupational data on neurotoxicity. Nitromethane in a fluoroalkane exposure has been associated with a partially reversible parkinsonian-type syndrome. Sandyk and Gillman (1984) report the case of a 20-year-old woman who had been a cleaner in an electronics factory, using a mixture of 94% trichlorotrifluoroethane, 5% methanol, and 25% nitromethane. Progressive 4- to 7-Hz resting tremor was noted, along with bilateral horizontal nystagmus, brief vertical nystagmus, cogwheel rigidity in upper limbs and neck, increased tendon jerks in lower limbs. Glabellar tap and head retraction reflex were positive. Gait showed parkinsonian features. Contribution of methanol was ruled out as were other causes of parkinsonism. Symptoms partially reversed on Sinemet and Symmetrel and the patient was taken out of the workplace.

The authors hypothesized that nitromethane reduction to methylisocyanide was responsible for symptoms. However, the influence of chronic exposure to freon (trichlorotrifluoroethane) cannot be ruled out, particularly since at least one other case report implicates freon as a neurotoxin.

Raffi and Violante (1981) describe the case of a 42-year-old female laundry worker who used liquid Freon 113 for 7 years. Contact was dermal and through inhalation. Complaints of pain and leg paresthesia were accompanied by EMS signs of neuropathy. Clinical improvement after rest and cessation of Freon use was accompanied by improved motor nerve conduction velocity. Metabolic, endocrine, paraneoplastic, and vascular disorders were ruled out.

Methyl Chloride (Chloromethane)

Conversion factor	1 ppm = 2.10 mg/m^3
IDLH	10,000 ppm
Odor threshold	> 10 ppm
OSHA	50 ppm (105 mg/m^3)
STEL	100 ppm (210 mg/m^3)
NIOSH	Not given
(*Quick Guide*, 1993)	

Methyl chloride is used as an industrial solvent, a refrigerant, and in the production of synthetic rubber, methylcellulose, tetramethyl lead, and as a blowing agent for plastic foams (Anger & Johnson, 1985). It is also used as an insecticide propellant. The most common uses of methyl chloride are in the production of methyl silicone compounds and as an antiknock fuel additive. Industrial leaks and defective home refrigerators account for many cases of methyl chloride neurotoxicity (Repko, 1981).

Neurological Effects

The amount of literature on neurotoxic effects of methyl chloride has been described as "meager," even though neurotoxic effects have been stated to resemble encephalitis (Repko, 1981).

Repko (1981) has summarized the neurological effects of methyl chloride exposure as consisting of "profound CNS depression and diffuse toxic damage" (p. 426), with reports of headache, confusion, double vision, and damage to the spinal cord and brain. Frontal and parietal atrophy, hyperemia, edema, and various degenerative cortical changes have been reported. Symptoms of neurotoxicity may remain latent for several hours after exposure. Patient symptoms include headache, somnolence, euphoria, visual disturbances, and chronic personality changes (Repko, 1981). Ataxia, vertigo, tremor, weakness, and loss of coordination have also been reported. There are few data on the long-term neuropathology of methyl chloride exposure, but one case study found EEG evidence of frontal and parietal atrophy in a patient exposed for 10 years (Rodepierre, Truhaut, Alizon, & Champion, 1955).

Neuropsychological Effects

Neuropsychological symptoms include depression, emotional lability, psychomotor impairments, ataxia, confusion, and incoherent speech (Repko, 1981). Perhaps the most complete neuropsychological investigation of methyl chloride was a NIOSH-sponsored study performed by Repko and his colleagues. They examined 122 workers engaged in the manufacture of foam products and compared them with 49 nonexposed controls (Repko et al., 1976). Mean exposure over all plants tested was 34 ppm, and mean after-work breath concentration of methyl chloride was 13 ppm. Subjects received an extensive neurological evaluation, including EEG, and were also administered eight neuropsychological tests and a personal data questionnaire. Neuropsychological tasks included measures of intellectual abilities, vigilance, visual acuity, equilibrium, strength, tremor, coordination, feeling state, and personality.

Test results revealed decrements in vigilance, reaction time, and complex parallel processing. Exposed subjects performed significantly more poorly when walking on two rails of different width, and did significantly less well on the Michigan Eye–Hand Coordination Test. Neurological examination and EEGs did not differentiate between exposed and control subjects. The authors also failed to find effects of methyl chloride on the Edwards Personal Preference Schedule and the Multiple Affect Adjective Checklist (MAACL).

The effect of chronic methylene chloride exposure was also investigated in retired airline mechanics who routinely used methylene chloride-based paint stripper. Within a battery of neuropsychological tests, only Paced Auditory Serial Addition was significantly worse than control performance (Lash, Becker, So, and Shore, 1991).

n-Hexane and Methyl Butyl Ketone

n-Hexane
Conversion factor   1 ppm = 3.58 mg/m^3
IDLH   5000 ppm
Odor threshold   65–248 ppm
OSHA   50 ppm (180 mg/m^3)
NIOSH   50 ppm (180 mg/m^3)
(*Quick Guide*, 1993)

Methyl butyl ketone
Conversion factor   1 ppm = 4.17 mg/m^3
Odor threshold   0.068–0.085 ppm
IDLH   5000 ppm
OSHA   5 ppm (20 mg/m^3)
NIOSH   1 ppm (4 mg/m^3)
(*Quick Guide*, 1994)

These two hexacarbon solvents are classified together because their neurotoxic metabolite, 2,5-hexadione "is responsible for much, if not all, of the neurological effects" that follow their exposure (Spencer, Couri, & Schaumburg, 1980, p. 456; Perbellini, Brugnone, & Gaffuri, 1981).

n-Hexane is an inexpensive solvent included in gasoline, glues, lacquers, and glue thinners. Its use is as varied as the industries in which it is employed; the food industry uses n-hexane to extract vegetable oils, and it is used in the manufacture of perfumes, pharmaceuticals, and cleaning agents (Jorgensen & Cohr, 1981). Experimental animal studies have shown n-hexane to cause giant axonal degeneration of the CNS and PNS (e.g., Frontali, Amantini, Spagnolo, Guarcini, & Saltari, 1981; Spencer *et al.*, 1980), and significant reductions in neuron-specific enolase, a marker protein in distal sciatic nerves (Huang *et al.*, 1993). The lowest dose of n-hexane that has been reported to cause damage is 194–720 mg/m^3 (54–200 ppm) (Jorgensen & Cohr, 1981).

In humans, short-term effects include "narcosis, coma, and eventually, respiratory arrest," possibly caused by infusion of this solvent into CNS cell membranes (Jorgensen & Cohr, 1981, p. 160). Long-term effects are caused by oxidation to 2,5-hexanedione. In chronic solvent abuse of hexane-containing products, cranial nerve and autonomic function may also be affected (Spencer & Schaumburg, 1985). Clinically, workers exposed to *n*-hexane first report headache, nausea, and loss of appetite. Symptoms that precede neuropathy include paresthesis and stocking glove numbness (Chang, 1990). Two to six months after initial exposure, victims have been shown symptoms of *n*-hexane neuropathy including "symmetrical, predominantly distal, motor deficit, and frequently sensory symptoms" which can progress for up to 3 months postexposure. Symptoms have been reported to persist on 2- to 3-year follow-up (Bravaccio, Ammendola, Barruffo, & Carlomagno, 1981, p. 1369) and can accompany positive neurological findings. However, Spencer and Schaumburg (1985) report that "partial or complete recovery [after exposure to *n*-hexane ceases] is the rule" (p. 54). This proposition received some support in a study of nine patients with *n*-hexane-induced sensorimotor neuropathy, and two with motor neuropathy (Chang, 1990). All patients except the most severe case (grade 5 polyneuropathy) "were able to return to daily activities, including running, within 12 to 30 months" (p. 486). Sensory disturbances in patients with polyneuropathy disappeared much more quickly than motor disorders, "usually within three or four months," while symptoms of motor disorders lasted as long as 4 years. Notably, five of nine patients continued to experience deteriorating motor function for 2–3 months *after* cessation of hexane exposure. Two patients who were diagnosed with mild dyschromatopsia on initial examination that failed to improve 4 years later. Subsequent ophthalmological examination confirmed maculopathy in these patients.

About 200 cases of work-related *n*-hexane neuropathy have been reported in Naples, Italy (Bravaccio *et al.*, 1981). Chang (1991) described an additional outbreak of *n*-hexane neuropathy in Taipei printers, who showed evidence of CNS damage in abnormal evoked potential studies. On initial evaluation one patient was confined to a wheelchair, two were able to stand but not walk, one was able to walk with support, and the remaining seven were unable to walk on heels for ten steps. All patients returned to normal activities "including running" within 1–4 years, with earlier remission of sensory than motor impairments. In the initial study, there was "noticeable fall of NCV, profound amplitude reduction of [compound muscle action potentials] and [sensory action potentials] and obvious prolongation of [distal latencies]" (p. 12). An unusual feature was delayed worsening of several electrophysiological measures. Five to ten months after the initial evaluation, further reductions were found in muscle action potentials in 6 of 11 patients, "mainly in the peroneal nerve." Decrease in sensory action potential amplitude and a fall in sensory nerve conduction velocity also occurred in 5 patients. Residual abnormalities were detected even after workers regained full functional motor capacity. In addition, many final EPs "were still significantly different from the normal controls,' (p. 17).

Industrial outbreaks notwithstanding, the most common cause of neuropathy is said to be recreational abuse of compounds containing this solvent (Spencer *et al.*, 1980). Recreational abusers have also reported mild euphoria and, in one case, hallucinations (Spencer *et al.*, 1980).

MBK is a colorless liquid with a pungent odor resembling acetone. It is often mixed with other ketones (e.g., MEK) as a solvent of nitrocellulose, resins, oils, fats, waxes, lacquers, ink thinners, and varnish removers (Bos, Gerrit, & Bragt, 1991). One of the best known epidemics of MBK poisoning occurred in 1973 when 48 workers at a Columbus, Ohio, plastic-coated fabrics plant developed peripheral neuropathy. Their symptoms included mixed motor and sensory findings identified by EMG, and were highly correlated with introduction by the factory of MBK in a print-shop solvent mixture (Landrigan et al., 1980). This episode halted the increasing use of this solvent, which today is recognized for its serious neurotoxic potential (Spencer et al., 1980). Peripheral neuropathy from MBK exposure has also been noted in workers exposed at a fabric plant (Billmaier et al., 1974; Allen, Mendell, Billmaier, Fontane, & O'Neill, 1975) and in spray painters at a dam construction site (Mallov, 1976). Highest levels of MBK are found in the blood and liver, but the CNS is the organ critically affected by toxicity.

Neurotoxic potency of MBK is approximately 12 times greater than n-hexane, but 3 times less than its metabolite, 2,5-hexanedione (Bos et al., 1991). Neurotoxic exposure profile is that of a central–peripheral distal axonopathy and workers may develop this profile after exposure to "only a few ppm of MBK vapor" (Bos et al., 1991, p. 190). Large, long fibers are affected first, followed by shorter, smaller, myelinated fibers, and eventually, unmyelinated fibers; resulting in motor impairments, paresthesias, and sometimes, inability to walk. CNS effects include damage to long tracts of the ascending and descending spinal cord, cerebellar white matter, and distal optic nerve (Bos et al., 1991). Spastic gait, impaired vision, and memory dysfunction are possible (Bos et al., 1991). Recovery may be incomplete.

Methyl Isobutyl Ketone

Less commonly studied than MBK, methyl isobutyl ketone (MIBK) is also used in glues and paints. There are no studies of chronic MIBK exposure. Two experimental studies of brief exposures show generally negative results with regard to neuropsychological functions, but significantly increased CNS complaints at 200 mg/m^3 (Iregren, Tesarz, & Wigaeus-Hjelm, 1993).

Perchloroethylene (Tetrachloroethylene)

Conversion factor	1 ppm = 6.89 mg/m^3
IDLH	500 ppm
OSHA	Not given
NIOSH	25 ppm (170 mg/m^3)
(Quick Guide, 1993)	

Perchloroethylene (PCE) is a chlorinated hydrocarbon solvent commonly used in the dry cleaning industry and as a metal degreaser. Occupations at exposure risk include dry cleaning plants, chemical and textile industries, and degreasing facilities. PCE is also used to dissolve pitches and waxes, adhesives, glues, printing inks and is frequently an ingredient in blended solvents. U.S.

production of PCE is substantial, but has decreased over the past decade from 765 million pounds in 1980, to 414 million pounds in 1986 (ATSDR, 1990). Importation of PCE, however, increased to 142 million pounds as of 1984 (ATSDR, 1990).

NIOSH estimates that 500,000 workers are at risk for PCE exposure in drying cleaning plants and other industrial sites (NIOSH, 1978). Exposure is also possible from household water contaminated with PCE. A shower running with PCE-contaminated water takes only 17 minutes to raise bathroom air PCE to one-third ACGIH TLVs (ATSDR, 1990).

The liver and CNS have been identified as target organs in inhalation exposure to PCE. Available case studies suggest that 100–200 ppm exposure over 5.5 to 7 h produces acute impairment effects, including headache, dizziness, sleepiness, and incoordination (ATSDR, 1990). Accidental, high-level acute poisoning may cause more severe loss of coordination, nausea, and loss of consciousness. Brain examination in animal exposure studies also indicates CNS effects of PCE (Table 4.9).

Further, since liver enlargement and hepatocyte vacuolization have been detected in mice after 30 days of exposure to only 9 ppm (Kjellstrand *et al.*, 1984), it is possible that other organ system effects may contribute to neuropsychological effects in humans.

Although PCE has widely replaced trichloroethylene as a dry-cleaning agent, it too has been reported to cause similar neuropsychological effects, including changes in recent memory and personality (Gold, 1969). There are few neuropsychological investigations of human PCE neurotoxicity in the literature, and these few studies have suggested that PCE exhibits less severe neurotoxic sequelae than trichloroethylene. Dudek (1985) failed to find effects of acute PCE exposure on reaction time. Similarly, Tuttle *et al.* (1977) tested 18 workers who were occupationally exposed to PCE. Neuropsychological tests were given before and after each work shift over 5 consecutive days. There were no significant findings

TABLE 4.9. PCE CNS Neuropathology in Animal Models[a]

Exposure	Effect	Subjects	Study
200 ppm/6hr/day-4days	reduced brain RNA content, increased non-specific cholinesterase content Increased nonspecific cholinesterase content	rats	Savolainene *et al.* (1977)
60 ppm/continuous/90 days followed by 4 month rest	decreased frontal lobe DNA	gerbils	Rosengren *et al.* (1986)
120 ppm/continuous/12 months	biochemical changes	gerbils	Briving *et al.* (1986, Kyrklund *et al.*, 1984)

[a]From ATSDR (1990).

for any of the tests after a correction for worker fatigue was made. Several problems with the Tuttle *et al.* (1977) study mitigate their conclusions. For example, while the experimental group consisted equally of males and females, the control group contained only females. In addition, exposure was relatively low and the number of subjects was small, suggesting that the definitive study on PCE exposure remains to be performed. Hake and Stewart (1977) experimentally exposed subjects to either 0, 20, 100, or 150 ppm PCE for up to 7.5 h. Subjects received 5 days of exposure per level of air concentration. Clinical EEG interpretation of subjects exposed to 100 ppm suggested cortical depression. Neuropsychological tests of higher cortical function were normal, but scores on the Flannigan coordination test showed significant impairment, on at least one day during the weeks of 100 and 150 ppm exposure.

Ferroni *et al.* (1992) compared 60 female dry-cleaning workers exposed to PCE for an average of 10 years with 30 controls from a plant that did not use solvents. PCE-exposed workers showed increased serum prolactin during part of the menstrual cycle compared with controls, an effect the authors attribute to possible PCE effects on pituitary function; however, exposure variables were not correlated with increased prolactin levels. Similar results were found for PCE-exposed workers on a reaction time measure. Reaction time was slower on several tests for exposed subjects compared with controls, but again, neither exposure duration nor air or blood concentrations correlated with test scores.

Case Study

The following case of chronic PCE exposure is consistent with prior studies in that higher cognitive functions did not appear to be significantly compromised. Instead, results appear to be consistent with central or peripheral neurotoxic "solvent syndrome" as a result of sustained exposure to PCE.

It is also an example of the need for "flexibility" in conducting such evaluations. Some of the most pathognomonic test results were on examinations decided upon on the basis of history and test results collected *during the evaluation,* and thus usually not typically administered as part of a neurotoxicity screening battery.

Case Observations and Conclusions. The patient was a college-educated male in his 30s who had worked with PCE for 5 years. He reported constant exposure via ambient air and by direct handling without gloves of PCE-soaked clothing. The patient's wife corroborated continuous exposure and noted that she continued to smell PCE on the patient's breath when he fell asleep at night. The patient also admitted to other symptoms that appeared to be related to chronic solvent exposure, including anosmia, fatigue, dizziness, loss of balance, frequent paresthesias of the face and hands, and a reported history of four or five blackouts.

The patient denied prior head injury or any illness or form of substance abuse that might influence neuropsychological findings. He did report that he

TABLE 4.10. Test Scores of a Patient with Chronic Perchloroethylene Exposure, with Sparing of Higher Cortical Function

1. Finger Tapping: RH mean: 48	Russell 1
(Dom. hand: L) LH mean: 45.5	Russell 2
2. Wechsler Memory Scale:	
(a) Logical	
Immediate recall: A: 9.5 B: 12 Total 21.5	Russell 2
Delayed recall: A: 8.0 B: 12 Total: 20	Russell 1
(b) Figural	
Immediate recall: A: 3 B: 4 C: 6 Total: 13	Russell 0
Delayed recall: A: 3 B: 4 C: 6 Total 13	Russell 0
3. Trail-Making Test: A: 15", B: 40"	Russell 0
	Russell 0
4. WAIS-R Digit Symbol Test: Raw score: 55	Scaled score 9
5. Stroop Test:	
No. of Words: 100 + age corr.: 0 Total: 100	T score 46
No. of Colors: 71 + age corr.: 0 Total: 71	T score 44
No. of Colored Words: 48 + age corr.: 0 Total: 48	T score 53
6. Luria–Nebraska Pathognomonic Scale: Critical Level 67.74	
No. of errors: 17	T score 51
7. Spatial Relations: (Greek Cross) Best: 1 + Worst: 1	Russell 1
8. Pegboard: (Grooved) RH: 71" T 44 LH: 80" T 29	
9. Grip Strength: RH: (1) 47 (2) 49 LH: (1) 45 (2) 58	
(kg) Mean RH: 48 Mean LH: 51.5	
10. IQ Estimate: Shipley Institute of Living Scale	
Raw score: (V: 38 A: 40) CQ: 115	
11. Rey Auditory Verbal Learning Test (Rey AVLT)	
12. Personality Test: MMPI: BDI: 39	
13. MMPI T scores (K corrected where appropriate)	
L t50, F t68, K t46, Hs t65, D t101, Hy t67, Pd t80, Mf t63,	
Pa t65, Pt t76, Sc t68, Ma t48, Si t82, A t77, R t70, Es t45,	
Mac t53	

Note: 0–5 Russell ratings are from Russell (1975) and Russell, Neuringer, and Goldstein (1970) with higher scores indicating greater impairment.

had been under acute job-related financial pressure. The patient admitted to increasingly severe feelings of depression, although these symptoms appeared to occur in the context of long-standing dysthymic symptoms. He was prescribed antidepressants and entered psychotherapy.

As can be seen from the summary table (Table 4.10), the patient displayed unimpaired higher cortical functioning. His only neuropsychological deficit in administered tests was significantly decremented fine motor coordination, as shown in Grooved Pegboard scores. Observation of the patient's performance revealed him to be bilaterally clumsy in the Finger Tapping and Grooved Pegboard tasks, missing the tapping paddle and dropping pegs, respectively.

After history-taking and administration of the neuropsychological battery, the patient was further examined with several "cerebellar" tests. He proved unable to sustain a tandem Romberg for more than 3 seconds and could not sustain his balance when lightly pushed. In addition, the patient displayed mild bilateral dysdiadochokinesia—being unable to perform rapid alternating palm-dorsum slaps on his thigh.

History and neuropsychological test results were interpreted as consistent with possible peripheral or cerebellar neurotoxic effects of PCE with sparing of higher cortical function. Because the patient was currently experiencing very severe psychosocial stressors, it was not possible to determine if his elevated Beck Depression Inventory and MMPI scores were directly solvent-induced, reactive, or unrelated. The patient was diagnosed as experiencing mild neurotoxic effects of solvent exposure, with rule-outs of other cerebellar or PNS disorder. He was subsequently referred for further neurological and occupational medical examination.

Styrene (Vinyl Benzene, Ethenyl Benzene, Phenylethylene, Styrene Monomer, Styrol)

Conversion factor	1 ppm = 4.33 mg/m^3
Odor threshold	0.017–1.9 ppm
IDLH	5000 ppm
OSHA/NIOSH	50 ppm (215 mg/m^3)
STEL	100 ppm (425 mg/m^3)
(*Quick Guide*, 1993)	

Styrene is one of the most widely used components in the production of plastics, where the highest levels of industrial exposure occur (*Styrene*, 1983). It is also a constituent of fiberglass and rubber items, and is widely used in the boat-building industry (Baker, Smith, & Landrigan, 1985). Styrene exposure may also occur in the home with the use of styrene-containing materials, using "floor waxes and polishes, paints, adhesives, putty, metal cleaners, autobody fillers, and varnishes" (*Occupational Exposure*, 1983, p. 16). In addition, the general population is exposed to this solvent through tobacco smoke, automobile exhaust, and through food wrapped in polystyrene containers (*Styrene*, 1983). In the United States, production of styrene was 6612 million pounds in 1981, and it has been estimated that from 30,000 to 300,000 workers are exposed to compounds containing this solvent. Exposure to small amounts of styrene is apparently ubiquitous in the United States, with tissue samples of nonoccupationally exposed persons containing 1.12–1.06 (mean–SD) ppm styrene, with environmental intake estimated to be 2.23 mg/h equivalent to an inhaled concentration of 1.96 mg/m^3 (Pierce & Tozer, 1992).

Styrene is very soluble in blood and tissues. It is taken up into the lung and less so through exposed skin. Small amounts are excreted in breath or urine, while about 85% is metabolized and eliminated as mandelic acid and 10% as phenylglyoxylic acid (Bardodej & Bardodejova, 1970). Styrene is detectable in breath samples and in urine metabolites for less than 24 h after exposure, but has been found in adipose tissue for up to 13 days after experimental exposure to the Swedish threshold limit value of 50 ppm (Engstrom, Bjurstrom, Astrand, & Ovrum, 1978). Chronically exposed workers may show much greater fat storage of this solvent (Engstrom et al., 1978). Rats exposed to 300 ppm have shown "marked styrene accumulation in the brain and perinephric fat" (Savolainen & Pfaffli, 1977, cited in Seppäläinen, 1978, p. 182). Styrene may affect dopamine neurochemistry, since rodent studies show increased dopamine receptor sensitivity and depletion of corpus striatum dopamine (Agrawal, Srivastava, & Seth, 1982). Agrawal et al. (1982) suggest that styrene may destroy dopamine-containing neurons, while Mutti, Falzio, Romanelli, and Franchini (1985) hypothesize that dopamine may be biotransformed into a metabolite that competes with dopamine at synaptic vesicle sites. Dose-related serum prolactin and TSH increases have also been reported.

General Effects

Common symptoms of styrene exposure include initial irritation of mucous membranes, nausea and dizziness, memory complaints, fatigue, giddiness, headache, and paresthesias in fingers and toes (Seppäläinen, 1978; Rosen, Haeger-Aronsen, Rehnstrom, & Welinder, 1978; Cherry & Gautrin, 1990). Vomiting, vertigo, and anemia have also been reported (*Occupational Exposure,* 1983). Headache, complaints of memory loss, feeling "drunk," dizziness, fatigue, irritability, and light-headedness increase with increasing blood styrene concentration (Checkoway et al., 1992). Workers with high exposure indexes are more likely than workers with low exposure to report headache, memory disturbance, sensory abnormalities, and excessive sweating. Balance and coordination problems were found in 26% of all workers but this symptom did not correlate with exposure indices (Matikainen et al., 1993).

Acute Effects

Human experimental study of acute styrene exposure has noted "listlessness, drowsiness and impaired balance" on initial exposure to 800 ppm, while a 1-h exposure to 376 ppm produced decrements in "balance, coordination and manual dexterity" (*Occupational Exposure,* 1983, p. 121). O'Donoghue (Table 4.11) summarizes acute effects of styrene exposure.

Chronic Effects

There is debate as to the neurotoxicity of styrene at levels below 100 ppm. Triebig, Lehrl, Wellie, Schaller, and Valentin (1989) failed to find significant neurobehavioral impairments in a matched cohort of styrene workers with an average exposure of less than 100 ppm, although a measure of short-term memory approached significance. The authors may have been overzealous in controlling for confounding variables. On the other hand, Anger and Johnson (1985) suggest that chronic styrene exposure produces "EEG abnormalities at 30 ppm, psycho-

TABLE 4.11. Acute Effects of Styrene Inhalation on Volunteers[a]

Styrene air concentration (ppm)	Length of exposure	Effects
0–100	1–6h	Odor not detected (NIOSH, 1973) Styrene odor strong but not objectionable. Transient eye irritation at 100 ppm; tests of coordination and a modified Romberg test were not affected (NIOSH, 1973; Hänninen et al., 1976)
200		Strong objectionable styrene odor and nasal irritation occur (NIOSH, 1973; Hänninen et al., 1976)
350		Continuous exposure to 50, 150, 250, and 350 ppm styrene for 30 min at each exposure level impaired reaction time only at the 350 ppm level; perceptual speed and manual dexterity were not affected (Cohr & Stockholm, 1979)
376	25 min	Unable to perform a modified Romberg test (Hänninen et al., 1976)
	50 min	Nausea present; manual dexterity and coordination decreased (Hänninen et al., 1976)
	60 min	Headache and inebriated feeling (Hänninen et al., 1976)
600		Very strong odor; strong eye and nasal irritation (NIOSH, 1973)
800	4 h	Immediate eye, nose, and throat irritation. Pronounced, persistent metallic taste, listlessness, drowsiness, and impaired balance, aftereffects included slight muscle weakness, unsteadiness, inertia and depression (Seppäläinen, Husman, & Mattenson, 1978)

[a]Reprinted from O'Donoghue, J. L. (1985). *Neurotoxicity of industrial and commercial chemicals* (Vol. II, p. 131). Copyright CRC Press, Boca Raton, FL.

motor impairments at 50–75 ppm and symptoms of fatigue, difficulties in concentration, and headaches at 50 to 100 ppm" (p. 68). Yokoyama, Araki, and Murata (1992) evaluated 12 male styrene laminating workers exposed to low levels of styrene for several months prior to testing, and 22–2.4 ppm of styrene in air on the day of testing. Workers took the Japanese version of the WAIS Picture Completion test, Digit Symbol, and the Maudsley Personality Inventory (MPI). Compared with matched controls, styrene workers were impaired on Picture Completion, but, surprisingly, not on Digit Symbol or the MPI. Since the workers were also exposed to acetone and MEK, it is unclear whether results reflect styrene exposure alone or in combination with these solvents.

CNS toxicity from styrene exposure greater than 100 ppm "is a well accepted phenomenon" (Checkoway et al., 1992, p. 560). Chronic PNS changes from styrene exposure have been noted by Rosen et al. (1978), who found mild sensory neuropathy in a group of older, more heavily exposed styrene workers. The authors hypothesized that the PNS may be synergistically affected by age and chronic exposure, possibly via increased production of intracellular free radicals. Cherry and Gautrin (1990) also found that 71% of workers exposed to 100 ppm styrene for longer than 4 weeks had deficits in sensory but not motor nerve conduction velocity (NCV). When exposure was less than 50 ppm, only 23% of workers showed this symptom. There was the suggestion that abnormal nerve conduction, but not reaction time impairments, take longer than 1 month to develop, since five men exposed to more than 100 ppm for less than 4 weeks had unimpaired NCVs but slower reaction times (Cherry & Gautrin, 1990). Murata, Araki, Kawakami, Saito, and Hino (1991) investigated NCVs of 11 male laminating workers employed in boat manufacture and who were not occupationally exposed to vibration, lead, or other neurotoxins. Mean exposure duration was 5 years, mean age was 40. V80 NCV was significantly slowed in these 11 workers compared with matched controls. Findings were consistent with styrene neurotoxicity to peripheral nerves, "especially the faster nerve fibers among the large myelinated nerves" (p. 779).

Psychological symptoms may continue after exposure ceases. Flodin, Ekberg, and Andersson (1989) studied 21 styrene workers who had been free from exposure for several months after the factory where they were employed had become bankrupt. Abnormal fatigue and forgetfulness, headache, and irritation were frequently reported, although it is unclear to what extent effects reflected job loss stress as opposed to continuing neurotoxic effects.

Abnormal EEGs have also been detected in several studies of styrene workers (reviewed in *Occupational Exposure*, 1983, p. 152; Harkonen, Lindstrom, Seppäläinen, Sisko, & Hernberg, 1978). In one such investigation, abnormal EEGs were more than twice as likely in a population of styrene workers exposed a mean of 5 years than in the general population (Seppäläinen, 1978).

Finally, several findings suggest that styrene exposure causes damage or dysfunction to the autonomic nervous system, including higher prolactin response to thyrotropin-releasing hormone (Arfini et al., 1987), significantly decreased ECG CV_{RR} (Murata et al., 1991), and vegetative nervous system disturbances (Araki, Abe, Ushio, & Fujino, 1971; *Styrene*, 1983). It is unclear whether autonomic effects are primarily an acute or a chronic effect.

Neuropsychological Effects

Lowered reaction times related to exposure intensity are the most commonly noted neuropsychological effect of styrene (Mutti et al., 1983; Gamberale & Kjellberg, 1982; Cherry & Gautrin, 1990). Behavioral effects of acute exposure have been suggested to depend on individual differences in clearing styrene and its metabolites from the body. Subjects who were slow to clear showed longer reaction times (Cherry, Rodgers, Venables, Waldron, & Wells, 1981; Cherry & Gautrin, 1990). However, a third study failed to replicate this finding in workers with an average of 2.5 years of exposure to levels below 110 mg/m^3 (Edling & Ekberg, 1985).

The lowest concentrations suggested to produce neurobehavioral symptoms are in the range of 10–25 ppm, noted by Flodin et al. (1986). Symptoms included reversible deficits in memory, concentration, irritability, and libido. Letz, Mahoney, Hershman, Woskie, and Smith (1990) performed a large scale study of five fiberglass boat building companies. The 105 fiberglass laminators and other employee participants were exposed to styrene an average of 4.6 years. Exposure levels varied greatly by company and season, but the grand mean over all companies was 13.5 ppm. Performance on the NES's computerized digit symbol test during afternoon testing was related to styrene exposure after adjusting for age and education. Digit symbol was also related to the natural logs of urinary mandelic acid, and personal air styrene samples corrected for respirator wear. Exposure to styrene levels of greater than 50 ppm, 8 h TWA account for most subjects' neuropsychological impairments.

Workers with an average of 5 years' exposure to styrene showed significant decrements in visuomotor accuracy, psychomotor performance, and vigilance as a function of mandelic acid concentration (Harkonen et al., 1978; Lindstrom, Harkonen, & Hernberg, 1976). Lindstrom and colleagues found that one test of visuomotor speed and one of visual memory correlated with exposure duration; however, these effects may have been confounded by significant relationships between age and exposure duration (Lindstrom et al., 1976).

Personnel exposed for an average of 8.6 years showed significant decrements relative to controls in reaction time, Block Design, Rey's Embedded Figures, and Wechsler Memory subtests including Logical Memory (immediate and 30 minute delayed) and Visual Memory (immediate) (Mutti et al., 1984).

An unpublished study reports chronic CNS changes in a group of seven patients who were exposed to styrene for a mean of 15 years (range 6-26 years) (Melgaard, Arlien-Søberg, & Brulin, cited in *Occupational Exposure*, 1983). "Intellectual impairment" was said to be found in six of the seven patients, with five showing moderate impairment and one severe impairment. Cerebral atrophy was detected in four of these patients by CT scan, and in one patient by pneumoencephalography.

Finally, a single case study suggests that styrene may interact with substance abuse and/or alcohol or Antabuse producing transient paranoid psychosis and neuropsychological deficit (White, Daniell, Maxwell, & Townes, 1990).

Overall results on styrene exposure suggest that exposure greater than 50 ppm or greater than 8.6 years is capable of causing neuropsychological impairments. Information about recovery from acute or chronic styrene exposure has not entered the literature.

Tetrabromoethane (Acetylene Tetrabomide)

Conversion factor	1 ppm = 14.37 mg/m^3
IDLH	7 ppm
OSHA	1 ppm (14 mg/m^3)
NIOSH	Not given
(*Quick Guide*, 1994)	

Tetrabomoethane (TBE) is a halogenated aliphatic hydrocarbon used to separate ore and waxes. Morrow, Callender, et al. (1990) report a case study of a 31-year-old oil worker who was exposed to TBE on the head and torso. Initial symptoms included nausea, vomiting, dizziness, and severe headache. Chronic symptoms included headache, parosmia, memory loss, depression, and irritability which caused the patient to be unable to return to work. Mildly elevated thyroid and liver function tests were the only unusual results in a complete medical and neurological workup. However, when PET scan using R-2-deoxyglucose was compared against controls, significantly decreased uptake was observed in the left superior frontal, left posterior medial frontal, right mid corpus callosum, right putamen, right medial and posterior thalamic, right anterior gyrus rectus, left amygdala, and left hippocampus. Topographic EEG was also abnormal, as were memory, learning, attention, and psychomotor speed factors on the POET neuropsychological battery. No evidence of malingering was found. The authors suggest that basal ganglia and parietal cortex may be especially sensitive to solvent effects as these structures were also found to be damaged in the autopsy of a chronic solvent abuser (Escobar & Aruffo, 1980).

Tetrahydrofuran (Diethylene Oxide, 1,4-Epoxybutane, Tetramethylene Oxide)

Conversion factor	1 ppm = 3.00 mg/m^3
Odor threshold	31 ppm
IDLH	20,000 ppm
OSHA/NIOSH	200 ppm (590 mg/m^3)
STEL	250 ppm (735 mg/m^3)
(*Quick Guide*, 1993)	

Tetrahydrofuran (THF) is a relatively rare solvent, used to dissolve plastics, epoxy compounds, and cellulosics. Use of the solvent is increasing in glues, paints, varnishes, ink, and wetting or dispersing agents in textile processing (Garnier, Rosenberg, Purssant, Chauvet, & Efthymiou, 1989).

Two cases of THF poisoning have been reported with neurological consequences. Both cases were plumbers working with THF-containing glues to repair plastic pipe in confined spaces. Symptoms included mucosal irritation, headache, dizziness, chest or gastric pain, and abnormal elevations of liver enzymes which cleared within 2 weeks. Mental status or neuropsychological performance was not specifically investigated. Mode of liver toxicity was suggested to be toxic hepatitis with serum aminotransferase elevation (Garnier *et al.*, 1989).

Toluene (Toluol, Methyl Benzene, Methyl Benzol, Phenyl Methane)

Conversion factor	1 ppm = 3.83 mg/m^3
Odor threshold	0.16–37 ppm
IDLH	2000 ppm
OSHA/NIOSH	100 ppm (375 mg/m^3)
STEL	150 ppm (560 mg/m^3)
(*Quick Guide*, 1993)	

Toluene is an aromatic hydrocarbon with primary uses as a cleaner and degreaser, and also as a solvent for paint, varnish, and rubber adhesive (Cavanaugh, 1985). Exposure is usually through inhalation. Animal studies note that toluene exposure produces a "U-shaped concentration/response function" (Benignus, 1981) initially excitatory at lower concentrations or short exposures but depressant at higher levels/concentrations. Absorption is via inhalation and ingestion. Eighty percent of oxidation is through the aliphatic chain, where absorbed toluene is metabolized from benzoic acid through glycine to hippuric acid, or with glucuronic acid to create benzoyl glucuronic acid. The aromatic ring is a minor route of oxidation, resulting in formation of o-, m-, and p-cresol which are then excreted in urine as glucuronide and sulfate conjugates (Meulenbelt, deGroot, & Savelkoul, 1990). These three cresols may be monitored as an index of exposure (Meulenbelt et al., 1990).

Acute Exposure

"Toluene may induce acute as well as chronic disturbances, especially in the central nervous system, not only at high concentrations but also at levels close to the TLV" (Arlien-Søberg, 1992, p. 97).

Neuropsychological impairment and neurological damage vary in severity and reversibility as a function of exposure level, chronicity, and individual differences that are as yet unresearched.

In humans, depending on the study cited, toluene approaches asymptote in the body between 10 and 80 min, with brain asymptote approximately 1.26–2.50 times that of blood (Benignus, 1981, p. 408). Acute effects observable without neuropsychological testing include those listed in Table 4.12, summarized by Von Oettingen et al. (1942) and listed by Benignus (1981).

Acute Neuropsychological Effects

Several studies have shown decrements in simple and choice reaction time from acute exposure. For example, Gamberale and Hultengren (1972) exposed 12 male subjects to increasing doses of 100–714 ppm of toluene within a single exposure session. They found decreases in concentration which caused reaction time and perceptual speed deficits at concentrations of 300 ppm and above. Echeverria et al. (1989) exposed 42 college students to 0, 75, and 150 ppm of toluene for 7 h over 3 days. At 150 ppm, a 6% decrease was found in digit span, 12% for pattern recognition latency, 5% for pattern memory. Decrements were

TABLE 4.12. Short-Term Exposure Effects[a]

Levels in air (ppm)	Duration	Effect
1.0	8 h	Minimal risk
100	8 h	Moderate fatigue
300	8 h	Severe fatigue, headache
600	8 h	Extreme fatigue, confusion, dizziness, intoxicated behavior

[a]ATSDR (1990).

TABLE 4.13. Acute Behavioral Effects of Toluene Inhalation[a]

Toluene level in air (ppm)	Duration (h)	Results
0–40	6	No effects
100	6	Impaired manual dexterity and visual perception
	7 1/2	Impaired dual task visual vigilance & tone detection
200	7	Impaired SRT
300	4 1/3	Impaired SRT
600	8	Severe staggering and incapacitation

[a]After Echeverria, Fine, Langolf, Schort, and Sampaio (1989).

also found for dexterity and tracking, along with increased fatigue on the Profile of Mood States. The most robust finding was a dose–response relationship between toluene level and the number of times subjects fell asleep, ranging from 7% at 0 ppm to 22% at 150 ppm. In general, short exposures and lower concentrations of toluene do not produce neuropsychological or neurological effects. For example, no effects were noted in subjects exposed for 4 h, either to toluene alone or in combination with xylene (Olson, Gamberale, & Iregren, 1985). No significant effects were noted after 6 h of 40 ppm exposure (Andersen et al., 1983). Benignus (1981) failed to find scalp EEG changes during a 6-h exposure of 200 ppm, and Winneke's (1982) review of the literature found no effect of acute, short-term exposure at levels below 300 ppm.

Somewhat longer exposure intervals may begin to produce observable effects; e.g., Andersen et al. (1983) found borderline significance levels of impairment after 6 h of 100 ppm toluene exposure on three neuropsychological tests (multiplication errors, Landolt's rings, and screw plate test). Pharmacokinetic properties of toluene also suggest that lack of short-term exposure effects may be misleading and fail to generalize to longer duration exposure effects. Because toluene blood levels are cumulative when exposure is extended, levels increase between 10 and 20 times their initial value over the course of 2 weeks' exposure to 184–332 ppm (Konietzko, Keilbach, & Drysch, 1980). Thus, short-term experimental exposures may not reflect typical workplace blood levels, even if air concentrations are equivalent. Accumulation of blood levels suggests the possibility of equally high concentrations in sensitive brain tissue; more research on what might be called "intermediate" toluene exposure seems warranted.

Chronic Exposure

Chronic exposure to toluene produces both neurological and neuropsychological impairment. Animal studies have shown toluene-induced nystagmus, vestibular and optico-oculomotor abnormalities, EEG disruptions in hippocampal theta wave activity, changes in brain neurotransmitter content, and structural damage to cerebrum and cerebellum (ATSDR, 1989). In rats, toluene is ototoxic, producing high-frequency hearing loss and cochlear damage (ATSDR, 1989;

TABLE 4.14. Long-Term Exposure Behavioral Effects[a]

Level (ppm)	Exposure	Effect
0.3		Minimal risk
200	1–10 years	Headache, dizziness, loss of coordination
300	1–10 years	Impaired memory and thinking ability

[a]ATSDR (1990).

Pryor, Dickinson, Howd, & Rebert, 1983). Hearing loss is permanent with damage to the outer hair cells of the cortical organ of the inner ear (Sullivan, Rarey, & Conolly, 1989). Sixteen weeks of exposure to toluene in rats produced "marked" changes in cerebellar and spinal cord proteins. Neuron-specific enolase (NSE), creatine kinase-B (CK-B), and β-S100 "were elevated by 15, 20, and 55% respectively. In spinal cord CK-B was reduced by 13%, while β-S100 increased by 35%" (Huang et al., 1993, p. 85). CK-B and β-S100 are marker proteins found primarily in astroglial cells and their increase may reflect the development of gliosis, a common indication of early CNS damage (Huang et al., 1993).

Toluene inhalation as part of a mixed solvent exposure is developmentally neurotoxic, producing neural tube defects and other neuropsychological abnormalities with *in utero* exposure. Microcephaly, growth retardation, craniofacial abnormalities, attentional impairments, and developmental delays have been reported in children of mothers who sniffed toluene in solvent mixtures while pregnant (Holmberg, 1979; Hersh et al., 1985; Goodwin, 1988). Pure toluene exposure produces developmental neurotoxicity in animal models.

Additionally, Svensson et al. (1992) found a reversible, dose-dependent decrease in gonadotropin hormones (HL and FSH) in printers exposed to toluene, suggesting additional effects on the hypothalamic–pituitary axis.

Neuropsychological effects of toluene inhalation include general intellectual decrement and increased emotional reactivity, both in the workplace (Hänninen et al., 1976) and as a product of solvent abuse. Toluene is said to be "by far the most widely abused solvent" (King, 1983, p. 76), perhaps because acute subjective effects include exhilaration, euphoria, and disinhibition (Cavanaugh, 1985). Toluene abuse shows toxic effects throughout the body—including liver, heart, kidneys, and bone marrow—although "the main toxic impact is on the nervous system" (King, 1983, p. 77). In rat studies, toluene damages brain structure; rats exposed to toluene continuously for 30 days showed lower total brain weight and lower cortical weight. Data also suggested breakdown of phospholipids producing gray matter loss (Kyrklund, Kjellstrand, & Haglid, 1987).

Female electronics workers exposed to a time-weighted average (TWA) of 88 ppm over the course of a work shift were compared with controls exposed to a TWA of 13 ppm (Foo, Jeyaratnam, & Koh, 1990). The study population was reported to be exceptionally devoid of confounding health influences; all subjects were stated to be nonsmoking teetotalers in good medical and psychiatric health. Mean age was in the mid-20s; quite a bit younger than typical populations studied. Toluene-exposed subjects performed significantly more slowly in Trails A and B, Digit Span, Digit Symbol, and Grooved Pegboard. From the study design, it is not clear how to partition acute versus chronic effects, since workers were tested midweek. Because subjects were tested before they began work that day, cumulative effects of acute toluene exposure would not have had time to diminish before testing.

A dose–response relationship of toluene exposure to symptom complaints was reported by Ukai et al. (1993). A statistically significant relationship was found in subjective symptom prevalence when subjects were grouped into increasing categories of exposure intensity, including dose-related increases in dizziness, floating sensations, feelings of heaviness in the head, headache. The off-work symptom "inability to concentrate" was significantly elevated in exposed but not control populations, and was linearly related to toluene dose.

Neurological Effects—Abuse

As of 1982, the literature on toluene abuse consisted of case studies rather than large-scale group research. In a review of these case reports, King (1983) reported diffuse cerebellar encephalopathy to be a common finding, with or without accompanying cognitive deterioration or dementia. In another case study, toluene was apparently able to cross the placenta and produce cerebellar damage in an unborn infant (Streicher, Gabow, Moss, Kono, & Kaehny, 1981). Other neurological findings from chronic toluene abuse include optic and peripheral neuropathy. CT scans of toluene abusers have shown widened sulci, enlargement of the subarachnoid cisterns, and dilated ventricles (Cavanaugh, 1985). For example, Fornazzari et al. (1983) found 7 of 14 patients to have abnormal CT scans with significantly more prominent cerebellar sulci, and larger overall ratings of ventricular size. In the same study, 46% of the sample showed cerebellar test abnormalities including gait ataxia and intention tremor. Similar findings have been reported for a group of 20 primarily Hispanic patients who had inhaled solvents (mostly toluene) for at least 2 years (Hormes, Filley, & Rosenberg, 1986). The subjects inhaled an average of one 12-ounce can of toluene-based spray paint per day, and their mean duration of exposure was about 12 years. Neurological dysfunction was found in 65% of these patients, and included pyramidal, cerebellar, cranial nerve, or brain stem abnormalities and tremor. Cerebellar signs of gait or leg ataxia or leg spasticity were detected in 9 of these patients. Dysarthria, nystagmus, and arm incoordination were found in 5 patients. Miscellaneous other abnormalities were found in several patients, including anosmia, bilateral hearing loss, ocular flutter, and opsoclonus. A case study of an 18-year-old girl with a 2-year history of sniffing toluene vapor showed her to have defective color perception (Malm & Lying-Tunell, 1980).

The underlying neuropathological process induced by toluene is unknown, although volatile hydrocarbons like toluene are all highly lipophilic and are easily absorbed into the lipid-rich nervous system (Hormes et al., 1986).

Neuropsychological Effects–Abuse

Neuropsychological studies suggest that toluene inhalation may cause more severe neuropsychological consequences than any other form of drug abuse, including polydrug abuse (Korman, Matthews, & Lovitt, 1981; Berry, 1976). A representative study found inhalant (toluene) abusers to perform significantly worse than polydrug abusers on 20 of the 67 neuropsychological measures employed (Korman et al., 1981). Fornazzari et al. (1983) found that of 24 patients, 15 (63%) were classified as impaired on a variety of neurological and neuropsychological measures. Abusers' Verbal IQs were less impaired than their Performance IQs, which were in the borderline range. The pattern of impairment led the authors to conclude that toluene abuse causes "profound impairment of motor control ... intellectual and memory capacity" (Fornazzari et al., 1983, p. 327). Toluene abuse appears to produce more serious neuropsychological consequences than other types of drug abuse, including cocaine abuse. Compared with chronic users of other drugs, toluene abusers "performed worse than controls on tests of short-term memory, card sorting, free recall, verbal memory, verbal IQ, word learning and comprehension of complex language." Groups did not differ on performance IQ, although both groups had lower performance IQ's than normal controls (Grigsby et al., 1993, p. 229).

Neuropsychological impairment was also detected in a majority of the Hormes et al. (1986) subjects: seven were demented, and another five had mild to moderate degrees of impairment, with abnormalities of "apathy, poor concentration, memory loss, visuospatial dysfunction, and complex cognition" (p. 701). Personality or emotional dysfunction was noted for all patients, who showed characteristically flattened affect and apathy.

Finally, Tsushima and Towne (1977) also investigated combined acute and chronic effects of paint sniffing on neuropsychological performance. While they were unable to specify the precise paint solvent under investigation, it was most probably toluene. Compared with controls matched on age, education, sex, and peer group, paint sniffers performed more poorly on 11 of the 13 neuropsychological measures employed. Sensitive tests included Grooved Pegboard, WISC-R Coding B, Stroop Color-Word Interference Test, Memory for Designs, and the Peabody Picture Vocabulary Test. Because the latter test is typically seen as tapping premorbid abilities, the study's results could also suggest that paint sniffers may self-select for sniffing because of lower premorbid abilities and intelligence. However, a trend toward significant duration effects suggests that neurotoxic effects may occur over and above self-selection factors (Tsushima & Towne, 1977).

Chronic Workplace Effects

Chronic workplace effects of toluene exposure are frequently reported. For example, Ørbaek and Nise (1989) studied two groups of Swedish rotogravure printers with a mean exposure of 29 years to toluene. Workers reported a variety of neurological and psychological complaints significantly more frequently than controls, including fatigue, recent memory and concentration impairment, paresthesias, depression, irritability, and sexual problems. Neuropsychological test scores were not found to be different from controls when results were adjusted for verbal ability in a synonym test, but it has been questioned whether this test is an appropriate index of premorbid function (Mikkelsen, Jørgensen, Brown, & Gyldensted, 1988).

Female workers with high exposure to toluene have significantly higher rates of spontaneous abortions than controls (Ng, Foo, & Young, 1992).

1,1,1-Trichloroethane (Chlorothane)

Conversion factor	1 ppm = 5.55 mg/m^3
Odor threshold	390 ppm
IDLH	1000 ppm
OSHA	350 ppm (1900 mg/m^3)
NIOSH	350 ppm (1900 mg/m^3)
STEL	450 ppm (2450 mg/m^3)
(*Quick Guide*, 1993)	

Overall, 1,1,1-trichloroethane had been considered to be one of the more benign solvents with regard to experimentally produced neurotoxic effect. However, recent clinical studies have documented neurotoxic effects from exposure. Several days of exposure to 500 ppm of 1,1,1-trichloroethane produced no impairment in psychomotor abilities (Stewart, Gay, Schaffer, Erley, & Rowe, 1969). Two 4 1/2-h exposures of 450 ppm likewise failed to show impaired functioning on neuropsychological measures (Salvini, Binaschi, & Riva, 1971a). Alternatively, Gamberale and Hultengren (1973) found simple and choice reaction time to be impaired as a function of exposure to 350 ppm of 1,1,1-trichloroethane.

Mackay *et al.* (1987) point out that none of the previously cited studies have correlated neuropsychological performance with blood or breath solvent level (a deficiency common to much solvent research). They performed such a study on 12 subjects in a counterbalanced design. Tests included simple and four-choice reaction time, Stroop Color-Word Test, as well as tests of verbal reasoning, motor tracking, and stress-arousal. Acute exposure to 1,1,1-trichloroethane significantly impaired tracking, Stroop, and both reaction time tests. Exposure concentration, duration, and the interaction of the two produced different patterns of performance decrement.

1,1,1-trichloroethane was the solvent investigated in one of the few studies to utilize the Luria–Nebraska Neuropsychological Battery (Berg & Kelafant, 1993). Thirty workers with mean exposure of 17.62 years were evaluated, with significant elevations found on the Rhythm, Memory, Intermediate Memory, and Speed scales. Trails B approached significance. Personality evaluation using the Personality Assessment Inventory showed scores in the normal range, although mean Somatization and Depression scales approached significance.

1,1,1-Trichloroethane is able to produce toxic encephalopathy with egregiously high dermal exposure, e.g., where workers routinely wash their hands and arms in this solvent (Kelafant, Berg, & Schleenbaker, 1994). Peripheral neuropathy due to workplace 1,1,1-trichloroethane exposure has also been reported (Liss, 1988; House, Liss, & Wills, 1994, Liss & House, 1995). Progressive liver disease as a function of 1,1,1-trichloroethane exposure has also been reported (Cohen & Frank, 1994), which may produce indirect central nervous system and neuropsychological abnormalities.

Trichloroethylene (Trichloroethene)

Conversion factor   1 ppm = 5.46 mg/m^3
Odor threshold   82 ppm
IDLH   1000 ppm
OSHA   50 ppm (270 mg/m^3
STEL   200 ppm (1080 mg/m^3)
NIOSH   Not given
(*Quick Guide*, 1993)

Trichloroethylene (TCE) has been recognized as a neurotoxic solvent for over 50 years. Most commonly used to degrease machine parts, TCE is also employed in the dry-cleaning industry, as a household cleaner, and as a constituent of lubricants and adhesives. It has been used as a short-acting anesthetic in dental and obstetric procedures (Annau, 1981). More recently, TCE has become one of the most extensively used solvents in Silicon Valley's (USA) electronics industry for cleaning, stripping, or degreasing (Baker & Woodrow, 1984; Feldman *et al.*, 1985).

Given this variety of applications, it is not surprising that about 3.5 million workers are exposed to TCE, with 100,000 of these exposed on a full-time basis. It has also been estimated that 67% of these workers work with TCE under inadequate safety conditions (NIOSH, 1978).

Commonly noted neurological effects of TCE exposure include sensory disturbances and trigeminal anesthesia, which can be permanent and spread to other cranial nerves (Solvent Neurotoxicity, 1985). The neurotoxic effects of TCE on the trigeminal nerve actually once led to its use as a treatment for tic douloureux. Cranial neuropathies have also occurred as a consequence of TCE's use as a dental and obstetrical anesthetic (Firth & Stuckey, 1945).

While consistent behavioral deficits have not been noted at TCE levels below 300 ppm, "subtle alterations of brain wave-activity" have been observed in subjects exposed to just 50 ppm for 3.5 h (Winneke, 1982, p. 126).

Neuropathological changes caused by TCE exposure have been inconsistently reported. Peripheral neuropathy and cranial nerve damage have been found (Cavanaugh, 1983; King, 1983; Annau, 1981), although several autopsy reports of TCE inhalation fatalities fail to cite neurological damage. In at least one subject, however, the brain section showed "striking alterations" in the brain stem, nerve roots, and peripheral nerves with "extensive" myelin degeneration

(Annau, 1981, p. 418). Case report inconsistencies in both human and animal exposure studies have led to the suggestion that peripheral and central lesions may not be produced from TCE *itself,* but rather by its decomposition product, dichloroacetylene (Annau, 1981). It has also been argued that TCE destroys neural tissue indirectly by acting as a release for an otherwise dormant virus (Solvent Neurotoxicity, 1985). With threshold limit value exposure levels, Ruijten, Verbeck and Salle (1991) found only slight effects of chronic trichloroethylene exposure on the sural and trigeminal nerve.

A possible relationship between TCE and symptoms of systemic lupus erythematosus (SLE) has been noted in individuals exposed to small amounts of solvent via well water (Kilburn & Warshaw, 1992). Kilburn and Warshaw evaluated symptoms of SLE in a population exposed to 6–500 ppb of TCE for up to 25 years. Using the American Rheumatism Association SLE questionnaire and blood testing, among other symptoms, it was found that the incidence of arthritis was 2–5 times normal, Raynaud's symptoms were 2–40 times controls, and neurologic disorders including seizures were 3–6 times more frequent in exposed subjects. Increased prevalence of antinuclear antibodies among the exposed group also suggested a link to SLE. The authors suggest that their results are analogous to drug-induced lupus, seen in patients treated with procainamide, hydralazine, isoniazid, chlorpromazine, methyldopa, quinine, and diphenyl hydantoin. Similarities were also cited with workers exposed to a substance similar to TCE, vinyl monomer, which was cited to cause scleroderma-like reaction in exposed workers (Arlien-Søberg, 1992.)

Neuropsychological Effects

One study of short-term, low level, experimentally induced exposure to TCE has shown decreases in manual dexterity and visuospatial accuracy (Salvini, Binaschi, & Riva, 1971b); however, these results have not been replicated and are not consistent with other studies using longer exposures and higher concentrations that failed to find similar effects (Annau, 1981).

Subjects' reports of other nonspecific CNS difficulties seem to be more frequent than actual neuropsychological changes. Complaints reported include headache, dizziness, fatigue, and diplopia (Steinberg, 1981), as well as tremor, giddiness, increased tear production, decreased skin sensitivity, alcohol intolerance, "neurasthenia and anxiety, bradycardia and insomnia" (Spencer & Schaumburg, 1985).

A 2-year follow-up of a 56-year-old male acutely exposed to TCE showed neuropsychological deficits suggesting posterior cortical lesions. In addition to decreased general intelligence, the patient showed losses in efficiency, concentration, dyspraxia, dysgraphia, and visual concept formation (Steinberg, 1981).

Field studies of workers exposed to TCE fail to agree on clear patterns of neuropsychological deficit. At least two studies failed to find chronic, low-level neuropsychological effects of TCE. In the first, no effects were found in workers chronically exposed to 50 ppm (Triebig, Schaller, Erzigkeit, & Valentin, 1977).

In the second, female subjects exposed for 6 years to air concentrations ranging from 110 to 345 ppm showed no effects of exposure (Maroni et al., 1977). However, both studies used a small number of subjects, and the absence of vigilance and reaction time tests in the second study mitigates its conclusions.

In several other studies, TCE-exposed workers did show significant impairment in choice reaction time (Gun, Grysorewicz, & Nettelbeck, 1978) and several types of psychomotor weakness (Konietzko, Elster, Bencsath, Drysch, & Weichard, 1975). Grandjean (1960) examined a group of 50 workers exposed to TCE for intervals of 1 month to 15 years. Nine of these workers displayed the "organic psychosyndrome" of "memory disturbances, difficulties in understanding, and changes in affective state" (Lindstrom, 1982, p. 133).

Finally, the risk for developing an "organic psychosyndrome" among metal degreasers who worked primarily with TCE for an average of 11 years, increased with increasing exposure (Rasmussen, Jeppesen, & Sabroe, 1993). Diagnosis was made on the basis of symptom complaints and neuropsychological test performance. Effects remained significant after adjusting for age, premorbid intellectual level, alcohol abuse, and other confounders.

To summarize, the literature on TCE tends to suggest that short-term acute exposure under the current threshold limit value does not produce neuropsychological deficits. Alternatively, acute exposure to toxic levels of TCE can cause trigeminal anesthesia as well as long-lasting neuropsychological abnormalities.

The effects of chronic exposure to TCE are unclear. CNS dysfunctions and neuropsychological deficits have been inconsistently demonstrated. TCE does not appear to cause overall increases in mortality among exposed workers (Shindell & Ulrich, 1985).

1,1,1-Trichloromethane (Chloroform, Methane Trichloride)

Conversion factor	1 ppm = 4.96 mg/m^3
Odor threshold	133–276 ppm
IDLH	1000 ppm
OSHA	2 ppm (9.78 mg/m^3)
NIOSH	TWA not given
STEL	2 ppm (9.78 mg/m^3) (60 min)
(*Quick Guide,* 1993)	

1,1,1-Trichloromethane is a chlorinated hydrocarbon with wide use as a cleaning solvent, aerosol propellant, dry-cleaning solvent, and in various industrial processes. Berg and Kelafant (1993) examined 15 workers exposed to 1,1,1-trichloromethane and who reported difficulties in memory, attention, concentration, balance, and affective disturbance. Somewhat unusual for a solvent study, the authors employed the Luria–Nebraska Neuropsychological Battery (LNNB), in addition to Trails, Symbol Digit, and the Personality Asessment Inventory (PAI). LNNB elevations were found on Rhythm, Memory, Intermediate Memory, and Speed, slower performance on Trails B, and PAI indications of somatization and mild depression. Data were not reported in this preliminary report.

Xylene (m-, o-, and p-Xylene)

Conversion factor	1 ppm = 4.41 mg/m^3
IDLH	1000 ppm
OSHA/NIOSH	100 ppm (435 mg/m^3)
STEL	150 ppm (655 mg/m^3)
(*Quick Guide*, 1994)	

Approximately 2 million tons of this aromatic hydrocarbon are produced in the United States annually (Arlien-Søberg, 1992). Xylene is used in paint solvents, varnishes, glues, printing inks, and in the rubber and leather industries (NIOSH, 1975, cited in Savolainen, Riihimaki, & Linnoila, 1979). Histology technicians who prepare tissues for staining and mounting are also exposed through their use of xylene and other solvents to clear fat and paraffin from tissue samples (Kilburn, Seidman, & Warshaw, 1985). Pure xylene isomers are used in the production of polymers (Seppäläinen *et al.*, 1991).

Xylene vapor is easily absorbed through the lungs and in its liquid form easily penetrates skin. It is mainly metabolized in the liver, but some metabolization may take place in the olfactory bulbs and nasal mucosa (Arlien-Søberg, 1992). Chronic, low-level exposures of between 160 and 320 ppm in animal models produced long-lasting gliosis in frontal and posterior cerebellar vermi (Arlien-Søberg, 1992).

In high concentrations, acute exposure to xylene is anesthetic and can be fatal. One report details one fatality among three painters exposed to xylene in a confined space. Xylene concentration was estimated at 10,000 ppm and autopsy revealed petechial brain hemorrhages and anoxic neuronal damage (Morley *et al.*, 1970).

The effects of short-term xylene exposure on the EEG patterns of young healthy experimental subjects were investigated by Seppäläinen *et al.* (1991). Subject exposure was either high (200 ppm) or low (135 ppm), during rest or exercise. Exposure increased alpha wave frequency and percentage during exercise, but overall, effects were minor and not considered deleterious.

Lower concentration acute exposure of xylene was examined by Gamberale, Annwall, and Hultengren (1978). In the first experiment, subjects inhaled either ordinary air or xylene at high or low concentration for 70 min. No neuropsychological effects were found. Another group first inhaled the higher concentration of xylene (12.3 mol/liter) while exercising on a bicycle ergometer to increase uptake of xylene. These subjects did become significantly impaired on three neuropsychological tests: reaction time addition, visually presented Digit Span, and choice reaction time. Another study (Savolainen *et al.*, 1979) investigated acute effects of xylene on sedentary subjects during more extended inhalation periods (i.e., several successive days for 6 h each day). Unadapted subjects showed initial problems with equilibrium when exposed to 4.18 mol/liter of xylene in inhaled air; however, they appeared to adapt when tested at double the dose of xylene several days later. The following week, when exposure increased to 16.4 mol/liter, reaction time and equilibrium were both impaired. No effects were found on critical flicker fusion, the Santa Ana Dexterity Test, or the Maddox Wing Test. The study suggests that in acute exposure to xylene, both rapid temporary adaptation and longer-term impairment are possible outcomes of inhalation. Another study suggests that lower doses of xylene in short-term exposure may actually improve performance on some neuropsychological tests (Savolainen, Riihimaki, Laine, & Kekoni, 1981). Effects of higher doses, consistent with earlier studies, produced impaired functioning.

Acute exposure to an estimated 700 ppm xylene accidentally distributed through a hospital ventilation system caused headache, nausea, dizziness, vomiting, and irritation of mucous membranes within 30 min of exposure. Symptoms lasted from 2 to 48 h (Klaucke, Johansen, & Vogt, 1982).

There are no neuropsychological studies of workers chronically exposed to pure xylene; however, one study that details subjectively experienced symptoms among chronically exposed laboratory technicians includes reports of impaired work performance and confusion (Hipolito, 1980). Workers exposed to solvent mixtures containing xylene have shown a variety of neuropsychiatric symptoms, including depression, fatigue, paresthesias, headache, nervousness, dry throat, increased thirst, alcohol intolerance, and gastric symptoms (Arlien-Søberg, 1992).

Thus, while there have not been many validating studies, xylene does appear to affect neuropsychological functioning, although at lower, short-term exposure, behavior may not be worsened and may even be facilitated. Longer, high-dose experimental exposures appear to reduce initial adaptive compensation, and produce consistent impairment of reaction time and short-term memory, especially after uptake is increased via exercise.

PROGNOSIS FOR CHRONIC SOLVENT EXPOSURE

Several studies address the longevity of neuropsychological impairment secondary to chronic long-term exposure to solvent mixtures. Bruhn, Arlien-Søberg, Gyldensted, and Christensen (1981) followed up their earlier study (Arlien-Søberg et al., 1979) by reexamining 26 of their original 50 painter subjects chronically exposed to solvent mixtures for an average of 28 years. Mild to moderate cerebral atrophy was found in two-thirds of these painters during their initial examination (Arlien-Søberg et al., 1979). While subjective complaints of headache and dizziness had declined in most subjects, no subjects showed improvement in neuropsychological, neurological, or neuroradiological status after 2 years without solvent exposure. The authors concluded that a syndrome of chronic exposure to solvents does exist, and that once cerebral atrophy or intellectual changes are observed, these symptoms are irreversible.

Another study assessed the recovery of 86 patients exposed to TCE, PCE, or solvent mixtures (Lindstrom, Antti-Poika, Tolla, & Hyytianinen, 1982; Antti-Poika, 1982b). Mean follow-up period was 5.9 years and the average age of patients at the time of the initial diagnosis was 38.6 years. Their mean exposure was 10.7 years. Fifty-three patients stopped working on initial diagnosis, the remainder continued to work for varying lengths of time after diagnosis.

Group means for Similarities and Picture Completion increased over time, suggesting possible learning effects, but two tests of motor functions showed decline. Prognosis was better among younger patients, those who did not use medicines with neurological effects, and those with longer follow-up (Lindstrom *et al.*, 1982, p. 585). Some relief of subjective symptoms was found in 52% of patients who had no other neurological condition. Symptoms of "abnormal fatigue, headache, dizziness, sleep disturbance, nausea and emotional lability" were reported to be significantly improved, although reports of memory disturbances increased somewhat (Antti-Poika, 1982a, p. 81).

When Seppäläinen and Antti-Poika (1983) examined the EEG recovery of these patients, they found slow recovery of EEG in about 59% of the cases, a finding they equated with the recovery course for closed-head injury. The authors noted, however, that a "substantial" number of patients "still had abnormal EEG's three to nine years after their diagnosis" (Seppäläinen & Antti-Poika, 1983, p. 23).

In a "case–control" or matched subject design study carried out in Denmark, Rasmussen, Olsen, and Lauritsen (1985) found a borderline significant difference in risk of development of late-in-life "encephalopathias" as a function of job classifications that involved solvent use. The greatest exposure odds ratios were found for development of psychosis, hypertensive cerebrovascular disease, and senile dementia. There was no increased risk of cerebral and cerebellar atrophy or ischemic cerebral atherosclerosis, a finding that is difficult to understand, considering that cerebral atrophy is itself a common concomitant of senile dementia.

Chronically exposed Finnish solvent workers have an increased risk of being "prematurely pensioned due to neuroses" (Lindstrom, Riihimaki, & Hänninen, 1984). Other studies have shown Swedish and Danish workers to have increased risk for neuropsychiatric disability (Axelson, Hane, & Hogstedt, 1976; Olsen & Sabroe, 1980) and for "presenile dementia" (used in these countries without the connotation of Alzheimer's disease) (Mikkelsen, 1980).

The most comprehensive neuropsychological follow-up of chronic solvent exposure involved the collaborative study of six Swedish departments of occupational medicine (Edling *et al.*, 1990). Data from 111 patients were divided into two groups: one with symptoms but no neuropsychological findings, and the second with diagnosed toxic encephalopathy involving positive neuropsychological tests and symptoms. Most subjects were exposed to solvent mixtures; 47 were house or industrial painters, 24 were car or industrial spray painters, 9 had exposure primarily to petroleum fuels and 3 to styrene. Exposure duration was 23 ± 11 years for the Type 1 group and 26 ± 10 years for the Type 2B group.

Follow-up reevaluation occurred not less than 5 years post initial evaluation and included clinical interview and a flexible battery shown to be sensitive to toxic encephalopathy (TUFF battery, see Chapter 2). Expressed symptoms of depression, concentration difficulties, and lack of initiative improved for more subjects of the Type 1 (symptom complaint only) subjects than Type 2B (neuropsychologically impaired) subjects. When the authors combined scores on the five "core" symptoms (memory, depression, concentration, lack of initiative, and fatigue), "more subjects in the type 2B group had deteriorated and fewer . . . had improved . . . between examinations" (Edling et al., 1990, p. 78).

Psychometric test results showed that most patients retained the same diagnosis over time, but 12 Type 2B subjects improved to Type 1 and three Type 1 deteriorated to Type 2B. Number of reported symptoms for both groups remained stable over time. Individual comparison on the Type 2B patients over time showed deteriorated performance on Block Design, while remaining stable (and low) in psychomotor speed and memory. Workers' quality of life was negatively related to diagnosis of Type 2B solvent encephalopathy. Seventy-four percent of those men had ceased work to receive disability or early retirement benefits, compared with 35% of the Type 1 group.

Shortcomings of the study include the fact that the second comparison was conducted on a follow-up group where 37% of the patients remained in solvent-related occupations. Thus, it is not clear whether noted recovery of 12 subjects took place in the group that continued working with solvents, or that had changed occupations. Another possibility could be that improvements in industrial hygiene might correspond with recovery of function in this group.

The relationship between neuropsychological performance and prognosis received recent attention by Morrow, Ryan, Hodgson, and Robin (1991) who longitudinally evaluated 27 workers with "mild toxic encephalopathy" on the MMPI and a standard neuropsychological battery. Approximately 50% of the group showed neuropsychological improvement on second testing, while 50% were rated as stable or worse. A history of peak exposure requiring medical attention accounted for 36% of the variance determining group status. Workers with peak exposure histories were more likely to be in the unimproved/deteriorated group. Although there was no difference in litigation status between groups, none of the workers in the poor-outcome group had returned to work, while nine from the good-outcome group had resumed employment. It was unclear whether deterioation of cognitive function postexposure in some workers, was neuropathological or psychosocial in origin.

Other studies have not demonstrated neuropsychological recovery after chronic exposure (e.g., Bruhn et al., 1981; Gregersen, 1988), so this particular finding must be considered preliminary.

In conclusion, prognostic studies generally indicate that once chronically solvent-exposed patients have developed cortical atrophy or other brain abnormalities, they recover either very slowly or not at all. Exposed workers who are otherwise healthy and who do not present with actual structural abnormalities appear to be more likely to recover. This conclusion concerns the long-term effects of solvent mixtures or for solvents as a class. Prognostic studies are not available for individual solvents but it may be expected that prognosis will vary, at least, as a function of the solvent's *a priori* toxicity, exposure levels, and chronicity of exposure.

CONCLUSION

Evidence continues to build for neurotoxicity across a wide range of solvents. The need for additional research persists, however, as very basic questions remain unanswered, including neurophysiological mechanisms of solvent toxicity, the sites of brain effects, the influence of individual physiological differences in metabolism, and the interactive effects of age, medications, alcohol, and other toxins. As is true in the field of neurotoxicity generally, current attempts to answer these questions appear to lag far behind the promulgation of new industrial toxins. As solvent effects become more widely known in science and government, perhaps increased multidisciplinary research participation, better governmental funding, and enlightened regulatory policies will give researchers a better chance to finally answer these questions.

5

Alcohol

INTRODUCTION

Alcohol has been described as "the intoxicant of choice in the Judaeo-Christian culture" (Parsons, 1986). Although the average amount of absolute alcohol consumed per person each year in the United States is 2.77 gallons, Parsons (1986) points out that this estimate includes one-third of the population who are abstainers. Thus, the remaining two-thirds of the population must have correspondingly greater per capita consumption rates. Furthermore, 10% of the population is said to consume 50% of the available supply of alcohol (Parsons, 1986, p. 101). Estimating numbers of problem drinkers in the United States depends largely on whether one subscribes to a categorical, disease model of alcohol abuse, and the strictness of the criteria used to divide alcoholics from nonalcoholics. Current estimates range from 5.2 to 13 million alcoholics, with somewhat over twice as many males than females (Hilton, 1989).

As is the case for metal and solvent neurotoxicity (Chapters 3 and 4), neuropsychological evaluation of alcohol-related deficits has been recommended for its "demonstrated efficacy in detection of brain dysfunction" (Eckardt & Martin, 1986, p. 123). Neuropsychological abnormalities are found in approximately 50–70% of detoxified alcoholics *without* organic brain syndromes, while 10% of alcoholics exhibit stable and severe cognitive dysfunctions (Adams *et al.*, 1994). Neuropsychological assessment is of value cataloguing the continuum of severity in alcoholic patients, and how deficits in cognitive and emotional functioning may influence social and vocational plans (Eckardt & Martin, 1986; Wilkinson & Carlen, 1980).

Alcohol and lead are the only neurotoxins whose properties have been extensively debated within the neuropsychological community. In contrast to the dearth of neuropsychological information available about most other neurotoxins, the utility of neuropsychological methods in the evaluation of alcoholism has been widely recognized, with hundreds of research studies extant investigating the neuropsychological concomitants of alcohol. This abundance of alcohol-related neuropsychological research may be related to the availability of specialized federal funding (Grant & Reed, 1985) or may simply be a reflection of alcohol's popularity and visibility as an intoxicant. In either case the volume of available literature precludes an exhaustive survey in the context of a single book chapter. This chapter will instead attempt a general update of neurological and neuropsychological information related to this ubiquitous neurotoxin.

ACUTE EFFECTS

Alcohol produces obvious cognitive and behavioral effects on human behavior that have been chronicled by saint and scientist alike. As far back as the third century, St. Basil presaged modern studies implicating alcohol-induced prefrontal impairments in observing that "drunkenness is the ruin of reason" (Byrne, 1982). Modern research investigations of alcohol's acute effects have generally validated such impressionistic observations of impairment by finding decrements in a range of psychological functions. Impairments in these basic functions, in turn, produce the realworld impairments in driving or performance of any occupation requiring these functions (Mitchell, 1985). Given recent estimates of alcohol-using drivers on the road, the question of acute neuropsychological impairment becomes more than just of academic interest. For example, a 1986 survey found that 3.1% of late-night weekend drivers have blood alcohol concentrations (BACs) of 0.10% or above, and 8.3% had BACs of 0.05% or more (Lund & Wolfe, 1991). While numbers have declined somewhat from those of the previous decade, 39% of sampled traffic fatalities had BACs of 0.10% or more (Lund & Wolfe, 1991).

Acute impairment is reasonably well correlated with blood alcohol level (BAL). BAL is computed using the following formula (Segal & Sisson, 1985):

$$BAL = \frac{80 \times g}{f \times w} = mg/100 \, mL$$

The constant 80 is used because blood absorbs 80% of alcohol from body tissue water; g is the number of grams of absolute alcohol consumed (13.6 g × No. of standard alcohol drinks); f is an estimate of the patient's body water content as a fraction of body weight or type (0.7=lean/muscular, 0.6=average, 0.5=fat); and w is the patient's weight in kilograms.

Acute neuropsychological dysfunction is demonstrable when BAL exceeds 50 mg/dl. At 60 mg/dl, reaction time, anticipation accuracy, and depth perception are impaired (Nicholson et al., 1992). BAL above 100 mg/dl elicits marked impairments in almost all behavioral skills (Mitchell, 1985). Table 5.1 lists acute behavioral effects from increasing BALs.

For any given blood level, neurobehavioral deficits are more serious when alcohol is being initially absorbed into the body (the ascending limb of the blood alcohol curve) than after blood alcohol levels have peaked and begun to decline (the descending limb) (Nicholson et al., 1992; Jones & Vega, 1982). Metabolic breakdown of alcohol in adults is in the range of 7–11 g of ethanol per hour, equivalent to 0.5–1 ounce of 50 proof drink per hour (Klaasen et al., 1986);

TABLE 5.1. Acute Effects of Incremental Blood Alcohol Level[a]

BAL	Behavioral effect
50–150 mg/100 ml	Incoordination, slow reaction time, blurred vision
150–300 mg/100 ml	Visual impairment, staggering, slurred speech, marked hypoglycemia, especially in children
300–500 mg/100 ml	Marked incoordination, stupor, hypoglycemia convulsions
500 mg or greater/100 ml	Coma and death, except in alcohol-tolerant individuals

[a]Klaasen, Amdur, and Doull (1986, p. 889).

however, women may lack a stomach amino acid needed to metabolize alcohol and show correspondingly slower alcohol metabolism (Frezza et al., 1990).

Alcohol is vasoactive, inducing vasodilation in small doses, but vasoconstriction in large amounts (Mathew & Wilson, 1991b). Cortical and cerebellar areas show greater declines in glucose metabolism than callosal or basal ganglia regions (Volkow et al., 1990). Alcoholics show these changes to a greater degree than nonalcoholics (Volkow et al., 1990).

Acute ingestion of ethanol has been linked to reductions in ability across a range of neuropsychological functions; a typical constellation of impairments in a recent study included losses in attention, concentration, auditory-verbal and visuospatial areas. Fine motor coordination was impaired at low and high doses, although motor speed was impaired at high doses only (Minocha, Barth, Roberson, Herold, & Spyker, 1985).

Tarter, Jones, Simpson, and Vega (1971) tested medical student volunteers when their BAL reached 1.1 mg/kg body wt on a neuropsychological battery that included the Stroop Test, Purdue Pegboard, Digit Span, a Dichotic Listening task, Critical Flicker Fusion, simple and choice reaction time, and the Shipley Institute of Living Scale. All tests except the Stroop were affected by alcohol ingestion, although only the abstract reasoning portion of the Shipley was worsened by alcohol. Interestingly, test practice eliminated all neuropsychological differences by the second testing session. This suggests the possibility of acute neuropsychological adaptation to alcohol, at least with bright, presumably nonalcoholic subjects.

Acute effects on perception, balance, and attention were found by Baker, Chrzan, Park, and Saunders (1986). The authors tested 21 nonalcoholic adult males on several measures, including Critical Flicker Fusion, Tachistoscopic Perception, Visual Afterimages, Digit Span, and standing stability. Besides applying relatively uncommon psychophysical measures to the study of alcohol intake, this study is notable for its careful methodology, including the use of a breathalyzer to measure alcohol levels, a MANOVA analysis, and the selection of a double-blind, crossover design where subjects served as their own controls. Participants received either 0 or 1.4 ml/kg body wt of vodka to produce breathalyzer readings of 0.004 or 0.054% (prebehavioral testing). Subjects performed significantly more poorly in the intoxicated condition. Significant decrements were found in patients' tachistoscopic perception and one of the stabilometry measures. Trends toward significance were found in Critical Flicker Fusion and cerebellar tests.

Alcoholics abstinent less than 3 weeks in an inpatient program were examined on the WAIS-R and parts of the WMS-R by O'Mahony and Doherty (1993). Similarities, Digit Span, and Vocabulary scores were significantly greater than Picture Completion, Block Design, Memory subtests, and Trails. The impaired scores did not significantly differ among themselves. Impairments were characterized as diffuse, rather than selectively targeting visual or visuospatial, "right hemisphere" functions.

SOCIAL DRINKING

There has been considerable controversy about the effects of moderate, social, nonalcoholic drinking on neuropsychological performance. While animal studies have noted changes in dopamine receptor sites with low-level alcohol intake (e.g., Lograno et al., 1993), generalization of these results to humans remains controversial and premature. The controversy is probably consistent with the potential impact of a finding that drinkers who average one or two drinks or more each day could suffer alcohol-related brain and behavioral changes. Parsons (1986) estimated that there are millions of such social drinkers in the United States alone.

In an extensive review of the subject, Parsons (1986) concluded that current understanding of the relationship between social drinking and neuropsychological performance is inconclusive, and that the investigation is still in its infancy. Existing studies have suffered from a variety of methodological deficiencies, including failing to match for education in alcohol and control groups; failure to measure or control for acute versus chronic alcohol effects; and not including the influence of relevant historical, medical, educational or sociodemographic variables. Considering the potential magnitude of the problem, Parson (1986) recommends large random sample prospective studies with long-term recording of drinking practices.

Reviewing the research support for the hypotheses shown in Table 5.2, Parsons found little support for hypothesis 1, and suggested that hypothesis 2 is worthy of further exploration. There appears to be a small but growing amount of evidence that links genetic makeup with poorer neuropsychological performance in the families of alcoholics (e.g., Schaeffer, Parsons, & Yohman, 1984; Drejer, Theilgaard, Teasdale, Schulsinger, & Goodwin, 1985). The "alcohol threshold" hypothesis is potentially interesting, since variation in individual thresholds could account for differences among research findings. However, there has been no method identified for establishment of such thresholds in the existing literature. Finally, this hypothesis does not have any direct evidence that would prove that amount consumed per occasion rather than frequency of drinking is related to brain impairment.

TABLE 5.2. Possible Relationships between Social Drinking and Neuropsychological Performance[a]

1. Alcohol–causal	Alcohol directly produces brain dysfunction as a consequence of amount of alcohol drunk per occasion
2. Cognitive–causal	Those with poorer intellect drink more than individuals with better cognitive abilities
3. Stress emotional–causal	Subjective stress or emotional problems lead to higher alcohol intake and poorer neuropsychological functioning
4. Genetic–causal	Genetic heritage leads to increased drinking and poorer neuropsychological performance
5. Alcohol–causal–threshold	Same as alcohol–causal but damage occurs only after a threshold has been achieved
6. Alcohol–causal–transient	Alcohol ingestion causes "limited" changes accompanied by neuropsychological and emotional disturbances

[a]Adapted from Parsons (1986).

CHRONIC EFFECTS

Acute intoxication and social drinking have typically been of less interest to neuropsychological researchers than the chronic effects of alcohol on brain and behavior. These effects could be conceptually divided into at least three areas of research interest that span the lifespan continuum: (1) prenatal teratogenic effects, including fetal alcohol syndrome (FAS), (2) the long-term consequences of chronic alcohol abuse uncomplicated by Werniche–Korsakoff syndrome, and (3) specific neuropathological and neuropsychological deficits induced by alcoholic Wernicke–Korsakoff syndrome.

Prenatal Effects of Alcohol Exposure

There is an assumed continuum of prenatal alcohol effects from minimal brain dysfunction to severe FAS. Determination of FAS in low-income or impoverished populations may be particularly difficult, since environmental stressors, low parent IQ, and other factors may be difficult to covary out from alcohol effects (Boyd, Ernhart, Claire, Greene, Sokol, et al., 1991). Nonetheless, the effect of alcohol on the fetus is real and multidetermined. Alcohol may interfere with protein synthesis as a consequence of (1) diminished placental transport of amino acid and nutrients, (2) alcohol or acetaldehyde effects on ribosome content, tRNA content, aminoacylation of tRNA, or transfer enzyme function in protein synthesis, and (3) malnutrition (Michaelis, 1990). Rat studies suggest that alcohol interferes with brain protein synthesis especially in hippocampal, amygdala, and cerebellar regions (Peters & Steele, 1982). Some potential effects of prenatal alcohol exposure are listed in Table 5.3.

Alcohol's neurotoxicity to the developing fetus can produce both subtle alcohol-related birth defects (ARBDs) as well as the severe deficit constellation

TABLE 5.3. Pathophysiological Mechanisms Contributing to Fetal Alcohol Syndrome[a]

Ethanol-induced embryotoxicity
Acetaldehyde-induced embryotoxicity
Chromosomal abnormalities
Ethanol-induced placental dysmorphology
Zinc deficiency
Ethanol-induced blood flow reduction inducing hypoxia
Alcohol-induced prostaglandin release
Concomitant drug use, smoking, malnutrition, other factors

[a]From Randall, Ekblad, and Anton (1990) and Michaelis (1990).

of FAS. Jacobson *et al* (1994) studied the former using the Bayley Scale to analyze the behavior of black infants born to inner-city mothers who were moderate to heavy drinkers during pregnancy. Clinically significant impairment on the Mental Development Index (MDI) was seen when mothers drank half an ounce or more of absolute alcohol per day, although the women surveyed resembled binge drinkers more than daily drinkers, with a median intake of six drinks per occasion. Subtle decrements on the MDI were seen at lower levels but the risk of having a low score on the MDI or McCall Index more than doubled above this level. The Psychomotor Development Index (PDI) on the Bayley was impaired only in the heaviest drinkers (2 ounces or more of absolute alcohol a day, or 4 ounces at the time of conception) suggesting that "gross motor coordination is less sensitive to prenatal alcohol exposure that [is] early mental development" (Jacobson *et al.*, 1994, p. 181). Adverse effects on the Bayley were more common in infants born to women older than 30 years of age and alcohol effects in the second and third trimester were worse than those from exposure at the time of conception.

FAS consists of biochemical, behavioral, and psychological abnormalities that include craniofacial and dysmorphic anomalies, retardation of intrauterine growth, and CNS involvement as well as tremulousness, fine-motor dysfunction, and poor sucking behavior (Rawat, 1979). FAS is understood to be the principal cause of mental retardation the United States (Abel & Sokol, 1987). Global incidence is estimated to be 0.4–3.1 per 1000 births, but may be as high as 190/1000 in rural, native communities with a high incidence of maternal alcohol abuse (Conry, 1990).

Diagnosis is made with positive history of maternal prenatal alcohol and postnatal symptoms that include (1) growth deficiency of prenatal origin (2) minor physical abnormalities, including short palpebral fissures, midface hypoplasia, smooth and/or long philtrum, thin upper lip, and CNS impairment. The latter may manifest as microcephaly, developmental delay, hyperactivity, attention deficit, learning disability, lowered IQ, or seizures. Growth and physical development may be impaired. FAS may be the endpoint of a continuum of prenatal alcohol effects, with less severe symptom constellations characterized as fetal alcohol effect (FAE) (Streissguth, Barr, & Sampson, 1990).

Neuropsychological Effects

Neuropsychological evaluation of FAS children has been advocated to increase understanding of alcohol's neuroteratogenicity and its correlation with functional aspects of the nervous system (West, Goodlett, & Brandt, 1990). Conry (1990) administered a neuropsychological and psychoeducational battery to 13 FAS and 6 FAE children, comparing them with age-matched controls. Multivariate analysis indicated significant main effect of alcohol and borderline significant interaction the latter testing for whether "the difference between affected and control children was greater for FAS than for FAE groups" (p. 652).

Streissguth et al., (1991) examined a cohort of 61 FAS patients ranging from 12 to 40 years old. Subjects received WAIS-R or WISC-R IQ, WRAT-R achievement tests, and underwent a physical and clinical record review at an interval of between 5 and 12 years. Average IQ for all subjects was 68, but IQ ranged from 20 through 105. Fifty-eight percent of patients had IQs in the range of mental retardation. Mean IQ of full FAS patients was 66, while FAE children showed a slightly higher mean IQ of 73. Educational functions were similarly impaired with fourth grade reading, third grade spelling, and second grade arithmetic. Intellectual and educational impairments were accompanied by social and interpersonal impairments. Vineland maladaptive behavioral scale showed that 62% of patients had "significant" levels of abnormal social behaviors, including poor attention, dependency, social withdrawal, and anxiety. However, as the authors correctly note, it is difficult to partial out pre- and postnatal effects. Their FAS population came from severely dysfunctional homes; almost one third were never raised by their biological mother, but were abandoned or given up for adoption at birth. On average they had lived in five different homes. The patients' mothers may have been an especially pathological sample, as 69% were dead at the time of the study. The extent to which extremely pathological upbringing influenced IQ and social competence in this FAS population is unknown. Studies of higher SES/higher education populations may be necessary to better partial out these factors.

Chronic Adult Alcoholism without Wernicke–Korsakoff Syndrome

Neuropathology of the Central and Peripheral Nervous System

The adult nervous system is also a principal target of this well-known neurotoxin (Juntunen, 1982a) with neurochemical, behavioral, and neuropathological changes reported as consequences of chronic exposure. Animal studies suggest that alcohol decreases the viscosity of biological membranes, which in turn causes neurotransmission abnormalities (Shoemaker, 1981). Neurochemical correlates of acute alcohol administration also include changes in acetylcholine (ACh) release and Pelham, Marquis, Kugelmann, and Munsat (1980) found that chronic alcohol use alter brain cholinergic neurons. They suggest that such effects may be permanent, or at least persistent. Such findings may help explain the genesis of some alcohol-related forms of memory disorder. Other neurotransmitters, including norepinephrine, are affected by chronic alcohol intake, as is neurotransmitter turnover (Hoffman & Tabakoff, 1980). Neuropathological changes in neuronal cell membrane structure are also associated with chronic use (Hoffman & Tabakoff, 1979). Walker, Heaton, Lee, King, and Hunter (1994) have hypothesized a role for alcohol-mediated reductions of nerve growth factors that are required for the "survival, function and maintenance of basal forebrain cholinergic neurons" (p. 13) suggesting that neurons in the hippocampus and septal regions which depend on such neurotrophic influences, would be especially compromised. Excitotoxicity from excessive glutamatergic nerve transmission caused by alcohol abuse may destroy selective neurons and underlie mechanisms of cortical atrophy and gray matter loss (Lovinger, 1994).

There is growing evidence that the action of alcohol is neurotoxic in itself, without accompanying avitaminosis. Recent animal studies have found a variety of alcohol-induced neuropathological changes in animals maintained on a nutritionally complete ethanol diet. For example, Rosengren, Wronski, Briving, and Haglid (1985) sought to investigate cortical changes in gerbils fed alcohol for 3 months in an otherwise balanced liquid diet. Neuropathological studies were conducted after 4 additional months when the animals were placed on an ethanol-free diet to allow recovery from acute toxic effects. The authors found results suggesting permanent changes in the composition and volume of brain cells which varied according to the cortical areas studied. They viewed such changes as consistent with gliosis found in the cerebral frontal and temporal cortex and the cerebellar vermis of human alcoholics (Rosengren *et al.*, 1985). Two other studies suggest that mice maintained on a nutritionally complete ethanol diet for 4 months show losses of hippocampal neurons, hippocampal pyramidal cell dendritic spines, and dentate granule cells (Walker, Hunter, & Abraham, 1981; Riley & Walker, 1978).

Cerebral and cerebellar damage are common consequences and cortical neuropathology is typified by diffuse and uniform atrophy, with both diffuse and patchy loss of neurons (Tarter & Van Thiel, 1985). Loss of subcortical white matter is proportionally greater than declines in cerebral gray matter (Charness, 1994). Cortical white matter loss is found in alcoholics with and without liver disease and degree of white matter loss (6–17%) is sufficient to account for documented ventricular enlargement (Charness, 1994).

Chronic alcohol abuse may reduce brain weight (and presumably cell mass). Harper and Blumberg (1982) found autopsied alcoholics to have brain weights below age-matched controls. In a later study, Harper and Kril (1985) studied the autopsied brains of controls, alcoholics, alcoholics with Wernicke's encephalopathy, and alcoholics with liver disease. The authors estimated brain tissue loss to occur for all alcoholic groups, with progressively greater losses in the Wernicke's and liver disease groups. However, a recent comprehensive review by Charness (1994) concluded that brain weight in alcoholics "is reduced only slightly as compared with that in non-alcoholics and some workers find no differences at all" (p. 8).

Atrophy of the cerebellar vermis occurs in about 30% of otherwise healthy alcoholics under age 35, with two thirds of this group showing more extensive, bilateral cerebellar atrophy (Tarter & Van Thiel, 1985). Older alcoholics show similar percentages of cerebellar atrophy (Lindboe & Løberg, 1988). Cerebellar damage is usually found after 10 or more years of alcohol abuse, and hypothesized causes for this particular brain abnormality include thiamine deficiency or electrolyte disturbance, rather than direct neurotoxicity (Charness, 1994).

Alcoholism is also one of the most common causes of peripheral neuropathy; this etiology accounts for almost 30% of peripheral neuropathy cases (Kemppainen, Juntunen, & Hillbom, 1982). Faster large myelinated fibers are more sensitive to chronic alcohol abuse than slower large myelinated fibers. Neurophysiological testing shows moderate slowing in motor and sensory nerve conduction (Fujimura, Araki, Murata, Yokoyama, & Handa, 1993).

Indirect Toxic Effects

The gradual supplanting of nutritive calories for ethanol may produce avitaminosis, and therefore the elimination of thiamine (B_1) from the diet. A complete thiamine-deficient diet, whether from alcoholism or other factors, can, in a matter of weeks, precipitate permanent structural damage to the CNS in the form of Wernicke's encephalopathy (Hartman, Sweet, & Elvart, 1985). If uncorrected, the syndrome can be fatal. Neuropsychological dysfunction resulting from this indirect effect of alcoholism is discussed later in the chapter.

Alcohol may also indirectly damage the CNS by producing cerebrovascular spasm and contraction of cerebral microvessels. Altura and Altura (1984) demonstrated that blood alcohol concentrations as small as may result from a single drink are capable of producing significant contractions of blood vessels. With alcohol blood levels high enough to produce stupor or coma, "intense spasm . . . and often rupture" of cerebral microvessels can occur (Altura & Altura, 1984, p. 328). This phenomenon may partially account for the Hillbom and Kaste (1983) findings that 40 of 100 patients presenting with ischemic brain infarction had used alcohol heavily within the previous 72 h. A possible relationship of ethanol-induced vascular spasms and hypertension has also been postulated (Altura & Altura, 1984).

Ethanol impairs neural recovery of function after injury (Charness, 1994), and individuals with histories of alcoholism show poorer postoperative neurosurgical course and higher postoperative morbidity when there is additional head trauma (Sonne & Tonnesen, 1992; Laatsch, Hartman, & Stone, 1994). A history of chronic alcohol abuse is also associated with poorer prognosis following severe closed head injury. Ruff *et al.* (1990) found greater prevalence of head injury-induced mass among chronic alcohol abusers, and poorer outcome using the Glasgow Outcome Scale (GOS). The association remained significant when the effects of age and BAL were controlled.

Similarities and Interactions with Other Neurotoxic Exposures

Since alcohol is itself a solvent, it would not be surprising if the neurological and neuropsychological features of chronic alcohol exposure resembled those resulting from chronic solvent intoxication. Acute symptoms are very similar, although ethanol may be more potent at lower doses (Echeverria, Fine, Langolf, Schort, & Sampaio, 1991). There would appear to be many similarities in symptom development. For example, both disorders have insidious onset and are difficult to diagnose in the early stages. Second, each syndrome is usually caused by repeated intoxicating exposures over a long period of time. Third, chronic exposure to either toxin may produce deteriorations in occupational and interpersonal coping (Juntunen, 1982a). Finally, there is preliminary evidence that impairments of central and peripheral nerve conduction are larger in solvent workers who are also alcoholics, suggesting the additive or synergistic effect of the two substances (Massioui *et al.*, 1990). Exposure to certain solvents preceded by alcohol ingestion slows solvent metabolism and clearance, allowing solvents to reside in the tissues of heavy drinkers longer and produce correspondingly more damage. Workers with chronic solvent exposure who are also heavy alcohol abusers are more likely to be hospitalized with organic brain damage than for other psychiatric diagnoses (Cherry, *et al.*, 1992). Cherry (1993) reviewed three studies suggesting that solvent workers are at particular risk for developing alcohol-related disabilities, including dementia, alcoholic psychosis, or alcohol dependency. As most studies do not find painters to differ from controls in alcohol consumption, Cherry suggests that either solvent-injured workers cannot tolerate larger amounts of alcohol without damage, or that solvents and alcohol potentiate one another. Since experimental exposure to certain solvents preceded by alcohol ingestion slows solvent metabolism and clearance, solvents may reside in the tissues of heavy drinkers longer and produce correspondingly more damage.

Alcohol may cause changes in bioavailability or toxicity of other neurotoxic substances. For example, alcohol may increase absorption of lead in the gastrointestinal tract, change its targets, and mobilize bound lead from tissues (Cezard, Demarquilly, Boniface, & Haguenoer, 1992).

There remains a comparative dearth of information concerning the ways in which alcohol and other neurotoxic substances might interact. The topic is more complex than might first be assumed. Questions that would need to be addressed include analyzing how drinking habits affect or potentiate solvent symptoms, whether tolerance to alcohol changes after solvent exposure, and even how prior alcohol intake influences choice of a neurotoxin-related occupation (Baker & Seppäläinen, 1986). Unfortunately, these questions are not the subjects of investigation at the present time. It remains for alcohol or neurotoxicity researchers to extend their investigation to these effects.

Disease Pathology

While comparative and interactive effects of alcohol with solvents have yet to be studied in detail, the neurological and neuropsychological effects of alcohol alone have been widely studied. Alcohol has been linked to an extensive list of disease pathology, and is toxic across the entire human life cycle, beginning with prenatal exposure, and possible before that, with direct alcohol-related damage

TABLE 5.4. Primary or Secondary Neurological Disorders of Alcoholism[a]

Alcohol intoxication	Dementia associated with alcoholism
Alcohol idiosyncratic intoxication	Wernicke-Korsakoff syndrome
Alcohol withdrawal	Alcoholic polyneuropathy
Alcohol withdrawal delirium	Alcoholic cerebellar degeneration
Alcohol epilepsy	Retrobulbar neuropathy
Alcohol hallucinosis	Alcohol myopathy
Alcohol amnestic disorder	Hepatic encephalopathy and hepatocerebral degeneration
Central pontine myelinolysis	Fetal alcohol syndrome

[a]Reprinted, by permission of the publisher, from Juntunen, J. (1982a). Alcoholism in occupational neurology: Diagnostic diffculties with special reference to the neurological syndromes caused by exposure to organic solvents. *Acta Neurologica Scandinavica*, 66, 89-108.

to sperm or egg prior to conception. Adults are subject to a diversity of diseases resulting from excessive alcohol consumption (Table 5.4).

Hepatic Dysfunction

Liver disease is a common concomitant of alcoholism, and as many as 30% of alcoholics develop cirrhosis (Tarter, Moss, Arria, & Van Thiel, 1990). Hepatic damage from alcohol can be considered an indirect cause of neurotoxicity. Full hepatic encephalopathy is characterized by impaired sensorium, asterixis, frontal release signs, hyperreflexia, extensor plantar response, and occasional seizures (Charness, 1994, p. 6). Brain damage is produced in the basal ganglia, thalamus, red nucleus, pons, and cerebellum, but neuronal loss or glial changes are not observed (Charness, 1994). A more progressive and severe variant of the syndrome produces patchy cortical, cerebellar, and basal ganglia neuronal loss with "cavitation of the cortico-subcortical junction and superior pole of the putamen" (Charness, 1993, p. 6).

Less severe hepatic dysfunction has also been shown to be highly correlated with both neurological and neuropsychological impairment. Acker *et al.*, (1982) found that cerebral atrophy covaried with liver disease. Two studies suggest that EEG abnormalities are worse in cirrhotic alcoholics than those without liver disease (Kardel & Stigsby, 1975; Kardel, Zander, Olsen, Stigsby, & Tonneson, 1972). Cirrhosis without alcoholism may also cause neuropsychological difficulties that are similar to those of alcoholics (Tarter *et al.*, 1983).

In what may eventually prove to be a landmark study in the neuropsychology of alcoholism, Tarter, Hegedus, Van Thiel, Gavaler, and Schade (1986) examined the neuropsychological performance of 15 subjects whose alcoholism was confirmed by the most direct evidence possible; all subjects received a diagnosis of Laennec's cirrhosis after liver biopsy. Subjects had been detoxified before their most recent hospital admission and all had been abstinent for 2 months or more. The authors found that abnormalities of hepatic biochemistry in a large neuropsychological battery accounted for between 23 and 56% of the variance depending on the particular combination of biochemical measure and neuropsychological test score (Tarter *et al.*, 1986; Table 5.5). In a later study, Tarter *et al.* (1988) compared the neuropsychological performance of alcoholics with and without cirrhosis, to controls. While both alcoholic groups showed impairments, cirrhotic alcoholic performed more poorly in short-term memory, visual tracking, and eye-finger coordination (Tarter *et al.*, 1990).

Finally, Richardson *et al.* (1991) showed that extremely elevated liver enzyme γ-glutamyl transferase (GGT) was associated with neuropsychological impairments in Digit Symbol, Block Design, Category Test, TPT total time and localization, Trails B, and Russell's modification of the Wechsler Memory Scale. GGT elevations during acute alcohol withdrawal predicted residual deficits. None of the alcoholics were clinically cirrhotic, although biopsies were not performed. No alcohol was consumed for 21 days before neuropsychological testing.

Without longitudinal data and corresponding imaging (e.g., CT, MR), it is not possible to determine whether liver and other biochemical abnormalities *cause* neuropsychological impairment, or whether abnormal liver enzymes simply correlate well with concurrent neuropathology. This area of study also awaits larger replications and follow-up research designs to examine subjects over time and determine reversibility. Current findings do suggest caution in automatically attributing the neuropsychological effects of alcoholism to brain abnormalities on the basis of psychometric test data alone.

Other Diseases

Two other diseases that present with neuropathological abnormalities associated with alcoholism are *Marchiafava-Bignami syndrome* and *central pontine myelinosis*. Marchiafava-Bignami syndrome is rare, but occurs principally in malnourished alcoholics. The disease is characterized by corpus callosum demyelination or necrosis. Pericallosal white matter may also be damaged. Characteristic symptoms include dementia, dysarthria, inability to ambulate, and spasticity. Central pontine myelinosis typically affects alcoholics with liver disease, but may occur in nonalcoholics having diseases affecting the liver, e.g., Wilson's disease. Lesions are produced at the base of the pons, with 10% of cases having lesions in other locations, including the striatum, thalamus, cerebellum, and cerebral white matter (Charness, 1994). "Locked-in" syndrome may be observed with normal level of consciousness, but limited ability to move the face or limbs.

Evidence of Chronic Effects: Imaging and Electrophysiological Studies

CT Scan

CT scan has been shown to be a reliable and valid method of measuring brain changes of alcoholic etiology. Depending on the study, the criteria, and the location of atrophy, from 33 to 95% of alcoholics show some form of atrophy on CT scan (Begleiter, Porjesz, & Tenner, 1980). There is general accord that CT scan is able to discriminate alcoholics from nonalcoholic controls (e.g., Bergman,

TABLE 5.5. Correlation between Liver Function Variables and Neuropsychological Test Performance[a]

Neuropsychological test	Liver function								Fasting plasma ammonia level
	Alanine transaminase	Aspartate transaminase	Alkaline phosphatase	Bilirubin direct	Albumin	Globulin	Prothrombin time	Indocyanine green serum (level at 20 min)	
Digit Span Plus One					−0.51*				
Fluency							−0.49*	0.62*	0.66*
Confrontation Naming	−0.50*	−0.55*	0.48*						
TPT—Memory	−0.72**								
TPT—Location	−0.54*					−0.74*			
Raven's Matrices IQ								−0.71*	−0.84**
Visual Memory		−0.53*							
Logical Memory—Delayed				−0.49*			−0.71**		
Trail Making Test	0.57*						0.51*		
Star Drawing Test—Errors									0.59*
Purdue Pegboard			0.62**						

[a]Tarter et al. (1986, p. 76).
*p <0.05; **p <0.01.

Borg, Hindmarsh, Idestrom, & Mutzell, 1980b). For example, Bergman, Borg, Hindmarsh, Idestrom, and Mutzell (1980a) examined 130 alcoholics with CT and neuropsychological tests. Diagnosed alcoholics and heavy drinkers from the control group showed greater widening of cortical sulci, larger ventricles, and cerebellar atrophy. While these abnormalities also correlated with aging, sulcal widening was common in alcoholics below age 40. Similar results have been found by Ron (1983).

CT also appears to be useful in differentiating alcoholics with Wernicke's encephalopathy from nonamnesic alcoholics. Wilkinson and Carlen (1980c), for example, compared these two groups who were diagnosed by Wechsler Memory Scale performance and neurological evaluation. Alcoholics with amnesic memory and positive neurological examination had significantly larger sulcal width and ventricle and sulcal scores on CT scan.

In contrast, CT alone may be less able to discriminate among non-Korsakoff alcoholics with different drinking histories. Ron (1983) surveyed 100 alcoholics using CT, and collected information on alcohol intake in terms of peak daily amount (g/day) and duration (number of years during which the patient had drunk more than 150 g of alcohol per day, "several" times per week). Measures of drinking pattern and abstinence duration were also collected (Ron, 1983, p. 12). CT measures included ventricle size, and width of sulci, Sylvian and interhemispheric fissures. Within older and younger age groups, alcoholics showed significantly greater structural abnormalities on all four variables. However, when Ron divided her groups into patients of similar age but different drinking histories (mean = 10 versus 26 years), no significant CT differences were observed between the two groups.

While a majority of CT studies appear to demonstrate some form of cortical atrophy in chronic alcoholic patients, the lack of well-controlled studies, and continuing development of more accurate machines, make definitive conclusions impossible at this time. There are also several problems common to CT research design, including possible selection bias in referral of alcoholics for CT; it may be that only more visibly impaired patients are labeled as alcoholic and referred for CT. Second, many CT studies have not required or controlled for abstinence periods prior to CT, potentially confounding acute and chronic effects. Third, some studies have used mixed populations of alcoholics who have other forms of neurological or physical illness, or have not controlled for the effects of cirrhosis. Fourth, the mere presence of CT deficits in alcoholics does not necessarily suggest a casual relationship between alcohol and atrophy. The study of Bergman *et al.* (1980a) found significant CT abnormalities in *all* age groups of alcoholics relative to nonalcoholic controls. This may indicate global, acute toxic influence of alcohol across all age groups, but may also suggest preexisting neuropathological abnormalities among these subjects. Support for the latter hypothesis can be extrapolated from studies of alcoholic families (e.g., Alterman, Tarter, Petrarulo, & Baughman, 1984; Alterman & Tarter, 1985). Finally, age and population differences may account for some CT effects (Ron, 1983).

CT Relationship to Neuropsychological Test Results. The relationship between CT scan results and neuropsychological tests has been examined in several studies. Wilkinson and Carlen (1980a) found that Wernicke's alcoholic had both poorer CTs and greater impairment on the WAIS Digit Symbol Test and the memory subtest of the Tactual Performance Test. Gebhardt, Naeser, and Butters (1984) analyzed the relationship of CT scans to neuropsychological performance in 24 chronic alcoholics who were administered a verbal Paired Associate Learning Test, a four-word short-term memory test, the Digital Symbol Substitution Test, and the Symbol Digit Paired Associate Learning Test. CT density in the thalamic dorsomedial nucleus correlated with verbal paired associate learning scores and with Symbol Digit performance. Third ventricle intracranial width also correlated with verbal paired associate learning scores. The authors suggest a relationship between the integrity of midline thalamic structures and the ability to form new associations in long-term verbal memory. Alcoholics who show the lowest mean CT density numbers are also the most impaired on a verbal paired associate learning task. In contrast, tests of short-term memory or their corresponding brain structures did not show impairment. Hill (1980) found that the ventricle/brain index correlated with Category Test errors in a sample of 42 subjects using alcohol and/or heroin.

Finally, Bergman *et al.* (1980c) surveyed a mixed group of binge and daily drinkers on CT with a median 11.7 years of alcohol abuse. The Halstead–Reitan Impairment Index, Block Design, Category Test, and TPT-location correlated significantly with an index of sulcal width. Verbal reasoning, Synonyms, Block Design, the Category Test, Trails A and B, as well as several tests of memory correlated with indices of ventricular size. The authors suggest that the Halstead–Reitan Impairment Index is sensitive to cortical changes on CT, while neuropsychological tests of learning and memory reflect subcortical integrity.

Not all studies have confirmed the relationship between CT and neuropsychological impairment. For example, Ron (1983) examined 100 male alcoholics and compared them with 50 volunteers on CT and neuropsychological tests. Ventricles, sulci, and interhemisphere fissures were wider in alcoholics, and cerebellar sulci were "visible only in the alcoholic group" (Ron, 1983, p. 13). However, other significant correlations between CT scan results and neuropsychological measures went in the opposite direction; some measures of atrophy correlated with *better* performance on a visuo-perceptual task as well as measures of immediate recall and partial cued recall. The lack of correlations between CT results and psychometric data in the direction of impairment was termed "disappointing" by the author, who hypothesized that the discrepancy between immediate and delayed recall of verbal material may have been better correlated to ventricular abnormalities than sulcal or Sylvian fissure width (p. 26). However, Wilkinson and Carlen (1980b) found no significant relationship between the Halstead–Reitan Impairment Index and ventricle size or sulcal width. CT results were unrelated to impairment index scores, whether or not alcoholic had neurological abnormalities. A wide variation (between 2 and 20 weeks) in abstinence among patients before receiving CT scan may have obscured a relationship between neurological and neuropsychological results in this study.

Lusins, Zimberg, Smokler, and Gurley (1980) studied 50 alcoholic patients with CT and a battery consisting of the WAIS and Trail Making Tests A and B. Duration of drinking, rather than neuropsychological test results, was the only factor that correlated with cerebral atrophy as imaged by CT. However, this study can be faulted for its use of the WAIS, a measure that is relatively insensitive to chronic alcohol intoxication. The authors correctly point out the need for a more comprehensive neuropsychological battery in future studies. Melgaard, Danielsen, Sorensen, and Ahlgren (1986) examined the CT scans of 46 male, chronic alcoholics and compared CT abnormalities to neuropsychological test performance. They were unable to show alcoholism history (as sampled by the Modified Missouri Alcohol Severity Scale) to be significantly correlated with CT or neuropsychological data. It may be, however, that questionnaire surveys of alcohol history were not valid for the particular populations surveyed, particularly since several neuropsychological tests also failed to correlate with the two indices of alcohol exposure. Cala, Jones, Wiley, and Mastaglia (1980) were also unable to confirm a general relationship between neuropsychological performance and CT results. The disparity among various studies remains unresolved, although variations in test sensitivities, population differences, willingness to disclose drinking histories, and the multifactorial nature of chronic alcohol damage undoubtedly contribute to the disparity among results.

PET Scan

Positron emission tomography (PET) scan measures local tissue concentration of positron-emitting radioactive isotopes. Unlike more conventional CT scans, PET scans can image active regional brain metabolism. PET apparatus is expensive and relatively uncommon; thus, few investigations of alcoholic patients using this methodology have thus far been conducted. Some preliminary PET research has detected lower glucose metabolism in the medial prefrontal and temporal cortex, thalamus, and basal ganglia of Korsakoff syndrome alcoholics (Kessler *et al.*, 1984). Non-Korsakoff alcoholics are reported to show more inconsistent PET abnormalities, although two studies showed significantly lower than normal metabolism "predominantly in the frontal and parietal cortexes" (Volkow *et al.*, 1994, p. 178). Similar results were found by Adams *et al.* (1994) who found that older alcoholics showed abnormal local cerebral metabolism on PET scan that localized to a sagittal strip of the medial frontal cortex.

Volkow *et al.* (1994) used PET to examine the recovery curve of recently detoxified, presumably non-Korsakoff, alcoholics. Brain metabolism showed significant increases during detoxification with most recovery occurring between 16 and 30 days. At the end of detoxification, alcoholics continued to show lower basal ganglia metabolism compared with their own scans and that of controls, along with lower parietal cortex metabolism relative to their own scans.

Event-Related Potentials

Alcoholics generally produce two types of deficits in these electrophysiological evaluations: N2 delay, which reflects defective stimulus evaluation, and low P3 amplitude, suggesting defective cognitive processing involved in the determination of stimulus matching. Parsons, Sinha, and Williams (1990) compared event-related potentials (ERPs) and neuropsychological performance in middle-aged alcoholics and community controls. Deficits interacted with gender, and family history of alcoholism. Only male alcoholics were impaired on ERP, showing reduced amplitudes for P3, N1, and Nd, and increased latency for N2. ERPs also differentiated alcoholics and controls in a later study from the same lab, where controls showed higher amplitudes and shorter latencies than alcoholics (Glenn, Sinha, & Parsons, 1993). Alcoholics performed significantly worse on all four factor analytic groupings of neuropsychological tests, including verbal tests, visuospatial, perceptual motor, and verbal story paragraph memory. Both male and female alcoholics were impaired relative to controls on these groupings.

ERP abnormalities persist in recently abstinent alcoholics. Realmuto, Begleiter, Odencrantz, and Porjesz (1993) examined 63 alcoholics with mean abstinence of 31 days. Alcoholics showed a significantly lower amplitude for N2, P3, and N2-P3. Results were hypothesized to reflect defective automatic processing "because of a defect in the mnemonic template necessary to match with an infrequent deviant stimuli" (p. 594). In addition, alcoholics' significantly lower amplitude on the Fz electrode was proposed by the authors to reflect "relative weakness of frontal cortical organization" (p. 598).

Positive correlations were obtained between ERPs and neuropsychological test scores; higher ERP amplitude was reflected in better neuropsychological performance. In the alcoholic sample, amplitude of visual and auditory P3's correlated significantly with the perceptual motor test cluster at Pz. Visual P3 latency negatively correlated with perceptual motor performance: faster latencies were associated with better neuropsychological performance. Gender and family history variables also produced significant correlations with ERP and neuropsychological performance. Positive family history of alcoholism in the male sample correlated with several variables, suggesting that positive family history subjects "have less efficient signal identification and processing than [negative family history] peers," a result the authors link to attention and concentration difficulties (p. 753). Among individual tests, Symbol-Digit was the most highly correlated with P3 latency.

ERPs may also have some ability to predict recidivism in sober alcoholics. Glenn et al. (1993) examined alcoholics at the end of an inpatient treatment program of 21-45 days. When patients were reexamined 13 months later, N200 latencies were significantly longer in the original ERPs of resumed alcoholics and a discriminant function of N2L, P3A, and N1A correctly identified 61.5% of abstainers and 65.6% of resumers. Somewhat higher rates of discrimination were found when functional and affective variables were considered. Depressive symptoms, psychosocial maladjustment, and N2L produced a prediction rate of 71%, supporting the construct of alcoholism as best understood within a biopsychosocial framework (Glenn et al., 1993).

Finally, visual discrimination ERPs may also predict likelihood of becoming alcoholic in the first place, since children with high familial risk for alcoholism appear to show delayed maturation via developmentally reduced slope of P300 amplitude in visual and auditory modalities (Hill & Steinhauer, 1993; Steinhauer & Hill, 1993).

Regional Cerebral Blood Flow

Cerebral blood flow (CBF) and regional cerebral blood flow (rCBF) studies directly measure cerebral activity and show high correlation with brain oxygen consumption (Berglund, 1981). Human studies of rCBF during acute alcohol consumption generally find that small amounts of alcohol cause vasodilation, while larger doses induce vasoconstriction (Mathew & Wilson, 1991b), although regional differences make this effect more complex. For example, in one study, small doses (0.7 g/kg) of alcohol increased mean hemisphere blood flow by 12%, while larger doses (1.5 g/kg) produced a 16% increase in CBF over the entire hemisphere (Sano et al., 1993). However, temporal regions increase rCBF in a linear fashion over dose, while prefrontal regions cause increased rCBF at the lower dose but decreased blood flow at the higher dose (Sano et al., 1993).

Chronic alcoholic damage to white matter is inadequately measured with xenon-133 inhalation rCBF technique, but two studies that injected ^{133}Xe found reduced white matter flow in this population (Berglund & Ingvar, 1976; Johannesson, Berglund, & Ingvar, 1982). The utility of CBF and rCBF to localize the effects of alcoholism has been reviewed by Berglund (1981), who notes the effects of ethanol toxicity (Table 5.6).

Wernicke-Korsakoff Syndrome

Wernicke-Korsakoff syndrome is a brain disorder produced by a thiamine-deficient diet, a condition that produces destruction of brain tissue in the midbrain surrounding the third ventricle. The syndrome is believed to result from the

TABLE 5.6. Review of CBF and rCBF Effects of Alcoholism[a]

1. CBF decreased in Wernicke's encephalopathy
2. CBF shows large variability of flow in Korsakoff's psychosis
3. Normal CBF in "average alcoholic" but alcoholics with DTs and physical illness may have lowered CBF
4. Possible accelerated reduction in CBF as a function of aging in alcoholics over 50 years old
5. Reduced mean hemispheric blood flow does not improve after the first week of abstinence in most alcoholics; after 3 weeks of abstinence "there are hardly any changes"
6. rCBF has also been found to positively correlate with neuropsychological test scores

[a]Berglund (1981).

alcoholic's increasing reliance on ethanol as a primary, nonnutritive calorie source. While only a small percentage (under 10%) of hospitalized alcoholics will present with neurological and neuropsychological symptoms of Wernicke–Korsakoff syndrome, the syndrome is of theoretical interest because it is one of the few consequences of alcoholism that is well understood; extensive neuropsychological and neuropathological investigations have been performed on victims of this syndrome.

Wernicke–Korsakoff syndrome is almost certainly more common than hospital statistics would suggest. Harper (1983) found that 80% of patients with features of Wernicke–Korsakoff syndrome were diagnosed during autopsy rather than during life. Harper, Giles, and Finlay-Jones (1986) argue that many cases are missed because conventional diagnostic criteria for this disorder are too strict, and would identify only 30 out of 200 potentially diagnosable cases in their city each year.

Wernicke's Encephalopathy

The clinical course of Wernicke–Korsakoff syndrome occurs in two stages, the initial acute phase of which is called "Wernicke's encephalopathy," a disorder that can be fatal if left untreated. Carl Wernicke first observed this neurological syndrome in 1881. Diagnosis, then as now, is made by history and clinical observations of pathognomonic symptoms, the most common of which are related to impairments in higher mental functions and consciousness (Harper *et al.,* 1986). The classic "Wernicke's triad" of ataxia, ophthalmoplegia, and memory deficit has been cited as diagnostic by Victor, Adams, and Collins (1971), although Harper *et al.* (1986) suggest that all three signs are "by no means invariably present" in these patients (p. 344). Other common PNS and CNS abnormalities have been reported, including polyneuropathy of the arms and legs (Victor *et al.,* 1971), a confusional and/or obtunded state, apathy, and an inability to maintain spontaneous conversation. If treated with large doses of thiamine at this point, the patient may show some evidence of improvement after several weeks, although only about 25% of Wernicke's patients are said eventually to recover to premorbid levels of cognitive functioning (Victor, Adams, & Collins, 1971). Wernicke's encephalopathy has been induced by other thiamine-deficient etiologies, including hyperemesis gravidarum (prolonged vomiting during pregnancy), anorexia nervosa, and gastrointestinal disorders that interfere with vitamin absorption. Alcoholics, however, are by far the group most at risk for developing the syndrome (Hartman *et al.,* 1985).

Wernicke's encephalopathy is probably more common among alcoholics than diagnostic statistics would indicate. Torvik, Lindboe, and Rogde (1982) found that only 3 of 68 cases of Wernicke's encephalopathy diagnosed at autopsy had been suspected prior to death. The neuropathology of Harper's (1979) patient group was similarly underestimated. Only 17 of 51 cases of Wernicke's encephalopathy were diagnosed before autopsy.

Korsakoff's Syndrome

About 6 years after Carl Wernicke's research had been published, S. S. Korsakoff published a number of papers that detailed amnesic and confabulatory symptoms of the disease which came to bear his name. In current use, a "Korsakoff syndrome" denotes the chronic residual symptoms of Wernicke's encephalopathy, most commonly confabulation and a severe anterograde amnesia, or the inability to learn new information, dating from the time of illness.

Neuropsychological Investigations of Wernicke–Korsakoff Syndrome

Severe anterograde amnesia has also been extensively documented in neuropsychological research. Reviewing the research of his colleagues, Butters (1985) found that Korsakoff patients are virtually unable to learn short lists of paired associates, or retain even three words in memory for 9 seconds if a distractor task intervenes. Susceptibility to distractors during rehearsal was particularly characteristic of Korsakoff patients.

The extensive neuropsychological investigations that have been performed on Wernicke–Korsakoff patients have been reviewed by Brandt and Butters (1986). These authors note characteristic patterns of neuropsychological functioning in Korsakoff alcoholics. For example, overall IQ scores of these patients are stated to be "indistinguishable" from controls matched on education, age, and socioeconomic class (Brandt & Butters, 1986, p. 456). While the WAIS-R Digit Symbol subtest score is low, Digit Span is normal.

Other neuropsychological tests detect the patterns of neuropsychological dysfunction in Korsakoff's syndrome far more sensitively than IQ tests. For example, Wechsler Memory Scale (WMS) scores are severely depressed and may be 20–30 points below IQ score (Butters & Cermak, 1976). Logical Memory, Figural Memory, and Associated Learning are the most sensitive measures, within the WMS, to Korsakoff's syndrome. Further, given the susceptibility of Korsakoff's patients to distractors, Russell's (1975) modification of the WMS to include incidental delayed 1/2-hour recall of the Logical and Figural Memory subtests would make the WMS even more sensitive to Korsakoff's syndrome memory deficit.

Other neuropsychological impairments of Korsakoff patients may be found on tests having constructional or visuospatial components, including the Tactual Performance Test, and tests containing embedded figures. Tests of planning, categorization, and rule learning (e.g., the Wisconsin Card Sorting Test and the Category Test) are also impaired.

Comparing Chronic Alcoholics with and without Wernicke–Korsakoff Syndrome

What is the relationship between chronic alcoholism with and without Wernicke–Korsakoff syndrome? Butters (1985) found similarities in both groups in their poor performance on problem-solving tasks and on tests involving abstract reasoning. Visuoperceptual deficits are also common to both types of patients. However, Korsakoff patients show much more severe anterograde amnesia than chronic alcoholics without this syndrome. The quality of recall in experimental tests is also different between the two groups. While Korsakoff patients tend to produce intralist intrusions of words (proactive interference) from previous lists, chronic alcoholics without Korsakoff's make errors of omission.

Wilkinson and Carlen (1980c) compared the neuropsychological performance of Wernicke's alcoholics with other, nonamnesic alcoholics on the WMS, the WAIS (IQ), and a modified Halstead–Reitan battery. The two groups were most consistently differentiated by the WMS, all of whose subtests except Mental Control were significantly different between groups. Digit Symbol from the WAIS and TPT Memory were the only other tests significantly different between groups, although several other scores, including Category Test and Trails B, were above standard impairment cutoffs.

Such results cast doubt on the validity of the "continuity hypothesis" originally proposed by Ryback (1971), which implied that Korsakoff's syndrome could be the end-product of chronic alcoholism. The continuity hypothesis suggested that neuropsychological deficits shown by Korsakoff patients were part of a continuum of cognitive decline as a function of alcohol quantity, frequency of use, and years of intoxication. Thus, for the continuity hypothesis to be supported, a similar, though less severe, pattern of "pre-Korsakoff" deficits should be seen earlier in chronic alcoholics' drinking histories.

Butters (1985) argues that such pre-Korsakoff patterns have *not* been found in alcoholics without Korsakoff's syndrome. While certain cognitive abilities do show continual decline during the course of chronic alcoholism, these functions have been localized to the *association cortex* rather than to the *subcortical* areas involved in memory and learning. Neuropsychological studies do not indicate that subcortical structures gradually decline to the degree of Korsakoff impairments; chronic alcoholics without Korsakoff's syndrome are not nearly as severely impaired in memory functions as Korsakoff patients. The impairments of these latter patients may best be understood as the consequence of acute trauma (vitamin B_1 deficiency) superimposed on the chronic traumata of long-term alcohol abuse. Thus, available evidence does not support the continuity hypothesis. Instead, separate but possibly additive effects of chronic alcoholism with and without Wernicke–Korsakoff syndrome have been found.

Chronic Alcoholism without Wernicke–Korsakoff Syndrome

Alcohol exerts direct toxic effects on the brain beyond those caused by vitamin depletion. These effects have been shown on autopsy, CT scans, CBF studies, and neuropsychological test results. For example, Harper and Kril (1985) autopsied the brains of controls, alcoholics, alcoholics with Wernicke's encephalopathy, and alcoholics with liver disease. Using an index of cerebral atrophy the authors estimated a mean loss of 41 g brain tissue in the alcoholic group,

89 g in the Wernicke's sample, and 109 g in the group with both alcoholism and liver disease. Alcoholics who show loss of cerebral tissue may do so in the absence of other alcohol-related organ pathology. A CT examination of alcoholics with no clinical evidence of liver disease revealed significantly greater cerebral atrophy than was present in an age-matched control group (Carlen et al., 1981).

The strictest matched sample control in a neuropsychological study of alcohol abuse would be a comparison of monozygotic twins, only one of whom was an alcohol abuser. Research thus far in twin studies has been mixed; an early monograph by Kaij (1960) found no deficits between abusing and nonabusing twins, while a more recent investigation suggested impairment in field dependence among alcohol-abusing twins (Bergman, Norlin, Borg, & Fyro 1975). Recently, another group of such twins received a computerized neuropsychological battery called the Maudsley Automated Psychological Screening Tests. Subtests included symbol digit, visuospatial ability ("mannikin"), verbal and spatial recognition memory, "visual perception analysis," vocabulary, tactual performance, and a computerized category sorting test. Alcohol-abusing twins performed significantly worse in TPT memory and localization, made more perservative errors on the category sorting test, and made more errors in the most difficult condition of the mannikin test. TPT localization correlated with marginal significance to CT ventricular area, and number of years drinking correlated significantly with TPT memory, and marginally with TPT localization (Guirling, Curtis, & Murray, 1991).

Locus of Damage—Hypotheses

The three main hypotheses advanced to explain the neuropathological effects of non-Wernicke-Korsakoff alcoholism are (1) diffuse injury, (2) lateralized damage to the right hemisphere, or (3) injury to the frontal-limbic-diencephalic area (Loberg, 1986). Recent clinical research shows increasing neuropsychological support for frontal lobe axis abnormalities, but it must be emphasized that alcohol neuropathology is clearly multifactorial. Different combinations of alcohol use, nutritional status, and other factors may produce any one or combination of the three so-called "syndromes" in clinical cases.

Diffuse Injury. The diffuse injury hypothesis is controversial. Loberg (1986) considers the evidence for diffuse injury to be the weakest of the three hypotheses and criticizes the ambiguity of the concept; "diffuse" could mean either distributed focal lesions or global, overall damage. In addition, neuropsychological tests that have shown consistent deterioration as a result of alcoholism (e.g., constructional tasks, Performance IQ subtests) have typically been thought of as "right hemisphere" rather than diffuse injury tasks. Verbal abilities that are strongly localized to the left hemisphere do not typically evidence alcoholic impairment.

Alternatively, Goldman (1983) suggests that neuroradiological evidence *favors* the diffuse injury hypothesis, including chronic alcoholism studies that show bilateral enlargement of the lateral ventricles, the third ventricle, and cerebral sulci. Altura and Altura's (1984) theory that heavy alcohol use may cause constriction and rupture of cerebral microvessels, along with hypertension, could also be construed as support for a form of diffuse alcohol-related injury. Goldstein and Shelly (1979) suggest that generalized CNS damage would explain why their alcoholics' performance did not resemble either anterior- or posterior-brain-injured controls.

Right Hemisphere Injury. Alcoholic neuropsychological deficit patterns, including intact Verbal IQ, impaired Performance IQ, visuospatial difficulties, and constructional deficits, are commonly thought to represent right hemisphere injury (e.g., Jones, 1971; Parsons, Tarter, & Edelberg, 1972; Jones & Parsons, 1972). There are several reasons why these findings may not provide as strong support for right hemisphere localization as has been thought. First, many of these tasks may not localize as clearly to one hemisphere as has been previously assumed (Loberg, 1986). Second, since such tasks, unlike linguistic abilities, are not practiced day-to-day, the failure to perform well could also reflect the novelty or difficulty of that task (Loberg, 1986). Finally, since most of these tests are timed, any affective or neuropsychological impairment interfering with arousal or cognitive efficiency may cause deficit patterns looking much like right hemisphere injury.

One of the most elegant tests of the right hemisphere hypothesis was performed by Ellis (1987) who failed to find differences between alcoholics and normals in dichotic listening tasks; both showed right ear superiority for verbal phonemic tasks and left ear preference for a musical dichotic listening task. Visually presented dichotic tasks have been inconsistent in showing alcoholics to be impaired with one study suggesting right hemisphere impairment (Ellis & Oscar-Berman, 1984) and another failing to find the effect (Oscar-Berman, Weinstein, & Wysocki, 1983).

Frontal-Limbic-Diencephalic Injury. This hypothesis predicts alcohol damage to the anterior and basal areas of the brain. This theory is consistent with PET scans that have identified diminished glucose metabolism in the frontal lobes as a function of alcoholism (Kessler *et al.,* 1984) and neuropsychological studies that show frontal lobe and memory deficit. Recent support for this hypothesis was reported by Adams *et al.* (1994), who found that older alcoholics showed abnormal local cerebral metabolism (LCMR) on PET scan that localized to a sagittal strip of the medial frontal region of the cortex. Performance on the Wisconsin Card Sorting Test (but not the Halstead Category Test) was significantly related to medial frontal LCMR. This finding suggests that abstract reasoning deficits found on the Category Test by earlier researchers (e.g., Fitzhugh, Fitzhugh, & Reitan, 1960, 1965; Smith, Burt, & Chapman, 1973; Long & McLachlan, 1974) may not be true examples of focal frontal abnormalities.

Other studies that support frontal lobe localization of neuropsychological abnormalities include those that find alcoholic deficits on the Finger Tapping

Test (Long & McLachlan, 1974), cognitive impersistence when attempting to solve unsolvable anagrams (Cynn, 1992), and motor impersistence on a Knob Turning Task (Parsons *et al.,* 1972). Commonly reported memory impairments like those found on the Russell-modified Wechsler Memory Scale delayed recall tasks (Russell, 1975) also support this axis of injury. Poorer frustration tolerance or cognitive persistence may be an index of subclinical abnormalities to this axis in that they may precede abnormalities in set-shifting and problem-solving in younger, healthier alcoholics (Cynn, 1992).

Hypothalamic–Pituitary–Gonadal (HPG) Axis Injury. Though not considered a primary locus of damage, HPG axis neuropsychological impairments can be inferred from alcohol's known toxicity to the HPG axis, and the knowledge that sex hormone dynamics may affect neuropsychological function, especially visuospatial abilities. The relationship between sex hormones and the neuropsychological functioning of alcoholics was recently investigated by Errico, Parsons, Kling, and King (1992). While normal male controls replicated the finding that visuospatial performance correlates with serum testosterone and follicular stimulating hormone levels, alcoholics failed to show this correlation. Curiously, alcoholic subjects showed significant correlations between *verbal* memory/problem-solving skills and testosterone levels. Controls showed significantly better performance on tests of visuospatial ability compared with alcoholics.

WHICH TESTS SHOW IMPAIRMENT?

The neuropsychological complexities of alcoholism make it understandable that testing methods be similarly multifactorial. Almost every neuropsychological test or test battery has at one time been investigated for its sensitivity to the effects of alcoholism. As is the case with other neurotoxins, not all neuropsychological tests are equally sensitive to the effects of alcohol. Those that have been found to be impaired typically assess a common set of functions. These include abstract reasoning, visuospatial abilities, memory, sustained attention, and cognitive flexibility. Not surprisingly, test batteries that load heavily on these abilities have also been shown to be sensitive indicators of alcoholic effects.

Individual Tests

Abstract Reasoning

Loss of ability to form hypotheses and reason abstractly has been a frequent finding among alcoholic subjects. One test of abstract, inductive problem-solving that is consistently impaired among alcoholics is the Category Test. For example, Jones and Parsons (1971) found alcoholics to perform significantly more poorly on the Category Test but not on the Shipley–Hartford test of verbal abstraction. Alcoholics had "normal and surprisingly good performance" on the Purdue

Pegboard, Finger Tapping, and Grip Strength, leading the authors to speculate about a prefrontal cortical or subcortical locus. An elaboration of this study published a year later successfully replicated the findings of Jones and Parsons (1971) on the Category and Shipley–Hartford (Jones & Parsons, 1972). Alcoholics also performed more poorly than controls on the Raven's Progressive Matrices, but not on the Embedded Figures Test. The authors suggest their results are interpretable as a deficit in visuospatial concept formation, possibly coupled with defective visual scanning. A more recent investigation suggests that Category Test results are sensitive to duration, mean and maximum alcohol consumption, alcoholism onset, and abstinence duration (Braun & Richer, 1993). Studies using another test of abstract reasoning, the Wisconsin Card Sorting Test (e.g., Klisz & Parsons, 1977), have also shown alcohol-related impairment.

Visuospatial Tests

There is ample evidence that alcoholics, with and without Wernicke–Korsakoff syndrome, are impaired in tests of spatial abilities. Block Design, Object Assembly, and Picture Arrangement subtests of the Wechsler Adult Intelligence Scale (WAIS) were impaired in 70% of the studies reviewed by Parsons and Farr (1981).

Tests of General Intelligence (IQ)

IQ test results of alcoholic subjects typically suggest that overall Verbal and Full Scale IQ scores are insensitive to alcohol effects. Performance IQ is much more frequently shown to be impaired, because that section of the WAIS or WAIS-R loads on subtests requiring spatial and constructional functions, nonverbal reasoning, cognitive efficiency, and flexibility. Similarity in the pattern of impairments between WAIS subtests and parts of the Halstead–Reitan Neuropsychological Battery have also been noted by Parsons and Farr (1981).

Memory Tests

Memory deficits in alcoholic patients are common findings, both for Korsakoff patients and for non-Korsakoff chronic alcoholics. Alcohol-based impairments in memory are found on a wide variety of tests. In one example of the genre, Ryan and Lewis (1988) found deficits on the Wechsler Memory Scale Revised in recently detoxified alcoholics, with alcoholics performing more poorly on all five index scores and 5 of 12 subtests.

IQ–Memory Discrepancy

It has been common practice for neuropsychologists to examine the discrepancy between IQ and scores on a memory index (most commonly the Wechsler

Memory Scale MQ, or the WMS-R-GM) to determine the severity of alcohol-related or other memory impairment. A study by Oscar-Berman, Clancy, and Weber (1993) suggests that this technique is only sensitive to Korsakoff patients, whose IQ–Memory score discrepancy averaged -19.7 points for MQ and -33.9 for GM. Older, non-Korsakoff alcoholics showed a much smaller discrepancy score (5.5 MQ; -9.8 GM), not significantly different from scores of age-matched nonalcoholics.

Efficiency Analysis

Glenn and Parsons (1992) have proposed a method of test analysis that cuts across different testing methodologies and cognitive domains. They propose use of Accuracy/Time as a measure of cognitive efficiency that significantly distinguishes alcoholics from controls in the neuropsychological domains of problem-solving, abstracting, verbal functions, perceptual-motor skills, learning, and memory. Alcoholics in their study were found to show not only main effects of slower speed and reduced accuracy, but also in the ratio of the two. The authors propose further research to determine whether poorer efficiency found in alcoholics is the result of increased impulsivity, or a difference in cognitive strategy common to alcoholics. An additional explanation may be that lowered cognitive efficiency reflects the reduced capacity of a cognitive system damaged by alcohol. The authors suggest that reduced efficiency ratios in alcoholics have ecological validity in explaining alcoholics' increased incidence of work-related injuries, auto accidents, and other situations where their efficient response to a rapidly changing situation is impaired and insufficient to avoid injury.

Neurosensory Tests

Work by Mergler *et al.* (1991; Mergler, Blain, Lemaire, & Lalande, 1988) has validated the use of desaturated color perception accuracy as an effective test of neurotoxic exposure in general, and chronic alcoholism specifically. In one such study, chronic alcoholics made more than twice as many errors on this color sorting test (i.e., Lanthony D-15 Desaturated Panel) as matched controls, reflecting a neurotoxic effect of alcohol on the retina (Braun & Richer, 1993).

Test Batteries

Halstead–Reitan Battery (HRB)

A great number of neuropsychological investigations with alcoholics have used the HRB (Loberg, 1986). Deterioration of HRB performance with respect to controls or to relevant age norms is a usual finding. Parsons and Farr (1981) observed that chronic alcoholics share a pattern on the WAIS and the HRB that is similar to, albeit not as great as, that of brain-damaged and aged non-alcoholics.

Parsons and Farr (1981) conclude that alcoholics exhibit consistently poor performance on three HRB subtests. Impairment reached 87% on the Category Test, 73% on Trails B, and 62% on Tactual Performance Test-Localization in the studies they reviewed. Fabian, Jenkins, and Parsons (1981) further elaborated a specific pattern of performance on the Tactual Performance Test that appears to be specific to alcoholics. Unlike controls, alcoholic subjects failed to improve their performance times on the second trial using the nondominant hand. In addition, alcoholic subjects improved more than controls from the second to the third trial (both hands). The authors hypothesize that right hemisphere dysfunction, or perhaps disruption of less well-established behaviors, may explain this pattern of TPT performance (Fabian et al., 1981).

Luria–Nebraska Neuropsychological Battery (LNNB)

Only a few studies have used the LNNB (Golden et al., 1980) to investigate the effects of alcoholism. One such study was performed by de Obaldia, Leber, and Parsons (1981), who examined 30 male alcoholics. Half were in the acute phase of detoxification between 2 and 3 weeks after hospital admission, the other half between 10 and 12 weeks postadmission. The alcoholic groups were compared with a group of age-and education-matched controls recruited from the community. Controls were significantly superior to both alcoholic groups on every LNNB scale except Reading. Only the Rhythm scale (Scale 2) differentiated the alcoholics in the acute phase of detoxification from those 8–10 weeks later.

Despite the ability of the LNNB to differentiate the two groups statistically, it was apparently not an especially sensitive clinical measure. For example, the mean T-score of the *alcoholics* on the LNNB was about T50. Using Golden's cutoffs for "brain damage," only 50% of the alcoholics were identified, suggesting either that the alcoholic groups employed by de Obaldia et al. (1981) were not especially impaired, or that the LNNB may not be as sensitive as the HRB in detecting alcohol-related impairment.

Results from the few other studies using the LNNB on an alcoholic population suggest the latter possibility. For example, Chmielewski and Golden (1980) found six impaired scales in alcoholics relative to controls: Visual, Receptive Speech, Arithmetic, Memory, Intelligence, and the Pathognomonic Scale. However, just as in the de Obaldia et al. (1981) study, Chmielewski and Golden (1980) differentiated controls from alcoholics on several scales, but failed to show significant clinical elevations in themselves; with the exception of one scale (Pathognomonic) all scales were below T60. Also, as in de Obaldia et al. (1981), Chmielewski and Golden (1980) were led to conclude that "the brain is fairly resilient to the effects of alcohol" (p. 103), a supposition that does not seem to be supported by the more extensive HRB alcoholism literature. Finally, a third research report (Rebeta, McAllister, Gange, Jordan, & Riley, 1986) described testing of 100 male VA alcoholics using the LNNB and finding no association between drinking variables and LNNB scale scores when age, education, and SES effects were factored out.

Thus, available studies do not support the clinical use of the LNNB to test for impairment of alcoholic etiology. The HRB appears to have a clear advantage over the LNNB for testing alcoholics for several reasons. First, the few available studies using the LNNB have reported inconsistent results and pattern elevations (e.g., Rhythm scale in one study, six scales but *not* including Rhythm in another, and no elevations in a third). In contrast, the extensive HRB literature shows fairly consistent impairments on the Category, TPT, and Trails tests. Second, while alcoholics may be clinically differentiated from controls on the LNNB, their actual level of performance is not generally in the range of clinically meaningful impairment using Golden's cutoffs. HRB-tested alcoholics more often fall into a range of clinical impairment using widely accepted cutoffs, both for specific tests and for the battery as a whole (i.e., Impairment Index).

In conclusion, although hit rates for detecting structural brain abnormalities have been suggested to be comparable between the LNNB and the HRB, the two batteries nevertheless appear to show differential sensitivity to alcoholism. The HRB loads more heavily on neuropsychological functions that are difficult for alcoholics, including sustained attention, abstraction, and complex perceptual processes. By contrast, the LNNB emphasizes discrete simple tasks that are heavily mediated by verbal functions that have been shown to be resistant to alcohol effects.

Factors Influencing the Neuropsychological Performance of Alcoholics

Recent neuropsychological investigations have strongly emphasized the multifactorial nature of alcohol-related neuropsychological impairments (Tarter & Edwards, 1986; Eckardt & Martin, 1986; Alterman & Tarter, 1985). Many previously unidentified subject variables have been found to influence the neuropsychological performance of alcoholic individuals; variables that strongly suggest that familial, genetic, and indirect environmental and physical concomitants of alcohol abuse may be as likely to be associated with neuropsychological impairments as is actual alcohol-caused brain damage. Grant's (1987, p. 320) flowchart captures the complex interplay of influences related to the neuropsychology of alcoholism (Fig. 5.1).

For the neuropsychological researcher, the knowledge that alcohol effects are multifactorial may cause reevaluation of existing research; many early studies which equated impaired neuropsychological performance with alcohol use may have been influenced by one or more of these preexisting or concomitant variables, and not necessarily by the direct toxic effects of alcohol on the nervous system. For the clinician involved in the assessment of alcoholism, the many new variables found to be related to neuropsychological performance should spur an increase in data collection of these variables, and at the same time should temper any tendencies to make direct causative inferences about any single influence of alcohol on observed neuropsychological performance.

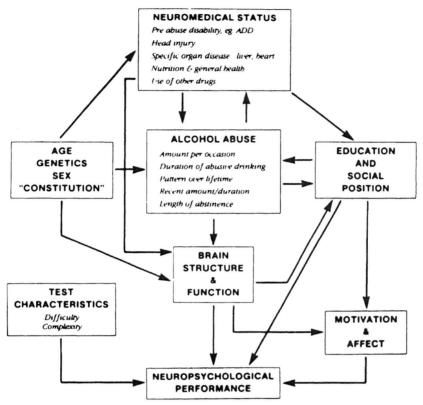

FIGURE 5.1. Variables to consider in any causal model of alcohol-associated neuropsychological deficit. ADD, attention deficit disorder. (Reprinted, by permission of the author, from Grant, I. Alcohol and the brain: Neuropsychological correlates. *Journal of Consulting and Clinical Psychology*, 55, 310–324. Copyright 1987 by The American Psychological Association.)

Age

"Premature Aging" versus Independent Decrement. The effects of age on neuropsychological performance are inherently complex, and made more so by the chronic assault on the nervous system by alcohol. Several studies have noted that chronic alcoholics tend to perform neuropsychological tasks at a level more appropriate to the norms of older intact subjects. The basic theory, called "premature aging," makes ambiguous performance predictions and other authors (i.e., Noonberg, Goldstein, & Page, 1985) have suggested alternatives. The following alternatives have been proposed.

Premature aging. The basic theory, proposed by Ryan and Butters (1980), suggests that the neuropsychological performance of chronic alcoholics will resemble older nonalcoholic controls.

Accelerated aging. This variant of the premature aging theory hypothesizes that alcoholic performance and underlying neuropathology may occur at any age following an appropriately long and intensive drinking history. Thus, alcoholics will perform more poorly than controls at any point across the life span.

Increased vulnerability. This second variant of the premature aging hypothesis predicts that younger alcoholics are less vulnerable or temporarily protected from the neuropsychological dysfunctions of alcoholism. Deficits may occur in these subjects with increasing age and consequent neuronal depletion (Noonberg *et al.,* 1985). This hypothesis would predict a fan-shaped interaction between aging and alcoholism. "Novice" alcoholics would initially look like their neuropsychological controls, but the two groups would gradually diverge, with alcoholics becoming increasingly poorer performers compared with controls as a function of age and drinking history.

Independent decrement. Alcohol and age affect cognitive functioning differently. These effects may or may not overlap. Support for "accelerated aging" or independent decrements in alcoholics could be extrapolated from several studies, including the original study performed by Ryan and Butters (1980). The authors administered a battery of neuropsychological tests to detoxified younger (34–49) and older (50–59) alcoholics and controls. In several learning and memory tests, including the Symbol Digit Test, a four-word short-term memory test, and a paired associate learning test, the younger alcoholics performed more like the older controls than their age-matched nonalcoholic peers. No interactions were observed between age and alcohol impairments. A study by Brandt, Butters, Ryan, and Bayog (1983) also found effects of alcohol and of age, but no interaction, suggesting independent decrements from each variable.

Independent decrement effects could be differentiated from premature aging if there was evidence of different patterns of functioning in old alcoholics compared with old controls. One study found such differences in behavior between normal old adults and alcoholics; the former but not the latter showed psychomotor and attentional difficulties (Grant, Adams, & Reed, 1984). Reige, Tomaszewski, Lanto, and Metter (1984) also found results consistent with independent effects of age and alcoholism. In tests of verbal and nonverbal memory, older alcoholics did not perform to the same pattern as controls, but were distinguished by deficits in auditorily presented memory stimuli. Younger alcoholics were statistically indistinguishable from their nonalcoholic "social-drinking" controls.

Finally, other evidence suggests the independence of aging and alcoholism effects. First, severe memory deficits of alcoholic etiology (e.g., Wernicke–Korsakoff syndrome) do not have any equivalent in the normal aging brain. Second, alcoholics make more frequent self-reports of cognitive errors than do nonalcoholics, suggesting potentially different introspective capacities in the two groups. Brandt *et al.* (1983) suggest that older nonalcoholic patients show naming difficulties not found in middle-aged alcoholics, while older alcoholics have much more pronounced nonverbal learning impairments than elderly control subjects. Finally, Oscar-Berman et al. (1993) examined individuals from age 24 to 74 for discrepancies between IQ and Wechsler Memory Scale scores and found no interaction between age and alcoholism. They found that patients without clinical evidence of severe memory disorder showed no significant discrepancy between IQ (WAIS or WAIS-R) and a Global Memory Index (WMS MQ or WMS-R GM). This remained true both for normal controls and for non-Korsakoff alcoholics. There was no difference in rate of performance decline between the younger alcoholics compared with controls or their older peers.

To summarize, most current evidence appears to favor the effects of alcohol and age as independent decrements on neuropsychological performance. That these factors separately impair behavior does not, however, obviate the possibility of additive or synergistic effects.

Onset of Impairment

The age at which neuropsychological deficits begin to appear is an issue with serious social, political, and clinical implications. For example, when Grant, Adams, and Reed (1979b) investigated alcoholics in their 30s with a 6-year drinking history, and found no deficit relative to controls, their results were criticized for minimizing the serious effects of alcohol abuse and other concerns (Eckardt, Ryback, & Pautler, 1980). In addition, the study has been criticized for its overly strict selection criteria, since subjects with mild residual organic brain syndrome were eliminated (Parsons & Farr, 1981).

Other investigations of alcoholics in their 30s have found results contrary to those of Grant et al. (1979). Noonberg, et al. (1985) evaluated alcoholics in their mid-30s, with the Brain Age Quotient (BAQ) battery, a collection of neuropsychological tests that included the Category Test, Tactual Performance Test, and Trails B from the Halstead–Reitan Battery, and Block Design and Digit Symbol tests from the WAIS. Even the youngest group of alcoholics performed more poorly on the BAQ than did their age-matched peers. Eckardt et al. (1980) also compared neuropsychological performance of alcoholic males in their mid-30s on 24 neuropsychological measures. Subjects' primary diagnosis was alcoholism with no secondary psychiatric disorder. Those who were tested from 2 to 6 days after their last drink showed impaired levels of functioning on 13 of the 24 tests administered. A second group examined after 14–31 days of abstinence were impaired on 11 of the 24 tests. An all-possible-subsetsregression analysis on the first (acute) group showed, when drinking variables were removed, that amount of alcohol consumption was related to 70% of the variance. Age contributed an additional 20% and education another 10%. Patients in this study had significantly longer drinking histories with significantly greater alcohol intake than those in the Grant et al. (1979) study, which may account for the differences between the two studies.

Drinking Pattern

There are conflicting findings about the influence of drinking pattern on neuropsychological performance. Specifically, the debate revolves around whether neuropsychological decrements are a function of amount consumed per sitting, recent drinking pattern, or lifetime alcohol consumption. Svanum and Schladenhauffen (1986) examined 40 inpatient detoxified alcoholics to determine whether recent drinking pattern or lifetime consumption accounted for Halstead–Reitan testing scores. Drinking pattern proved to be as important as duration in accounting for neuropsychological deficits. Tarbox, Connors, and McLaughlin (1986) compared daily drinkers (5 or more days a week) to "bout" drinkers (drinking for days, weeks, or months with intervening abstinence). Bout drinkers scored significantly higher on age-corrected performance IQ measures as well as Wechsler Memory Scale Mental Control and Digit Span tests. Post hoc comparisons also found that older bout drinkers performed better than older daily drinkers across neuropsychological tests.

Drinking History

Early drinking onset has been associated with greater social maladjustment (Lee & DiClimente, 1985) as well as neuropsychological impairment. Pishkin *et al.* (1985) investigated the influence of both age of onset and years of drinking. Older controls and late-onset alcoholics performed significantly better than younger, early onset alcoholics on the Shipley–Hartford test. The authors concluded that early onset alcoholism is particularly likely to result in eventual cognitive impairment. It is equally likely, however, that younger alcoholics may be more predisposed to neuropsychological deficits because of the same premorbid factors that elicited early use of alcohol.

Familial Factors

It is generally well established that vulnerability to alcoholism is intergenerational. About 50% of alcoholics have a family history of alcoholism (Berman, Whipple, Fitch, & Noble, 1993). Recent genetic, neurophysiological, and neuropsychological studies support the hypothesis that vulnerability to alcoholism is multidetermined and may be associated with genetic patterns as well as structural and functional brain differences that predate the onset of alcohol abuse. For example, Blum *et al.* (1993) have located a dopamine receptor gene that is found significantly more frequently in severe alcoholics and may play an etiologic role in that disorder. Several studies found that P3 (P300) abnormalities are present in sons of alcoholics and may even predict the likelihood of adolescent alcohol and substance abuse (Whipple, Berman, & Noble, 1987; Whipple, Parker, & Noble, 1988; Berman *et al.*, 1993). Berman *et al.* (1993) found that P3 or P300 event related potential latency, amplitude, and subject age significantly predicted the development of alcohol and drug abuse 4 years later. P3's with the lowest amplitude were correlated with the highest levels of substance abuse. Results are consistent with data suggesting that both alcoholics and their sons show increase P3 latency and decreased P3 amplitude relative to controls (Whipple, et al., 1987, 1988).

Several studies likewise suggest that relatives and other family members in alcoholic families may have neuropsychologic deficits regardless of their sobriety. Peterson, Finn, and Pihl (1992) found that sons of male alcoholics performed significantly worse than controls on four neuropsychological tasks: self-ordered pointing (pointing to a familiar object in arrays of other objects), Rey–Osterreith copy, WMS Difficult Paired Associates, and WAIS-R Information; tasks requiring or reflecting the ability to categorize and organize new information. Schaeffer et al. (1984) found that males with an alcoholic close relative performed significantly less well on verbal and nonverbal abstraction and perceptual-motor tasks than did matched controls without an alcoholic relative. Another group of papers by Alterman and colleagues suggest that other familial factors should be considered. For example, Alterman and Tarter (1985) reviewed possible genetic factors related to alcoholism and concluded that histories of childhood hyperactivity (with presumed attentional deficits and low impulse control) may be a factor in poorer neuropsychological performance among alcoholics.

Several studies suggest that children of alcoholic fathers may exhibit learning and neuropsychological difficulties compared with children of nonalcoholic parents. For example, Tarter, Jacob, and Bremer (1990) contrasted cognitive and behavioral performance of school-age sons of early and late-onset alcoholics with sons of normal social drinking fathers and depressed fathers. The authors found that sons of early onset alcoholic fathers showed lower verbal IQs compared with other groups, and had lower performance IQs compared with sons of depressed fathers. Neuropsychological test factor analysis suggested that sons of early onset alcoholics were impaired relative to other groups on tests of general attention from the Detroit Learning and Aptitude Test. Unlike earlier studies, no differences in visuospatial abilities, abstraction or praxis were found, suggesting that community-recruited subjects were less severely impaired than typical clinical referrals (Tarter et al., 1989).

A second study by Tarter and colleagues provides additional evidence of impairment in sons of male alcoholics. Sons of alcoholics showed deficient performance relative to controls on verbal tests (Peabody Picture Vocabulary, Peabody Individual Achievement Test, and a Peterson Memory Test) (Tarter et al., 1990). Finally, studies that find impairments among young alcoholics may be consistent with either preexisting cognitive deficit or subtle neuropsychological insult. The study of Alterman et al. (1984), which examined 30-year-old male alcoholics, would fall into this category. Subjects showed significantly impaired performance on tasks that required organization, sequencing, and sustained goal-directed behavior (Alterman et al., 1984).

Handedness

Handedness is another variable that has been suggested to correlate with alcoholism. Studies by London (1985) suggest that larger percentages of alcoholics are left-handed than is true for the general population (17–39 versus 10%). In addition, significantly fewer left-handed alcoholics were classified as "improved" after 1 year (29 versus 56%). While other factors, such as small sample size or having an alcoholic father, may equally explain London's results, other studies not specifically investigating the effects of alcoholism have linked certain kinds of left-handedness to minimal brain dysfunction or subtle neurological disability (e.g., Geschwind & Behan, 1982). It is possible that whatever mediating variables create this form of pathological left-handedness may also increase the likelihood of developing alcoholism.

Nutritional Factors

The relationship between thiamine deficiency and its contribution to the neuropsychological impairments of Wernicke–Korsakoff disease is common knowledge. Recently, it has been proposed that another nutritional component, vitamin E, may also mediate a subset of neuropsychological performance in the area of psychomotor function. Patients with primary biliary cirrhosis (PBC) and vitamin E deficiency showed significantly worse performance than normals on six different motor and visuomotor tests, while patients with PBC alone showed significant impairment in only one such test (Arria *et al.*, in press, cited in Tarter *et al.*, 1990). These results suggest the need to monitor vitamin E levels in alcoholics under neuropsychological study.

Psychopathology

Neuropsychological deficits shown by alcoholics are often clinically paired with emotional disturbance and serious psychosocial problems. The relation between this toxic brain disturbance and emotional functioning remains obscure. Loberg's (1986) review of the literature concluded that there is "no uniform alcoholic personality" (p. 431), although it suggests the possibility that demographically similar groups of alcoholics may also have consistent profiles and that these profiles, in turn, may be related to neuropsychological measures. Future research may also extend the investigations of depression and neuropsychological deficit (e.g., Fisher, Sweet, & Pfaelzer-Smith, 1986; Newman & Sweet, 1986) to alcoholic populations. It is unclear how other forms of affect alter neuropsychological performance; for example, the anxiety states so commonly noted in alcoholics. There is very little research that addresses these potential sources of variance in neuropsychological performance (Alterman & Tarter, 1985).

Head Trauma

Recent studies suggest that alcoholic individuals with a family history of alcoholism are twice as likely to have a history of significant head trauma than nonfamilial alcoholics (Tarter & Edwards, 1986). Moreover, the social and environmental context of alcoholism leads to increased risk of head injury. Thus, some observed neuropsychological deficits may be in part or whole the effect of exogenous brain injury rather than endogenous toxic assault.

Gender

Gender-based differences in alcohol metabolism suggest that women may not metabolize orally ingested alcohol as efficiently as men and that women may be more vulnerable to acute and chronic consequences of alcohol abuse (Frezza et al., 1990). Neuropsychological support for this difference may be found in a study by Acker (1986), who compared performance of male and female alcoholics on a variety of tests including Logical Memory, Benton Visual Retention, Trails, Digit Span, and an automated neuropsychological battery. She found female alcoholics to show greater impairment than males on immediate recall, psychomotor speed, abstract reasoning, and visuoperceptual tasks. It was suggested that females may be more sensitive than males to neuropsychological effects of alcohol for an equivalent amount of drinking; however, since the study lacks statistical comparison with female nonalcoholic controls, findings cannot discriminate between alcohol-based and premorbid sex differences in the abilities Acker suggested are impaired.

Support for a nonalcoholic explanation of Acker's results could be extrapolated from the results of Fabian et al. (1981), who administered the Tactual Performance Test (TPT) to four groups of subjects; male and female alcoholics were compared with male and female nonalcoholic controls matched for age and education. There were main effects for both gender and memory, as women remembered more shapes but localized fewer of them than men. There was a main effect for alcoholism but not for gender in TPT completion time, with the alcoholic group performing more slowly than controls. Thus, the study indicates effects of alcoholism and effects of gender, but no interaction between the two. The lack of interaction supports the existence of premorbid processing differences.

One study that provides some support for differential vulnerability to alcohol in women is that of Hannon et al. (1985) who found that amount of alcohol consumed was significantly related to an increased number of total errors on the Category Test, an effect not found for men. The Hannon et al. (1985) study awaits replication. However, results suggesting an opposite conclusion were obtained by Silberstein and Parsons (1979), who found that alcoholic women were *less* impaired on visuoconstructive tasks than were men.

A factor to consider in gender studies is the Nicholson *et al.* (1992) finding that acute neuropsychological differences between males and females disappeared when corrections were made for body fat composition, suggesting that this correction should be made in future gender comparison studies.

In summary, theoretical arguments for differential vulnerability of women to alcohol can be made, including differing efficiency of alcohol metabolization, different degrees of cerebral lateralization, lower body mass in women (and hence higher blood alcohol levels for a given number of drinks), or otherwise unspecified sex-related differences in processing alcohol. However, available studies on gender differences have thus far failed to consistently verify differential neuropsychological impairments tied to any of these hypotheses.

RECOVERY OF FUNCTION

Tracking recovery of neuropsychological functions is difficult, not only because improvement is not uniform across abilities, but for the same reasons that determining precise etiology of neuropsychological impairments in alcoholics is difficult. Alcoholism, in its etiology as well as its pattern of recovery, is multifactorial syndrome (Goldman, 1986). Drinking pattern and severity, degree of abstinence, age, education, emotional status, and drug use are just a few of the factors capable of influencing recovery. Other variables related to alcoholic recovery include one's particular definition of "recovery" and the time at which recovery is assessed.

Neurological Recovery

Carlen and Wilkinson (1980) studied the recovery of 122 chronic alcoholics with at least a 10-year history who were admitted to an inpatient treatment center. Over the course of a 6-week study, the authors found continuing neurological improvement on "simple standardized neurological testing" with a "rapid drop during the first 5 weeks" (p. 109). The authors note that practice may have improved performance somewhat, but cite other studies that controlled for practice that have also shown significant improvement. Carlen and Wilkinson's results may not apply to the most severely impaired alcoholics since these patients were excluded from the study.

CT Scan

Initial assessment of recently abstinent alcoholics reveals cortical atrophy (Wilkinson & Carlen, 1980a,b). Carlen, Wortzman, Holgate, Wilkinson, and Rankin (1978) demonstrated some reversal of atrophy in four of eight patients examined with CT. A later extension of the study, using more patients, did not confirm results of the initial study. However, alcoholics who claimed abstinence during the intertest interval had a significant negative correlation between age and sulcal measurements, while nonabstinent alcoholics showed no correlation between age and sulci, suggesting the possibility that younger abstinent alcoholics show acute recovery while older alcoholics do not.

Ron (1983) performed follow-up CTs on 56 of her original 100 alcoholic patients at periods ranging from 30 to 152 weeks. Patients were divided according to their drinking history into abstinent and nonabstinent groups. Initial CT examination showed no significant radiological differences between these subgroups. On follow-up no significant CT differences were found between these groups, although "trends toward lessened sulcal and Sylvian fissure widths" were found in the abstinent group (Ron, 1983, p. 20). In addition, the alcoholics rated as abstinent showed significantly smaller ventricles at follow-up, in contrast to the nonabstinent group who failed to improve. The meaning of this finding is obscured by the fact that the abstinent group showed a trend toward larger ventricles on initial testing. Thus, the "improvement" may be artifactual. When Ron analyzed the results of individual patients, five of the seven rated to be improved at follow-up were abstainers, while only one of five in the deteriorated-at-follow-up group claimed abstinence.

The evidence for alcoholic recovery on CT is complicated by questionable alcohol history self-reports, inability to exactly reproduce prior head positions in follow-up CT evaluation, sample attrition, and an unknown proportion of variance contributed by other factors correlated with alcoholism. Perhaps the best conclusion that can be made about CT data at this time is that while abstinence will not *necessarily* produce structural recovery that is viewable on CT, such recovery is far more likely to be observed during abstinence than if drinking continues. There is no evidence to suggest that CT scans of nonabstinent alcoholics improve with time.

PET Scan

Volkow et al. (1994) used PET to track the recovery curve of recently detoxified alcoholics. Brain glucose metabolism showed significant increases on PET during detoxification with most recovery occurring between 16 and 30 days. At the end of detoxification, alcoholics continued to show lower basal ganglia compared with their own scans and that of controls, along with lower parietal cortex metabolism relative to their own scans.

Regional Cerebral Blood Flow

Berglund, Bliding, Bliding, and Risberg (1980) studied recovery during the first 7 weeks of alcoholism using measurements of global and regional cerebral blood flow. Mean cerebral blood flow remained stable after the first week of abstinence. Within the first 2 days of abstinence an average 20% decrease in cerebral blood flow was found, a result Goldman (1983) associates with ionic imbalance. Significance increases in regional cerebral blood flow subsequently occurred in the upper frontal and parieto-occipital regions of the right hemisphere. The right temporal region showed significantly decreased blood flow compared with the left. The authors found recovery of rCBF in the frontal and anterior temporal cortex to be associated with improved performance on the Koh's block and Trails B neuropsychological tests. Berglund (1981) summarizes rCBF relationships to recovery as follows:

1. Mean hemispheric blood flow in most patients is not reversible after one abstinent week, although a minority of patients will show partial recovery during the first few weeks.
2. Relative reductions of blood flow in lower frontal regions without changes in mean hemispheric flow may be "related to reversibility of impaired spatial ability" (p. 298).
3. Older alcoholics show larger reductions of blood flow in lower frontal regions compared with younger alcoholics.
4. Wernicke–Korsakoff alcoholics may also show recovery of cerebral blood flow "but ... the flow normalization probably takes more than a few weeks in many patients" (p. 298).

Neuropsychological Recovery

Many types of neuropsychological deficits have been identified in alcoholics, including impairments in motor abilities, perception, verbal and nonverbal abstract reasoning, conceptual shifting, and memory. Recovery from such impairments is dependent on many factors. First, the pattern of neuropsychological recovery is not uniform; some abilities return completely while others may fail to improve significantly above initial levels of impairment. Second, recovery of neuropsychological functions appears to depend on remaining abstinent; neuropsychological improvements noted in alcoholics who have maintained sobriety do not occur in alcoholics who continue to drink (Guthrie, 1980).

Acute Neuropsychological Recovery

Recently detoxified alcoholics show deficits on a wide range of neuropsychological functions, including global impairment in memory and most other neuropsychological skills (e.g., Ryan & Lewis, 1988). Short-term (under 6 months) recovery has been summarized by Goldman (1986), who finds that alcoholics continue to show impairments on the Halstead–Reitan during their first week of treatment. They show some improvement over the next 23 weeks and then remain at that level. They do not become "fully normal" even at 6 months after the last drink (Goldman, 1986, p. 137).

Interaction of Recovery with Age

There are several reasons to believe that alcoholics' neuropsychological recovery may interact with age. First, older drinkers may simply have a longer history of alcoholism. Second, aging alcoholics may have developed other direct and indirect consequences of alcohol abuse, including malnutrition effects, head injuries, and hypertension or related vascular injuries (e.g., Altura & Altura, 1984). Finally, loss of brain mass from chronic alcoholism may limit the adaptive capacity of the aging brain to recover completely from chronic toxic assault.

Several studies have suggested that alcoholics over 40 recover neuropsychological function more slowly and less completely than younger alcoholics. For example, in alcoholics whose mean age was 37, Grant, Adams, and Reed (1979) showed that those abstinent a minimum of 18 months scored equivalently to matched controls on the Halstead–Reitan battery. Normal performance was maintained on follow-up 1 year later (Adams, Grant, & Reed, 1980). In contrast, Yohman, Parsons, and Leber (1985) studied alcoholics whose mean age was approximately 11 years older (48) than the Grant *et al.* study. Yohman *et al.* (1985) found neuropsychological differences in abstracting, perceptual-motor, and learning-memory tests to persist 13 months after detoxification.

Goldman, Williams, and Klisz (1983) monitored the visuospatial recovery of alcoholics over the course of 3 months. Alcoholics over age 40, while demonstrating some recovery over repeated testings, remained significantly impaired relative to controls or younger alcoholics on Digit Symbol Substitution, Trails B, and Grooved Pegboard dominant hand. Three months of 14 test sessions "were not sufficient to permit full recovery" (p. 373) in the older group. Age, and not years of drinking, was the predominant predictor of neuropsychological test scores for alcoholics. However, as the authors point out, since older controls failed to show impairment, chronic alcoholism must be added to normal aging to account for the results of the study.

Long-Term Recovery

Obvious practical difficulties maintaining contact with an experimental population and verifying their drinking histories make studies of long-term recovery from alcoholism quite difficult. One of the longest periods to have been investigated was 5 years, in a study by Brandt *et al.* (1983). They compared three groups of alcoholics with progressively longer periods of abstinence, including a "short-term abstinence" group (sober between 1 and 2 months), a "long-term abstinence" group (abstinent from 1 to 3 years), and a "prolonged abstinence" group (who remained alcohol-free for 5 years or more) (p. 439). Age, years of drinking, and WAIS vocabulary score were covariates in the analysis.

Time of abstinence had an effect on some neuropsychological measures but not on others. Recovery of function in the prolonged abstinence group equal to the performance level of matched nonalcoholic controls was demonstrated in verbal and visual short-term memory tasks (Four Word STM and Benton Visual Retention Test), and one test of cognitive efficiency and flexibility (Digit Symbol). Another test of efficiency (Symbol Digit Substitution Test) improved over time but failed to return to the level of controls, and no recovery was observed in the Symbol Digit Paired Associate Learning Test or the Embedded Figures

Test (Brandt et al., 1983). The authors concluded that, while psychomotor and short-term memory may recover with prolonged abstinence, certain neuropsychological functions in the realm of long-term memory may be refractory to recovery.

A more optimistic view of alcoholic recovery was provided in a study that assessed memory, learning, and neuropsychological functioning in non-Korsakoff alcoholics who abstained from alcohol for an average of 7 years (Reed, Grant, & Rourke, 1992). The authors compared a cohort of these individuals with recently detoxified alcoholics, 2-year abstinent alcoholics, and matched nonalcoholic controls. Tests included the "Anna Thompson" study from the Wechsler Memory Scale initial recall and trials to criterion of 15 story elements, a Brown–Peterson interference test, WMS 30-minute delayed free recall, WMS Figures initial trial and trials to criterion of 15 elements, Face Learning, WMS Figures 30-minute delayed recall, Symbol-Digit Pair Associate Learning, Verbal Pair-Associate Learning, and seven standard neuropsychological tests from the Halstead–Reitan and the WAIS (Category Test, TPT, Trails, Digit Symbol, Block Design, Digit Span, and Vocabulary). A hypothesized relationship between length of abstinence and learning was only demonstrated for visual learning, where it was revealed that recently detoxified alcoholics performed significantly more poorly than the other three groups on WMS Visual Trial 1, and worse than controls or long-term abstinent alcoholics on Figural Memory trials to criterion. Most important was the fact that nonalcoholic controls and long-term abstainers did not differ, either on memory tasks or on the other neuropsychological tests. The fact that the alcoholic group was in their 40s when they began to abstain also contradicts conventional clinical wisdom that recovery is impossible or at least less complete after age 40. Some boundaries of generalization must still be observed, however, since the Reed et al. alcoholics were medically "quite healthy," had an average of 13 years' education, and apparently did not begin drinking until their late 20s. The authors also observed that it is possible that the "cognitively best endowed" alcoholics were most likely to utilize treatment to become long-term abstainers, and only the most intact individuals would volunteer to participate in research of this kind. Future studies may sort out whether long-term abstinence allows neuropsychological functions to normalize, "if those who escape impairment in the first place are more able to achieve long-term sobriety," or if there is recovery in this medically healthy group of abstaining alcoholics, whether it can be generalized to other alcoholic groups.

Other Factors

Alcoholic recovery interacts not only with age, but also with abstinence or other factors. This question was addressed by Grant et al. (1984), who compared 4-week abstinent alcoholics with 4-year abstainers and nonalcoholic controls. Results of their factor analysis suggested the existence of independent, noninteractive factors consisting of recency of active drinking and aging. Regression analysis further suggested that age, education, and head injury scores contributed to the variance in several factors, although the "number of weeks since the last drink" predictor contributed only about 36% of the variance in the various factors. Multivariate analysis of factor scores suggested that recently detoxified alcoholics had significant difficulty in problem-solving and learning new material. Their findings led Grant et al. (1984) to suggest that, while alcoholics continue to show neuropsychological impairments at approximately 1 month of abstinence, somewhere between 18 and 48 months, many seem to recover neuropsychological functions.

ISOPROPYL ALCOHOL ABUSE

Drinking of so-called "rubbing alcohol" is occasionally seen in hospitalized alcohol abusers with no access to ethanol. Acute symptoms appear significantly greater than those resulting from ethanol abuse. Kratz, Holliday, and Donahoe (1994) report the case of patient who had been hoarding bottles of rubbing alcohol who would become periodically dysarthic and ataxic and would show out-of-the-ordinary deficits in attention, speed of processing, executive functions, word finding, memory, and time orientation. The patient was described as normally well groomed and showing good articulation.

CONCLUSIONS

Parsons and Farr (1981) concluded in an earlier review of alcohol's neuropsychological effects that it was "impossible to escape" the many variables that can influence the neuropsychological performance of alcoholics. Recent alcoholism research has borne out the validity of that observation. The days are long gone when a direct causal relationship could be assumed between the intake of alcohol and neuropsychological impairment. That simplistic relationship has been replaced by the concept of multiple causation in alcohol-related neuropsychological impairment. Familial, genetic, nutritional, sociodemographic, and biochemical factors have been shown to contribute, sometimes substantially, to the variance tapped by neuropsychological tests. Direct toxicity to the brain, indirect hepatic and circulatory abnormalities, thiamine deficiency, genetic predisposition, and even an injury-prone alcoholic life-style, have all been shown to affect neuropsychological functioning. Such diverse etiologies await resynthesis in future models of neuropsychological function in alcoholism.

The individual clinician must, by extension, also be aware of this diversity of factors when evaluating single cases. Detailed clinical history of alcoholic patients may benefit from inclusion of these new factors. These include information on developmental history (e.g., hyperactivity history or family alcoholism), demographic factors, preexisting or concurrent psychopathology, age, physical disorders (e.g., liver disease), history of head injuries, and nutritional status (Alterman & Tarter, 1985).

Future neuropsychological research on alcoholism will undoubtedly become more multidisciplinary as psychologists utilize increasingly accurate brain-imaging and metabolic scans to better validate patterns of impairment shown on neuropsychological tests. The interaction of alcohol with drugs and other toxic substances (e.g., solvents) on neuropsychological performance is another area that deserves increased research attention.

The goal of understanding each and every interaction between alcohol and all of its cofactors is an idealistic one, to say the least. However, the extensive existing corpus of neuropsychological data and ongoing research in these areas can be viewed as a paradigmatic example of the neuropsychological contribution to toxicology. Neuropsychologists who wish to become neuropsychological toxicologists need to look no further than alcoholism research to appreciate the very real current and potential contributions of neuropsychological methods to the understanding of neurotoxic effects.

6

Drugs

Cognitive and emotional side effects of prescription drugs are essential clinical information in the context of neuropsychological evaluation. Some drugs effect very specific neuropsychological functions, e.g., benzodiazepines impact on short-term memory consolidation. Other medications may produce or exacerbate depressive disorders.

PRESCRIPTION DRUGS

Anticholinergics

Many types of drugs and neurotoxins have anticholinergic properties. Anticholinergic drugs have been used to control tremor in Parkinson's disease, and today are common adjuncts to neuroleptic therapy. Neuroleptic medications, without adjunctive antiparkinsonian agents, have anticholinergic effects. Finally, pesticides have potent anticholinergic properties.

Neuropsychological symptoms of anticholinergic substances can include amnesia and short-term memory deficit. High doses induce toxic states of delirium, agitation, confusional state, and hallucinations. Dysarthria and ataxia may also be present (Goetz, 1985). Herschman, Silverstein, Blumberg, and Lehrfield (1991) report a case of headache, blindness, agitation, and disorientation induced by the anticholinergic atropine sulfate, administered in nebulized form to control bronchospasm. Chronic use of multiple anticholinergic drugs may induce dementia that could be incorrectly ascribed to psychiatric disorder (Moreau, Jones, & Banno, 1986). Anticholinergic drugs appear to show a dose–response-related impairment of free recall, and when schizophrenic subjects already receiving anticholinergic drugs have their dose increased to the point of dry mouth or blurred vision, significantly decreased memory was observed (McEvoy & Freter, 1989). Unregulated herbal and patent medicines may have anticholinergic properties. For example, Chan (1995) reports on nine cases of anticholinergic poisoning in individuals being treated by herbalists for various ailments. *Datura Metel L* and *Panax ginseng* have both produced anticholinergic poisoning, although it is not clear from the report whether poisoning episodes were from the herb itself or adulterants (scopolamine, reserpine) that have been found in U.S. ginseng preparations. Commercially available Thorn Apple tea *(Datura Stramonium)* has precipitated anticholinergic poisoning in adolescents attempting to obtain a hallucinogenic "high" (Coremans, Lambrecht, Schepens, *et al.*, 1994). Neuropsychological reactions to anticholinergic substances receive additional discussion in the sections on neuroleptic medications and pesticides.

Anticonvulsants

Anticonvulsant drugs, administered as antiepileptic agents, are commonly prescribed for long periods, if not throughout the lifetime. Neuropsychological studies of anticonvulsant effects may be difficult to interpret, since they are, *a priori*, administered to patients exhibiting or at risk for significant brain dysfunction. A selected sample of neuropsychological and neuropsychiatric effects of anticonvulsants are described below.

Carbamazepine (Tegretol)

Carbamazepine is commonly prescribed for childhood seizure disorders. Carbamazepine has also been found to have several applications in psychiatry and may prove useful in treating depression, bipolar disorder, and as treatment for chronic schizophrenia that has been refractive to neuroleptics. Other possible, although not yet approved, indications for carbamazepine include episodic dyscontrol syndrome and miscellaneous behavioral disorders (Evans & Gualtieri, 1985).

Neurological side effects of carbamazepine can include mild "drowsiness, vertigo, ataxia, blurred vision and diplopia" which occur in about one-third of patients on this medication (Evans & Gualtieri, 1985, p. 223). Less common effects can include nystagmus, tics and various forms of dyskinesias (Evans & Gualtieri, 1985), or reversible dystonias (Reynolds & Trimble, 1985). Diplopia, an early sign of toxicity, occurs when carbamazepine blood levels exceed the normal range of 59 g/cm^3 (Goetz, 1985). Overdose may cause ataxia, nystagmus, seizures, and myoclonus with asterixis. Serum levels $\geq$ 170 μmol/liter (40 mg/liter) are associated with high risk for serious complications including "coma, seizures, respiratory failure and cardiac conduction defects" (Hojer, Malmlund, & Berg, 1993, p. 449).

Reviewing the neuropsychological effects of carbamazepine, Evans and Gualtieri (1985) suggest that both children and adults show higher levels of alertness with carbamazepine than with other anticonvulsants (i.e., phenytoin or phenobarbital). One study tested patients on a single medication, in a crossover design, and found that patients performed significantly better on the Stroop Color-Word Test and the Trail-Making Test during a Tegretol trial than during a period of Dilantin (phenytoin) use (Thompson & Trimble, 1982). Patients also performed more efficiently on a digit cancellation task when taking Tegretol, a finding the authors attribute to increased ability to sustain attention. The same study found significantly improved recall for pictures and words in both immediate and delayed recall conditions (Thompson & Trimble, 1982). Evans and Gualtieri (1985) failed to find any well-documented neuropsychological studies of carbamazepine influence on psychomotor performance, although high levels of carbamazepine have been associated with impaired concentration, motor speed, reaction time, decision-making speed, and threshold detection (Gillham, Williams, Wiedmann, Butler, Larkin, & Brodie, 1988; Massagli, 1991).

Carbamazepine is not without risk for adverse effect in a minority of patients. Silverstein, Parrish, and Johnston (1982) retrospectively reviewed seven cases were carbamazepine administration in children was associated with insomnia, aggression, paranoia, severe irritability, and developmental regression.

Developmental neurotoxic effects of carbamazepine include microcephaly, early developmental delay, and meningomyelocele (Adams, Vorhees, & Middaugh, 1990).

Phenytoin (Dilantin)

Phenytoin has been prescribed for over 50 years to treat seizure disorder and its efficacy in that regard is undisputed. It has also been associated with a variety of neurologic abnormalities, neuropsychological deficits, and developmental defects.

Neurological Effects. Reynolds and Trimble (1985) reviewed evidence for long-term phenytoin toxicity and described both PNS and CNS effects. Phenytoin has been linked to peripheral neuropathy, or at least electrophysiological abnormalities, in about 18% of seizure patients, a clinical result not found with other anticonvulsants (e.g., carbamazepine or sodium valproate). Permanent cerebellar syndrome has been noted in several case studies of phenytoin intoxication, although this outcome appears to be rare. Other less common neurological sequelae of phenytoin include various kinds of dystonia, reversible involuntary movement disorders, and asterixis (Reynolds & Trimble, 1985).

Neuropsychological Effects. Toxic blood levels of phenytoin have produced encephalopathy, along with common concomitants of nystagmus and ataxia. More recently, researchers have become aware of neuropsychological deficits produced by therapeutic or even subtherapeutic blood levels of phenytoin. One study investigated the effects of phenytoin on normal, nonepileptic volunteers who received either placebo or phenytoin 200 mg. TID for 2 weeks, in a double-blind, crossover design. Subjects receiving phenytoin showed significantly poorer delayed recall of pictures, made more errors on the Stroop Color-Word Test, and showed impaired ability to decide color or category membership in a choice reaction time task (Thompson, Huppert, & Trimble, 1981). In addition, the authors report significant correlations of serum phenytoin level with scores on five neuropsychological tests: immediate and delayed picture recall, reaction times to correctly identify words and pictures, and decision-making about color categories. In all cases, performance deteriorated as a function of increasing serum phenytoin level (Thompson *et al.,* 1981). The authors rightly caution that results may not necessarily be applicable to chronic users of Dilantin, since short-term administration precludes the investigation of possible tolerance or adaptation. Results are, however, consistent with a later study by some of the same authors. They investigated a group of 28 patients who had been prescribed a variety of anticonvulsants, including phenytoin. All patients except one had subtoxic blood levels of anticonvulsant. Significant decrements were noted in measures of immediate recall of words and pictures, although no differences were found in delayed recall or recognition. Other neuropsychological functions found to be affected included concentration in a visual scanning task under auditory distraction, and category decision-making. Similar results were found when the authors reanalyzed their data and included only the 16 patients who did not have toxic or near-toxic levels of anticonvulsant. These patients also showed poorer immediate and delayed word recall, poorer visual scanning under distraction, and a poorer category membership decision-making. Unfortunately, the results of phenytoin users were not analyzed separately, since some of the patients in the study received phenytoin, while others received carbamazepine, an anticonvulsant with fewer neuropsychological side effects (Reynolds & Trimble, 1985; Evans & Gualtieri, 1985; Thompson & Trimble, 1982).

Pulliainen and Jokelainen (1994) compared the effects of phenytoin with carbamazepine on young adult, newly diagnosed epileptic patients in an interesting study design that included pre- and post-drug administration neuropsychological testing, along with a control group that allowed comparison of practice effects. An interaction between group and time was seen on the Benton VRT immediate and delayed conditions. Marginally significant interactions were also observed with Purdue Pegboard and Symbol Digit; with the carbamazepine group showing faster performance in the drug condition and the phenytoin group performing more poorly. Drug levels of phenytoin were significantly correlated with slowed pegboard in the dominant hand and women showed greater slowing then men. While both drugs decreased practice effect compared with controls, phenytoin patients benefited much less from practice than all other groups, leading the authors to hypothesize that this drug interferes with procedural learning to a much greater degree than carbamazepine.

Developmental Effects. Phenytoin is capable of producing a wide range of teratogenic effects, ranging from growth deficiency to microcephaly and intellectual impairments. Craniofacial anomalies are commonly noted, and include short nose, low nasal bridge, low-set or abnormal ears, hypertelorism, epicanthic folds, eyelid ptosis, and prominent lips (Adams *et al.,* 1990). The constellation has been called *fetal hydantoin syndrome* (FHS) by Hanson and Smith (1975). No large-scale epidemiological studies of developmental neuropsychological effects have entered the literature; however, increased risk for learning disability may be present. IQ may be reduced relative to controls, but mental retardation has not been substantiated (Adams *et al.,* 1990). The relationships between dose and subsequent neuropsychological dysfunction have not been established in humans (Adams et al., 1990).

Sodium Valproate (Depakote)

Depakote was compared with phenobarbital for neuropsychological effects in children by Vining et al. (1983). Using a double-blind, crossover design, the authors examined the effects of both medicines on 21 children. While seizure control did not differ between drugs, significant neuropsychological differences were noted between valproate and phenobarbital. When children received phenobarbital they performed significantly more poorly in Full Scale IQ on an (unspecified) Wechsler Intelligence Scale. Phenobarbital subjects also showed decreased performance on Block Design, Picture Completion, and Vocabulary subtests. Two tests of attention span were also impaired, and every item but one on a behavioral questionnaire filled out by parents showed less dysfunction with sodium valproate. Sodium valproate use is not, however, completely free of neuropsychologically related side effects. Valproate use has been linked to symptoms of mild sedation, ataxia, dysarthria, and nightmares (Goetz, 1985). High levels of valproic acid have been linked to impairments of concentration, decision-making speed, and recall (Thompson & Trimble, 1983). Neurological developmental deficits associated with valproate include developmental delay and spina bifida (Adams et al., 1990).

Ethosuximide

Another anticonvulsant drug, ethosuximide, does not appear to have been the subject of neuropsychological study as of this writing. Psychosis has been reported as an effect of ethosuximide use, but many of the reported patients have a prior psychiatric history (Reynolds & Trimble, 1985).

Phenobarbital

Phenobarbital, a barbiturate used in childhood febrile convulsions, has been shown to produce a reversible behavior disorder in children, characterized by hyperactivity. Irritability, lethargy, insomnia, and oppositional behaviors are also sometimes seen (Reynolds & Trimble, 1985). Phenobarbital for childhood febrile seizures has shown unfavorable neuropsychological side effects relative to other anticonvulsants (e.g., carbamazepine). Byrne, Camfield, Clark-Tovesnard, and Hondas (1987) conducted intellectual and behavioral assessments on a monozygotic twin pair, one of whom had developed febrile seizures treated with phenobarbital from 17 to 30 months of age. While normal intelligence for both twins was found on the Bayley and McCarthy scales, the twin receiving phenobarbital performed consistently less well than the unmedicated twin. Posttreatment evaluation suggested that the medicated twin had improved in endurance and showed a better emotional tone, although he remained less cooperative than his brother. The originally medicated twin also remained significantly lower in the McCarthy's Quantitative and Memory subtests. The authors discount the possibility that febrile seizure and not phenobarbital may have been responsible for the medicated twin's lasting impairments, citing research to the contrary. They suggest that administration of phenobarbital "during infancy may have mildly attenuated general as well as specific intellectual ability, with such attenuation more evident later in the preschool period" (Byrne et al., 1987, p. 397).

Neuropsychological Effects in Relation to Anticonvulsant Serum Levels

Thompson and Trimble (1983) examined the effects of anticonvulsant serum levels on memory, concentration, perceptual and motor speed, and mood in 28 seizure patients who used a variety of single and dual anticonvulsant medications. While mean serum levels were characterized as subtoxic, 12 patients did have toxic serum levels. Relative to their own low-serum-level performance, high-level anticonvulsant patients were impaired on immediate recall for pictures, concentration under a distracting stimulus, speeded decision on a color and category membership task, and a related motor latency measure when no decision was required. No differences were found in motor speed, mood, or delayed memory recall. The authors then removed date culled from the 12 patients with serum levels above the therapeutic range, leaving 16 patients. Fewer neuropsychological tests were found to be impaired at therapeutically high but nontoxic serum levels. These included immediate word recall, visual scanning, decision-making on categories (the more difficult task) but not color membership.

In conclusion, anticonvulsants have a variety of neurotoxic and neuropsychological effects that vary according to dosage and drug type. Neuropsychological function decrements may be observed even when serum levels of anticonvulsant are subtoxic. Certain common medications like phenytoin have been linked to a variety of neuropsychological deficits including sustained attention, perceptual and motor performance, and speeded decision-making tasks. Other anticonvulsants appear to produce fewer neuropsychological deficits compared with phenytoin, especially carbamazepine (Tegretol) and sodium valproate. Administration of single anticonvulsants is associated with better neuropsychological performance than polydrug administration.

Future neuropsychological research is needed to continue investigation into the newer anticonvulsants like valproate and carbamazepine, and it is hoped that researchers will utilize the crossover, double-blind design favored by Trimble and associates. Such a design would obviate many of the difficulties inherent in performing neuropsychological tests on populations with preexisting brain dysfunction.

Norms for individuals using anticonvulsants would also be of use to clinicians and neuropsychological researchers. In the meantime, neuropsychologists engaged in testing individual patients need to take into account alterations in behavioral and cognitive functions produced by these drugs.

Antidepressants

Tricyclic antidepressants (TCA) as a class (e.g., amitriptyline, imipramine, desipramine, doxepin) are capable of producing overt CNS toxicity with high plasma levels, as well as more subtle side effects of normal dose administration capable of affecting neuropsychological performance. True tricyclic CNS toxicity occurs in 6% of patients and mostly typically presents with symptoms of delirium, disorientation, memory impairment, and agitation. Approximately one-third of cases reviewed included symptoms of psychosis (Preskorn & Jerkovich, 1990). Toxicity risk correlates positively with TCA plasma levels and risk is ten times greater when plasma levels are greater than 45 ng/ml (Preskorn & Jerkovich, 1990).

Commonly reported symptoms at lower plasma levels include tremor, blurred vision, and postural hypotension. Neuropsychological studies of antidepressant medication administration show mixed results. For example, several neuropsychological examinations of subjects receiving tricyclics have failed to find significant neuropsychological impairments, either after a single dose (Heimann, Reed, & Witt, 1968) or after extended (6 and 12 weeks) therapy (Kendrick & Post, 1967). In studies finding deviations from the baseline performance, there appear to be approximately equal numbers showing positive and negative effects. For example, Liljequist, Linnoila, and Mattila (1974) found that nortriptyline slightly worsened Paired-Associate and Digits Backward scores. Legg and Stiff (1976) examined subjects given clinical trials of imipramine for 3 weeks and found significant impairments relative to controls on the Benton Visual Retention Test, the Wechsler Memory Scale Logical Memory subtest, the WAIS Digit Symbol and prorated Performance IQ. In contrast to Liljequist *et al.* (1974), however, paired-associate test performance was not impaired.

Since the elderly are at an enhanced risk for both depression and memory deficit, the influence of antidepressants on memory in that group is an important question. Confusional states in patients over age 45 who receive amitriptyline or imipramine have been reported to be as high as 35% (Deptula & Pomara, 1990). Marcopulos and Graves (1990) compared 27 elderly subjects on a variety of antidepressants with an equal number of unmedicated controls matched for levels of depression. The drug group showed mild, but significant impairment on all variables. The authors properly point out that their study does not necessarily detect anticholinergic memory impairment, especially since they failed to find an anticholinergic dose–effect relationship. Sedation, or other unmeasured factors may be contributory. Hoff, Ollo, and Helms (1986) reported impairment in verbal memory in 9 elderly inpatients receiving nortriptyline. Impairments increased with age and drug plasma concentration.

The relationship between antidepressants and memory is complicated by confounding effects of depression on memory. Mitigating depression with tricyclics may actually improve the performance of depressed patients on a battery of memory and learning tests (Sternberg & Jarvik, 1976).

Available results suggest a complex relationship between TCA medications and neuropsychological impairment. Individual differences in cholinergic response to antidepressant medications, severity of depressive effects counterbalanced against medication-based improvements in function, and the multifactorial neurochemical nature of depression all may prove relevant factors to future research. Deptula and Pomara (1990) correctly criticize current antidepressant research, pointing out problems that include small group size, vague, mixed, or multiple diagnoses among the "depression" group, lack of control subjects, and inadequate neuropsychological test batteries. They conclude that, to date, only amitriptyline and to a lesser extent imipramine show reasonable evidence of neuropsychological effects. In younger subjects these effects may be temporary as tolerance develops over a period of 1–2 weeks.

Selective Serotonin Reuptake Inhibitors

There is little neuropsychological research as yet on more recently developed selective serotonin reuptake inhibitors (SSRIs). Fluoxetine (Prozac) is the most well known of this group and has been extensively employed for depression management since being introduced in the United States in 1987. Common side effects include headache, insomnia, nausea, and nervousness, which occur in 15–23% of patients (Messiha, 1993). Other more uncommon neurological side effects include akathisia, dystonia, dyskinesia, and isolated case reports of parkinsonian-like syndromes (Messiha, 1993). So-called "serotonin syndrome" is another uncommon complication, understood to be caused by serotonergic stimulation. Symptoms include aggression, hypomania, tremor, shivering, myoclonus, diaphoresis, restlessness, and thought disturbances (Messiha, 1993). Combining fluoxetine and lithium has produced neurotoxic side effects in two reports (Noveske, Hahn, & Flynn, 1989; Sacristan *et al.*, 1991).

An unusual neuropsychological sequela of frontal lobe impairment was noted as a reversible side effect in a patient taking fluoxetine. The patient was a 23-year-old student who was administered the drug for symptoms of obsessive–compulsive disorder and social anxiety. Treatment was gradually increased to a prodigious 100 mg/day over an 8-week period, at which time the patient developed apathy, decreased attention, perservation, and memory difficulties. SPECT scan revealed decreased blood flow to the frontal lobes relative to premorbid scan. Neuropsychological screening showed decreases in word fluency, production of complex novel designs, and copying of a complex design. Four weeks after discontinuation of medication, the patient reported that perceived symptoms had cleared, but were accompanied by resumption of obsessive–compulsive symptoms (Hoehn-Saric *et al.*, 1991).

Antihypertensives

CNS impairment from adrenergic-inhibiting antihypertensives has been noted since reserpine was found to cause depression and impairment in psychomotor coordination. Neuropsychological effects of currently available adrenergic-inhibiting antihypertensives are similar and have been noted primarily in the domains of depressive affect, lowered attention/arousal, and impaired memory. Highly lipophilic beta blockers (e.g., propranolol) produce high brain concentrations and are considered to be more neuropsychologically impairing than hydrophilic beta blockers like atenolol (Neil-Dwyer, Bartlett, McAinish, & Cruickshank, 1981). Neuropsychological symptoms appear to correlate with plasma brain concentrations (Neil-Dwyer et al., 1981).

There have been several anecdotal accounts of "forgetfulness" attributed to medications that lower and stabilize blood pressure (e.g., Ghosh, 1976; Fernandez, 1976). These observations received preliminary validation in a recent study that used Russell's (1975) modification of the Wechsler Memory Scale (Solomon et al., 1983). Their study compared hypertensives on propranolol or methyldopa with hypertensive subjects on diuretic alone, as well as normotensive subjects prescribed propranolol. Subjects receiving either propranolol or methyldopa displayed significant impairment on the Logical (Verbal) Memory subtest of the Wechsler Memory Scale, both in immediate and 20-minute recall. No other deficits were found, and results could not be explained by variations in patient blood pressure. Because the mean age of normotensive subjects (60.5) differed by about 8 years from other subjects (early 50s), the conclusion of a main effect of propranolol on verbal memory is weakened somewhat because age-related changes in normotensives' memory may have contributed some of the variance the authors attributed to propranolol. Nonetheless, the study presents a potentially significant neuropsychological side effect of antihypertensive agents, and is deserving of replication.

Benzodiazepines

Benzodiazepines are a group of medications commonly prescribed for their antianxiety, hypnotic, and muscle-relaxing effects. Prescription use of these products is truly extensive, with recent estimates of over $400 million spent annually on these drugs. A survey taken in the Boston area during the 1970s found that 30% of all hospitalized patients were prescribed diazepam and 32% received flurazepam (Greenblatt & Shader, 1974; Hall & Zisook, 1981). Almost 1.25 million adults in the United Kingdom have taken tranquilizers (primarily benzodiazepines) for more than a year (Balter, Manheimer, Mellinger, & Uhlenhuth, 1984). Nearly 1 in 10 American adults takes at least one benzodiazepine a year. One to 1.5 *billion* benzodiazepine prescriptions were filled in the United States between 1965 and 1985 (Salzman, 1992).

Benzodiazepine neurochemistry involves the stimulation of GABA receptors in the ascending reticular activating system, which inhibits cortical and limbic system arousal. Anxiolytic, sedative, muscle relaxant, ataxic, and amnestic effects may be traced to neuronal inhibition in diverse brain regions affected by these receptors (Roy-Byrne et al., 1993). Limbic system discharges are also suppressed (Hall & Zisook, 1981). Benzodiazepine can also decrease catecholamine uptake in parts of the brain, and increase acetylcholine concentrations (Hall & Zisook, 1981). Cerebral blood flow decreases significantly in the right hemisphere and borderline significantly in the left with the most prominent reduction occurring in the right frontal lobe (Mathew & Wilson, 1991a). Benzodiazepines are considered euphorigenic and euphoric intoxication effects may be correlated with rapidity of peak plasma levels, with large doses causing greater euphoria than smaller doses administered at intervals (de Wit, Dudish, & Ambre, 1993).

Valium (diazepam) is the most well known of the benzodiazepine derivatives, a group of minor tranquilizers of varying half lives and potency. Other benzodiazepines include chlordiazepoxide (Librium), alprazolam (Xanax), lorazepam (Ativan), oxazepam (Serax), flurazepam, and clobazam.

Neurophysiological Effects

The relationship between benzodiazepine administration and brain activity is not completely worked out. One study (Buchsbaum et al., 1987) showed PET scans with decreased cerebral glucose metabolism in frontal and occipital cortex, especially in the right hemisphere. Another found that the frontal lobes, in particular the right frontal lobe, showed decreased cerebral blood flow (CBF) (Mathew, Margolin, & Kessler, 1985). However, these results were not replicated in a follow-up study by the same authors using a 20% higher dose injected more rapidly (Mathew & Wilson, 1991a). Roy-Byrne et al., (1993) also failed to find localized CBF reductions with benzodiazepine but noted reduced whole brain CBF (25–30% decrease) after initial dose of alprazolam. Tolerance effects were found with respect to whole brain CBF, memory performance, and plasma epinephrine over the course of 1 week of continuous treatment with alprazolam.

Acute Neuropsychological Effects

Acute subjective effects of low-dose (7–10 mg) benzodiazepine are quite similar to those of alcohol, but significantly greater impairment is produced by benzodiazepines on measures of memory, body sway, and time estimation (Schuckit, Greenblatt, Gold, & Irwin, 1991). Higher doses (12 mg average) produce clumsiness, sleepiness, and concentration impairments of a degree that would make it highly dangerous to operate machinery or a motor vehicle (Schuckit et al., 1991). Clinical abnormalities on tests for drunken drivers are positively correlated with blood benzodiazepine concentration (Kuituren, Aranko, Nuotto, et al., 1994).

The most commonly reported acute neuropsychological effect of benzodiazepine has been memory dysfunction. Acute amnestic reactions have been produced in subjects receiving these drugs immediately prior to surgery. Several studies suggest that benzodiazepines may impair recall of words presented after drug administration, but not influence prior word learning, a finding interpreted as consistent with impairment in encoding but not of retrieval (Lister, 1985).

Block and Berchou (1984) used the Buschke Selective Reminding test to examine the acute effects of lorazepam (Ativan) and alprazolam (Xanax). Both drugs impaired total recall compared with placebo. Impairment in total recall was greater over successive trials. Both drugs also significantly reduced the imagery advantage inherent in remembering high-imagery words.

In the same day, benzodiazepine administration also impaired performance in a discriminative reaction time task and reduced the frequency for which Critical Flicker Fusion could be perceived. Main effects of impaired learning and retrieval found in this study are consistent with reductions in memory and/or motivation to encode information seen in other studies. However, the authors' use of multiple ANOVAs over the same data makes it difficult to vouch for their many significant findings.

Lucki, Rickels, and Geller (1986) examined acute benzodiazepine effects on 22 chronic users of these medications. Subjects received a fixed dose of their usual benzodiazepine and were tested 60–90 min later. Compared with their predrug status, subjects showed significantly reduced Flicker Fusion but improved their performance on the Digit Symbol substitution test. Memory was tested for immediate and 20-min recall using a 16-item word list of noncategorized nouns. While immediate recall was not affected, delayed recall was significantly reduced.

Bornstein, Watson, and Kaplan (1985) failed to find neuropsychological effects of triazolam on normal university volunteers given single doses of flurazepam or triazolam. Other studies that used higher doses of benzodiazepines and more complex tasks found significant impairments. For example, Hindmarch and Gudgeon (1980) found that lorazepam impaired a four-letter cancellation task in doses of 1 mg.

Finally, a recent study examined memory decrements as a function of increasing doses of diazepam (Wolkowitz et al., 1987). Subject volunteers received cumulative intravenous diazepam at 8.8, 35.1, and 140.1 μg/kg. This would correspond to doses of 0.625, 2.5, and 10 mg in a 70-kg subject. Each subject was read a list of 18 words in the same semantic category at a rate of one word every 3 seconds. Six of the words were read twice, and six were presented only once. Subjects were asked to notify the examiner if a repeat was heard, as a test of attention. A distractor task followed in which subjects were given a semantic category and asked to generate a word list from that category. A recall and recognition task followed this distractor trial, and subjects were also asked whether the words they recognized had been presented once or twice, which the authors characterized as a measure of "automatic processing" (Wolkowitz, et al., 1987). Main effects were found for diazepam on attention, recognition, and automatic processing, with decreasing performance as a result of diazepam administration. Incorrect word intrusions into free recall were also significantly increased. While diazepam did not affect free recall *per se,* when the authors corrected for intrusions as guessing, they did find significantly decreased recall in the diazepam conditions. The authors noted anecdotally that the amnestic effects of diazepam were so marked that some subjects "did not even recall having heard a given memory list" (Wolkowitz *et al.,* 1987, p. 27).

In contrast to these impairments of episodic memory, subjects showed no deficits in their ability to generate semantic category words in the interference task. The authors interpreted their results as consistent with a model in which diazepam interferes with acquisition or consolidation of new memories, but does not affect memory or retrieval of previously acquired information. The authors further suggest that this pattern of memory dysfunction is similar to that found in Korsakoff's patients, and that diazepam may be a pharmacological model for certain organic amnesias. Unfortunately, since subjects' memory abilities were highly correlated with subjective ratings of sedation, the possibility cannot be ruled out that attention rather than memory function is impaired.

Temazepam and flunitrazepam were compared with one another and with controls. Drug administration to otherwise unmedicated subjects was found to change EEG and neuropsychological functions (Saletu, Grunberger, & Seighart, 1986). Compared with placebo, temazepam caused increases in EEG beta and decreased alpha activity. Delta/theta activity was increased. Decreases in neuropsychological tests of attention, concentration, general cognitive function, and memory were found, although low blood levels produce some beneficial effects in reaction time and several other tests.

Several studies have looked at acute benzodiazepine effects in a more "real world" task of driving an automobile. Hindmarch and Gudgeon (1980) compared clobazam and lorazepam administration on five car handling tasks, including parking, three-point turn, slalom, width estimation between obstacles, and brake reactions. Examiner ratings showed significant driving impairments for lorazepam but not clobazam. Regrettably, simultaneous neuropsychological testing was not administered to ascertain the existence of more subtle impairments.

O'Hanlon and Volkerts (1986) administered nitrazepam and temazepam in a double-blind crossover design, using more exacting measures of driving impairment than examiner rating; Doppler radar for measuring velocity and an electro-optical lane tracker to compute the vehicle's position within its lane. Eleven adult females who had previously received benzodiazepines for insomnia were assessed while administered medication the night before and tested on the following day for their ability to drive in actual traffic. Results suggest that shorter-acting benzodiazepines (i.e., Temazepam) do not impair driving (i.e., ability to maintain accurate lane position) the day after drug administration, while benzodiazepines with long half lives will cause significant impairment in lane tracking. The authors also note that two other benzodiazepines, flurazepam 30 mg and loprazolam 2 mg, had an effect greater than a blood alcohol concentration of 1.0 mg/ml, a BAC associated with seven times the risk of traffic fatality.

There are few studies that assess the comparative neuropsychological effects of benzodiazepines on the same measures. One such study compared effects of alprazolam 1 mg/2 mg, bromazepam 3 mg/6 mg, clobazam 10 mg/20 mg, oxazepam 30 mg/50 mg, lorazepam 1 mg/2 mg, and placebo on 28 female volunteers (ages 21–45). Statistical design was a 7 × 7 Latin square. Neuropsychological tests were critical flicker fusion and choice reaction time. The only benzodiazepine to impair critical flicker fusion was oxazepam. In contrast, choice reaction time was affected by higher doses of bromazepam, alprazolam, lorazepam, and oxazepam; the latter three also produced notable "hangover" effects that included difficulties awakening the next morning and coordinating behavior. Bromazepam 6 mg showed reduced hangover response, while clobazam and bromazepam 3 mg were free of hangover effects.

Benzodiazepines tend to show greater neuropsychologically impairing effects relative to anxiolytics which do not belong to the benzodiazepine family. One such contrasted diazepam and buspirone on tests of memory, motor speed and coordination, and cognitive efficiency (Unrug-Neervoort, Luijtelaar, & Coenen, 1992). Diazepam produced significant decreases in motor coordination, motor speed in the nonpreferred hand, delayed recall of a complex figure, and verbal learning compared with buspirone.

Chronic Effects

There has been mixed support for chronic benzodiazepine-induced neuropsychological impairment. In what might be thought of as "experimental chronic" benzodiazepine administration, Bornstein, Watson, and Pawluk (1985) gave normal college student volunteers without history of neurological or psychiatric illness nightly capsules of either triazolam, flurazepam, or placebo for 7 days. Baseline, intermediate, and final neuropsychological tests were administered, including Grooved Pegboard, Trails, Finger Tapping, Hand Dynamometer, Wechsler Memory Scale Logical and Figural Memory, the Knox Cube Test, and a choice reaction time task. Subjects were also given a questionnaire to uncover possible side effects. Although no significant differences on any neuropsychological measures were found when baseline differences were controlled, many more side effects were reported by the experimental groups. While this study suggests that low-dose, intermediate-term use of certain benzodiazepines does not produce neuropsychological deficit, the study does not really address long-term sequelae of chronic use or abuse of minor tranquilizers.

Several more recent studies investigating longer-term use of benzodiazepines produced mixed results. The first used a patient population that had been prescribed benzodiazepines for a mean of 60 months (Lucki *et al.,* 1986). Thirty-three of the experimental subjects were single-benzodiazepine users, four additional patients received two (flurazepam and triazolam), and one patient regularly received clorazepate dipotassium (Tranxene) and lorazepam (Ativan). Six of the total group received additional medications—four taking antidepressants and two receiving antihypertensives. Daily intake was in the therapeutic range "in nearly every case" (Lucki et al., 1986, p. 427). Twenty-six drug controls were recruited from a patient group presenting with anxiety disorders at the same clinic. Control subjects were matched for sex, age, and education with the norms of the experimental group. Tests administered included subjective mood scales, Digit Symbol Substitution Test, as well as tests involving symbol copying, letter cancellation, Critical Flicker Fusion, and free recall of word lists. Only performance on Critical Flicker Fusion differentiated experimental subjects from controls, and since multiple comparisons were made without adjusting the alpha level, this single significant result may be artifactual. The same authors retested 17 of their subjects 48 days after termination of benzodiazepine. Again, only Flicker Fusion thresholds showed significant increases. Patients also reported significantly greater sedation and anxiety during withdrawal.

Two subsequent investigations by Golombok and her colleagues suggest that chronic use of benzodiazepines does produce structural and functional brain changes. The first study (Golombok, Moodley, & Lader, 1988) compared 50 patients presently using benzodiazepines for at least 1 year with a group that had stopped taking benzodiazepines, and a matched control group that had never taken the drug. Tests administered included the National Adult Reading Test (NART), Cancellation, Reaction Time, Digit Symbol, Symbol Copying, Block Design, Verbal Memory Recall, New Learning, Visual Spatial Recognition Memory, a "Manikin" spatial orientation task, Trails, the Bexley–Maudsley Category Sorting Test, Controlled Oral Word Association (COWAT), a test of visuospatial analysis, a state anxiety inventory, and a questionnaire assessing frequency of self-reported cognitive failures.

Several of the neuropsychological tests were significantly related to a measure of benzodiazepine intake in subjects presently taking the drug: Cancellation, Digit Symbol, Symbol Copying, Block Design, New Learning, the "Manikin" test, and visual perceptual analysis. Subjects with high benzodiazepine intake were significantly impaired on these tests after controlling for state anxiety. The authors suggested that their pattern of results reflect parietal, posterior temporal, and occipital effects, a pattern different from acute effects, particularly in the display of visuospatial impairments. The lack of memory impairment in chronic users also provides support for tolerance for that effect in long-term users. The pattern of impairment may be reversible, as a group that had withdrawn from long-term benzodiazepine use could not be distinguished from controls. However, the group that had withdrawn had generally lower lifetime benzodiazepine intake, indicating that the critical comparison—former high-dose, long-term benzodiazepine users against controls—has not yet been investigated.

A more recent study by the same group recorded CTs of individuals with chronic benzodiazepine users against similar comparison groups (Moodley, Golombok, Shine, & Lader, 1993). Reduced density was noted in several regions, especially in the frontal, occipital lobes, and the caudate nuclei. Most of the results were found in lorazepam users. The authors suggest that lorazepam and other high-potency anxiolytics may have greater neuropathological correlates;

however, they could not rule out the possibility that more severely anxious patients (with presumably greater premorbid neuropathology) may have been prescribed lorazepam.

The results of the Lucki et al. (1986) study suggest that chronic use of benzodiazepines within therapeutic guidelines does not have neuropsychological effects on the functions tested by the authors. However, this lack of effect may be related to tolerance and drug adaptation in very long-term users, or else the possibility that more difficult neuropsychological tasks (e.g., those requiring divided attention or parallel processing) might be necessary to exhibit subclinical effects in this population. That several such "human performance" tasks *were* found to be impaired in the Golombok study, suggests that particular types of tasks, and perhaps specific benzodiazepines are required to produce abnormal neuropsychological and neurophysiological effects in chronic users.

Chronic use of benzodiazepines in combination with other medications may produce severe neuropsychological impairment. Brooker, Wiens, and Wiens (1984) report the case study of a 53-year-old male attorney who showed severe neuropsychological impairment on the Halstead–Reitan Battery after long-term use of diazepam (Valium) and meprobamate (Equanil/Miltown). The patient showed almost complete symptom reversal at 1 year post-initial assessment, when he had been abstinent for 6 months.

Comparison with Alcohol. Although the class of benzodiazepines have been colloquially viewed as "powdered alcohol," their acute neuropsychological profiles are distinct. While both produce psychomotor impairment at high doses, benzodiazepine (i.e., triazolam) produced memory impairments on both a digit recall task and a picture recognition test, particularly in delayed recall conditions, whereas alcohol did not (Roache, Cherek, Bennett, Schenkler, & Cowan, 1993).

Tests Used. Wittenborn (1980) reviewed tests that discriminated benzodiazepine users from controls. He suggests that the following tests are most capable of discriminating the two groups: Critical Flicker Fusion, learning and memory tests, manipulation tests, time estimation, Digit Symbol Substitution Test, and cancellation tasks. In general, test batteries that load heavily on memory and attentional processes would appear to be the most sensitive to benzodiazepine-induced impairments.

Psychological and Psychiatric Effects. While the expected therapeutic effect of benzodiazepines is usually the reduction of anxiety, paradoxical changes in affect and behavior have also been reported. Hall and Zisook (1981) reviewed these paradoxical reactions, which include depression, tremulousness, apprehension, insomnia, severe anxiety, suicidal ideation, hatefulness, and rage. Gross cognitive disturbances linked to benzodiazepine use include psychosis, manic or hypomanic behavior, confusional states, and visual hallucinations. Korsakoff-like syndromes have also been reported (Hall & Zisook, 1981).

Chemotherapy and Antineoplastic Drugs

By definition, antineoplastic drugs are cytotoxic, and CNS complications are frequently observed in children and adults undergoing therapy for cancer. Children may be at higher risk for intellectual and neuropsychological damage from these drugs because of incomplete myelination (Brown *et al.*, 1992).

In adults, intrathecally (via lumbar puncture or indwelling Ommaya reservoir) administered antineoplastic therapy has been seen to cause transient complications of aseptic meningitis, acute myelopathy (transient or permanent), and, rarely, seizures. High-dose intravenous drug administration can cause acute focal cerebral dysfunction, e.g., disseminated necrotizing encephalopathy induced by methotrexate therapy alone (Asada *et al.*, 1988).

Cyclosporin

Major neuropsychiatric side effects include tremor, anxiety, seizures, ataxia, visual hallucinations, and disorientations. Less common reactions include cortical blindness, paresthesias, dysarthria, and paresis (Trzepacz, DiMartini, & Tringali, 1993).

Methotrexate

CNS damage is also associated with weekly low-dose methotrexate treatment in patients who did not have cancer (Wernick & Smith, 1989). Of 25 consecutive patients receiving methotrexate for controlling rheumatoid arthritis, 5 patients spontaneously reported subtle cognitive dysfunction, mood alterations, or unusual cranial sensations. Symptoms included "shaky," "head heavy," "strange or funny," "thick," "decreased memory," "crazy," "head in vise," "fuzzy," and "altered mood." Other symptoms produced by methotrexate therapy are hallucinations, ataxia, depression, and confusion. Nyfords (1978) noted that 11% of patients developed headache, 5% had dizziness, and 2% reported memory impairment. Complaints were thought to be underreported, since they were not considered serious in comparison to more life-threatening situations common to cancer-related illness. Wernick and Smith conclude that "because of its negative impact on overall quality of life, the possibility of CNS toxicity should be considered in patients receiving methotrexate therapy" (p. 774).

5-FU

Neurological side effects of 5-FU therapy include cerebellar ataxia, gait and limb ataxia, dysarthria, nystagmus and hypotonia. High doses may cause encephalopathy in up to 40% of patients. Symptoms vary from lethargy to coma (Shapiro & Young, 1984).

Vincristine

Vincristine is an alkaloid that has been used in cancer chemotherapy since 1962. It is "predictably and uniformly neurotoxic" (Le Quesne, 1984). Neurotoxic symptoms begin with paresthesias, with severe weakness ensuing. Extensor muscles of the fingers, wrists, and legs become weak. There is partial recovery with elimination of paresthesias; however, ankle jerks usually failed to show recovery and superficial loss of sensation can persist (Le Quesne, 1984).

Disulfiram (Antabuse)

Disulfiram is prescribed in the treatment of chronic alcoholism. Its daily use causes exaggerated and aversive sensitivity to ethanol. By blocking the enzyme aldehyde dehydrogenase, disulfiram causes an accumulation of acetaldehyde, with resulting autonomic reactions, including nausea, vomiting, headache, and hypotension. After 19–38 days of treatment, mean concentration of CS_2 in alcoholics taking Antabuse was 9482 ng/liter, approximately 35 times that of unmedicated subjects (Brugnone et al., 1992). Average mean value of total CS_2 was about 50 times that of normal subjects. Blood half-life for free CS_2 was 10.6 h and 6.9 h for total CS_2 (Brugnone et al., 1992).

Sterman and Schaumburg (1980) find distal axonal breakdown resulting in peripheral neuropathy to be associated with standard therapeutic doses (250–500 mg daily). The course of impairment begins several months after initiating treatment and takes another several months to recover after the drug is stopped. Paresthesias, weakness, and sensory impairment are characteristic, with rare episodes of optic neuritis. Disulfiram has also been linked to a reversible toxic encephalopathic syndrome with symptoms of delirium, psychosis, and cerebellar signs. These central and peripheral symptoms may be related to accumulation of carbon disulfide, since disulfiram in the body is catabolized into this potent neurotoxin (Sterman & Schaumburg, 1980).

Palliyath, Schwartz, and Gant (1990) expanded on Sterman and Schaumburg's findings, by showing that alcoholic patients receiving a lower dose of Antabuse (125 mg) did not reveal significant electrophysiological abnormalities, while replicating earlier findings of Antabuse neurotoxicity at a dose of 250 mg. Within 6 months of administration, the large dose produced significant declines in median motor latency, peroneal conduction velocity, median sensory velocity and amplitude, and sural sensory conduction velocity. Since patients were abstinent from alcohol during this period, declines were most likely associated with Antabuse.

It has been generally held that high dose or even overdose of disulfiram in adults is typically reversible and "rarely serious" (Ryan, Sciara, & Barth, 1993, p. 389; Brewer, 1984). However, a recent case study of a cirrhotic alcoholic taking up to 1500 mg daily noted a progression of confusion to lethargy and coma after 16 days of disulfiram administration (Ryan et al., 1993). Six years later, the patient received neuropsychological testing, the profile of which was characterized as diffusely impaired, with HRB Impairment Index of 0.7, 100 Category Test errors, a Digit Symbol scaled score of 3, and Trails B at 232 s. Given the patient's history of alcoholism, hepatic damage, and other difficulties, test results likely reflect both Antabuse overdose and premorbid factors. There was, however, was no negative exaggeration of symptoms noted on the patient's MMPI and the patient had been apparently functioning adequately in his job prior to a self-arranged hospitalization (Ryan et al., 1993).

Disulfiram overdose profile is more variable in children, where some recover, but others sustain moderate or severe brain injury (Benitz & Tatro, 1984).

There are few controlled studies on the effects of Antabuse on neuropsychological functioning, but available data suggest that further studies are needed. Prigatano (1977) compared the neuropsychological functioning of 22 inpatient, recidivist alcoholics treated with disulfiram against an equal number of inpatient alcoholics involved in a 90-day inpatient milieu treatment program. Patients were matched as closely as possible on age, education, years of alcohol abuse, and general intelligence. The subjects in the Antabuse group had been taking this medication somewhere between 14 and 21 days.

Disulfiram-treated alcoholics were significantly more impaired than the milieu treatment group in abstract reasoning (Category Test), speech perception (Speech Perception Test), and in their overall impairment indexes. Percentages of patients from each group who would be classified as impaired using Reitan's norms appear in Table 6.1. While it is tempting to ascribe such results to Antabuse administration, they are most likely a function of preexisting differences in the two groups, particularly since the Antabuse group was composed of recidivists with a history of previous treatment failure. Support for this hypothesis was found in a preliminary follow-up study reported by Prigatano (1977) in the same paper. When six slightly younger and better-educated alcoholics were tested on entry into treatment, and then again after 2 weeks of Antabuse treatment, they failed to show significantly increased impairment. The Category Test of these

TABLE 6.1. Neuropsychological Impairment as a Function of Disulfiram Treatment[a]

Test	Percentage of subjects rated as impaired	
	Antabuse alcoholics	Milieu alcoholics
Category Test	91	59
TPT Total Time	82	68
TPT Memory	23	27
TPT Location	86	77
Seashore Rhythm Test	50	41
Speech Perception Test	91	36
Tapping Preferred Hand	91	55
Impairment Index	82	50

[a]Reprinted, by permission of the publisher, from Prigatano, G. P. (1977). Neuropsychological functioning in recidivist alcoholics treated with disulfiram. *Alcoholism: Clinical and Experimental Research, 1*, 81–86.

subjects was significantly impaired on entry into the program, and thus not as a function of Antabuse. In fact, when tested after 2 weeks of Antabuse administration, these subjects improved significantly on the Category Test (75.2 versus 48.5 errors), TPT total time, and the Impairment Index. There was a nonsignificant tendency for Speech Perception to be worse on the second testing, which the author speculates could achieve significance with a larger N. Thus, the issue of whether Antabuse causes neuropsychological impairment remains open at this time. Studies that test the effects of Antabuse on more completely matched groups controlling for recidivism would appear to be essential to answer the question.

Naloxone

Naloxone is an opioid antagonist used to block the emotional or toxic effects of opiate ingestion. Naloxone appears to be capable of inducing its own effects of human neuropsychological function. In a study of 12 hour naloxone infusion on normal volunteers, Del Campo, McMurray, Besser, and Grossman (1992) found increases in choice reaction time, lessened serial recall of numbers and letters, and lower accuracy of spatial orientation. Increased depression was found on the POMS, which may be related to naloxone-induced elevations in cortisol production.

Neuroleptics

Neuroleptics or "major tranquilizers" are prescribed to manage and control symptoms of confusional states, schizophrenics, late Alzheimer's disease, and other functional or structural disorders that cause a breakdown of cognitive and emotional functioning. Neuroleptic agents include the phenothiazines, thioxanthenes, and butyrophenone. Typical action is the blockade of dopamine D_2 receptors, although there is little detailed research on the effects of neuroleptics on the brains of healthy humans without psychiatric disease. An important advance in this area was recently published by Bartlett *et al.* (1994) who used PET scan to study regional glucose utilization of normal subjects who were administered haloperidol (Haldol). Twelve hours after administration of 5 mg of Haldol, significantly lower glucose utilization was measured in the cortex, limbic system, thalamus, and caudate nucleus, but not the putamen or cerebellum. Localized reductions were also seen in frontal, occipital, and anterior cingulate cortex.

Neurological Effects

Neuroleptics can have various deleterious effects on the CNS. Acute intoxication from neuroleptic medications can produce nonspecific CNS changes, including cerebral edema, reversible neuronal swelling and vacuolation, secondary anoxic and vasocirculatory lesions (Jellinger, 1977). Acute behavioral changes from neuroleptic administration include akathisias or states of intense restlessness coupled with anxiety, which are usually seen in the first days of neuroleptic therapy or when the dose of drug has been substantially increased. Symptoms can be reversed with anticholinergic treatment (Goetz, 1985) and can spontaneously resolve within several weeks. Dystonias are another potential side effect seen early in the course of neuroleptic administration, and may occur with just one parenteral dose. The head and neck show abnormal positions, and eyes may maintain a fixed upward gaze. Piperazine derivatives are most likely to precipitate dystonias (Goetz, 1985).

One of the more well-recognized symptoms of chronic neuroleptic treatment is the development of parkinsonian symptoms. Patients with a history of Parkinson's disease and/or essential tremor among immediate family or close relatives may be most at risk for developing these symptoms (Negretti, Calzetti, & Sasso, 1992).

Toxic confusional states are produced in some patients which may be related to the anticholinergic effects of these drugs.

Neuroleptic Malignant Syndrome (NMS)

The most serious acute reaction to neuroleptics is that of "neuroleptic malignant syndrome," a potentially fatal reaction to antipsychotic medication. The syndrome is characterized by extrapyramidal signs, including parkinsonian symptoms, muscle rigidity, dystonia, akinesia, and tremor. Hypothermia and other signs of autonomic dysfunction are also present, including very high fever, dysarthria, seizures, and deterioration of mental status.

NMS is thought to be rare and estimates vary according to strictness of diagnostic criteria. Larger prospective studies suggest frequencies between 0.03 and 0.12%. Including milder or atypical variants of NMS inflates that figure as high as 12.2% (Modestin, Toffler, & Drescher, 1992). More than 150 cases have been reported thus far, with a mortality rate of approximately 22% (Shalev & Munitz, 1986). NMS usually begins within a few days of drug administration and develops quite rapidly. Termination of neuroleptic treatment usually reverses symptoms "within 4 to 40 days" (Shalev & Munitz, 1986, p. 339). While etiology is unknown, patients with existing neurological abnormalities including OBS, mental retardation, and substance abuse (opiates and alcohol) are "overrepresented among fatal cases, comprising 32% of fatalities" (Shalev & Munitz, 1986, p. 340). The syndrome is apparently not dose-related.

Neuropsychological Effects of NMS. There is little literature addressing the specific neuropsychological effects of NMS. In this regard, Rothke and Bush (1986) examined a 28-year-old female with postpartum psychotic depression. She developed NMS after receiving increasing doses of haloperidol over the course of 8 days at a final, prodigious dosage of 70 mg/day. Neuropsychological examinations were performed approximately 4 months and 10 months after her initial hospitalization. Initial evaluation suggested significant deficits in visuospatial functions and in memory. While immediate verbal memory was unimpaired, mild to moderate deficits in delayed recall were found. Severe impairments in immediate and delayed visual memory were also noted. The patient's overall memory quotient on the Wechsler Memory Scale was only 69. Her score on the Category Test was 103, suggesting significant impairment in abstract spatial reasoning. MMPI scores were not reported, but were described as unremarkable.

On the follow-up examination the patient showed some improvement in attentional and overall intellectual abilities. Her Memory Quotient increased to 101. In contrast, delayed verbal recall was severely impaired, more so than in the initial testing, and delayed visual recall continued to be severely impaired. The patient's Category Test score continued to be in the range of clinical impairment. Since the patient's attentional abilities and nonpathological MMPI appear to rule out the influence of psychopathology on test scores, the authors suggest that NMS may have caused the patient's lasting neuropsychological impairment. Neuropsychological investigation of other cases of NMS appears warranted to determine the typicality of Rothke and Bush's (1986) results.

Chronic Effects

Jellinger (1977) performed neuropathological postmortem examinations on patients with long-term use of neuroleptics. He found neuronal swelling, glial satellitosis and neuronophagia in patients' caudate nuclei with more frequently encountered changes in patients with symptoms of tardive dyskinesia. Christensen, Moller, and Faurbye (1970) performed postmortem neuropathological examinations on schizophrenics with oral dyskinesia and found degeneration of the substantia nigra in 27 of the 28 patients, and gliosis of the midbrain in 25. Only 7 of 28 unmedicated control subjects showed similar degeneration of the substantia nigra, and only 4 of 28 showed midbrain and brain stem abnormalities. Since neuroleptic drugs block both acetylcholine and dopamine in the CNS, it is possible that some of the neuropathological effects cited above are the result of prolonged alterations in levels of these neurotransmitters. It is unclear, however, whether premorbid neuropathology found in a subset of schizophrenics could also be responsible for autopsy findings like these.

Chronic behavioral effects of neuroleptic medication include extrapyramidal syndromes, behavioral and autonomic effects (Sterman & Schaumburg, 1980). The most well-known extrapyramidal effect of neuroleptic administration is tardive dyskinesia (TD), a syndrome comprising involuntary choreiform movements of the face, lips, and tongue. The syndrome is usually considered to be a potential side effect of chronic neuroleptic administration, although short-term trials of medication have also been found to evoke symptoms. TD may be reversible on termination of the causative medication, although long-term sequelae have also been noted.

Neuropsychological Effects—Tardive Dyskinesia

Response to neuroleptic drugs varies greatly among patients. Very little research has been performed on neuropsychological differences as a function of neuroleptic response. An obvious research topic is the investigation of normal and abnormal (TD) neuroleptic responders. Struve and Willner (1983) performed such a study and examined the scores on an analogical reasoning task of all patients who began neuroleptic treatment. Patients who would later develop TD were found to score significantly lower on this task than patients who would eventually become part of the "normal" non-TD controls (Struve & Willner, 1983). Additionally, TD patients' scores did not change as a function of medication over time, suggesting either abnormal but stable initial medication response, or else structural brain differences between the two groups. While this study's findings are obscured by failure to control for educational differences among patients, the goal of discovering a neuropsychological predictor of TD is worthy of further exploration.

Other Effects

The same neuroleptic-induced alterations of dopamine and acetylcholine that may be related to long-term neuropathology may concomitantly produce neuropsychological side effects. A review of relevant studies by Heaton and Crowley (1981) suggested that neuroleptics with little cholinergic activity, including the piperazine phenothiazines and haloperidol, "do not seem to impair cognitive function or motor speed. . . . However these agents are more likely to produce acute extrapyramidal and other side-effects and may thereby impair performance on some motor coordination tasks" (p. 501).

Those neuroleptics that do reduce CNS cholinergic activity have been associated with other types of neuropsychological deficits. Tune *et al.* (1982) tested the recent memory of 24 outpatient chronic schizophrenics, using a ten-item free recall word list. Subjects were taking the equivalent of 200 mg/day of chlorpromazine and also received anticholinergic medications (benztropine or trihexyphenidyl) for extrapyramidal symptoms. The authors report a "highly significant inverse correlation between increasing serum levels of anticholinergics and recall scores." Recall did not correlate with an estimate of IQ or the overall severity of schizophrenic symptoms (Tune *et al.*, 1982, p. 1461).

The relationship between memory deficit and serum anticholinergic levels was further explored in a study that examined both free recall and recognition memory as a function of anticholinergic activity (Perlick *et al.*, 1986). The authors examined 17 chronic schizophrenics receiving either mesoridazine, haloperidol, chlorpromazine, thioridazine, or fluphenazine. Subjects were administered a neuropsychological battery that included WAIS-R Vocabulary, Similarities, Block Design, and Picture Completion Tests, the Associate Learning subtest of the Wechsler Memory Scale, Benton's Revised Visual Retention Test (BVRT), and the Mattis–Kovner Memory Inventory (a test of verbal recall with periodic recognition memory "probes"). Replicating the Tune et al. (1982) results, the authors found that, after controlling for IQ, subjects with anticholinergic serum levels (ASLs) above the group mean showed significantly impaired free recall compared with subjects below the mean ASL. Recognition memory for the "probes" was not affected by ASL. There was also no relationship between ASL and the other neuropsychological measures, or between serum neuroleptic levels and memory.

While patients in the Tune et al. (1982) study had anticholinergic effects induced with antiparkinsonian drugs, the Perlick et al. (1986) subjects had the majority of their anticholinergic reductions produced by the neuroleptic medications themselves. The results of these studies taken together suggest that either anticholinergic neuroleptic medications or antiparkinsonian agents may produce memory impairment that is correlated with serum anticholinergic level. Further investigation of anticholinergic memory deficits seems warranted since unimpaired recognition memory in the Perlick et al. subjects suggests that the anticholinergic "memory" deficits may not be in the encoding phase. Instead, retrieval may have been impaired in some way by faulty attentional, motivational, or planning processes (Perlick et al., 1986). Clinically, it is noted that cognitive impairments induced by anticholinergic drugs can further impair the already compromised mentation of patients with Parkinson's disease or psychosis who receive these drugs (McEvoy & Freter, 1989).

Not all neuropsychological functions of schizophrenics are impaired by neuroleptic medications. Strauss, Lew, Coyle, and Tune (1985) showed beneficial effects of neuroleptic levels on distractibility. They examined 28 outpatients with a diagnosis of chronic undifferentiated schizophrenia on reaction time and an audio-presented digit span test. This latter test was presented using a female voice alone, or with a male voice listing extraneous numbers between presentations by the female voice (distracting condition). In either case, subjects were asked to attend only to the numbers presented by the female voice. Neuroleptic levels showed significant negative correlation with auditory distractibility. Serum anticholinergic levels were not significantly correlated with test scores.

Similarly, Braff and Saccuzzo (1982) found improved speed of information processing in a tachistoscopic task for medicated schizophrenics compared with unmedicated schizophrenics. Since both groups performed significantly more poorly than depressed controls, the authors concluded that schizophrenic disorder, rather than medication, was responsible for slower information processing, and neuroleptic medication may even reverse certain information processing deficits.

The problem of assaying the effects of medication apart from the schizophrenic disease process itself was addressed in a study design that performed repeated measures testing on schizophrenics, first at their normal medication doses and then at a mean interval of 29 weeks later at 80-90% dose reductions (Seidman et al., 1993). Eleven matched, normal controls underwent testing once to establish testing baselines. The only neuropsychological test score to change was a measure of auditory dichotic listening, in which the left ear accuracy significantly improved. Interpretation of results is complicated by the age of the schizophrenic group (mean=52, SD=11.3), the fact that 9 of 11 had a history of moderate to severe alcohol abuse, the control population that was almost 13 years older than the patients, and that Bonnferroni corrections for multiple tests were not performed. The general design, however, bears replication with better matched, somewhat younger patients, and a control group that is tested twice to determine practice effects.

Clozapine

Clozapine has been used to treat symptoms refractory to more common neuroleptic medications, with one study finding that 30% of patients refractory to Haldol improved with clozapine (Kane *et al.*, 1988). There are two neuropsychological studies on clozapine administration at this writing. The first failed to find differences between clozapine and Haldol or flupenthixol on several sensorimotor tests (Classen & Laux, 1988). A more recent study found that clozapine significantly worsened WMS Figural Memory, a likely function of the potent anticholinergic properties of the drug (Goldberg *et al.*, 1993). The authors also note that while clozapine is 17 times less potent than the anticholinergic, benztropine, it is administered in doses that are 100 times greater.

Neuropsychological Effects—Children

Children are much less frequently prescribed neuroleptic medications, and there are correspondingly fewer studies of neuropsychological dysfunction in neuroleptic-medicated children. One example is Platt, Campbell, Green, and Grega (1984) who found mild effects on cognitive functions in 61 children with a diagnosis of undersocialized aggressive conduct disorder. A mean dose of 2.95 mg/day of haloperidol caused slowing in simple reaction time and worsened Porteus Maze scores. No effects were found in tests of memory, concept attainment, short-term recognition memory, or the Stroop test.

Lithium

Lithium was discovered in 1817 and is classified as a group I alkali metal. Before its current use as a psychiatric medication, it was first tried as a gout treatment, and then later as a salt substitute, the latter having toxic or fatal results. Lithium's utility in normalizing the psychiatric symptoms of mania and bipolar disorder was not known until 1949 when the first case reports were published (Sansone & Ziegler, 1985). Today, lithium is commonly accepted as the preferred treatment for bipolar disorder and unipolar manic disorder and is also used to treat chronic cluster headaches (Sterman & Schaumburg, 1980). Individuals at risk for lithium neurotoxicity are those being treated for the above-mentioned psychiatric disorders. Although lithium is used in several industrial capacities, including plastics and ceramic production, and in lubricants (Dempsey & Meltzer, 1977), lithium toxicity is not reported to be an industrial problem (Hamilton & Hardy, 1974).

Neurological Effects

Lithium shows preferential toxicity for the CNS over the heart or kidney. Neurotoxic levels of lithium can induce abnormal EEG and dysarthria, as well as gait and limb ataxia; these appear to abate without permanent sequelae as lithium levels decline. Acute lithium overdose can precipitate coma, and long-term deficits from lithium-induced coma may be more a function of the coma itself than lithium (Sterman & Schaumburg, 1980). Overdose of lithium can be fatal and immediate hemodialysis is indicated to prevent death or permanent brain damage (Sansone & Zeigler, 1985).

PNS Effects. Girkem, Krebs, and Muller-Oerlinghausen (1975) found 6 of 17 bipolar patients on long-term lithium to show significant slowing of motor nerve conduction velocity (NCV); however, half of these patients were simultaneously taking other medications and one patient had cancer. In the same report the authors replicated their finding of lower NCVs in a group of 7 normal volunteers who received lithium for 1 week. NCVs remained significantly decreased 7 days after lithium withdrawal and may have been the cause of five volunteers' complaints of weakness or fatigue (Sansone & Zeigler, 1985; Girkem *et al.,* 1975). Other PNS effects of maintenance therapy with lithium include "fine rapid tremor of mild to moderate intensity" (Sansone & Zeigler, 1985, p. 242). Parkinsonian symptoms and cogwheel rigidity have been infrequently observed. Reversible lithium neurotoxicity has been observed clinically, even in the context of normal serum levels (Bell, Cole, Eccleston, & Ferrier, 1993). Symptoms can include confusional state, cerebellar abnormalities, delusions, and neuromuscular symptoms. Patients with premorbid EEG abnormalities, concurrent organic impairment, and those given a high starting dose have elevated risk for neurotoxic reactions (Bell *et al.,* 1993).

CNS Effects. Lithium encephalopathy caused by overdose can be accompanied by "flaccid paralysis, proximal muscle weakness, fasciculations, and areflexia" (Sansone & Zeigler, 1985, p. 243). EEG patterns show dose-related slowing that may persist "for at least 11 days after the last dose of lithium" (Sansone & Zeigler, 1985, p. 245).

Severe and sometimes fatal reactions to lithium in the context of normal blood lithium have been linked to toxic brain accumulations of the metal. Animal studies suggest that brain lithium levels may be higher than serum concentration after steady-state blood levels are produced. Sansone and Zeigler (1985) report two studies in which autopsies showed elevated brain lithium compared with serum lithium. In one of these, the pons showed 1.5–2 times the levels of serum lithium, while the second study found overall brain toxicity with highest accumulations in the white matter and brain stem (Sansone & Zeigler, 1985). A third study reports permanent damage to the basal ganglia and cerebellar connections as a consequence of lithium toxicity (Hartitzsch et al., 1972). Smith and Kocen (1988) note that lithium toxicity can resemble a Creutzfeldt–Jakob-like syndrome.

Neuroleptics in Combination with Lithium. Lithium, when used in combination with neuroleptics, may be more likely to produce neurological abnormalities than use of lithium alone. Fetzer, Kader, and Danahy (1981) report three cases of organic brain syndrome produced when these two medications were combined. Withdrawing lithium ameliorated symptoms in two of these patients, although the authors cite a potential risk of irreversibility. Prakash, Kelwala, and Ban (1982) reviewed 39 cases of lithium/neuroleptic neurotoxicity and observed that the syndrome was characterized by confusion, disorientation, and unconsciousness. Extrapyramidal signs were often present and included tremor, akathisia, dyskinesia, and dystonia. Also occasionally reported were cerebellar signs and slow-wave EEG. The authors speculate that neuroleptics may somehow raise brain uptake of lithium to toxic levels.

Neuropsychological Effects

Patients taking lithium typically do not enjoy the effects of the medication, and give subjective accounts of cognitive and especially memory impairment. Various studies have demonstrated neuropsychological impairment in short-term memory, long-term recall, visual recall, and psychomotor speed (Kocsis et al., 1993). Results across studies tend to be inconsistent, probably because of the many potential confounds inherent to this type of research. For example, neuropsychological and structural brain impairments are frequently found in individuals most likely to receive the drug, i.e., those diagnosed with bipolar disorder (e.g., Savard, Rey, & Post, 1980; Pearlson et al., 1984). Other potential confounds that limit the interpretation of early neuropsychological studies include failure to control for effects of age, inadequate assessment of pre- and postmorbid mood, not assessing serum lithium levels with consequent failure to rule out lithium toxicity, educational variation among subjects, and inadequate statistical treatment (e.g., Lund, Nissen, & Rafaelsen, 1982).

Lithium has been examined for its effects on memory in several studies. Some studies have not found lithium to significantly affect memory. For example, Ghadirian, Engelsmann, and Ananth (1983) attempted to determine whether duration of lithium treatment was related to memory function. When age, serum lithium, psychopathology and physical illness were controlled, there were no significant effects of lithium duration. However, when Christodoulou, Kokkovi, Lykouras, Stefanis, and Papadimitriou (1981) tested the memory of euthymic patients maintained on lithium after receiving placebo for 15 days, they found memory improvements in word list memory and two measures of visual retention.

Another study eliminated affective or psychiatric disorder as a source of variance by examining the neuropsychological effects of lithium administration on normal college males (Weingartner, Rudorfer, & Linnoila, 1985). Subjects were maintained on standard lithium doses of 1225 300 mg/day until steady-state serum concentrations of 0.82 0.17 mEq/liter by the 8th day. Double-blind, crossover experimental design was employed, with all subjects participating in both conditions and each subject matched as his or her own control. Subjects were asked to listen to sets of categorically related words, presented once or twice. They were required to identify word repetitions, attempt free recall, and then identify correct words among distractors in a recognition task. They were also asked to judge whether each identified word occurred once or twice.

No differences in free recall or recognition were found regardless of the number of word presentations. However, subjects produced more intrusion errors in the lithium condition than the control condition, both in recall and in recognition. The authors suggest that while lithium does not affect number of items recalled or recognized, it may affect the "clarity" of what is remembered and therefore may be an analogue of the "cognitive blurring" that occurs in patients on lithium, and which may be responsible for their noncompliance (Weingartner *et al.*, 1985).

A third study used psychiatric patients as their own controls and found no overall effects between patients before and after they began lithium treatment (Smigan & Perris, 1983). However, the authors also did not find effects of other drugs, age, and serum lithium levels on performance; factors that have been implicated in neuropsychological impairment by many other studies. The overall sensitivity of their methodology is therefore somewhat suspect.

Examining the effect of age and lithium administration, Friedman, Culver, and Ferrell (1977) found that patients below 55 years old did not exhibit neuropsychological impairment. Eight patients under age 55 who received prescription doses of lithium for bipolar affective disorder showed an average impairment index of 0.26 on the Halstead–Reitan Battery. Five older patients (mean age 62.4) receiving lithium for the same amount of time (mean 3.6 years) had an average impairment index of 0.72, suggesting that age alone, or a possible interactive effect of age, illness, and lithium use, may have produced neuropsychological impairment.

Other neuropsychological functions may be more consistently affected by lithium than memory, particularly those involving motor and/or perceptual speed. For example, Squire, Lewis, Janowsky, and Huey (1980) tested a mixed group of psychiatric patients in a double-blind, crossover design. Subjects who took an average dose of 0.94 mEq/liter did not show learning or memory deficit, but were impaired in several speeded tasks, including Digit Symbol and a clerical copying task. Serum lithium levels correlated significantly with time to complete Trails B, suggesting a loss of cognitive efficiency or flexibility. This effect was also found by Small, Small, Milstein, and Moore (1972), who administered parts of the Halstead–Reitan Battery to a mixed group of affective and schizophrenic disorder patients. The authors reported high lithium levels to be correlated with poor performance on Finger Tapping, Block Design, Trails B, Digit Symbol, the Minnesota Paper Form Board Test, and a verbal test. In addition, lithium also produced a significant decrement in scores on the Tactual Performance Test. These effects may be related to PNS abnormalities or yet-to-be-understood cortical mechanisms. Judd, Hubbard, Janowsky, Huey, and Takanashi (1977) studied normal subjects receiving lithium and found that subjects' general slowing on information processing tasks but not on tests of motor speed may indicate that slowing is of central rather than peripheral origin. Further research will better clarify the nature and etiology of lithium's neuropsychological effects.

One of the better controlled studies of lithium discontinuation tested psychiatric outpatients (mean age 54) repeatedly on three neuropsychological measures: finger tapping, Buschke Selective Reminding Test, and a test of word association productivity and idiosyncrasy (Kocsis *et al.,* 1993). The authors first obtained a testing baseline, then subjected patients to blinded placebo for 1–2 weeks, reassessing neuropsychological functions at the end of placebo administration. Subjects were assessed for a third time after 2 weeks of resumed lithium administration. Age, sex, plasma lithium, thyroid function, duration of lithium administration, and depressive symptoms were analyzed as intervening variables. Compared with baseline performance, significant improvement was found on measures of memory and motor speed at the end of the placebo washout period. Improvements were said to range from 8 to 44% above baseline. Arguing against practice effects, all of the measures of memory and word association productivity declined significantly again after lithium resumption. Mood ratings did not change during the course of the study. Age and depression did not interact with results. While the study can be criticized for data massaging (50 analysis, 10 dependent measures, and 5 grouping variables), data are consistent with neuropsychological effects of lithium on memory, word association, and motor speed.

In conclusion, data appear to be accruing that implicate lithium in mild to moderate impairment of memory, psychomotor functions, and general information processing productivity. High serum lithium levels can be neurotoxic and can be correlated with acute effects on several neuropsychological tests, especially those that measure attention, cognitive efficiency, and flexibility. It is not known whether these acute effects cause chronic, cumulative neuropsychological damage.

Continued neuropsychological studies using double-blind, crossover, counterbalanced designs would seem to be necessary to resolve remaining questions. Subject population variables must be addressed; individual differences in brain structure among bipolar patients must be controlled; subjects who receive lithium first must be allowed sufficient intervals for drug clearance and practice effects must be addressed. Neuropsychological interactions, if any, of lithium and aging also require further exploration.

Muscle Relaxants

Baclofen has been linked to both toxic and withdrawal-related encephalopathy syndromes. Toxic encephalopathy is characterized by confusion, and somnolence and coma may ensue. Seizures, myoclonus, and "combative hallucinatory agitation" during unconsciousness can occur (Goetz, 1985, p. 205). Abrupt withdrawal may also precipitate psychiatric and neurological symptoms, including hallucinations, seizures, and mania. Dantrolene is another muscle relaxant cited by Goetz (1985) to have neurological effects. Muscle weakness, "euphoria, lightheadedness, dizziness and fatigue" have been reported (Goetz, 1985, p. 206).

Steroids

Corticosteroids (e.g., dexamethasone, prednisone) are employed on a spectrum of diseases requiring their immunosuppressant or anti-inflammatory properties. Common symptoms of steroid-treated patients are emotional lability, depression, insomnia, poor concentration, and psychosis (Wolkowitz et al., 1990). Major psychiatric side effects of corticosteroids include delirium, depression, and mania, often in patients receiving doses greater than 40 mg/day (Trzepacz et al., 1993). Dementia may also occur.

Corticosteroids may affect the nervous system directly and indirectly. Corticosteroid receptors are found in the hypothalamus, pituitary, hippocampus, septum, and amygdala. The full set of interactions between corticosteroids and body systems is not yet known but includes influence between steroids and "genomically related events i.e., transcription," nucleotide metabolism, alteration of neuronal membrane ion conductances, central carbohydrate, protein, and lipid metabolism (Wolkowitz et al., 1990). Steroids alter the metabolism of monoamine neurotransmitters and neuropeptides.

Steroid administration can be both a cure for and a cause of dementia. Dementias resulting from inflammatory processes with resulting increased intracranial pressure may show temporary or permanent remission with steroid administration. Paulson (1983) lists several dementing illnesses that respond positively to steroids, among them global encephalopathies, e.g., subacute sclerosing panencephalitis and Kreutzfeldt–Jacob disease. Lupus erythematosus, sarcoid meningitis, and dementias with fluctuating mental states are also suggested to improve with steroids.

Steroids may also worsen CNS functioning and thus may induce, rather than ameliorate, dementing illness (Varney, Alexander, & MacIndoe, 1984; Paulson, 1983). Varney et al. (1984) studied six patients who received 60–125 mg/day of prednisone for a variety of steroid-treatable diseases. Two of six patients had long-lasting cognitive sequelae subsequent to resolving steroid psychosis. The other four patients showed global, but reversible, deficits on follow-up (which varied from 80 days to 2 years). All patients were tested with a neuropsychological battery that included WAIS, parts of the Wechsler Memory Scale, several of Benton's tests (e.g., BVRT, Facial Recognition Test, 3D Constructional Praxis scale, and other tests). The authors characterized the deficits as disturbances in "memory retention, attention, concentration, mental speed and efficiency, and occupational performance" (Varney et al., 1984, p. 372). The authors note that the patients were not overtly toxic, intoxicated, or demented, but rather that each was impaired relative to his own prior and subsequent functioning.

TABLE 6.2. Selected Drugs with Neurotoxic Effects[a]

Drug	Use	Effect
Acyclovir	Antiviral agent	Tremor, confusion, lethargy, psychotic depression, seizures, agitation, abnormal EEG
Aminoglycosides	Antibiotic	Eighth cranial nerve damage; ototoxic, vestibular toxicity, neuromuscular blockade
Chloramphenicol	Antibiotic	Distal symmetrical neuropathy, optic neuropathy
Ciprofloxacin[b]	Antibacterial agent	Disorientation, restlessness, headache, hallucinations
Clioquinol	Amebicide	CNS degeneration of optic tracts and long spinal cord tracts
Dapsone	Treatment of leprosy	Reversible motor neuropathy
Ethionamide	Treatment of tuberculosis	Symmetrical sensory polyneuropathy, paresthesias of the feet
Glutethimide	Sedative/hypnotic	Rarely—distal sensory axonopathy (reversible)
Isoniazid	Treatment of tuberculosis	Sensory paresthesias (numb fingers and toes); slow but complete recovery of peripheral neuropathy
Metronidazole (Flagyl)	Antibacterial/protozoan	Peripheral distal neuropathy
Misonidazole	Radiosensitizing agent	Peripheral neuropathy, CNS pathology similar to thiamine deficiency syndrome (Wernicke's encephalopathy
Nitrofurantoin	Antimicrobial	Peripheral neuropathy with paresthesias, rapid deterioration can occur with severe sensory loss
Nitrous oxide	Anesthetic	Mild, distal, symmetrical polyneuropathy, high levels of abuse may cause myelopathy
Perhexiline maleate	Angina pectoris	Peripheral neuropathy with 300–400 mg/day, 4 months to 1 year
Pyridoxine	Vitamin B_6	Distal sensory neuropathy
Sulfonamide	Treament of urinary tract infections	Peripheral neuropathy, CVA, headache, fatigue, tinnitus, psychosis, focal CNS symptoms, e.g., aphasia (rare)

[a]Schaumburg et al. (1983) and Goetz (1985).
[b]Trzepacz et al. (1993).

Behavioral and cognitive impairments from corticosteroid administration are dose related, with one study finding increased intrusion (commission) errors in verbal recall, and diminished corrected free recall (Wolkowitz, Reus, Weingartner, Thompson, et al., 1990). Plasma prednisolone levels were directly correlated with commission error rates, leading to the hypothesis that corticosteroids suppress the brain's ability to filter out extraneous information via reduced excitability of certain hippocampal neurons as well as influencing levels of cortisol, ACTH, β-endorphin, norepinephrine, and somatostatin which may secondarily affect cognitive function (Wolkowitz et al., 1993).

Anabolic steroids include a class of synthetic testosterones that have limited, albeit legitimate treatment applications of breast cancer, aplastic anemia, angioneurotic edema, and growth or sexual development retardation in young males (Daigle, 1990). However, the principal use of these drugs at present is to enhance athletic performance. In this capacity, steroid use has reached epidemic proportions, with professional and amateur athletes at particular risk. More than 80% of body builders and 50% of professional football players have taken steroids, and track and field, cycling, and other sports have experienced widely publicized steroid abuse. The spread of steroid into high school and junior high school is of special concern, with recent estimates of 6–10% of male high school students admitting steroid use to "improve" athletic performance or physical appearance (Katz & Pope, 1990). Individual case studies have noted increased aggressivity, depression, anxiety, and sometimes psychosis with visual and auditory hallucinations, religious and paranoid ideation. Approximately one-third of individuals surveyed in a 1988 study indicated symptoms close to or fulfilling the DSM-IIIR criteria for a manic syndrome (Katz & Pope, 1990). Systematic studies of neuropsychological effects of anabolic steroid abuse have not entered the literature.

Other Drugs

The medications listed in Table 6.2 have not been investigated with neuropsychological measures. However, their effects might be expected to produce abnormalities on such tests, especially in sensory and motor functions.

ABUSED AND NONPRESCRIPTION DRUGS

Drug abuse continues to be a problem of international scope. In comparison with earlier decades, currently available illicit drugs tend to be more potent, more addictive, and perhaps more damaging to the nervous system. "Crack" cocaine, "ice" methamphetamine, marijuana, and lysergic acid diethylamide (LSD) are available in new or more potent variations than were possible previously (Smith, Ehrlich, & Seymour, 1991). Drug abuse plagues adolescents in particular, with 25% of children age 12–17 reporting illicit drug use (National Institute on Drug Abuse, 1988).

Abused drugs are substances that have a high reinforcement potential in human beings. To some extent, this category overlaps previously described neurotoxins (e.g., toluene, benzodiazepines); however, other abused drugs are prescription medicines with legitimate uses but which also induce physical and/or psychological addiction. Although not all highly reinforcing drugs are chronically neurotoxic, all produce acute neuropsychological impairment.

Amphetamines

Amphetamines and related stimulants are prescribed appropriately for narcolepsy and for childhood hyperactivity. They are also highly reinforcing and thus have a high abuse potential. Neurological consequences of amphetamine abuse include arteritis, vasculitis, and intracranial hemorrhage (Carlin, 1986). Like cocaine, amphetamines are vasoconstrictive on small artery branches, putting the user at risk of stroke and hypertension. In human subjects, amphetamine appears to reduce cerebral blood flow, leading Mathew and Wilson, (1991b) to speculate that amphetamine-induced cortical blood flow may produce subcortical disinhibition, leading to elation or euphoria.

Psychiatric consequences of prolonged use of amphetamines include paranoid-schizophrenic-like psychosis. There is very little neuropsychological literature on the effects of amphetamine use or abuse *per se*. Medical complications with neuropsychological import include hypertension, stroke, retinal damage, hyperpyrexia, pancreatitus, renal failure, and neuropathy (Kroft & Cole, 1992). The long-term psychological effects of amphetamine use have yet to be addressed in a research study (Carlin, 1986).

Developmental neurotoxicity of amphetamines (i.e., methamphetamine) is a more definite finding. Several studies of infants exposed to methamphetamine prenatally have noted effects very similar to those found in so-called "cocaine babies," including lethargy and tremor. Increased incidence of intraventricular hemorrhage and other structure damage is also found (Vorhees, 1994).

Amyl Nitrite

Amyl nitrite has legitimate, although infrequently prescribed, medical application as a coronary vasodilator. Psychological effects include vasodilation, smooth muscle relaxation, and reductions in blood pressure. The drug is reported to be in wide recreational use by male homosexuals for orgasm enhancement and relaxation of the anal sphincter to permit anal intercourse. Amyl nitrite significantly increases bilateral cerebral blood flow and pulse rate (Mathew, Wilson, & Tant, 1989). There do not appear to be any neuropsychological case studies of drug use, but cerebrovascular accident and consequent neuropsychological deficits are risks consistent with the properties of the drug.

Caffeine

Caffeine is said to be the most widely used psychoactive substance in the world. Caffeine increases plasma catecholamines in humans, and also affects central norepinephrine and central dopamine turnover (Stoner, Skirboll, Werkman, & Hommer, 1988). Performance-enhancing effects of caffeinated beverages have been assumed to provoke tolerance. However, a recent, large-scale survey suggested that psychomotor and cognitive effects of drinking coffee or tea may outweigh tolerance effects. Jarvis (1993) reviewed the relationship between cognitive performance and habitual coffee and tea consumption in a cross section of 9003 British adults. After controlling for social, age, health, and demographic variables, significant trends for better performance as a function of higher coffee intake were observed on choice reaction time, incidental verbal memory, and visuospatial reasoning. Compared to coffee teetotalers, individuals who drank more than six cups of coffee per day were faster by 6% in simple reaction time, 4% in choice reaction time, and 4–5% better in memory and visuospatial tasks. Improved performance in reaction times did not occur at the price of increased error rate. Tea drinkers showed somewhat weaker improvements in cognitive function. The largest improvements appeared to occur in the oldest subjects. Survey results must be considered tentative since coffee and tea consumption information was obtained without determining type of coffee or tea (instant, ground decaffeinated) or strength. The author correctly suggests the need for a follow-up study utilizing blood levels of caffeine at the time of testing.

Cocaine

History

Cocaine is derived from an alkaloid of the coca plant, *Erythroxylon coca,* and has been used since at least 3000 B.C. for various sacramental, medical, and recreational purposes. Cocaine was thought to be in common use before the Inca civilization, and the Incas themselves believed the plant to be divine, restricting its use to nobles and priests (Mulve, 1984). They may also have used cocainized saliva to prepare for trephining (Nicholi, 1984).

The psychological and pharmacological effects of cocaine were rediscovered by the medical community in the 1880s. Sigmund Freud enthusiastically recommended the drug for its anesthetic and mood-elevating properties. Freud prescribed cocaine for several of his patients and also self-medicated to relieve his own migraines and "neurasthenia."

Freud's experience with cocaine was probably similar to that of countless present-day users: initial moderate use and enthusiastic recommendations, followed by addiction, deterioration of affect, memory, and attention. There is at least circumstantial evidence linking Freud's rapidly written, cryptic, and opaque *Project for a Scientific Psychology* (1895) to a period of cocaine use, a year before he finally put "the cocaine brush aside" in 1896 (Freud, in Masson, 1985, p. 201).

The early 20th century saw the production of many cocainized patent medicines, the most popular of which was Coca-Cola. The Pure Food and Drug Act of 1906 forced the removal of cocaine from Coke and the Harrison Narcotic Act made cocaine illegal in 1914 (Mulve, 1984).

There are still several legitimate medical uses of cocaine, primarily as an anesthetic in nasal operations (Nicholi, 1984).

All available data indicate that cocaine use has reached epidemic proportions in the United States, with estimates of 33 million individuals who have used cocaine, up from about 9 million in 1972 (Nicholi, 1984). An increase of 10 million users in this decade alone suggests that it has become a major national health problem (Mulve, 1984). There has been more than a 200% increase in cocaine-related deaths and emergency room visits, and a 500% increase in admissions to federally funded cocaine treatment programs from 1976 through 1981 (Washton & Tatarsky, 1984). Cocaine use now spans all demographic groups with 9% of all professionals having tried cocaine (Siegel, 1982). With the production of "crack" cocaine, and a lowering of price per dose, there has been wide adoption of cocaine by the rest of society as well, with estimates of approximately 14% of the U.S. population having tried cocaine at least once (Washton &

Tatarsky, 1984, p. 247; Kain, Kain, & Scarpelli, 1992). In inner-city hospitals 18–20% of pregnant women have urinary metabolites indicating cocaine use (Kain *et al.*, 1992). Even high school students have discovered cocaine; one recent study estimated that one in every six high school seniors had tried cocaine, and that almost 5% had used this drug in the month prior to the survey (O'Malley, Johnston, & Bachman, 1985, cited in Melamed & Bleiberg, 1986).

Initially thought to be fairly benign and non-addictive, cocaine has become increasingly recognized for its addictive potential and its toxicity. Rats allowed to self-administer cocaine injections, and that in consequence developed severe tonic–clonic seizures, would reinitiate cocaine administration "as soon as the convulsions subsided" (Bozarth & Wise, 1985, p. 83). In the same study, rats were placed on a 30-day protocol of either cocaine or heroin self-administration. Group mortality of the cocaine-injecting rats was 90% compared with 36% for the heroin group (Bozarth & Wise, 1985). Cocaine is reinforcing regardless of administration route.

Cocaine can be inhaled (snorted), smoked, or injected. The drug is commonly available in a variety of forms: as cocaine hydrochloride salt with various levels of purity, or by mixing the salt with an alkali (e.g., baking soda) to produce a lipid-soluble alkaloid that is far more potent (free base or *crack*). Kain *et al.* (1992) estimate that 40% of cocaine addicts use the drug intranasally, 30% are free base smokers, 20% inject the drug, and 10% use some combination of the above.

The smoked form of cocaine reaches the brain more rapidly than other preparations, with production of tachycardia and "intense euphoria" within 5–10 min (Kain *et al.*, 1992). It is also associated with increased risk of seizure activity and stroke (Smith *et al.*, 1991). Being lipophilic, cocaine crosses the blood–brain barrier easily and produces brain concentrations of about four times peak plasma concentration (Kain *et al.*, 1992).

Low to average dose of conventionally inhaled cocaine crystals is about 25–150 mg. Toxicity depends on route of administration. Reported toxic reactions to topical cocaine anesthesia in plastic surgery are uncommon, and a conservative estimate of maximum safe dose is from 100 to 300 mg (Jones, 1984). Fatal dose is approximately 1 g.

Neurochemical Effects

Cocaine is lipid-soluble and capable of penetrating the blood–brain barrier. It is normally broken down quickly by plasma and liver cholinesterases into water-soluble metabolites, but breakdown is slowed in persons with less active blood cholinesterases, including babies, pregnant women, elderly men, and individuals with liver disease (Dial, 1992). Cocaine has effects on all major neurotransmitter systems. Brain dopamine stimulation is apparently responsible for the brief, but highly rewarding, effects of cocaine administration; when animals were treated with a drug causing destruction or blockage of dopamine neurons (e.g., 6-hydroxydopamine, pimozide) they no longer self-administer cocaine (de

Wit & Wise, 1977). Cocaine stimulates dopamine production, but blocks reuptake. In this case, a failure to recycle places additional demand on dopamine production, which soon fails to produce adequate replacement supplies. Limited supplies of dopamine induce dopamine receptor supersensitivity and increased receptor binding (Daigle, Clark, & Landry, 1988). Cocaine's stimulatory effects also include the PNS. Effects on the latter include vasoconstriction, pupil dilation, sinus tachycardia, and ventricular arrhythmias (Goetz, 1985).

Cocaine seems to exert its potent reinforcing properties by blocking the reuptake of dopamine, increasing the availability of dopamine at receptor sites and producing increased neurotransmission. Further, since reuptake and reuse of dopamine cannot occur, dopamine production must be stepped up. This suggests that depression-depletion states following continued cocaine administration are the result of exhaustion of dopamine production sites (Dackis & Gold, 1985). Further support for cocaine's action on the dopaminergic system is provided by animal and human studies showing reduction in reinforcement potential of cocaine and amphetamine when dopamine antagonists are administered (Wise, 1984).

Depletion states in thyroid axis hormones may also partially account for psychological symptoms of cocaine-induced depression, producing a state of drug-induced hypothyroidism (Dackis & Gold, 1985). Cocaine also affects other neurotransmitter systems. Cocaine blocks norepinephrine reuptake, facilitates central and peripheral norepinephrine release, and activates inhibitory presynaptic norepinephrine neuron autoreceptors (Gold, Washton, & Dackis, 1985). Serotonin release and reuptake may also be blocked (Taylor & Ho, 1977, 1978).

Neurophysiological Effects

While topical anesthetic use of cocaine in modern plastic surgical procedures has been described as relatively safe, cocaine abuse has several frequently reported and potent neurotoxic effects. First, cocaine is a powerful CNS stimulant that mobilizes massive adrenergic discharge and stimulates the neurophysiological response of a "fight or flight" reaction. EEG and ECG show general desynchronization after cocaine administration (Gold & Verebey, 1984). Acute cocaine exposure in chronic drug abusers is associated with increased EEG alpha in frontal and temporal structures and increased beta in frontal and central region. Herning, Glover, Koeppl, and colleagues (1994) suggest that such a pattern may reflect reduced cortical neural activity, since the EEG effect is similar to that obtained with barbiturates and benzodiazepines. Cocaine use is associated with CVAs, cerebritis, and hyperpyrexia.

Second, seizures, possibly tonic–clonic type (Cohen, 1984), have been reported by about one-quarter of free base and intravenous cocaine users interviewed on a cocaine hotline (Washton & Gold, 1984) and have been documented in neurological research reports (see Pascual-Leone, Dhuna, & Anderson, 1991a, for review). Seizures caused by repeated cocaine administration may be related to the phenomenon of "kindling," an outward spread of electrical activity from the limbic system as a result of repeated stimulation (Melamed & Bleiberg, 1986). Experimental kindling paradigms in animals have resulted in reduced seizure threshold. Two other acute neurotoxic effects of cocaine include vasospasms and subsequent cerebrovascular damage (Altura, Altura, & Gebrewold, 1985), and multifocal tics in previously asymptomatic individuals (Pascual-Leone et al., 1991a). Vessel vasoconstriction and transient increases in blood pressure may induce tissue ischemia and hemorrhage or cerebrovascular accident in human cocaine abusers (Volkow, Mullani, Gould, Adler, & Krajewski, 1988). Continued resting hand tremor after 12 weeks of abstinence may reflect subclinical extrapyramidal dysfunction similar to, but more mild in character than is found in patients with Parkinson's disease (Bauer, 1993); further evidence for dopamine depletion in this population that may predict early development of full parkinsonian syndrome. Cerebral atrophy severity on CT correlates with duration of chronic abuse (Pascual-Leone et al., 1991b).

Acute effects of cocaine administration in eight experienced IV users were graphically demonstrated in a recent PET scan study (London et al., 1990). Each subject received two PET scans, accompanied by either 40 mg of cocaine hydrochloride or saline placebo in a double-blind crossover design. Tracer was fluorine-18 deoxyglucose (FDG). Cocaine administration significantly lowered glucose utilization by 14% over the whole brain, but up to 26% in higher cortical areas. Caudate nucleus, putamen, amygdaloid nucleus, hippocampus, and parahippocampal gyrus also showed significantly reduced cerebral glucose metabolism. Only the pons and cerebellum failed to show metabolic reductions. The authors suggested that dopaminergic action may account for decreased glucose metabolism. Euphoriant effects of cocaine, and reductions of glucose metabolism may be related, as similar reductions have been found in other euphoriants, including benzodiazepines, barbiturates, and amphetamine.

Chronic Effects

Chronic CNS effects of cocaine abuse have likewise been demonstrated by electrophysiological and imaging techniques. EEG in cocaine abusers showed markedly reduced alpha power in frontal and temporal regions, along with significantly increased average relative power in delta, theta, and beta (Pascual-Leone et al., 1991a). Longitudinal EEGs run on five patients showed gradually increasing theta power over the temporal regions (Pascual-Leone et al., 1991a).

Imaging studies have also validated abnormalities associated with chronic cocaine abuse. Pascual-Leone and colleagues (1991b) found CT evidence of cerebral atrophy in chronic cocaine abusers. Volkow et al. (1988) measured cerebral blood flow with PET scan and oxygen-labeled water. Subjects were 20 young adult mixed IV or free base cocaine users. Scans were performed within 72 h of hospital admission and 10 days subsequent to initial scan. Seven of twenty-four normal controls had one area of abnormal CBF, while fourteen of the cocaine users were abnormal prior to detoxification. Nine of twelve cocaine users reassessed at 10 days continued to show at least one area of abnormal CBF. Differences between normals and cocaine users were most apparent in the left frontal and parietal cortex and the occipital lobe with cocaine users showing significantly lower CBF in those areas (Volkow et al., 1988). Mental status showed concentration deficits and a neuropsychological screening (that did not assess attention or executive functions) was largely normal except for mildly decremented Scale scores on WAIS Arithmetic and Digit Span. Possible explanations include cocaine-induced small vessel occlusions, neurotransmitter changes, hemorrhagic reactions, allergic vasculopathy, or microembolisms from cocaine adulterants.

Baxter et al. (1988) scanned cocaine users for glucose utilization with FDG, a tracer whose transport is proportional to glucose. While these authors, unlike Volkow et al., failed to find hypometabolism in frontal lobe areas, very small numbers of subjects weaken this finding. Preliminary PET study of catecholamine systems suggests damage to presynaptic dopamine neurons similar to Parkinson's disease.

Recent studies examining the longevity of cocaine-induced neuropsychological impairment have produced disturbing findings. For example, Berry, van Gorp, Herzberg, and colleagues (1993) found that cocaine-induced neuropsychological impairments in memory, visuospatial abilities, and concentration persist for at least two weeks in abstinent hospitalized cocaine users.

Finally, Strickland et al. (1993) examined eight free base cocaine abusers with SPECT and MRI and neuropsychological measures after 6 months' residence in a drug-free closed treatment facility. Subjects were young (mean age 32 years) but had used 2–6 g of cocaine daily for 3–6 years and were classified as moderate to heavy cocaine abusers. Subjects were compared with age-matched controls. SPECT data showed that all cocaine-abusing subjects had "regions of significant cerebral hypoperfusion in the frontal, periventricular and/or temporal–parietal areas" that were associated with "striking" neuropsychological deficits (p. 419). Impairments in attention, concentration, and new learning were seen in seven of eight patients, with memory impairment seen in five of seven. MRI data were considered unremarkable in all subjects, with no demonstrable evidence of stroke or focal brain atrophy. One subject displayed enlarged ventricles and diffuse atrophy, and two others showed a single small white matter lesion. Suggested etiology, according to Strickland (personal communication), was that free base cocaine users experience microinfarction or other vascular or neurochemical damage that apparently does not recover during 6 months of abstinence.

Damage shown by these studies is consistent with recent surveys that indicate elevated risk for cocaine-induced stroke. When consecutive admissions for 214 young adult patients admitted for ischemic or hemorrhagic stroke were compared against a control population with unrelated illness, 34% of the stroke patients were drug abusers while only 8% of the controls were so identified (Kaku & Lowenstein, 1990). In patients younger than 35, drug abuse was the most commonly identified etiology, occurring in 47% of stroke patients. Cocaine was the drug most frequently implicated (Kaku & Lowenstein, 1990). Recent neuroimaging studies of chronic cocaine abusers are particularly disturbing for the suggestion of long-term brain injury during enforced abstinence. Behavioral, cognitive, and emotional correlates of these subtle brain injuries may be mistaken for functional problems or poor motivation in rehabilitation settings where neuropsychological and neuroimaging studies are not typically performed. Results also suggest caution at returning chronic cocaine abusers to employment that requires sustained attention and/or memory (e.g., air traffic controller).

Prenatal Effects

Cocaine crosses the placenta and causes CNS stimulation, peripheral vasoconstriction, tachycardia, and elevated blood pressure (Schneider & Chasnoff, 1992). A great deal of developmental impairment is explained by vasoconstrictive effects of cocaine, since vessels like the middle cerebral artery show early muscular development and thus are particularly vulnerable to drug-induced contraction and subsequent ischemia. Maternal cocaine during pregnancy is associated with increased rates of spontaneous abortion, *in utero* cerebral infarction, hypoperfusion, hypoxia, and intracranial hemorrhage (Schneider & Chasnoff, 1992; Kain *et al.*, 1992). Neural tube, cardiac, gastrointestinal, and genitourinary defects may also occur. Cocaine-induced hyperthermia may also produce brain hemorrhage. Used throughout pregnancy, cocaine significantly affects fetal growth parameters, with particular impact on birth weight and head circumference (Coles, Platzman, Smith, James, & Falek, 1992). Infants who did not survive show a variety of histological abnormalities, especially in prefrontal and temporal regions.

Prenatal neurotoxicity may depend on both severity of abuse and factors extrinsic to cocaine abuse (prenatal care, polydrug abuse). There is mixed evidence for effects of cocaine on the Brazelton Neonatal Behavioral Assessment Scale with some authors finding significant impairments (Chasnoff, Burns, Burns, & Schnoll, 1986; Chasnoff, Griffith, MacGregor, Dirkes, & Burns, 1989; Schneider, Griffith, & Chasnoff, 1989) and others who do not (Coles *et al.*, 1992; Eisen *et al.*, in press; Neuspiel, Hamel, Hochberg, Greene, & Campbell, 1991; Richardson & Day, 1991). Four-month-old infants exposed to cocaine and polydrug abuse *in utero* perform significantly worse than controls in muscle tone, primitive reflexes, and volitional movement (Schneider & Chasnoff, 1992).

Postnatal behavior in babies born to cocaine-addicted mothers includes tremor, sleep disturbance, hypotonia, high-pitched crying, irritability, hyperreflexia, poor feeding, increased muscle tone, motor impairment, seizures, apathy, and lowered orientation. Some clinical signs diminish or disappear with reduction and elimination of cocaine metabolite from the body (Kain *et al.*, 1992). Infants have improved somewhat after 1 month, but 42% of children in one study exposed to cocaine *in utero* had delayed fine motor skills, 32% had problems with gross motor skills, and 24% had neuromuscular abnormalities (Belcher & Wallace, 1991).

Psychological Effects

The profile of psychological symptom onset varies with type of administration; inhalation or injection of cocaine produces almost instantaneous mood changes. Oral cocaine requires from 10 to 20 min to fully express its effects.

Clinical observations on the continued use of cocaine suggest systematic symptom progression. Initial acute reactions of low to average doses (25–150 mg) include euphoria, increased feelings of alertness, mental acuity, and energy. Anorexia and decreased need for sleep are common. Elated mood slowly changes to irritation, restlessness, depression, and psychomotor retardation. These are reversible with continued administration of cocaine, setting a cycle of positive and negative reinforcement that becomes a potent addiction (Gold & Verebey, 1984).

More serious psychological symptoms occur with increased use, including panic attacks in half of the users reporting to a cocaine hotline (Washton & Gold, 1984). Paranoid ideation, depression, irritability, anxiety, and loss of motivation are reported by more than half of other subjects, and memory difficulties are described by a third of all respondents (Washton & Tatarsky, 1984).

Cocaine psychosis occurs with toxic levels of cocaine use, and is characterized by delusions of persecution and auditory, visual, or olfactory hallucinations. Parasitosis and formication, the sensation of snakes or insects crawling under the skin, are common, and addicts have been known to insert needles into their skin, looking for "cocaine bugs" (Nicholi, 1984). Similarities between the toxic psychosis produced by cocaine and that resulting from amphetamine use are said to be "far more striking than their differences" (Fischman, 1984, p. 86).

Withdrawal

Depressive disorders of various types are common and prolonged in cocaine withdrawal. An "abstinence" syndrome has been postulated by Gawin and Kleber (1986), who noted an initial depressive "crash" with lethargy, extreme dysphoria, suicidal ideation, intense cocaine craving, paranoia and hypersomnia, lasting from 1 to 40 h after termination of cocaine administration. These symptoms were followed by a subphase of hypersomnolence, intermittent awakenings, and hyperphagia, lasting 8–50 h.

Phase two, withdrawal, consisted of 15 days of normalized affective response, which progressed to anxiety, increased cravings, and anhedonia. These symptoms lasted an additional 110 weeks until cravings ceased. If subjects remained abstinent, baseline emotional functioning returned in this third phase, characterized by more limited cyclic cravings, possibly triggered by environmental cues. These cravings may be indefinite (Gawin & Kleber, 1986). Lithium and antidepressants may increase ability to abstain.

Interactions with Alcohol

Diagnosis of cocaine abuse in persons with family histories of alcoholism is apparently common; approximately 70–90% of inpatients receiving cocaine treatment and 50% of those treated on an outpatient basis for cocaine dependence are simultaneously dependent on alcohol (Higgins *et al.*, 1992). It may be the case that genetic vulnerability and/or alcohol "priming" increase predilection for cocaine abuse (Dial, 1992). Both alcohol and cocaine act on dopaminergic neuronal activity, while cocaine in combination with alcohol may deplete catecholamines and increase acetylcholine turnover (Cocores, Miller, Pottash, & Gold, 1988). Use of cocaine in combination with alcohol attenuates the effects of either substance alone (Higgins *et al.*, 1992). This suggests that cocaine users utilize alcohol in part to lessen undesirable side effects of extended cocaine use.

Neuropsychological Effects

Neuropsychological studies have lagged behind neuroimaging investigations. Initial data on specific neuropsychological concomitants of cocaine use were garnered from subjective symptom questionnaires without accompanying neuropsychological validation. For example, a survey of cocaine users suggested that 57% experience memory problems (Washton & Gold, 1984). Press (1983) studied the neuropsychological effects of either cocaine or opiate use on the Luria–Nebraska Battery. Sixteen subjects had either a 2-year history of regular heavy cocaine use or a 3-year history of semiannual bingeing. All cocaine users had been abstinent for at least 2 months prior to testing. This group was compared to 16 opiate abusers with a 5-year history, 5-month abstinence user profile. An equal number of matched non-drug-using subjects formed the control group. Polydrug abusers and heavy alcohol users were excluded from the study.

Results on the Luria–Nebraska approached but did not achieve significance ($p=0.068$), with the opiate users performing more poorly than the cocaine users. In post hoc comparisons the controls appeared to perform better than either drug group on the Rhythm and Expressive Speech Scales. Drug users also showed a trend toward having more scales on the LNNB in the impaired range. Factor-analytic studies further suggested that verbal memory was significantly more impaired in both heroin and cocaine abusers. In reviewing the records of several subjects the author suggests that high-dose chronic users of free base cocaine may be more likely than snorters to show "severe generalized deficits in memory" (Press, 1983, p. 119).

Melamed and Bleiberg (1986) studied free-basing cocaine users after psychiatric nursing staff in a drug treatment hospital noted an anecdotal impression that free base cocaine users seemed far more likely than other patients to injure themselves while participating in patient baseball games (Melamed, 1986, personal communication). The authors subsequently examined a group of these young adult subjects who reported an average of 32.5 months' use of free base cocaine. For their initial assessment, subjects were tested 48–72 h after their last reported use of free base cocaine. Design of the experiment involved the use of multiple t tests; however, Bonferroni's test corrected for multiple comparisons.

Results of the study, although somewhat inconsistent, suggested impaired neuropsychological functioning. Both Trails A and B were within the range of impairment, with Trails B showing "serious" impairment on Reitan and Wolfson's (1985) norms. On the Paced Auditory Serial Addition Test (PASAT), free base users showed impairment only on the easier 2.4- and 2-s, but not on the shorter, more difficult 1.6- and 1.2-s trails. This latter result is difficult to understand as a function of neuropsychological impairment, although Melamed (personal communication) hypothesizes that the easier subtest PASAT may be a more sensitive measure than the more difficult version. Otherwise motivational or functional factors, rather than the authors' hypothesis of neuropsychological factors, may provide a better explanation for the authors' data. A second assessment conducted after 10 days of abstinence showed significant improvement in Trails tests and in the PASAT 2.4 and 2.0 conditions. However, Trails B and PASAT 2.4-s trial continued to be performed at a significant level of impairment relative to normative data. Findings are consistent with what might be expected in patients with symptoms noted by Gawin and Kleber (1986) following withdrawal from cocaine.

The Melamed and Bleiberg (1986) study is not without methodological problems, including the use of published norms in lieu of a matched control group. It is unclear whether Reitan and Wolfson's (1985) neuropsychological norms are an appropriate control for the performance of Melamed and Bleiberg's mostly upper-middle and middle-class subjects.

Newer studies continue to find mild neuropsychological impairments related to cocaine abuse. O'Malley *et al.* (1988, cited in O'Malley & Gawin, 1990) compared 20 age- and education-matched controls with chronic abusers. Cocaine abusers scored significantly lower on WAIS-R arithmetic, symbol digit modalities, story memory, Halstead Category Test, and the impairment index for their neuropsychological screening battery. Deficits were mild, did not exceed normative cutoffs for clinical diagnosis of impairment, and would likely have disappeared with Bonferroni multiple comparison correction. Cocaine users also appear to show slower reaction times than controls, similar response accuracy on visual and auditory stimulus discrimination tasks which persisted over a period of 3 months (Roberts & Bauer, 1993). Such results are consistent with those of Strickland *et al.* (1993) and suggest longer-lasting changes in brain function with cocaine abuse than heretofore thought.

Evidence of lasting verbal memory impairment as well as attentional impairments have been discovered in cocaine abusers who were also polydrug abusers (Ardila, Rosselli, & Strumwasser, 1991). With an average of 16.6 months of cocaine abuse, subjects were impaired relative to matched controls on the Wechsler Memory Scale MQ, Logical Memory, and Associative Learning tests. Subjects had been abstinent for 27 days on average. Mittenberg and Motta (1993) investigated 16 cocaine abusers in a similar study, but with stricter exclusion criteria than Ardila *et al.* (1990). Subjects who were hospitalized at a private inpatient substance abuse facility had been abstinent an average of 20 days, and were screened to exclude subjects with a history of alcohol or other substance abuse. Controls were recruited from hospital staff and matched for age, race, gender, and IQ. On the California Verbal Learning Test (CVLT), cocaine-abusing subjects learned fewer words over five list repetitions, and recalled fewer words after a distractor list was interposed. They also produced fewer words under immediate and delayed cued recall, after delayed recall, and during a recognition trial. Cocaine abusers learned words significantly more slowly than controls. Significant negative correlations were found between money spent on cocaine and IQ and delayed recall but not on age or education.

The present state of neuropsychological literature in cocaine abuse does not permit differential diagnoses of brain abnormalities. Findings such as Melamed and Bleiberg (1986) or O'Malley, Gawin, Heaton, and Kleber (1988), as well as the results of Strickland *et al.* (1993) could be consistent with either neurotransmitter depletion, diffuse microinfarct process, or other vascular or metabolic mechanism. Given the visible effects of neurotransmitter depletion and/or microinfarct damage from cocaine abuse, the relative lack of neuropsychological validation tends to suggest that most research has used methodologies relatively insensitive to subtle, diffuse deficit (see Hartman, 1991). In addition, recent results suggest that "crack" cocaine abusers may constitute a neuropsychologically distinct class with more readily identifiable impairments in brain structure and function (see Strickland *et al.,* 1993).

Designer Drugs

The proposed Analogue Enforcement Act defines *designer drugs* as substances that have a substantial similarity in either structure or effect to substances already regulated by the Controlled Substances Act (Ziporyn, 1986). In practice, designer drugs are usually narcotics that have been synthesized in clandestine laboratories to imitate or "improve upon" existing drugs of abuse. Toxicity and potency are often vastly increased. For example, 3-methyl fentanyl, a synthetic heroin analogue, requires only 2 μg per dose; an amount placed on the head of a pin is sufficient to kill 50 people (Ziporyn, 1986).

Synthesis of these compounds has been lucrative and also initially legal, since the Food and Drug Administration cannot currently ban synthetic drugs in advance of their creation. The consciousness-enhancing drug MDMA (3,4-methylenedioxymethamphetamine) is of this class, as are the fentanyls, a group of drugs widely used in anesthesiology. Both are now controlled substances; the former was recently banned because of its similarity to another compound shown to cause brain damage in animals.

Perhaps the most striking observation of neurotoxicity resulting from designer drug use was the recent appearance of a severe Parkinson's disease syndrome in users of the synthetic heroin substitute MPPP (methyl-4-phenyl-4-propionoxy-piperidine). When distilled at the wrong temperature or pH, another compound, "MPTP" (1-methyl-4-phenyl-1,2,4,5-tetrahydropyridine), is also created (Campagna, 1986). MPTP selectively destroys dopamine-producing neurons in the substantia nigra. The unfortunate victims of this drug, most of whom were under 35, developed symptoms characteristic of advanced Parkinson's disease, including rigidity, stooped posture, and speaking difficulty. Pardoxically, the irreversible consequences visited on these drug abusers may prove valuable to researchers looking for causes of Parkinson's. Since this is the first time that Parkinson's has been traced to an exogenous poison, researchers are now investigating cases of Parkinson's disease that have occurred where certain herbicides similar in composition to MPTP (e.g., paraquat) have entered the water supply (A bad drug's benefit, 1985).

The designer drug, 3,4-methylenedioxymethamphetamine (MDMA; also known as "Adam," "X," or "Ecstasy") is a synthetic amphetamine that has been implicated in causing convulsions, psychosis, cerebral hemorrhage, subarachnoid hemorrhage, and cerebral sinus thrombosis (Rothwell & Grant, 1993). Animal studies show that MDMA damages brain serotonin neurons and primate studies suggest that typical human doses of MDMA are neurotoxic (McCann & Ricaurte, 1993). Neurotoxicity studies of MDMA and its analogue, MDE, show serotonin depletion and axonal damage. MDE is somewhat less neurotoxic than MDMA (Hermle, Spitzer, Borchardt, Kovar, & Gouzoulis, 1993). There is some suggestion that fluoxetine (Prozac) may prevent MDMA-induced serotonin release while failing to inhibit other prominent psychoactive effects (euphoria, sense of well-being, garrulousness). No neuropsychological evaluations of patients taking MDMA or MDE have yet entered the literature.

The future of designer drug use is "bleak," according to one source, with predictions of increasing occurrences of neurodegenerative disease resulting from use of these synthetic neurotoxins (Roberton, cited in Ziporyn, 1986). In addition, devastatingly addictive variants of existing drugs are already being developed in the laboratory (e.g., cocaine analogues that are 3–34 times more potent than cocaine at altering response schedules in animals) (Boja *et al.*, 1992). The possibilities for neurotoxic damage commensurate with addictive potential appear very real. While no neuropsychological studies of designer drug effects have entered the literature thus far, such investigations would seem strongly indicated, from both a theoretical standpoint, and for the possibility of further delineating the dangerous and toxic effects of these substances.

Glue Sniffing

See Chapter 4, pp. 190–191 and 214–215.

LSD and Psychedelic Drugs

Substances that induce "altered states" of consciousness, which distort cognitive, sensory, and perceptual awareness, are ancient intoxicants. Naturally occurring hallucinogenic substances have been used for thousands of years, and are mentioned in ancient Vedic scriptures. Despite an ancient lineage, abuse of these compounds is reported to have been relatively rare, perhaps because they served religious or cultural functions that strictly delimited proper use (Strassman, 1984).

LSD was discovered to be a hallucinogen in 1943 by researcher Albert Hoffman, who accidentally ingested it. It is synthesized from lysergic acid found in ergot, a rye fungus, and diethylamine. The drug is taken orally and absorbed well from the gastrointestinal tract. Peak LSD levels occur with 60 min of ingestion. Doses as low as 1 μg are capable of inducing noticeable effects (Gold, 1994). A 10-μg dose induces euphoria, 50 μg produces psychotomimetic effects, and 400–500 μg produces frank hallucinatory experiences (Miller & Gold, 1994; Gold, 1994). Hallucinatory effects of LSD are correlated with the drug's effect on 5-HT_2 receptors, but "marked effects" have also been noted in the 5-HT_{1a} or synaptic autoreceptors in the raphe nucleus and elsewhere (Gold, 1994, p. 126). LSD's primary effect appears to be on the serotonin system, and it has a high affinity for both types of serotonin receptors (5-HT_{2a} and 5-HT_{2c}) (Miller & Gold, 1994; Aghajanian, 1994). This suggests that 5-HT_2 antagonist drugs like clozapine may be effective in blocking adverse "bad trip" reactions experienced by some LSD users (Aghajanian, 1994).

LSD's effectiveness as a psychedelic and its ease of synthesis stimulated widespread use, both among scientists and abusers. Research applications of LSD included experimental trials with advanced cancer patients to expand their awareness and increase acceptance of their condition. Such research was terminated in the late 1960s as the drug's abuse potential came to national attention.

While other psychedelics have been widely available and abused (e.g., psilocybin, mescaline, STP), only LSD has been neuropsychologically investigated. Evidence of neuropsychological impairment is mixed in the available literature, and is hampered by lack of premorbid test data, poor methodology of existing studies, and the fact that LSD use rarely occurs in isolation from abuse of other drugs.

One of the more unusual reports is Abraham's (1982) finding of chronic color vision deficit in a group of 46 individuals who averaged 88 LSD experiences. The authors suggest their result is consistent with abnormal color persistence along the neuro-ophthalmic pathway caused by LSD's competitive inhibition of serotonin. The duration of color vision deficit is apparently prolonged, although it is not known whether it is permanent. Replication and validation of the Abraham study seems indicated. Another type of possibly related perceptual disorder associated with LSD use has been termed *posthallucinogen perceptual disorder* (PHPD). The disorder is considered infrequent but disabling when it occurs. PHPD individuals report living in a perceptual haze, seeing trails of light or images following hand movement, and various psychological symptoms of anxiety, depression, panic, or frank phobic reaction (Smith & Seymour, 1994). SSRI and/or benzodiazepines have been prescribed for treatment.

McGlothlin, Arnold, and Freedman (1969) studied LSD users with a median of 75 LSD experiences. The authors employed an elaborate battery of neuropsychological tests including the Halstead–Reitan Battery. The sole significant finding was that, compared with controls, LSD users were more impaired on the Category Test. However, absolute level of impairment was not, in itself, in the impaired range and did not correlate with number of exposures.

Cohen and Edwards (1969) studied LSD users who took LSD at least 50 times and controls matched on age, education, IQ, and SES. They excluded glue sniffers, but not other types of drug-taking from the experimental group. Drug subjects performed significantly more poorly on Trails A and a test of spatial abilities. Impairment on Trails A and The Raven Progressive Matrices Test correlated with number of LSD "trips." Unfortunately, lack of information on other drug use, and the exclusion of an unknown number of obviously impaired patients, mitigate the conclusions of this study.

Wright and Hogan (1972) studied a small group of LSD users with less LSD experience (average number of experiences 30). No differences between experimental and control groups were found on the Halstead–Reitan battery and the Aphasia Screening Test. It is unclear whether negative results simply reflect inadequate exposure history. Culver and King (1974) found that users of LSD performed somewhat less well on Trails Tests, though their scores were not in the impaired range.

In a review of studies prior to 1975, Grant and Mohns (1975) concluded that there was mixed evidence for neuropsychological impairment in LSD users but that there was no consistent evidence to suggest that LSD caused permanent neuropsychological disturbance. Strassman (1984) concurred in a more recent review, stating "the most carefully performed studies to date do not . . . support the contention that frequent LSD use is associated with permanent brain damage" (p. 588). Carlin (1986) has criticized all prior LSD researchers for their reliance on univariate statistical procedures, and for failing to correct for multiple t or F tests. He also speculates that central tendency research designs may wash out individual-specific patterns of impairment (Carlin, 1986).

While there were few LSD studies during the 1980s, this may change with the resurgence of LSD use among high school students. In a survey of 522,328 junior and senior high school students, Gold (1994) found that hallucinogen use increased from 4.9% to 5.3%. In 1990, the Drug Enforcement Administration confiscated over 500,000 doses of LSD, the third highest number of drug seizures that year. So-called "raves" or nonstop dance parties including the use of LSD and MDMA and/or methamphetamine have contributed to a resurgence of psychedlics.

MDMA is a drug with both psychedelic and stimulant properties. Clear evidence of MDMA neurotoxicity was found in animal studies, where MDMA reduced brain 5-HT by 90%, with chronic decreases in hippocampus and cortex for at least 12–18 months. Typical human doses of MDMA are also neurotoxic (McCann & Ricaurte, 1993). Human studies show that MDMA produces a 30–35% drop in 5-HT metabolism, with effects in women tending to be much greater, closer to 50% (Miller & Gold, 1994). Side effects of MDMA can include convulsions, psychosis, cerebral hemorrhage, subarachnoid hemorrhage, and cerebral sinus thrombosis (Rothwell & Grant, 1993). There is some suggestion that fluoxetine (Prozac) may prevent MDMA-induced serotonin release while failing to inhibit other prominent psychoactive effects (euphoria, sense of well-being, garrulousness). No neuropsychological evaluations of patients taking MDMA or MDE have yet entered the literature.

Marijuana

History

An intoxicant consisting of the leaves and flowers of the female *Cannabis sativa* plant, citations of marijuana use date back to 2737 B.C. in China, where it was termed the "liberator of sin" (Abel, 1976a). O'Shaughnessy (1839) introduced cannabis intoxication effects to Western medical audiences, with descriptions of the "delirium which the incautious use of the Hemp preparations often occasions...." During the 1800s, tinctures of cannabis were employed to treat neuralgias, although variable potency was a problem in dose determination.

Marijuana was one of the most popular intoxicants for the counterculture of the 1960s but recent estimates suggest that its use may have declined somewhat. A 1984 *New York Times* poll suggested that the proportion of teenagers who had tried marijuana declined from 51% in 1979 to 42% by 1982. Daily use dropped from 10.7% in 1978 to 5.5% in 1983 (Murray, 1986); however, recent U.S. estimates suggest that over 29 million people continue to use marijuana, with more than 7 million daily users (United Nations, 1990).

The psychoactive ingredient in marijuana is Δ-9-tetrahydrocannabinol (THC). Presently available marijuana varies widely in THC content; potency ranges from 0.35% for a U.S. sample to 11% for Sinsemilla variety. Crude marijuana extract contains 20% THC, while hashish oil ranges from 10 to 60% THC (Mikuriya & Aldrich, 1988). Schwartz (1991) expressed the concern that currently available varieties of marijuana are 600% more potent than earlier strains. This hypothesis is disputed by Mikuriya and Aldrich (1988) who attribute low-potency estimates of the previous decade to the artifact of using old, degraded police samples when measuring THC content.

Experimental use of THC has suggested several beneficial medical applications, including treatment of open angle glaucoma and asthma. In addition, its antiemetic and antianorexic effects make it potentially useful in cancer radiation therapy and chemotherapy (Cohen, 1980).

Neurological Effects

In addition to its well-known euphoriant action, marijuana intoxication may produce tremor, brief periods of muscle rigidity, or myoclonic muscle activity. High doses cause "hyperexcitability of knee jerks with clonus" (Jones, 1980, p. 67). Surface EEG recordings typically do not reveal significant alterations, in sharp contrast with results from primate studies using deep electrode implant recordings. These latter studies showed "marked alterations" in the septal and amygdala regions; alterations that persisted for up to 8 months after inhaling smoke equivalent to three marijuana cigarettes per day for a period of 36 months (Jones, 1980, p. 68). CBF was found to increase in experienced marijuana smokers but not significantly more than in a placebo condition. However, CBF decreased among inexperienced smokers, significantly more than placebo (Mathew & Wilson, 1991b). The authors suggest that experiment-induced anxiety or other expectations may have influenced findings. Chronic heavy use of marijuana appears to be associated with a general reduction in CBF, which increases over a period of abstinence (Tunving et al., 1986).

A recent PET scan investigation of marijuana's active ingredient, THC, points to the possibility that marijuana's intoxicating effect is not localized to cortical regions at all. When human subjects were given acute IV administration of THC, there was no consistent change in global cerebral glucose metabolism. However, all subjects showed increases in *cerebellar* metabolism. Moreover, THC subjective intoxication was correlated with increasing cerebellar metabolic activity (Volkow et al., 1991). Results are consistent with the very high density of cannabinoid receptors in the cerebellum. In addition, THC intoxication may be related to cerebellar functions of affect modulation, attention, and reinforcement, as well as its anatomical connections to the limbic system and prefrontal cortex (Volkow et al., 1991). High concentrations of cannabinoid receptors are also located in the substantia nigra, pars reticulata, globus pallidus, and hippocampal dentate gyrus; however, PET scan correlations with neuropsychological or subjective intoxication effects have not been performed as of this writing (Herkenham, Lynn, Little, Johnson, et al., 1990).

Only a single study using neurologically compromised individuals found cerebral atrophy in marijuana users (Campbell, Evans, Thomson, & Williams, 1971). This finding has not been replicated in studies of healthy long-term marijuana users.

Neuropharmacological studies of cannabinoid agents suggest that they exert their effect primarily through cholinergic pathways. Cannabis appears to affect limbic turnover and synthesis of acetylcholine. THC has also been shown to decrease hippocampal turnover of acetylcholine. Principal neurotoxic mechanism involves "selective reduction in Ach synthesis in the muscarinic pathways of the limbic system" because of high-affinity uptake of choline (Miller & Branconnier, 1983, p. 448).

In addition to its psychoactive effects, inhaled marijuana smoke acutely boosts carboxyhemoglobin levels more than four times that of a tobacco cigarette. Although chronic carboxyhemoglobin levels are much lower than those of tobacco smokers, there may be repeated acute cardiovascular risk in chronic users (Tashkin, Wu, & Djahed, 1988).

Acute Neuropsychological Effects

Acute intoxication effects depend on marijuana potency, smoking technique, and individual variation in marijuana bioavailability. Psychoactive effects of THC include immediate memory/attentional impairment for digit span which peaks between 1.5 and 3.5 h after oral administration (Tinklenberg, Melges, Hollister, & Gillespie, 1970). THC-induced memory impairment appears to affect acquisition but not retrieval, as list learning subjects were impaired when THC was ingested prior to learning, but not in the interval between learning and recall (Darley, Tinklenberg, Roth, Hollister, & Atkinson, 1973). Reductions in attention, competition from intruding thoughts, or more rapid loss of information from short-term store were possible explanations (Darley et al,. 1973).

Recreationally smoked marijuana has a rapid (1-10 min) onset and duration of about 3-4 h (Abel, 1976a). Feelings of intoxication are affected by context of administration and are prone to placebo effects. Acute intoxication appears to induce global changes in neuropsychological functioning. Memory, speech, cognitive efficiency and flexibility, attention, and spatial abilities are all disrupted, and the degree of this disruption in dose-dependent (Ferraro, 1980). For example, Dornbush, Fink, and Freedman (1976) found measures of short-term memory and reaction time to be affected at an unspecified "high dose" but not at a similarly unspecified "low dose." Similarly, several studies by Abel and his co-authors (1976b,c) suggest acute memory changes immediately after smoking marijuana. In one study, Abel (1976b) had eight subjects serve as their own counterbalanced controls. Each was asked to read Bartlett's *War of the Ghosts* and recall it 15 minutes later in writing. While under the influence of marijuana, seven of the eight subjects recalled significantly fewer words overall, and fewer content words compared with their test results while sober. Similar results were reported by Miller, Drew, and Kiplinger (1976) in the same volume. However, since neither study had a recognition condition, the results could be explained as a function either of memory or of motivational impairment.

While recognition memory impairment is not inevitable (e.g., Darley, Tinklenberg, Roth, & Atkinson, 1974; Miller *et al.,* 1977) when impairments in recognition memory do appear they are primarily related to the tendency of marijuana-intoxicated subjects to make false-positive identifications of material not previously learned. Intrusion errors for both words and nonverbal stimuli are a common research finding (e.g., Miller, Cornett, & Wikler, 1979; Miller, Cornett, & Nallan, 1978). Abel (1976c), for example, found that marijuana-intoxicated subjects make more intrusion errors in a recognition task. Dornbush (1974) found a small decrease in memory and an increase in false alarms for subjects tested for recognition memory while using marijuana. The effect was found whether subjects learned material while intoxicated or "sober."

New learning, as well as memory, is impaired with marijuana use. Rickles, Cohen, Whitaker, and McIntyre (1973) found results suggesting that moderate doses of marijuana interfered with and increased the difficulty of learning new material.

Such acute impairments appear limited to material learned under the influence of the drug. As with THC administration alone, recall or recognition of material learned while "sober" is not affected when recalled or recognized later in an intoxicated state (Ferraro, 1980; Darley *et al.,* 1973; Darley, Tinklenberg, Roth, Vernon, & Koppell, 1977; Stillman, Weingartner, Wyatt, Gillin, & Eich,

1974). Miller and Branconnier (1983) have suggested that the acute memory deficit pattern resulting from marijuana intoxication resembles that which occurs in patients with impaired cholinergic limbic system functioning, including herpes simplex encephalitis, Korsakoff psychosis, and Alzheimer's disease. Attentional impairments are linked to disinhibition of septal–hippocampal inputs to the reticular activating system. This produces inability to habituate cortical arousal to novel stimuli and consequent failure to focus attention (Miller & Branconnier, 1983).

Marijuana has also been shown to affect reaction time, motor coordination in machinery operation, accuracy, and hand steadiness. In one German study, driving impairment was stated to be found for as long as 6 h following THC intake (Kielholz et al., 1972, cited in Yesavage, Leirer, Denari, & Hollister, 1985). Several ecologically relevant and disturbing studies have examined the effects of marijuana on the complex perceptual-motor and decision-making abilities of airline pilots. The first asked certified pilots to maintain holding patterns on a flight simulator after smoking marijuana. Significant effects on performance were found up to 4 h after smoking compared with placebo (Janowsky, Meacham, Blaine, Schoor, & Bozzetti, 1976).

A second study examined ten experienced licensed private pilots who were trained for 8 h on a flight simulator, allowed to smoke one marijuana cigarette, and then retested at 1, 4, and 24 h later on the simulator (Yesavage et al., 1985). Subjects were significantly less able to align and land at the center of the runway after ingesting marijuana; impairments that persisted over the 24-h testing period. Unfortunately, the authors did not include a control condition that would allow potential confounds (e.g., fatigue, regression toward the mean) to be addressed.

More recently, Yesavage and colleagues again tested pilots on a flight simulator with results that extend and clarify their previous study. This time, the experiment varied pilot age (young versus old), marijuana dose (0, 10, or 20 mg), and time (1, 4, 8, 24, 48 h after smoking) (Leirer, Yesavage, & Morrow, 1989). The important comparison for this review is the Dose × Time Delay interaction, which showed detrimental effects of a single 20-mg THC cigarette at 1 and 4 h, but not afterward, thus failing to replicate Yesavage et al. 1985). Neither the 10-mg THC-containing cigarette nor the placebo produced performance decrements. Age, dose, and task difficulty were considered to be cumulative in producing impaired simulator performance. Since blood serum THC levels in the 20-mg condition were considered "relatively moderate" and consistent with social use of the drug, results suggest impairment in complex cognitive task performance for as long as 4 h following drug use.

These simulator studies suggest that marijuana may subtly but significantly impair complex psychomotor and decision-making behavior long after subjective intoxication ends. Even highly trained subjects are apparently unable to compensate for marijuana-induced perceptual-motor dysfunctions. The results are particularly unsettling in light of several highly publicized train and plane accidents during this decade where operators have subsequently tested positive for marijuana.

Acute Interactions with Alcohol. Ubiquitous use of these two intoxicants makes combined personal use of alcohol and marijuana likely, but there are few controlled studies testing their interactive effects. Perez-Reyes *et al.* (1988) addressed the issue in a single-blind, three dose/Latin square crossover design. Neuropsychological effects were assessed with a complex, computerized decision-making task that

> displayed two-digit numbers in the central field of vision that changed approximately every half-second.... If these numbers exceeded a critical value (57), the subject was instructed to press button 3.... If these numbers fell below a critical value (53), then the subject was to press button 2.... Simultaneously, the peripheral displays ... were programmed to display one of four digits.... If either the left or right peripheral display changed from the steady value of 4 to a 5, the subject was instructed not to respond.... If the value changed to a 3, the subject was to press a corresponding left number 1 or right number 4 response button.... If the value changed to a 7 in either of the peripheral displays, the subject was to press [a] foot pedal. [p. 269]

Mean subjective ratings of marijuana intoxication were increased by prior alcohol ingestion in a dose-dependent manner. Computer response accuracy and speed showed main effects for alcohol, marijuana, or their combination, but no differential effects. All subjects showed monotonic impairment related to alcohol dose. Four of six subjects showed accuracy impairments, and five of six subjects displayed latency impairments under the influence of marijuana. Marijuana interacted with alcohol to produce greater impairments of accuracy in three subjects and in latency in four subjects. After 360 min no significant drug effects were observed.

Marijuana and Formaldehyde. A drug combination astounding for its neurotoxicity and for the naiveté of its users, this combination of marijuana and formaldehyde-based embalming fluid ("Wicky Sticks," "AMP") was reported by Hawkins, Schwartz-Thompson, & Kahare (1994). The first individual with no known psychiatric history was admitted to the hospital with auditory hallucinations, disorganized thinking, and other symptoms of severe psychosis. Neuropsychological examination conducted 6 weeks after admission showed complete disorientation for time and date, with severe impairments of new learning that did not improve with repeated trials. Severe memory deficits persisted in a repeat examination 7 weeks later. A second case had no signs of psychosis but also displayed disorientation and severe memory deficit. A test of malingering did not show evidence of deliberate exaggeration.

Chronic Effects

Researchers are beginning to reevaluate the chronic neurotoxicity of marijuana. Studies during the 1970s were inconsistent and largely failed to provide evidence of significant cognitive and personality impairment as a consequence of long-term marijuana use. Some studies found impaired complex processing and short-term memory impairment (e.g., Melges, Tinklenberg, Hollister, &

Gillespie, 1970a,b; Casswell & Marks, 1973b; Carlin & Trupin, 1977). However, a number of other studies failed to find consistent deficits as a function of chronic marijuana intoxication. For example, Schaeffer, Andrysiak, and Ungerleider (1981) examined a group of ten subjects who reported using between 2 and 4 ounces of marijuana mixed with tobacco for an average of 7.4 years. When tested with a comprehensive neuropsychological battery, including the WAIS, the Benton VRT, the Rey Auditory Verbal Learning Test, Trails, and other tests, subjects showed no impairment in cognitive functioning; in fact, IQ scores were all in the Superior to Very Superior range—all this despite the fact that most of the experimental volunteers continued to use marijuana during the period of neuropsychological testing. Similar negative results were also obtained by earlier researchers (e.g., Grant, Rochford, Fleming, & Stunkard, 1973; Rochford, Grant, & LaVigne, 1977). In particular, when large group studies in Jamaica (Beaubrun & Knight, 1973) and Costa Rica (Satz, Fletcher, Sutker, 1976) failed to detect significant effects of chronic use, the prevailing opinion became that marijuana produced acute, but not chronic neuropsychological sequelae.

Recent research has cast doubt on this assumption. In particular, heavy users of marijuana appear to show impairments in neuropsychological function related to drug use. For example, Block and Ghoneim (1993) found deficits in subjects who used marijuana seven or more times weekly in memory for high-imagery words. They also found deficits on mathematical skills and verbal expression on the Iowa Tests of Educational Development in subjects who were previously matched by fourth grade test scores. However, data do not address whether lowered educational scores and marijuana use were effects of other, non-drug-related variables. Varma, Malhotra, Dang, Das, and Nehra (1988) examined heavy cannabis use in India, where social sanction against prodigious use is weak, and high-potency forms of the drug are easily available. Subjects were 26 daily (at least 20 times per month) users with daily intake of 150 mg of THC. Controls were matched for age, education, and occupation. A variety of neuropsychological screening measures were used, including tests of memory, personality, reaction time, pencil tapping, Trails A, Standard Progressive Matrices, WAIS-R verbal scale, Indian adaptation, and others. Groups did not differ on basic intelligence measures, several tests of reaction time, motor and perceptual speed, and time estimation. A memory battery showed significant differences on recent memory only. Users also scored significantly higher on psychoticism, neuroticism, and rated themselves more impaired on measures of personal, social, and vocational disability.

Methodological problems of the study include the use of a specialized Hindi language memory test with unknown comparability to Western memory tests, and the use of multiple t tests without correction. In addition, the 12 h minimum abstinence period seems insufficient to partial out acute effects in these very heavy users. Further, while the authors suggest that users who are not part of a socially isolated drug subculture eliminate confounding demographic selection factors, their cross-sectional study cannot rule out preexisting personality or other factors leading to heavy drug use.

Two studies that address longitudinal follow-up of heavy marijuana users provide stronger evidence that chronic marijuana use degrades neuropsychological function. The first study followed up Costa Rican marijuana users who were previously tested in 1980. At that time, subjects, who smoked an average of 9.6 marijuana joints per day for 17 years, failed to show significant effects (Carter & Doughty, 1976). Following up on a nonsignificant trend for users to perform worse on learning and memory tests, Page, Fletcher, and True (1988) sought out the original 82 subjects for retesting. Fifty-seven subjects (30 users and 27 nonusers) were available for neuropsychological follow-up and presented with a large battery of neuropsychological tests. The authors attempted to replicate the original test battery, but also added several other tests to allow more specific investigation of memory, learning, and attention. The full screening battery included tests for intelligence, memory, motor skills, sustained attention, and personality.

Results were notable in that the original 1980 neuropsychological tests failed again to discriminate between users and nonusers. However, three tests added to the later study significantly discriminated between the two groups: "underlining" (a cancellation test), Buschke Selective Reminding, and a computerized continuous performance test. The authors characterized resulting marijuana-induced deficits as occurring during sustained attention, effortful processing, and long-term storage word retrieval. Results from Page *et al.* (1988) suggested that earlier studies that failed to demonstrate drug effect may have been using insensitive neuropsychological measures.

Even heavier use of cannabis was investigated by Mendhiratta *et al.* (1988), who examined charas/ganja smokers and bhang (a cannabis-containing beverage) drinkers. Daily average THC intake was estimated to be a prodigious 140–150 mg. Users were initially evaluated in 1972 and retested in 1982. Compared with controls, retested users showed poorer word association reaction time, pencil tapping speed, Bender Visual Motor Gestalt score, and backward digit span. Users also scored significantly higher on the Neuroticism scale of the Maudsley Personality Inventory. Over 10 years, users deteriorated significantly in word associated reaction time and BVMG score compared with controls. While the study was not without methodological problems, including minimal washout time (12 h) and failure to correct for multiple comparisons, the fact that effects were found on relatively insensitive measures suggests that extremely heavy, chronic use of THC-containing substances eventually produces obvious clinical effect. Taken together, Page *et al.* (1988) and Mendhiratta *et al.* (1988) indicate that moderately high levels of chronic marijuana use produce "subclinical" effects on highly sensitive tests, and that extremely high intake of marijuana over 10 years can produce more obvious clinical deficits on less sensitive measures.

Preliminary data from a National Institute on Drug Abuse (NIDA) study showed statistical relationships of between $p < 0.1$ and $p < 0.05$ on a variety of neuropsychological measures on two groups of chronic (15 or more years) users (Leavitt *et al.*, 1991). Age and education effects corrections were taken, but there is no indication whether coexisting drug use, cultural factors, or employment influenced results. Since the authors were unable to test a set of full-time-employed individuals who also chronically used marijuana, the latter is a significant concern. The authors properly warn against attributing effects with their relatively small sample sizes. Discovery of neuropsychological effects of chronic marijuana use paves the way toward reexamination of related questions, including whether "amotivational syndrome" or structural brain changes can be demonstrated in chronic users. Acute impairment research must address the findings of Yesavage *et al.* (1985) and determine subtle effects and time course of acute intoxication.

Another NIDA study used quantitative EEG (QEEG) to compare maps of brain electrical activity of 15 "exceedingly" long-term users of THC (mean use 19.6 years), with those of shorter-term users (mean 4.1 years) and two nonuser control groups (Struve *et al.*, 1991). The study found several interesting results:

1. Very long-term duration (VLD) users had significantly elevated Absolute Power (voltage) of theta activity over bilateral frontal–central cortex compared with shorter-use or nonusing groups.
2. Relative Power (Amount) of theta was also significantly elevated in VLD users compared with shorter-term users, but not with controls.
3. The VLD group displayed significantly elevated interhemispheric coherence of theta compared with shorter-term and nonusers. The authors also found that the VLD group displayed hyperfrontality of alpha bilaterally over the frontal cortex. Preliminary analysis of accompanying neuropsychological data suggests that VLD marijuana users show deficits greater than shorter-term users; however, age and education do not appear to be covariates of this preliminary analysis. Neuropsychological deficits would, however, support the authors' speculation that increasing amounts of theta in VLD users found by QEEG, "if confirmed, may suggest organic damage."

Prenatal Exposure, Postnatal Exposure, and Neurotoxicity

Marijuana use among pregnant women has been reported to range from 5 to 34% with effects of decreased birth weight, greater likelihood of early delivery, and increased complications during labor and delivery (Astley & Little, 1990). Fried (1980) found that mothers who smoked more than five joints per week gave birth to infants who displayed tremors, abnormal startle response, and altered visual responsiveness between 2 and 4 days postpartum. Symptoms decreased by 30 days and, in a follow-up study, no effects on motor development were found at 1 and 2 years of age (Fried & Watkinson 1988).

Since Δ THC collects in maternal milk at eight times the levels found in blood, the relationship between maternal marijuana use and neuropsychological performance of breastfed infants is a logical topic of exploration. Such a study was performed by Astley and Little (1990), who measured infant performance on the Bayley Scales of Infant Development as a function of maternal postpartum marijuana use in the first and third month postpartum. Confounding variables included age, race, pregnancy medical history, tobacco, coffee, alcohol, psychoactive drug use, use of marijuana during gestation, paternal alcohol and tobacco use, and infant age and sex. Both experimental and control groups were typically white, middle class, and college educated.

Five percent of the variance of infant motor skills was explained by postpartum marijuana use. After controlling for the effects of tobacco smoking, alcohol use, and cocaine during pregnancy and lactation, infants' exposure to marijuana in breast milk at 1 month postpartum was associated with a 14 ± 5 point decrease in Bayley PDI (psychomotor) functioning. An unfortunate confounding influence was marijuana use during the first trimester, although regression analysis showed that postpartum use during lactation was the stronger predictor. No significant effects were found for postpartum marijuana use in the third month. Other potential confounds include prenatal marijuana smoking, infant exposure to passive smoke, and unaddressed maternal–infant interactional variables (Astley & Little, 1990).

Opiates

Opiates, particularly heroin, do not appear to have chronic neurotoxic effects on adult cognition. Rounsaville, Jones, Novelly, and Kleber (1982) found few significant differences when using a brief battery to test 72 opiate addicts compared with epileptics and CETA (Comprehensive Employment Training Act) workers. Addicts actually outperformed CETA controls on finger tapping. The authors' choice of control groups must be questioned, since all groups performed in the mildly impaired range for Trails A and B, Digit Symbol, and a pegboard task. There was also no relationship between recent drug use and neuropsychological status, suggesting that the tests employed may not have been sensitive to drug-related impairments. However, Fields and Fullerton (1975), using the complete Halstead–Reitan Battery, also failed to find significant differences between addicts and controls.

More recent investigations of opiate neurotoxicity suggest that opiates may be selectively toxic to neuroendocrine structures, rather than cortical functions. Mutti *et al.* (1992) examined neuroendocrine changes in heroin addicts as a function of performance on a neuropsychological vigilance task. While reaction times did not differ between addicts and controls, plasma ACTH was "markedly depressed" in heroin addicts compared with controls. Normal subjects showed progressive increases in ACTH over testing time, while heroin addicts' ACTH did not change. A twofold increase in serum prolactin was also initially observed in the heroin addicts, but decreased to control levels with long-term abstinence. β-Endorphin levels of heroin addicts were also much higher than those of controls for at least the first 2 months of abstinence. Further research is necessary to determine whether these unusual neuroendocrine responses reflect neural abnormalities or endocrine dysfunction. In addition, abnormal values may have preceded drug abuse. Further investigation is clearly indicated.

While current literature does not suggest neuropsychological risk to adults from chronic opiate intoxication, intravenous opiate administration has well-defined dangers. Secondary infections that penetrate the blood–brain barrier are easily introduced via contaminated needles or conditions of injection. Acquired immune deficiency syndrome (AIDS) has been shown to be transmitted in this way to produce neurotoxic bacterial, fungal, and viral infection.

Fetal and Infant Neurotoxicity

Material narcotic administration appears to affect fetal and infant arousal systems. While there are many methodological difficulties in existing research (e.g., poorly defined population, nutritional and educational deficit, polydrug abuse, and poor parenting), Householder et al. (1982) conclude there is a "remarkable consistency across studies." Toddlers who had been narcotic-addicted neonates tended to exhibit brief attention span, immature fine motor coordination, hyperactivity, and low frustration tolerance (Lodge, 1976; Ting, Keller, Bergman, & Finnegan, 1974; Wilson, Desmond, & Verniaud, 1973).

Phencyclidine (PCP)

Phencyclidine was developed in 1956 and intended as a new analgesic (Maddox, 1981). Human clinical trials were conducted after animal studies suggested that the material had potent analgesic properties. Over 3000 research patients received PCP for local surgical procedures or preparatory to general surgery. The first side effect noted was that all patients experienced complete amnesia for the operation and postoperative phase of recovery, a consequence considered beneficial by some. The second, termed "emergence phenomena," were "excitation reactions most frequently encountered in young or middle-aged males"— reactions that, along with symptoms of hallucinations, agitation, violence and prolonged depression of consciousness, eventually caused the termination of human clinical studies (Maddox, 1981, p. 5; Baldridge & Bessen, 1990). PCP became a "street drug" in 1967 and was immediately put into wide usage by a polydrug abusing population that either injects, "snorts," eats, or smokes it on marijuana (Fauman & Fauman, 1981).

More specifically, the range of PCP reactions have been characterized by Burns et al. (1975) as causing agitation, excitement, incoordination, catalepsy, and mutism in "low dosage," moderate dosage producing light coma or stupor, and "high dosage" causing prolonged coma.

Neurotoxicology

PCP is highly lipid soluble and accumulates in fat and in the brain (Baldridge & Bessen, 1990). Animal studies demonstrated PCP remains in the brain for as long as 48 h after administration, even after blood levels become undetectable (Meibach, Glick, Cox, & Maayani, 1979). Metabolism of body fat may release PCP, causing recrudescence of psychoactive effect. Animal experiments have also shown PCP to cause "dramatic" changes in glucose metabolism in specific brain regions. It appears to have particular affinity for the limbic system, increasing glucose metabolism in the hippocampus, subicular cortex, and especially the cingulate gyrus (37% increase). Smaller increases were seen in the substantia nigra. Inferior colliculus metabolism was decreased by 54% after PCP administration (Meibach et al., 1979).

PCP is thought to bind with high specificity to a receptor "colocalized" with N-methyl-D-aspartate (NMDA) receptors, a type of glutamate receptor (Olney, Labruyere, & Price, 1989). Animal preparations have shown PCP to antagonize the actions of glutamate and aspartate, suggesting its possible prophylactic use in brain damage prevention from stroke, heart attack, or perinatal asphyxia; however, PCP apparently induces a neurotoxic effect of its own. Rats given subcutaneous injections of PCP showed dose-dependent neurotoxic damage consisting of acute vacuolization and mitochondrial destruction (Olney et al., 1989). The effect is apparently acute, noncumulative, and reversible, at least by a relatively insensitive light microscopy examination. Cingulate and retrosplenial cortex were found to be selectively vulnerable. Tolerance also develops and neurons lose sensitivity unless they are separated by a long drug-free interval (Olney et al., 1989).

Clinical Effects

Clinical presentation is quite variable, with CNS stimulation, depression, cholinergic, anticholinergic, and adrenergic effects possible. Immediate mental status changes can include confusion, disorientation, auditory and visual hallucinations, delusions, and may resemble an acute schizophrenic episode (Baldridge & Bessen, 1990). Behavior is variable and patients may cycle between alertness, violence, bizarre behavior, or agitation. PCP's analgesia may mask significant bodily injury that was sustained but not perceived while intoxicated, in one case, allowing a PCP user to bite through his forearms "almost to the bone" (McCarron et al., 1981, p. 211). Analgesia and loss of sensory integration have been proposed to explain why another PCP user was found nude and head down in a garbage dumpster, but appeared neither cold nor uncomfortable (McCarron et al., 1981).

Neonatal Effects

Assessment of neonatal PCP exposure is complicated by confounding variables, including poor or absent prenatal care, and polydrug abuse, including cocaine. Nonetheless, PCP is transferred through the placenta and infants whose mothers' drug use included PCP ingestion, have shown postnatal irritability, hypersensitivity to auditory stimuli, tremors, hypertonicity, facial twitching, bizarre eye movements, and staring (Howard, Kropenski, & Tyler, 1986; Chasnoff, Burns, Hatcher, & Burns, 1983). Howard et al., who followed 12 prenatally exposed infants for 18 months, found that in spite of being placed in nurturing postnatal environments, infants showed borderline functioning in fine motor, adaptive, language and personal-social development. The authors suggest that neonatal PCP exposure may produce greater developmental impairment than heroin or methadone.

Children under age 5 who become intoxicated on PCP, show lethargy, staring, or depression of consciousness, but aggression or violence is unusual (Baldridge & Bessen, 1990).

Neuropsychological Effects

Unfortunately, there is very little information on the long-term neuropsychological effects of PCP use. PCP-abusing populations are likely to be polydrug abusers, which makes the unique contribution of PCP to neuropsychological impairment difficult to determine. Further, one-third to one-half of sampled "street" PCP has been found to contain other chemical agents with dissimilar pharmacological structure but similar behavioral effects (Lewis & Hordan, 1986).

Acute effects of PCP, however, are clear. Impairments in judgment, logical reasoning, abstraction, and attention have been noted, as has "extreme" loss of ability to sustain organized thought (Domino & Luby, 1981, p. 405).

Chronic effects of PCP have been addressed by Carlin, Grant, Adams, and Reed (1979), who compared 12 chronic PCP users with normal controls and 12 polydrug abusers who did not use PCP. Both PCP users and polydrug abusers were significantly impaired, suggesting that chronic neuropsychological effects of PCP may be similar to those of polydrug abuse.

Another recent study used a screening battery approach to study 30 adolescents and young adults who were referred to drug abuse counseling (presumably PCP from the context of the article) (Lewis & Hordan, 1986). Mean Full Scale IQ of this sample was 92.5, with Performance IQ being several points higher (96.2) than Verbal IQ (91.7). The tests most sensitive to impairment in this population included Trails B, where 30% showed scores above the brain-damage cutoff, and the Category Test, where "more than 70 percent of the sample were in the mild, moderate or severe impaired range" (Lewis & Hordan, 1986, p. 198). Almost 60% of the sample also performed "below normal" on the finger tapping test, a finding suggestive of impaired fine motor performance or motivational deficit.

As in any noncontrolled study, conclusions cannot be drawn as to whether population effects or drug effects produce these neuropsychological impairments. Replication in a controlled design is indicated.

Emotional Effects

In one study, 19 of 24 PCP users reported having bad reactions to PCP, including loss of sensation or motor control, paralysis, confusion, sensory distortion, or unconsciousness. Twenty-three of the subjects admitted to losses of judgment and memory during PCP intoxication, sometimes disposing them to accidents. Unfortunately, these reactions were not discouraging to this group, almost all of whom continued to abuse PCP (Fauman & Fauman, 1981).

Violent behavior has been associated with PCP abuse, and several studies have supported violence-inducing properties of the drug. For example, Graeven, Sharp, and Glatt (1981) sampled self-reports from 200 PCP users with a 10-year history of PCP abuse. With factor analysis, the authors identified three clear factors, what the authors termed an "energy" factor, a "violence" factor, and a "negative ideation" factor. Controlling for age, heavy users reported more frequent violent effects than recreational users, a dose–response relationship that lends credence to drug effect rather than cultural influence; however, since 30% of the population was Hispanic and those results were not separately analyzed, the question is still open. Heavy users also reported increased energy, happiness, and sexual arousal, so it is possible that PCP influence on the limbic system is more general than specific to violence. Looking at comparisons between recreational and heavy users, the authors found recreational users to experience depression during PCP use, while "expert" users experienced mood elevations.

Violent behavior has been associated with PCP abuse, but it remains unclear whether PCP has unique violence-inducing properties or whether individuals who abuse PCP are more likely to have histories of violent acting out. Psychosis is a common concomitant of PCP abuse and is of a toxic confusional type that does not resemble schizophrenia.

Polydrug Abuse

Polydrug abuse is characteristic of many heavy drug users. Users will often simultaneously or sequentially use alcohol, cocaine, opiates, barbiturates, and other drugs in various combinations.

Of all abused drug effects, polydrug abuse seems to show the most research support for neurotoxicity. Two out of three studies for which an Impairment Index could be compiled were in the impaired range (Parsons & Farr, 1981). For example, Grant *et al.* (1978), reporting the results of a 3-month multisite study, found 37% of polydrug abusers to show neuropsychological impairment on the Halstead–Reitan Battery. Three-month follow-up testing saw impairment in 31 of the 91 subjects who completed the second battery (Grant *et al.*, 1978).

A later study compared the likelihood of neuropsychological impairment in young (mean age 30) alcohol versus polydrug abusers (Grant, Adams, Reed, & Carlin, 1979). Half of the polydrug abusers were impaired on a clinician's rating of Halstead–Reitan Battery scores, compared with only 20% of the alcohol abusers. Equivalent verbal IQ scores between the two groups argue against educational differences accounting for these results, but the relatively small sample size in each group ($n = 20$) limits the generalizability of these conclusions.

Sedative-Hypnotics

There are few neuropsychological studies on the effects of sedative-hypnotics. Bergman, Bord, Engelbrektson, and Vikander (1989) studied neuropsychological function in a follow-up study of 30 hypnotic-sedative abusing patients.

While significant improvement was observed on most tests, almost half of the group continued to show signs of neuropsychological impairment, leading the authors to conclude that their mixed group of antianxiety, sedative, and hypnotic abusing patients showed permanent neuropsychological deficit as the result of drug abuse.

CONCLUSIONS—NEUROPSYCHOLOGICAL TOXICOLOGY OF DRUGS

Neuropsychologists who work with medical patients, or who are asked to evaluate impairment in drug-abusing individuals, must be aware that acute neuropsychological effects are possible under almost all medications, and that chronic effects have been seen for many others. Some drugs apparently produce quite specific neuropsychological impairments (e.g., verbal memory deficit in patients receiving antihypertensives). Others may produce nonspecific encephalopathies (e.g., steroids). A careful medical, educational, and demographic history is a necessary addition to the neuropsychological test results in such cases. Sometimes it is only with such information that the clinical neuropsychological can evaluate test results in the context of potentially confounding influences.

It is unfortunate that drug abuse studies typically fail to include such information in their experimental designs, thereby lessening their utility for clinicians and researchers. The majority of neuropsychological studies on drug effects have methodological or statistical design flaws that make their conclusions difficult to interpret. Some of these flaws are within the control of the experimenter, including use of small numbers of subjects, inadequate or missing control groups, and failure to control for premorbid abilities. Other less than optimal conditions are beyond the control of the researcher. For example, some types of drug abuse do not exist in isolation from other factors, e.g., underclass membership, alcohol abuse, head injury, or poor nutrition.

For health researchers, both abused and prescription drug effects are increasingly understood to be multifactorial in nature. Alcohol researchers have come to a similar conclusion. As the clinical complexity of drug effects becomes apparent, experimental design considerations become salient, and neuropsychological researchers have been increasingly called upon to design drug experiments, develop and administer tests, and implement interventions based on test results. This should have the effect of increasing the quantity and quality of drug research in the next decades.

7

Pesticides

INTRODUCTION

Most commonly used pesticides, including the chlorinated hydrocarbons (e.g., DDT), the organophosphates (e.g., Malathion, Diazinon, Ronnel), and the carbamates (e.g., Baygon, Maneb, Sevin, Zineb), are lethal to insects via neurotoxic action (Morgan, 1982). Studies using higher mammals have also demonstrated pesticide neurotoxicity (e.g., Vandekar, Plestina, & Wilhelm, 1971; Aldridge & Johnson, 1971; DuBois, 1971). It is not surprising, therefore, that neurotoxic effects of pesticides are also found in human exposure victims, and that both cognitive and emotional functions are affected. What may be surprising is the extent of the potential problem, since of all neurotoxic substances produced by civilization, pesticides probably vie with lead for the widest distribution in the environment. In the United States, where there has been a government-mandated ban of leaded fuels, pesticides may well have taken the place of lead as the most ubiquitous neurotoxic material deliberately released into the ecosystem.

The actual amount of pesticides put into the environment is staggering. There are over 34,000 pesticides registered by the EPA. In recent years, approximately 1.1 *billion* pounds of pesticides were utilized in the United States alone (Lang, 1993); not appreciably different from estimates of 1.4 billion pounds in 1975 (Ecobichon & Joy, 1982). Over 4 billion pounds are applied worldwide, with the United States using 1.1 billion pounds of that total annually (Lang, 1993). Pesticides are applied to over 900,000 farms in the United States. Seventy-five percent of all cropland and seventy percent of all livestock are treated with pesticides (Lang, 1993). An estimated 340,000 workers are involved in some aspect of pesticide production in the United States (Moses, 1983). An additional estimated 2.5–2.7 million migrant or seasonal workers come into contact each year with pesticides (Moses, 1983). Pesticide handlers are an additional risk group, with an estimated 1.3 million certified pesticide appliers in the United States. Combining these figures would still underestimate the total number of exposed subjects, since they do not include those involved in the commerce, transportation, and distribution of pesticide. Concisely summarized, "workers exposed to pesticides are one of the largest occupational risk groups in the world" (Davies, 1990, p. 330).

Nonoccupational exposure to pesticides also involves large numbers of individuals. Considering the EPA's estimate that 69 million families store and use pesticides, it is perhaps not surprising that there are 60,000 to 70,000 estimated poisonings annually from organophosphates alone (Lang, 1993; Muldoon & Hodgson, 1992). The number of nonoccupational poisonings may be underestimated, as results from a South Carolina hospital survey found that half of all patients hospitalized for pesticide exposure had incurred exposure in non-work-related situations (Schuman, Whitlock, Coldwell, & Horton, 1989). The dominant risk of poisoning from specific pesticides may actually be greater for nonoccupational exposure than for work-related situations. For example, home, school, or public access building exposure to Malathion between 1966 and 1980 accounted for 92% of symptomatic exposure reports (Muldoon & Hodgson, 1992).

Considering the magnitude of potential exposure, it is all the more startling to find less neuropsychological research on pesticide effects than for almost any other neurotoxic substance in common use. "In the United States, neurotoxicologic and neurobehavioral sequelae of pesticides have been recognized more by serendipity than by design" (Davies, 1990, p. 328). There are probably several reasons why pesticide-exposed individuals and neuropsychologists do not encounter each other more often, reasons that depend on both individual and sociocultural factors. For example, many exposed workers in the United States are migratory, or seasonal transient, workers. Medical facilities are not always provided; California is the only state where medical surveillance of organophosphate workers is mandated by law (Coye et al., 1986). Even if access to treatment were available, seasonal workers without union representation might justifiably fear job loss if physical illness is reported. In addition, poisoning episodes may not be recognized as such by itinerant agricultural employees not apprised of local pesticide spraying schedules or expectable toxic exposure symptoms. Fourth, few industrial employees of pesticide manufacturers have access to, or awareness concerning the value of, neuropsychological examination as part of a health screening. Finally, neuropsychologists, as a group, are not yet an active presence in industrial settings or occupational medicine clinics.

For whatever reason, the limited numbers of clinical and research personnel engaged in neuropsychologically related pesticide investigations cannot even begin to address the complexity and magnitude of the problem, the potential dangerousness of these materials, or the frequency of human toxic reactions. Estimated pesticide-related illnesses in the United States alone are between 150,000 and 500,000 per year (Coye, 1985; Koloyanova & El Batawi, 1991) and worldwide annual poisoning estimates range from 500,000 to 3,000,000 with mortality in excess of 1% (Ecobichon & Joy, 1982; Jeyaratnam, 1985). As many as 40,000 persons may die each year from pesticide poisoning, with 75% of lethal cases occurring in developing countries (Koloyanova & El Batawi, 1991). However, cases in the United States are also common, with an estimated 113–295 poisonings per 10,000 field workers among Florida citrus field workers (Koloyanova & El Batawi, 1991). The United States EPA estimates that there are 45,000 people poisoned and 200 people killed by pesticides annually (Ecobichon & Joy, 1994). Chronic neuropsychological impairments from acute poisonings are estimated to occur in 4–9% of exposure victims (Holmes & Gaon, 1956; Tabershaw & Cooper, 1966; Hirshberg & Lerman, 1984). It seems obvious from these statistics that adverse consequences of pesticide use are a common and serious danger. It can be hoped that growing interest in toxicology by neuropsychologists will also be reflected in increased attention to the cognitive and affective consequences of pesticide exposure.

ROUTES OF EXPOSURE AND INDIVIDUALS AT RISK

Most patients who present with pesticide poisonings are involved in formulation of the compounds, or in their agricultural application. However, individuals not connected with agricultural industries can also be at risk. Accidental exposure has been reported as a result of eating pesticide-contaminated, unwashed fruit (Ratner, Oren, & Vigder, 1983), drinking contaminated residential water (Dean et al., 1984), or through the misuse of home insecticide products (Reichert, Yauger, Rashad, & Klemmer, 1977). Deliberate pesticide ingestion as a suicide attempt has also been reported (Lerman, Hirshberg, & Shteger, 1984). Hospital personnel may be a high-risk group for pesticide exposure because of frequent and routine applications of insecticide in the hospital's physical plant. Hospital staff exposed to professional applications of insecticide have developed mental confusion, nausea, and other symptoms (Biskind & Mobbs, 1972).

Even office workers are not protected against accidental exposure. One study describes subjective symptoms and neurochemical alterations experienced by a group of five office workers who were exposed to chlorpyrifos (Dursban) and methylcarbamic acid (Bendiocarb 1%) dusted on the outside of the building to kill termites, but presumably carried into the office by an air-intake vent (Hodgson, Block, & Parkinson, 1986). It is not known how many of such outbreaks go unreported, since office and professional staff may be unaware of routine insecticide applications and hence may attribute pesticide-induced symptoms to other factors (Biskind & Mobbs, 1972).

ORGANOPHOSPHATES

Malathion
   Conversion factor   1 ppm = 13.73 mg/m^3
   IDLH   500 mg/m^3
   OSHA/NIOSH   10 mg/m^3

Parathion
   Conversion factor   1 ppm = 12.11 mg/m^3
   IDLH   20 mg/m^3
   OSHA   0.1 mg/m^3
   NIOSH   0.05 mg/m^3
   (*Quick Guide*, 1993)

There has been more research on organophosphate (OP) compounds than any other pesticides, perhaps because they were originally developed for their neurotoxic properties as "nerve gases" during World War II. *Soman, Sarin, and Tabun* are OP products, differing from their insecticidal counterparts mainly in potency and function (Duffy & Burchfiel, 1980). An OP compound, tri-*o*-cresyl phosphate (TOCP), was responsible for the highly publicized late 1920s and early 1930s epidemic of severe peripheral neuropathy called "Ginger Jake" paralysis, a syndrome that was eventually linked with TOCP adulteration of the "Jamaica Ginger" beverage.

TABLE 7.1. Active Ingredients in Organophosphate Compounds[a]

acephate	edifenphos	oxydemeton methyl
Akton	endothion	parathion
Aspon	EPBP	parathion methyl
azinophosmethyl	EPN	phencapton
bensophos (phosalone)	ethion	phenthoate
bensulide	ethoprop	phorate
Bomyl	ethyl parathion	phorazetim
bromophos	famphur	phosalone
carbophenthion	fenamiphos	phosfolan
chlorfenvinphos	fenitrothion	phosmet
chlormephos	fensulfothion	phosphamide (dimethoate)
chlorphoxim	fenthion	phosphamidon
chlorpyrifos	fonophos	phoxim
chlorthiophos	formothion	pirimiphos-ethyl
Coumaphos	fosthietan	pirimiphos-methyl
crotoxyphos	IBP	profenfos
crufomate	iodofenfos (jodenfos)	propaphos
cyanofenphos	isofenphos	propetamphos
cyanophos	isofluorphage (DFP)	prothoate
DDVP (dichlorvos)	isoxathion	pyrazophos
DEF	jodfenfos	pyridaphenthion
demeton	leptophos	pyrophosphate
demeton methyl	malathion	quinalphos
demeton-*O*-methyl sulfoxide	mephosfolan	ronnel
dialifor (dialifos)	Merphos	schradan
dialifos	methamidophos	stirofos
Diazinon	methidathion	sulfotepp
dicapthon	methyl demeton (demeton-methyl)	sulprofos
dichlofenthion		temephos
dichlorvos	methyl parathion	terbufos
dicrotophos	methyl systox	tetrachlorvinphos
diisopropyl fluorophosphate (DFP)	Methyl Trithion	tetraethylpyrophosphate
	mevinphos	thiometon
dimefox	mipafox	timet (phorate)
dimephenthoate (phenthoate)	monocrotophos	triazophos
dimethoate	naled	trichlorfon
dioxathion	nephocarp (carbophenothion)	trichloronate
disulfoton		

[a]Hallenbeck and Cunningham-Burns (1985).

OP compounds are more acutely toxic than other types of pesticides (Stopford, 1985). Parathion is the most toxic with a lethal dose as low as 100 mg/kg, while Malathion (LD of 858 mg/kg) is one of the least toxic OP products (Namba, 1971). Most incidents of significant OP poisoning have involved exposure to parathion or methylparathion. OP poisoning has been less frequent in the United States than in Egypt, Mexico, and India, where mass epidemics have been reported (Goetz, 1985). Suicidal individuals in the latter country account for many acute OP poisonings, forming 67.4% of a cohort of 190 patients examined in India by Agarwal (1993).

In the United States, OP and carbamate exposures are responsible for most acute occupational poisonings and 20–40% of OP applicators have reduced cholinesterase levels (a marker of OP toxicity) (Lang, 1993).

OP insecticides may be absorbed from "all possible routes" including skin, lungs, gastrointestinal tract, and conjunctiva (Namba, 1971, p. 290). Three types of neurotoxic reactions have been catalogued thus far from OP pesticides, with "cholinergic illnesses [being] the most frequent and classical undesirable manifestations of organophosphate ... exposures" (Davies, 1990, p. 328). Production of muscarinic, nicotinic, and CNS effects reflects "the inhibition of acetyl cholinesterase at the parasympathetic, sympathetic and central nervous system receptor sites" (Davies, 1990, p. 328). Inhibition of acetylcholinesterase may be irreversible (Arian, cited in Gershon & Shaw, 1961). Lesion or change in CNS acetylcholine receptors may occur (Duffy, Burchfiel, Bartels, Gaon, & Sim, 1979), through "accumulation of acetylcholine at neuroeffector junctions and autonomic ganglia" (Morgan, 1982, p. 12). This has been hypothesized to be the major cause of the toxicological effects of OPs (Levin & Rodnitzky, 1976).

A second neurotoxic effect of progressive and irreversible peripheral sensorimotor neuropathy may follow exposure to certain OP pesticides. The effect occurs from 2 to 5 weeks after a single exposure, and has been tied to indirect phosphorylation and inhibition of an enzyme called "neurotoxic esterase" (Davies, 1990).

A third neurotoxic effect with an "intermediate" delay of 24–96 h has been recorded by Senanayake and Karalliedde (1987), which developed between 24 and 96 h postexposure to chlorpyrifos, trichlorfon, trichloronate, and methamidophos. Four of ten patients died of respiratory paralysis. Patients experienced cranial nerve, limb, and neck weakness that was not responsive to atropine or oxime.

Presumptive diagnosis of OP poisoning can be made when serum cholinesterase activity is reduced by 10–50% in the context of OP exposure (Goetz, 1985). Mild behavioral symptoms can include fatigue, headache, dizziness, and abdominal complaints. More severe symptoms include miosis and muscle fasciculations, although these signs may not be present "even in severe poisoning" (Namba, 1971, p. 295). Extreme weakness or paralysis may result from acute exposure (Stopford, 1985). Bulbar signs may be seen, including difficulty speaking and swallowing, and shortness of breath (Goetz, 1985). Other characteristic physical symptoms include sweating, emesis, and excessive tearing (Namba,

1971). Finally, a "garlic" odor on the patient's breath may be indicative of OP poisoning.

Neurological Effects

Neurological effects of OP poisoning can be divided into effects that are visible on casual physical examination, and symptoms that are detectable on laboratory or neuropsychological evaluation. Clinically observable symptom patterns were reported by Whorton and Obrinsky (1983), who examined or reviewed the records of 19 farm workers who entered a field too soon after application of phosphamidon (Dimecron) and mevinphos (Phosdrin), two highly toxic OPs.

Initial complaints included weakness, blurring of vision, and nausea. Visual complaints of eye discomfort in reading or watching television were the longest-lasting symptoms, persisting at least 5 months postexposure. One month postexposure, 11 of the subjects continued to report headaches and weakness. Anxiety was reported by approximately one-third of the subjects, but these symptoms might have been reactive to fear of income loss or future medical disturbance rather than a function of neurotransmitter abnormalities. Three children who were part of this group exhibited the same pattern of symptoms as exposed adults, including weakness, fatigue, and inability to keep up with peers in play activity or sports. Symptom constellation in children was said to resolve within 23 months (Whorton & Obrinsky, 1983).

A syndrome of delayed polyneuropathy, typically present at 8 to 14 days postexposure, is also seen, characterized by numbness and tingling in the extremities, weakness and foot drop in the lower limbs, balance problems, and reduction or absence of knee and ankle jerk (Otto, Molhave, Rose, Hudness, & House, 1990).

Subclinical effects "can be found with great regularity in exposed subjects, notwithstanding the absence of any clinically apparent signs and in the face of normal cholinesterase levels in these same individuals" (Rodnitzky, 1973, p. 165). Subclinical neurological symptoms of OP exposure include decreased sensory nerve conduction velocities, and increased fiber density (Stalberg, Hilton-Brown, Kolmodin-Hedman, Holmstedt, & Augustinsson, 1978). Lasting neurological effects may be produced. One study followed 77 industrial workers exposed to Sarin. Although the workers were clinically asymptomatic, and did not have reductions in cholinesterase activity during the year prior to the investigation, their EEGs were significantly different from controls from the same plant. Exposed subjects displayed increased high-frequency beta activity (12–30 Hz) on spectral analysis and visual inspection of the EEG record suggested increased amounts of slow activity, nonspecific abnormalities, and increased amounts of REM in the sleep EEG (Duffy et al., 1979; Duffy & Burchfiel, 1980). EEG in OP-exposed workers is characterized by "high incidence of low to medium voltage slow activity in the theta range, that is, 4 to 6 Hz activity . . . during light drowsiness in brief episodes of 2–4 seconds duration" (Kaloyanova & El Batawi, 1991, p. 18).

Recent investigations have also located small, but statistically significant, signs of subclinical neuropathy in otherwise asymptomatic individuals exposed to OP compounds, although peripheral and polyneuropathies reported following OP exposure have also been suggested to be unrelated to cholinesterase inhibition (Namba, 1971). For example, Kaplan, Kessler, Rosenberg, Pack, and Schaumburg (1993) report eight patients who developed clinical and electrophysiological symptoms of sensory neuropathy after exposure to chlorpyrifos (Dursban) applied by an exterminator.

Chronic visual impairments can result from acute OP intoxication and may include "visual field stenosis, progressive myopia, astigmatism, edema and atrophy of the optic nerve" and other visual changes that can be attributed to anticholinesterase effects (Kaloyanova & El Batawi, 1991, p. 19).

Neuropsychological Effects

Neuropsychological symptoms of OP poisoning include a variety of cognitive and affective symptoms including "impaired vigilance and reduced concentration, slowing of information processing and psychomotor speed, memory deficit, linguistic disturbance, depression, anxiety, and irritability" (Ecobichon & Joy, 1982, p. 171). Severe neuropsychological impairment and depression have been noted in the absence of CT or other laboratory abnormalities (Rosenstock, Daniell, Barnhart, Schwartz, & Demers, 1990). One study that injected volunteers with an OP found slowed responses to motor and intellectual tasks, inability to sustain attention and concentration, as well as poorer learning and memory. One subject described his experiences:

> Several things are strange. Anything I think goes from big to small ... My arms feel strange to touch ... I can't seem to keep my mind on what I'm trying to think about ... I just go blank ... What were we talking about? ... I just kind of forget everything. [Bowers, Goodman, & Sim, 1964, p. 385]

Rosenstock *et al.* (1990) found a mixed range of physical and psychological symptoms in a 60-year-old farm worker with two episodes of acute OP intoxication separated by 5 years. Symptoms 2 years after the latter exposure included headaches, memory loss, confusion, and fatigue. While subjects' descriptions are consistent with other clinical reports of pesticide intoxication, the authors regrettably did not include a placebo condition, or specify the OP dosage level.

Savage *et al.* (1980, 1988) employed neuropsychological testing on OP-exposed workers, using the Halstead–Reitan Battery. Results corroborated with 24% of the exposed individuals performing in the impaired range, twice the number of impaired controls. The authors reported main effects of pesticide exposure on 34 neuropsychological subtest scores. Finally, subjects showed deficits in coordination and fine motor speed after acute OP pesticide poisoning.

Visual retention and constructional deficits have also been reported (Jusic, 1974) and memory dysfunctions are frequently cited.

Metcalf and Holmes (1969) found impaired performance on the WAIS, the Benton Visual Retention Test, and a story recall task in a group of agricultural and industrial workers chronically exposed to pesticides. "Mental confusion" is also a frequently reported finding as a consequence of intoxication, and has apparently resulted in aerial accidents by crop sprayers who became unable to fly (e.g., Reich & Berner, 1968). Most of these findings, however, have not been validated with appropriate neuropsychological tests (Levin & Rodnitzky, 1976).

Korsak and Sato (1977) administered part of the Halstead–Reitan Battery to 59 male volunteers who had varying (unspecified) degrees of exposure to OP and other pesticides. High-OP exposure subjects performed significantly more poorly on Bender–Gestalt and Trails B compared with low-OP subjects. Unfortunately, the authors did not include information about subject matching procedures, if any—leaving the results potentially subject to educational or other confound.

Otto *et al.* (1990) investigated the neurological and neuropsychological status of male OP workers in Egypt. Testing was conducted in three plants where combined OP production included 31,800 metric tons of various OP pesticides produced from 1961 to 1986. Pesticide and fertilizer workers had more neurological impairment than textile workers, including "quite dramatic" impairments in coordination, involuntary tremor, and knee jerk (p. 312). Vibration sensitivity on the Opticon was the most sensitive index of pesticide exposure. Vibration sensitivity decreased more rapidly with age in the exposed workers than among controls. Two neuropsychological tests (Block Design and Santa Ana) did not vary significantly with LNTE (lymphocyte neuropathy target esterase) or cholinesterase inhibition.

Finally, Daniell *et al.* (1992) report a failure to find pre/postseason differences in neuropsychological performance on the NES in a group of pesticide applicators. However, the conclusions of the study are seriously undermined by the control group selected: 68% of controls had worked picking or trimming crops with 27% having worked with pesticides.

Individuals exposed to OP pesticides after home extermination treatments may also be at risk for neuropsychological impairments. Dursban (chlorpyrifos) proved to be the active pesticide in several cases reported by Kaplan *et al.* (1993) where both central and peripheral neurotoxic effects were reported. One family developed headaches, nausea, painful muscle cramps, numbness, and paresthesias (more prominent in the legs), along with school decline in two teenage children. Sural nerve conduction studies showed low amplitudes which reverted to normal in 6 months. A second patient was a physician who developed paresthesias, memory impairment, and abnormal neuropsychological functioning after Dursban was applied to her basement exercise room. Neuropsychological dysfunction was characterized as a "persistent dysfunction in memory, word-finding, and visual perception" (p. 2196). In all, five of the eight patients reported by Kaplan *et al.* were found to have deficits in word-finding, concentration, and memory, with symptoms reversible in four patients. The authors suggest that effects may persist even after cholinesterase levels return to normal.

Mearns, Dunn, and Lees-Haley (1994) summarize the neuropsychological effects of OP pesticides as follows: short-term effects include memory and concentration deficits, while individuals who have suffered clear poisoning from OPs may have problems that last for a decade or more.

Effects on Personality and Emotional Functioning

Many changes in personality variables have also been reported as a function of exposure to OP and other pesticides. Although early case studies describing "schizophrenic" and depressive reactions to OP poisoning (e.g., Gershon & Shaw, 1961) have been disputed for being impressionistic and nonquantifiable (e.g., Barnes, 1961; Bidstrup, 1961), recent reports continue to note increased tension, restlessness, anxiety, and apprehension as a function of exposure to OP and other pesticides (Russell, 1983; Ecobichon & Joy, 1982; Levin, Rodnitzky, & Mick, 1976; Levin, 1974; Namba, 1971). Dille and Smith (1964) reported the case of two crop duster pilots with chronic OP exposure; both had symptoms of depression and anxiety. A larger group of 16 workers were followed for 18 weeks after an acute OP exposure. Anxiety complaints were present in 44% of this group.

Emotional effects of OP exposure are consistent with what is known about the cholinesterase-inhibiting properties of pesticides, since percutaneous injections of anticholinesterase in human volunteers produced initial feelings of fatigue, followed by subjective feelings of "jitteriness or tenseness inside." Subjects became depressed, irritable and "listless" (Bowers et al., 1964, p. 384). Anxiety and irritability may occur with long-term exposure to low-dose OP exposure (Rosenthal & Cameron, 1991).

Levin et al. (1976) investigated the possibility of subclinical personality disturbance in 24 individuals exposed to OPs in agricultural or commercial settings. Controls included a mixed group of farmers tested out of spraying season, or those who did not participate in insecticide application. Controls were matched with experimental subjects for age and education. Compared with controls, pesticide-exposed subjects exhibited significantly higher anxiety on the Taylor Manifest Anxiety Scale, an effect determined by the scores of the 13 commercial pesticide applicators, rather than the farmers. Commercial pesticide workers also had lower plasma cholinesterase than controls. Other measures, e.g., the Beck Depression Inventory, did not show differential effects of pesticide exposure or type of employment.

It is possible, of course, that depressed plasma cholinesterase is simply coincidental with higher stress levels inherent to the job of commercial pesticide applicator rather than a direct result of pesticide exposure. The authors rightly suggest replication of their study using larger and more diverse subject populations.

Emotional symptoms may be better correlated with plasma cholinesterase levels than blood cholinesterase, though symptoms may persist beyond recovery of blood and plasma cholinesterase levels, possibly for 6 months or more following exposure (Levin *et al.*, 1976).

DURSBAN (CHLORPYRIFOS)

Dursban is an organophosphate that had been generally considered safe for use in home and industry. Recently, however, several reports indicate significant neurotoxicity and neuropsychological impairment developing from this "benign," commercially applied pesticide. Kaplan, Kessler, Rosenberg, Pack and Schaumburg (1993) detail case reports on 8 individuals who developed peripheral neuropathy, with or without cognitive impairments. Neuropathy symptoms included sensory loss of all modalities in a stocking-glove distribution with mild distal weakness and areflexia in the lower limbs. Cognitive impairments included mild short-term memory loss on routine mental status examination, a decline in school performance in 2 children that lasted 6 months, and complaints of slowed thinking. Neuropsychological examination of a physician with the latter complaint suggested impairments in an embedded figures test, impaired picture completion, computerized abstract form discrimination, and low-average prose memory and serial digit recall. The authors cite that, of the 5 individuals with concentration, word-finding, and memory deficit, impairments were reversible in 4. One patient showed persistent impairments. The authors suggest reassessing the assumed safety of this pesticide.

In another set of case studies, individuals exposed to Dursban expressed multiple organ complaints, including an "initial flu-like illness followed by chronic complaints of fatigue; central nervous system problems (headaches, dizziness, loss of memory); upper and lower respiratory symptoms; joint and muscle pain; and gastrointestinal disturbances" (Thrasher, Madison, & Broughton, pp. 90–91). All autoantibodies, including ANA, were elevated compared with controls. Approximately half were positive for antimyelin antibodies, twice the number of controls. Exposure also appeared to cause new allergies and worsen preexisting ones (Thrasher, Madison, & Broughton, 1993). Finally, Sherman (1995) reviewed the symptoms of 41 Dursban- and Dursban-with-other-pesticide-exposed individuals. Seventy-eight percent developed headache and central or peripheral nervous system symptoms. Central nervous system problems of memory loss, confusion, sleep disturbance, weakness, and fatigue were reported. Among those who developed peripheral neuropathy, 50% were unable to continue working.

Case Study: Dursban Exposure

The patient is a high school-educated female in her 40s who was exposed when her wooden home was treated with several pints of Dursban, applied by hand. Initial symptoms resembled flu, but progressed to "pins and needles" sensations in all limbs, dizziness, extremely elevated ANA titer, and deep muscle pain. The patient complained of attention and memory-related problems and

TABLE 7.2. Dursban-Exposed Patient's Scores on Tests of Neuropsychological Functions

Test	Raw score	T score (Heaton) or scaled score
Category Test	79 errors	t 36
Trails A	39"	t 41
Trails B	87	t 44
TPT Memory	6	t 37
TPT Localization	4	t 50
Finger Tapping Dom. Hand	39.75	t 42
Finger Tapping N.D.	40.13	t 51
Grooved Pegboard DH	61"	t 53
Grooved Pegboard NDH	68"	t 49
TPT DH	.5 min/block	t 52
TPT NDH	.975 min/block	t 36
TPT Both	.508 min/block	t 41
Fingertip Number Writing	DH 8 errors, NDH 1 error	
Gordon Diagnostic System III	Vigilance 27/30 correct	Distractability 0/30 correct
Stroop Color Word Test	Words t 35, Colors t 43, Color–words t 37	
KBIT IQ Estimate	Verbal (vocab) 98, Matrices (performance) 105, Composite (full scale) 102	
COWAT	28	30%ile
WCST	Perservative responses 29	9th %ile t 37
Hooper VOT	27.5	Normal
Visual Search/Attention (VSAT) (Cancelation Test)		Impaired
Multi Digit Memory Test (Malingering)	64/72 correct	Borderline
Memory Assessment Scales	Impaired visual span, immediate and delayed prose recall, and delayed/visual recognition	
Observations	Ataxic irregularity to arm movement	
Neurological	MRI atrophy: positive jaw jerk	

was unable to continue working. Initial administration of steroids failed to produce improvement and evaluations for lupus and multiple sclerosis proved negative. The patient was seen for neuropsychological evaluation approximately one year postexposure with persistent complaints in aforementioned areas. Table 7.2 shows the patient's scores on various tests of neuropsychological functions.

Intellectual, Language, and General Cognitive Functions. On a test that estimates IQ from picture vocabulary, word identification, and nonverbal analogical reasoning, the patient scored in the average range for verbal, nonverbal, and full-scale abilities. Verbal and nonverbal abilities did not differ significantly. The patient's present IQ estimate is consistent with her level of education and occupational attainments.

Attentional skills ranged from borderline to severely impaired, depending upon task. On a test of visual scanning, sequencing, visuomotor speed, and attention—which required the patient to sequentially connect with a pencil line 25 randomly-spaced numbers on a page—performance was at the border between low average and mild clinical impairment. Similar performance was seen in the more complex version of this same test, which requires rapid drawing of lines between successive alternating numbers and letters randomly spaced on a page (e.g., A-1-B-2-C-3, etc.).

The patient showed slightly more impaired performance on a test of attention under continuous distraction, where she was required to read a list of color names printed in a contrasting color and say the name of the colored ink, rather than the word (e.g., the word "RED" printed in blue ink—patient says "blue").

The patient was severely impaired on a test of rapid visual response and decision making, that, unlike the others, was not self-paced. This "continuous performance" test (GDS) required visual identification and motoric response to predefined sequences of numbers (e.g., 1-9) rapidly displayed on a computer panel. Performance remained severely impaired in the more difficult variation of this test, which measures visual attention under highly distracting conditions, i.e., requiring visual scanning of number sequences while distracting numbers flash on either side (GDS). In the latter condition, the patient was simply unable to perform the task at all, stating that she did not see any "1-9" number sequences.

Tests of executive functioning and complex processing were in the mild range of clinical impairment. For example, the patient was mildly impaired on a test of sustained inductive problem solving that is thought to be a measure of both overall cortical integrity and prefrontal cortical function (Category Test). In addition, she displayed mild perseverative impairments on a similar test that measures more highly localized prefrontal lobe capacity to inductively problem solve and shift sets (WCST).

A test of verbal fluency that typically activates the left frontal lobe was performed in the low average range.

Sensorimotor Skills. On a finger tapping test, the patient produced low average fine motor speed in the dominant hand and average speed for the nondominant hand. She did not show the expected speed preference for the dominant hand, and, in fact, was slightly slower in the dominant hand than in the nondominant hand. Fine motor coordination on a pegboard test was normal bilaterally.

Tactile sensitivity was abnormal, with the patient making 8 errors in detecting numbers traced on her fingertips of her dominant hand but only 1 in her nondominant hand. On cerebellar testing, and in any task requiring rapid motor movement, the patient showed an unusual, jerky "ataxic" movement of each arm when attempting to negotiate rapid finger-to-nose movements. She also showed considerable impairment attempting to rapidly touch her thumb to each successive finger.

Basic object identification and visual acuity appeared normal, but on gross testing of visual field, the patient appeared to display a consistent slight lower right (from the examiner's perspective) bilateral quandrantanopsia.

Memory and Learning. The patient's overall level of memory functioning on an extensive test of various memory domains was lower than estimated IQ when these scores are usually quite similar. Areas of significant clinical deficit included short term memory for spatial locations, ability to comprehend and retain prose, and the ability to recognize and discriminate previously seen designs from distractors. Memory for tactually perceived shapes was also mildly impaired.

A test of recognition memory that is sensitive to malingered memory deficit (MDMT) was performed well without any evidence of deliberate exaggeration.

Visuospatial Skills. On a test requiring the patient to recognize common objects presented as "exploded" puzzle-piece line drawings, the patient's visual identification was normal and unimpaired. Spatial perception of shapes perceived through tactile stimulation alone was normal for the dominant hand, but mildly impaired in the nondominant hand.

The patient denied significant depression on self-report inventories and did not appear to be influenced by emotional difficulties during testing.

Summary and Conclusions. This is a female who was referred to evaluate the nature, extent, and possible etiology of reported cognitive impairments. At present, the patient demonstrates a variety of neuropsychological impairments, most severely in rapid visual attention under distracting conditions. Visual memory, prose memory, and nonverbal inductive problem solving were also significantly impaired. A test of memory malingering demonstrated no evidence of deliberate exaggeration of memory impairment. In addition, the patient displayed an unusual, almost ataxic, movement to her arm when performing finger-to-nose testing or any motor task that required arm movement. Informal visual field testing suggested the possibility of a bilateral right lower quadrant field cut. Formal visual field testing to corroborate this finding is indicated.

In general, the patient's profile indicates diffuse, patchy impairments of cognitive and subcortical and possibly cerebellar function somewhat more localized to the right hemisphere. The patient's prior medical evaluations have apparently not found an adequate explanation for the patient's symptom complaints.

The patient's symptom development, present complaints and complaints, and neuropsychological impairments are consistent with case study descriptions of Dursban-poisoned patients, showing a constellation of neuropsychological and an immune system dysfunction. There is no premorbid history of psychiatric dysfunction, substance abuse, or somatization history to explain present findings and obvious medical alternatives have been ruled out. Continuing neurological workup is recommended.

CARBAMATES

Carbamates are synthetic pesticides, herbicides, and fungicides that have achieved large distribution within the past 40 years. Insecticides include: adoxycarb, allyxycarb, aminocarb, bendiocarb, BPMC, bufencarb, butacarb, carbanolate, carbaryl, carbofuran, cloethocarb, dimethalan, dioxacarb, ethiofencarb, formethanate, hoppcide, isoprocarb, trimethacarb, metolcarb, mexacarbate, pirimicarb, promacyl, promecarb, propoxur, XMC, xylycarb, aldicarb, methomyl, oxamyl, thiofonax, and thiodicarb. Herbicides include: asulam, chlorbufam, desmedipham, phenmedipham. Fungicides include: benomyl, carbendazim, thiophanate-methyl, thiophanate-ethyl.

Carbamates are absorbed easily through the skin and symptoms of intoxication are observable within several minutes after absorption. Poisoning effects are produced by acetylcholine accumulation at the nerve endings since carbamates, like OP pesticides, inactivate acetylcholinesterase. Acute intoxication effects include constricted pupils, salivation, epigastric pain, sweating, lassitude, vomiting, and in severe cases, seizures, unconsciousness, cardiac arrythmia, hypertension, and death. Depressed cholinesterase activity in the blood may be a marker of recent intoxication. Acute symptoms include lightheadedness, blurred vision, salivation, weakness, and muscle fasciculations (Stopford, 1985). Although there have been no reports of formal neuropsychological examination performed on carbamate-poisoned individuals, exposed workers are highly symptomatic. Symptoms displayed by workers in a methomyl production factory include headache, blurred vision, giddiness, cramps, diarrhea, chest discomfort, and weakness (Morse & Baker, 1979).

Ecobichon and Joy (1982) described a case of carbamate exposure in a 55-year-old farmer who hand-sprayed a vegetable garden with carbaryl. Acute symptoms included severe vertigo, visual impairments, paresthesias, fatigue, and memory loss. Photophobia, mild paresthesia and memory loss continued to be experienced 1 year after the exposure, with the patient reportedly needing to continually compensate for memory impairment with written reminders. Personality changes may also occur, with one case report noting uncontrolled aggressive behavior in both a man and his pet cat, to which the subject applied a commercial tick powder composed of 5% carbaryl (Bear, Rosenbaum, & Norman, 1986).

CHLORINATED HYDROCARBONS

Chlorinated hydrocarbons or organochlorines are a structurally diverse group of pesticides that include cyclodienes, halogenated aromatics, and others. The best known organochlorine pesticide is DDT, a pesticide that is very persistent in the ecosystem, and whose use has been banned for that reason in many

TABLE 7.3. OCP Values[a]

Organochlorine pesticide	Highest allowable air level (mg/m^3) (USA)	Safe whole blood levels
Aldrin	0.25	0.1 μg/ml
DDT	1.0	0.2 μg/ml
Dieldrin	0.25	0.1 μg/ml
Lindane	0.5	0.02 μg/l
Endrin	0.1	
Heptachlor	0.5	
Chlordane	0.5	

[a]From Kaloyanova and El Batawi (1991).

TABLE 7.4. Organochlorine Pesticides[a]

Name	Trade name
Endrin	Hexadrin
Aldrin	Aldrite, Drinox
Endosulfan	Thiodox
Dieldrin	Dieldrite
Toxaphene	Toxakil, Strobane-T
Lindane	Isotox, gammaBHC or HCH
Hexachlorocyclohexane	BHC
DDT	Chlorophenothane
Heptachlor	Heptagran
Chlordecone	Kepone
Terpene polychlorinates	Strobane
Chlordane	Chlordan
Dicofol	Kelthane
Mirex	Dechlorane
Methoxychlor	Marlate
Dienochlor	Pentac
TDE	DDD, Rhothane
Ethylan	Perthane

[a]Morgan (1989).

TABLE 7.5. Active Ingredients of Organochlorine Pesticides[a]

aldrin	DDT	isobenzan
benzene hexachloride	dicofol	lindane
γ-benzene hexachloride (lindane)	dicophane (DDT)	methoxychlor
	dieldrin	methoxy-DDT (methoxychlor)
BHC (benzene hexachloride)	dienochlor	mirex
γ-BHC (lindane)	endosulfan	octachlorocamphene (toxaphene)
camphechlor (toxaphene)	endrin	
chlordane	ethylan	polychlorinated camphene (toxaphene)
chlordecone	HCH (benzene hexachloride)	
chlorinated camphene (toxaphene)	γ-HCH (lindane)	TDE
chlorobenzilate	heptachlor	tetrachlorodiphenylethane (TDE)
	hexachlorocylohexane (benzene hexachloride)	
chlorophenothane (DDT)		toxaphene
chloropropylate	γ-hexachlorocyclohexane (lindane)	
DDD (TDE)		

[a]Hallenbeck and Cunningham-Burns (1985).

developed countries. Because organochlorines accumulate, chronic intoxications are more likely than acute poisonings. Acute intoxication is most likely with aldrin, endrin, dieldrin, and toxaphene, with acute gastrointestinal symptoms, headache, dizziness, ataxia, paresthesias, tremor, and seizures. Liver and kidney damage may result (Kaloyanova & El Batawi, 1991).

Chronic exposure to organochlorine pesticides may show characteristic damage to metabolic and neuroendocrine mechanisms, producing a syndrome resembling Itzenko–Cushing syndrome (Kaloyanova & El Batawi, 1991). Abnormal EEGs were found in 21.9% of workers chronically exposed to organochlorine pesticides from 1 to 2 years, with bitemporal spikes, low voltage, and diffuse theta (Mayersdorf & Israeli, 1974). Chronic neurological effects of exposure include polyneuritis, ataxia, nystagmus, disturbance in olfactory sensitivity, headaches, giddiness, and gastric pathology (Kaloyanova & El Batawi, 1991).

DDT

OSHA	1 mg/m^3 (skin)
NIOSH	0.5 mg/m^3

DDT (dichlorodiphenyltrichloroethane) is a chlorinated hydrocarbon that, until recently, was one of the most widely used pesticides. It was the first synthetic insecticide to achieve widespread use, and was extensively employed during the 1940s and 1950s (Stopford, 1985). In recent years, however, the use of DDT has tapered off in developing countries because it is very stable in the environment and increases in concentration along the food chain. Like all organochlorine pesticides, DDT is highly lipid-soluble and accumulates in fatty tissues. It is also a potential carcinogen. DDT is still sold in developing countries where politics, insect predation, or disease risks outweight environmental concern.

DDT's neurotoxic action is to interfere with nerve impulse transmission by blocking normal repolarization of the axon after impulse transmission. The result is "a more or less continuous train of impulses along the fiber following a simple impulse" leading to "severe disruption of nervous system function" (Morgan, 1982, p. 7).

DDT changes electrical activity in the brain, including "the cerebellum, cortex, limbic system and various subcortical structures" (Ecobichon & Joy, 1982, p. 107). Acute ingestion of DDT causes nystagmus, hyperesthesias, and motor impairment, in dose-dependent severity. In mild cases most symptoms disappear within several days, although severe, possibly allergic reactions have also been reported. Doses as low as 6 mg/kg per day are associated with moderate poisoning. Two common symptoms of such poisoning are apprehensive excitement and persisting weakness in hands and feet. Chronic DDT exposure cases are rare and neuropsychological examinations of exposure cases have not entered the literature. Polyneuropathies have been reported in individual cases, though DDT's role has been disputed.

There are little data on neuropsychological impairment related to DDT. Misra, Nag, and Krishna Murti (1984) found that Bender–Gestalt test performance was correlated with DDT serum levels in DDT sprayers, but IQ and memory functioning on the Wechsler Memory Scale were not affected. EEG abnormalities were found in 55% of the sprayers. Four of the sprayers showed no alpha activity.

CHLORINATED CYCLODIENES

Chlorinated cyclodienes, including chlordane, heptachlor, aldrin, dieldrin, endrin, and endosulfan, are among the highest toxicity organochlorine pesticides. They cause convulsions as a primary manifestation of exposure (Stopford, 1985). Other neurological effects are similar to those found for pesticides as a group, including headache, dizziness, fasciculation, EEG abnormalities, hyporeflexia, loss of coordination, cerebral edema, and toxic hepatitis.

Chlordane

Conversion factor	1 ppm = 17.04 mg/m^3
IDLH	500 mg/m^3
OSHA/NIOSH	0.5 mg/m^3 (skin)

In use for about 40 years, approximately 200 million pounds of chlordane were applied in the United States, before it was commercially banned on April 14, 1988. From 1983 to 1988, use of chlordane had been restricted to underground termite control. Prior to that time, however, chlordane had been widely used as an agricultural pesticide. Because chlordane persists in soil for over 20 years, and in treated homes for up to 15 years, exposure risk continues, despite the banning of the substance.

Residing in a chlordane-treated home is one of the most common routes of chlordane exposure. The EPA estimates that about 19.5 million homes have been treated with chlordane, causing exposure to 52 million people (ATSDR, 1989). Because chlordane degrades very slowly in the environment, elevated air concentrations of chlordane in the home may persist for as long as 15 years after application (Livingston & Jones, 1981).

Other nonworkplace exposures may occur via contact with chlordane-treated soil, or through contaminated foodstuffs.

Neurotoxic effects of chlordane exposure have been documented through inhalation, dermal, and oral routes. Chlordane has been demonstrated to be neurotoxic at acute high dose or long-term, lower-level exposure. After acute ingestion, chlordane migrates to body fat stores and within 48 h, over 300 times more chlordane resides in fat than is excreted in urine (Curley & Garrettson, 1969). Humans may be more sensitive to this pesticide than animals, as the estimated human lethal dose of 25 mg/kg is about 10 times lower than values reported for animals (ATSDR, 1989).

Neurological Effects

Chlordane and its metabolites are present in many human tissues, including brain tissue. Neurological abnormalities have been reported as a consequence of chlordane exposure from oral, dermal, or inhaled exposures. The most sensitive indicants of acute chlordane intoxication are symptom complaints, including headache, dizziness, irritability, muscle tremor, confusion, convulsion, and coma (ATSDR, 1989). Not enough information is presently available to predict dose–response relationships in humans, particularly when subtle neuropsychological effects are at issue.

Optic neuritis has been reported in human exposure, while animal literature does not appear to demonstrate neurological impairment.

Chlordane causes fetal neurotoxicity, either through *in utero* exposure or nursing, in mice. Symptoms include increased seizure threshold, delayed learning, and increased exploratory behavior (Al-Hachim & Al-Baker, 1973). There does not appear to be human data on chlordane developmental neurotoxicity.

Agent Orange and Dioxin

Dioxin exposure has a very long half-life in human tissue. In one case, a chemist who synthesized TCDD, one of the most toxic congeners of the dioxin family, continued to show elevated dioxin levels 35 years after exposure (Schecter & Ryan, 1992). Agent Orange, one of the chlorphenoxy herbicides, is a 1:1 mixture of 2,4,5-T and 2,4-D. Over 100 million pounds of this potent herbicide were applied by the U.S. Air Force as a defoliant in Vietnam from 1962 through 1970. Agent Orange has been linked to a variety of toxic effects, including chloracne, porphyria cutanea tarda, liver disorders, and immune system abnormalities. However, psychological and neuropsychological measures have thus far failed to validate neurotoxic effects on exposed Vietnam veterans. For example, Korgeski and Leon (1983) evaluated 100 veterans who participated in the Vietnam war. The authors used both neuropsychological and personality tests, including the Wechsler Logical Memory Subtest of the Wechsler Memory Scale with delayed recall, the Porteus Maze Test, the MMPI, a psychological problem self-report, and other (unspecified) tests of spatial memory, problem solving, and learning.

Korgeski and Leon (1983) analyzed their subjects' data twice; once to determine the relationship between subjects' *belief* of their exposure and test performance, and a second time to see whether Defense Department computer records of herbicide spraying correlated with subjects' performance. Veterans who believed they had been exposed reported significantly more subjective psychological difficulties, both on the MMPI and self-report inventory. Problems reported included depression, anxiety, rage attacks, and irritability. However, when subjects' proximity to actual sprayings were extrapolated from Defense Department records, all differences between exposed and unexposed individuals disappeared. No differences whatsoever were found in the veterans' neuropsychological performance, whether they were classified by objective or subjective criteria. Finally, while general medical problems were diagnosed equally in each group, skin problems were significantly associated only in the group that *believed* it was exposed. The authors suggested that the belief of exposure to Agent Orange may be more associated with psychological difficulties than actual exposure, although there is always the possibility that subject report is a more accurate index of exposure than Defense Department records.

Another study investigated dioxin exposure in a nonmilitary population and reported similar findings (Hoffman *et al.*, 1986). The authors examined a set of households exposed to 2,3,7,8-tetrachlorodibenzo-*p*-dioxin (the active ingredient in Agent Orange) when it was combined with waste oil and sprayed on residential, commercial, and recreational lands to control dust. Households in a mobile home park were the focus of study.

While depressed immune system reaction to dioxin was detected in the exposed group, no neuropsychological abnormalities were shown on Trails, Grip Strength, and reaction time. The exposed group did show significantly elevated tension/anxiety and anger/hostility scores on the Profile of Mood States (POMS), but the etiology of these symptoms is unclear.

One study that did find neuropsychological evidence of dioxin effects was conducted in southwest Germany by Peper, Klett, Frentzel-Beyme, and Heller (1993). Subjects were individuals living in the vicinity of a site contaminated with dioxins (PCDDs) and furans (PCDFs). Exposure was chronic with a mean duration of 21 years and the majority consumed vegetables grown in gardens contaminated with PCDDs/PCDFs (200–8000 ppt). Attic dust of private homes in the area contained 585,000 ppt. Subjects were grouped for analysis based on whether blood levels were above or below the group median. Several tests of verbal memory, as well as WAIS-R Similarities and Picture Completion were impaired in the above-median group. When age was partialed out, both dioxins and furans remained significantly correlated with Picture Completion, a recognition score, Stroop, Trails A, visual memory, and paired associates. The study does not mention whether individuals were aware of their median status prior to being tested and the number of individuals tested is small (19) so the study should be considered more of a pilot worthy of replication with a larger group.

Bolla (1994) suggests in a review of the neuropsychological effects of Agent Orange that, at the time of her review, there was insufficient evidence to link Agent Orange exposure with neuropsychological or neuropsychiatric abnormalities. However, the role of the Reagan administration in discouraging or distorting research in this area has recently been uncovered. Thus, legal and medical controversies continue unresolved, with a final verdict contingent on impartial, independent research.

Kepone

Chlordecone, trademarked as Kepone, is responsible for one of the most recent and severe outbreaks of pesticide poisoning. In 1975, workers at the Life Sciences Products Company in Hopewell, Virginia, showed an attack rate of up to 70%, since "workers walked regularly through a slurry of the pesticide without boots or protective clothing, ... a powdery residue ... soiled workers clothing [and] caked their skin ... [and] the few protective masks provided were rarely used" (Taylor, Selhorst, & Calabrese, 1980, p. 411). Exposed workers developed the "Kepone shakes" with tremor and ocular trembling. Irritability and memory loss were reported in 13 of 23 patients, although no tests or statistics are cited (Taylor *et al.*, 1980).

Telone

Telone (1,3-dichloropropene) is used as a pesticide and a nematocide, and is manufactured by Dow Chemical Company. Several neuropsychological and neuropsychiatric effects of Telone exposure have been described. One exposed individual developed personality changes in the course of several weeks, becoming anxious and fearful, and 4 years later developed weakness, inability to concentrate, and anhedonia (Peoples, Maddy, & Thomas, 1976). The same authors also recounted the history of 36 firemen who developed headaches, anxiety, and memory and concentration problems, as a probable result of exposure to a 1200-gallon spill of Telone on the roadway. Russell (1983) tested 11 California highway workers exposed to Telone and compared them with seven nonexposed workers on a screening battery consisting of the Wechsler Memory Scale, Trail-Making Tests, and the Minnesota Multiphasic Personality Inventory. Significant differences were found only for MMPI personality variables, with Depression, Maladjustment, and Manifest Anxiety scales elevated relative to controls.

OTHER PESTICIDES

Pyrethroids

Pyrethroids are synthetic derivatives of the natural flower extract pyrethrin. They show selective toxicity to the CNS by interference with ionic permeability of the cell membrane, thereby blocking production and conduction of nerve impulses. They also slow the closing of sodium channels. Neurotoxic effects may be fatal, and also include loss of coordination and paralysis. Sensory irritation similar to a paresthesia or dysesthesia is frequently reported (Kaloyanova & El Batawi, 1991). There do not appear to be any neuropsychological studies of this class of pesticide exposure in the literature.

Paraquat

Paraquat is a herbicide that gained recent public attention when it was chosen by the U.S. government to destroy clandestine marijuana crops. While there are no neuropsychological studies available, autopsy brain studies of two fatal paraquat poisonings suggest alterations in cell contents consistent with effects of aging, vitamin E deficiency, hypoxia, and other causes. Cerebrovascular changes and marked alteration of myelin were also found (Grcevic, Jadro-Santel, & Jukic, 1977). Investigations of paraquat continue since it was found to be chemically similar to an incorrectly manufactured "designer drug" linked to parkinsonian abnormalities (A bad drug's benefit, 1985).

FUNGICIDES AND FUMIGANTS

Chloropicrin

This fungicide and fumigant has produced kidney, liver injury in animals, as well as skeletal muscle necrosis with chronic exposure (Anger et al., 1986). Eye irritation, corneal injury, fatigue, vertigo, and headache have been found in humans (ACGIH, 1982; Goldman et al., 1987).

Methyl Bromide

IDLH   2000 ppm

Methyl bromide is one of the two major fumigants employed in the United States with annual use in the range of 35 million pounds (Alexeeff & Kilgore, 1983). Agricultural as well as industrial fumigators of commercial buildings are at risk for exposure, a risk that is intensified because methyl bromide is odorless and has no reliable warning properties. From 1953 to 1981, 60 deaths and 301 cases of poisoning were reported (Herzstein & Cullen, 1990).

In rats and mice, methyl bromide is highly neurotoxic, with rats showing necrosis in cerebral cortex, hippocampus, and thalamus, while mice displayed cerebellar damage. Female rats were less susceptible to methyl bromide than male rats (Eustis, Haber, Drew, & Yang, 1988).

Human exposure produces various neurologic abnormalities, including ataxia, reflex abnormality, visual impairments, muscle weakness, fatigue, frontal headaches, and tremulousness (Shield, Coleman, & Markesbery, 1977; Herzstein & Cullen, 1990). Depression, irritability, confusion, and personality changes have also been reported (Alexeeff & Kilgore, 1983). Mild exposure has been associated with chronic polyneuropathy while severe exposures have produced motor incoordination and memory impairment (Alexeeff & Kilgore, 1983; Greenberg, 1971). Loss of hand dexterity along with upper and lower extremity paresthesias were reported by two workers 3 weeks postexposure (Herzstein & Cullen, 1990). Bishop (1992) reports a case study of an individual exposed to the gas after greenhouse soil fumigation. The patient became confused, apractic, and lost both voice and hearing. He developed grand mal seizures and went into coma on hospital admission. Slight increases in the size of the patient's lateral ventricles and cortical sulci were noted over the course of the patient's hospitalization. Improvement was described as "slow but steady" over a 4-month period; however, the patient continued to show intention tremor and periods of aggression and depression. The patient was referred to a residential rehabilitation unit.

Few neuropsychological studies of methyl bromide are available. Anger *et al.* (1986) found that methyl bromide-exposed subjects tended to perform less well than controls. However, large between-group differences in ethnicity, number of medications consumed that day, alcohol use, and education make firm conclusions difficult.

Sulfuryl Fluoride

Seizure activity has been shown in animals and there is a single case study of a 30-year-old male exposed for 4 h to sulfuryl fluoride. The patient developed nausea with emesis, stomach cramping, pruritus, and pinprick paresthesia on the right leg (Taxay, 1966).

EXPOSURE TO COMBINATIONS OF PESTICIDES

There is very little literature to address the combined effects of neurotoxic pesticides. Reviewing what literature is available, Kaloyanova and El Batawi (1991) summarized combination exposure symptoms as including "asthenovegetative syndrome, vegetative polyneuritis, radiculitis and diencephalitis, vegetative vascular dystonia, and brain sclerosis" (p. 169).

HERBICIDES

Herbicides used to kill and control undesirable competitive plant species are generally more toxic to plants than to animals; however, many herbicides contain neurotoxic metals and compounds. Herbicide constituents may include arsenic compounds, chlorate salts, and other highly toxic substances. The number of reported U.S. herbicide poisonings is steadily increasing, from 1054 in 1983 to 5531 in 1989 (Smith & Oehme, 1991). While most herbicide fatalities are intentional (suicide) there is little data on chronic non-fatal human exposures from occupational use or groundwater contamination.

8

Other Neurotoxins

ACRYLAMIDE

Conversion factor	Not given
IDLH	Unknown
OSHA/NIOSH	0.03 mg/m^3
(*Quick Guide*, 1993)	

Acrylamide is a neurotoxic vinyl monomer whose effects have been characterized as similar to those of organic solvents (Tilson, 1981). Only the monomer is considered to be neurotoxic and this form of acrylamide is used to separate solids out of solutions (flocculent), as grouting agent, and as a paper or cardboard strengthener. It is also used in photography, adhesives, metal coatings, algicides, and bactericides (Tilson, 1981).

Animal studies suggest that acrylamide binds to plasma erythrocytes and that less than 1% of available dose is found in the nervous system. Half-life in blood is approximately 2 h, but half-life in the spinal cord may be as long as 24 days (Tilson, 1981). Acrylamide is fetotoxic and developmental neurotoxicity is exacerbated by low-protein diet (Khanna, Husain, & Seth, 1988).

The neurotoxic effect of acrylamide is thought to be related to an ability to disrupt protein transport along the axon (Pleasure, Mishler, & Engel, 1969). Tilson (1981) in a more recent and extensive review suggested that "acrylamide decreases fast axonal transport and interferes with energy production and neuronal metabolism" (p. 451). In rats acrylamide decreases brain concentration of norepinephrine, dopamine, and 5-hydroxytryptamine after 9 days' exposure at 50 mg/kg per day (Dixit, Husain, Mukhtar, & Seth, 1981).

In humans, acrylamide exposure has been categorized as to early, intermediate, and late stage effects by Tilson (1981), as shown in Table 8.1.

Effects on the CNS may be independent of PNS effects. CNS effects have variously been linked to cerebellar or midbrain abnormalities (Fujita, Shibota, Kato, *et al.*, 1961; Garland & Patterson, 1967). One recent model has proposed that the neurotoxic action of acrylamide is characterized by giant axonal swelling and that the most severe effects occur in larger-diameter axons and unmyelinated axons. The authors termed this effect a "central-peripheral distal axonopathy" (Schaumburg, Wisniewski, & Spencer, 1974; Spencer & Schaumburg, 1977).

TABLE 8.1. Human Acrylamide Poisoning by Stage[a]

Exposure	Symptoms
Early	Numbness of lower limbs
	Paresthesias
	Tenderness to the touch
	Coldness
	Excessive perspiration of feet/hands
	Desquamation of feet/hands
	Muscle weakness in extremities
Intermediate	Loss of body weight
	Lassitude
	Hypersomnolence
	Emotional changes
Late	Positive Romberg
	Loss of vibration and position sense
	Weak or absent tendon reflexes
	Ataxia
	Foot drop
	Muscular atrophy
	Occasional urinary and fecal retention

[a]Tilson (1981).

Acrylamide is used to make grouting agents or flocculents and is considered by the World Health Organization to be neurotoxic (WHO, 1985). Many of the recent acrylamide intoxication cases were found in the People's Republic of China because of its use in cottage industries run by townships or villages. He et al. (1989, 1993) investigated the neurological and electroneuromyographic effects on workers who were heavily exposed to acrylamide. Symptoms of heavy exposure included peeling skin and numbness, coldness and clumsiness in the hands, muscle weakness, lassitude, and sleepiness. Three patients were admitted to hospital with horizontal nystagmus, truncal ataxia, and hand clumsiness. The course of poisoning included improvement of cerebellar symptoms followed by loss of vibration sense and tendon reflexes.

ENMG study of exposed subjects noted partial denervation of distal muscles caused by axonal degeneration. H-reflex was nonresponsive in one-third of subjects and ankle tendon reflexes absent in 30.4%. The authors concluded that chronic acrylamide exposure damages PNS function with heavy exposure causing cerebellar impairment followed by polyneuropathy.

ANESTHETIC GASES

Introduction

The first anesthetic inhalant was diethyl ether, used by a U.S. physician in 1842 (Edling, 1980). Many compounds have been used since then for their anesthetic properties, including halothane (2-bromo-2-chloro-1,1,1-trifluoroethane), nitrous oxide, chloroform, trichloroethylene, and methyoxyflurane (Edling, 1980; Bach, Arbit, & Bruce, 1974; Patterson et al., 1985). There is no literature suggesting that patients retain significant neuropsychological sequelae of properly administered anesthetics. The primary exposure risk is thought to be to operating room personnel where as many as 200,000 persons are routinely exposed to waste anesthetic gases. Seventy-five thousand of these individuals work in hospitals (Patterson et al., 1985). The operating room itself may be chronically subjected to low-level ambient exposure of anesthesia: up to 500 ppm of halothane in the working field of the anesthetist, and from 1 to 85 ppm throughout the larger area of the operating room (Dudley, Chang, Dudley, Bowman, & Katz, 1977).

Since anesthetic gases, by definition, have been developed for their consciousness-altering properties, it is not surprising that acute exposure to anesthetics can alter neuropsychological test behavior. However, despite the *a priori* neurotoxicity of solvent/anesthetics, most available surveys of anesthetic-exposed personnel have been epidemiologic in nature and have relied on questionnaire data rather than neuropsychological test results (Patterson et al., 1985). One of the better studies of this type surveyed symptom complaints as a function of how frequently operating room air was scavenged (Saurel-Cubizolles et al., 1992). After controlling for various working conditions, the authors found that the risk of developing neuropsychological complaints was 3.6 times higher than controls where air was scavenged 10 times an hour or less. Adjusted risk was three times higher with air scavenged from 11 to 20 times per hour, and there was no additional risk compared with controls when air was scavenged more than 20 times per hour (Saurel-Cubizolles et al., 1992). There are few studies of exposed groups using neuropsychological measures, and no longitudinal investigations of groups at risk. No studies of chronically exposed operating room personnel are available; however, animal experiments have documented CNS damage after continuous exposure to halothane.

Acute Neuropsychological Effects

There is obviously no disputing the consciousness-altering effects in clinical applications of anesthesia. The effects of smaller, subclinical exposures typically found in operating theaters are more controversial. Several studies by Bruce and his colleagues support the notion that very low doses of anesthesia can produce certain neuropsychological impairments. For example, Bach et al. (1974) subjected volunteers to 500 ppm nitrous oxide mixed with 15 ppm halothane in air. Subjects showed significant decrements in choice reaction time, several subtests of the Wechsler Memory Scale and memory for tachistoscopically presented patterns. Bruce and Bach (1975) exposed subjects to 500 ppm nitrous oxide and 15 ppm enflurane for 4 hours and then administered a 35-minute neuropsychological battery. Two tests showed impairment: Digit Span and a divided attention decision-making task. Other tests were unaffected, including memory subtests of the Wechsler Memory Scale, a tachistoscopic task, and five subtests of the WAIS.

In subjects receiving only 500 ppm NSIO, only Digit Span was significantly decremented.

Bruce and Bach (1976) tested 100 male subjects who were exposed to either nitrous oxide alone, or in combination with halothane. There were five independent groups:

1. 500 ppm nitrous oxide + 10.0 ppm halothane
2. 50 ppm nitrous oxide + 1.0 ppm halothane
3. 25 ppm nitrous oxide + 0.5 ppm halothane
4. 500 ppm nitrous oxide alone
5. 50 ppm nitrous oxide alone

Each subject was tested once with air exposure alone and once after 2 hours' exposure to anesthetic. All subjects received a tachistoscopic test of visual short-term memory, the Raven's Progressive Matrices Test, the O'Connor Dexterity Test, an audiovisual choice reaction time task, a test of vigilance, and the Digit Span test. While logical reasoning tasks like the Raven's and manual dexterity were resistant to anesthetic effects, significant decrements in performance were found in all conditions. Subjects in the lowest-dose condition (50 ppm nitrous oxide alone) showed significant decrements only in the most complex task, the audiovisual divided attention test. Volunteers exposed to the highest amount of anesthetic also performed significantly more poorly on the tachistoscope and Digit Span tests. Digit Span proved sensitive to all but the lowest exposure (50 ppm N_2O, no halothane) group. The authors use their results to speculate that even very low levels of waste gas available in operating rooms may be capable of impairing the neuropsychological functioning of operating room staff.

There are, however, some inconsistencies in the Bruce and Bach (1976) study. Results include an unexplained (probably chance) improvement in vigilance in the 50 ppm nitrous oxide–1 ppm halothane group. Also, subjects who received high-dose nitrous oxide alone were impaired on more tasks than those receiving high-dose nitrous oxide plus halothane, a result that seems explainable only in terms of preexisting differences among subject groups.

The findings of Bruce and his colleagues have also been called into question by several studies that failed to replicate their results. Cook, Smith, Starkweather, Winter, and Eger (1978) exposed subjects to 20 times the amount of halothane of the former study—200 ppm halothane—and from 1000 to 4000 ppm of nitrous oxide. Though the dose of anesthetics was many times larger than in the Bruce and Bach (1976) study, no evidence of neuropsychological impairment was found.

In the study that most resembled Bruce's paradigm (naive subjects and similar tests), Cook *et al.* (1978) compared subjects' performance on air versus anesthetic on the same day, while Bruce and Bach waited a week between air and anesthetic trials. Thus, subjects in the Cook *et al.* study may have remembered their performance on the first trial more clearly than subjects in the Bruce and Bach study, and may have attempted to equalize performance between trials. Also, Bruce and Bach (1976) tallied 100 responses to complex reaction time stimuli per trial, while Cook et al. used 50 responses per trial, suggesting that fatigue or vigilance failures over trials could have made Bruce and Bach's subjects perform more poorly in the latter 50 trials, thereby degrading overall performance.

Smith and Shirley (1977) also failed to find neuropsychological effects of anesthesia exposure in subjects who were exposed either to halothane 100–150 ppm, or to a mixture of halothane 15 ppm and nitrous oxide 500 ppm. However, the latter study employed anesthetists or anesthetic technicians who spent "a major part of the working day in an operating theater" (Smith & Shirley, 1977, p. 65), suggesting the possibility that adaptation or selection for high tolerance for anesthetics may have occurred in this group.

Thus, neuropsychological effects of low-level anesthesia inhalation are equivocal. Hopefully, future studies will resolve the discrepant results reported in these studies. An observation by Cook et al. (1978) suggests that variations between subjects may be an important factor in researching the neuropsychological effects of anesthetic exposure. They noted wide variation in subject responses to *clinical* amounts of anesthesia, which suggests that perhaps the stronger responders to clinical doses may be the only subjects who show impairments with subclinical exposures.

Developmental Neurotoxicity

Dudley et al. (1977) documented unexpected and prolonged learning disability in rat pups exposed to trace levels of halothane *in utero*. Based on their findings, the authors warn female operating room personnel to transfer to other medical services at least 1 month before conception. There appears to be increased risk of spontaneous abortion and congenital abnormality in operating room and recovery room personnel (Guirguis, Pelmear, Roy, & Wong, 1990). There is no evidence at this time to link postnatal exposure to operating room anesthetic gases to permanent neuropsychological disability (Edling, 1980). Neither, however, is there a study that addresses the question. Longitudinal neuropsychological studies of operating room personnel would seem to be strongly indicated.

Nitrous Oxide

Nitrous oxide was discovered by Priestley in 1779 and was first used as an anesthetic in 1844. No time was lost among the recreational drug users of that era, who immediately experimented with nitrous oxide and made it a popular abused drug during the 1840s. Since nitrous oxide is approved by the Food and Drug Administration as a food additive, and is on the GRAS (Generally Recognized As Safe) list, it has been legally used as a propellant aerosol in whipped-cream cans. These cans contain approximately 3 liters of 87–90% N_2O. Nitrous oxide gas chargers for commercial whipped-cream contain 4.3–5 liters of 93–98% N_2O (O'Donoghue, 1985a). O'Donoghue (1985a) reported there have been at least 22 cases of myeloneuropathy and/or polyneuropathy from chronic use of N_2O. All but two of these cases involved voluntary, recreational abuse. The remaining patients were exposed to N_2O in badly ventilated areas. Early symptoms included sensory paresthesias, loss of balance, gait ataxia, and impotence. Personality alterations have also been reported. Grigg (1988) reported the case of a nitrous oxide abuser with sleep disorder, decreased appetite, racing thoughts, and "intrusive hypnotic suggestions." Neurological examination results were consistent with sensorimotor neuropathy, involving the spinal cord and possibly the cerebellum (O'Donoghue, 1985a).

Acute Neuropsychological Effects

Experimental exposure to 15% nitrous oxide was investigated in volunteer subjects who inhaled nitrous oxide in oxygen for 45 min (Tiplady, Sinclair, & Morrison, 1992). Small but significant slowing in Digit Symbol, Symbol Digit, along with speed and error rates of a sentence verification task were found in the nitrous oxide condition. The effects diminished almost immediately with administration of pure oxygen. Greenberg *et al.* (1985) tested the acute neuropsychological effects of 20% nitrous oxide on 12 male subjects who served as their own controls. Subjects received nine tests from an automated neuropsychological battery presented on a COMPAQ microcomputer (Baker, Letz, & Fidler, 1985). Three tests were significantly affected by nitrous oxide: Finger Tapping; a Continuous Performance Test which required subjects to press a button whenever they saw a large "S" that randomly appeared in a string of other letters; and a Symbol Digit test where numbers were matched to abstract symbols presented on the screen. Nitrous oxide slowed subject performance in each case.

ARTHRINIUM

A fungus that grows on sugarcane, *Arthrinium* produces toxic encephalopathy and late-onset dystonia in Chinese children (Spencer, Ludolph, & Kisby, 1993). Symptoms include sudden nausea, vomiting, diarrhea, and, in children, diplopia, nystagmus, somnolence, seizures, decerebrate rigidity, and coma. The picture for recovery is grim, with development of delayed dystonia 7–40 days after ingestion, with "facial grimacing, sustained athetosis of hands and fingers, torsion spasm, spasmodic torticollis, hermiballismus, and painful spasms of the extremities" (Spencer *et al.*, 1993, p. 109). CT scan of children with these symptoms reveals bilateral hypodensities in the putamen and somewhat less frequently, in the globus pallidus, caudate, and claustrum. Between 1972 and 1989, there have been almost 900 recorded cases, with 88 deaths.

ASPARTAME

Despite the fact that this widely distributed artificial sweetener metabolizes into aspartic acid, phenylalanine, and methanol, there has been no evidence for its behavioral toxicity in normal adults, probably because only phenylalanine and aspartic acid enter portal blood (Tollefson, 1993). In children, however, two concerns are related to its excitotoxic properties. First, Zorumski and Olney (1992) warn that "immature humans are being fed multiple excitotoxins, including Asp contained in Aspartame (Nutrasweet®)" (p. 284) and that silent hypothalamic damage may result from children's and adolescents' consumption of normal amounts of aspartame and other excitotoxic substances.

Second, several studies suggest that Aspartame may lower the threshold for chemically produced convulsions (Sze, 1989). Effects of Aspartame on children may be somewhat more evident, however, particularly in children already diagnosed with seizure disorder. Using a population of children with untreated generalized absence seizures, Camfield and colleagues (1992) found duration (but not number) of spike wave discharges to significantly increase with Aspartame compared to sucrose. While the dose required was somewhat large (equivalent to a 25-kg child drinking 1.6 liters of diet cola), Aspartame is present in many other sources of artificially sweetened foods, and thus, this amount could be possible for a child to consume. Results suggest avoidance of Aspartame in children already diagnosed with seizure disorder.

Adult seizure disorder from Aspartame has not been validated. In the 265 cases of epileptic seizures associated with aspartame ingestion logged with the FDA from 1986 through 1992, 50% had physician records or other data showing seizures were not associated with aspartame. Only 36 reports were logged as having occurred in connection with aspartame consumption consistently more than once. When a double-blind, placebo-controlled crossover design study was conducted, using continuous EEG monitoring, no seizures were seen during aspartame ingestion. Seizures were recorded in two subjects, but after placebo administration. Four headaches were reported after placebo consumption (Tollefson, 1993).

BROMINE

Conversion factor	1 ppm = 6.64 mg/m^3
Odor threshold	<0.0099–0.46 ppm
IDLH	10 ppm
OSHA/NIOSH	0.1 ppm (0.7 mg/m^3)
STEL	0.3 ppm (2 mg/m^3)
(*Quick Guide*, 1993)	

Bromine gas is a respiratory irritant and is capable of causing chemical burns to exposed skin. Carel, Belmaker, Potashnik, Levine, and Blau report a 4-ton release of bromine gas from an overturned tanker truck. Delayed cognitive and emotional complaints were reported after 6–8 weeks, but were not validated by psychological examination. No test data were reported.

BUTANE

Mathew, Kapp, and Jones (1989) report the case of a 16-year-old female who inhaled commercial butane gas for over a year. The patient became increasingly irritable, aggressive, and impulsive, leading to her rejection by a peer group. Etiology and sequelae remain unclear, as the patient refused detailed medical or neuropsychological evaluation. The frequency of butane inhalation abuse is not known.

CARBON MONOXIDE

Conversion factor	1 ppm = 1.16 mg/m^3
Odor threshold	100,000 ppm
IDLH	1500 ppm
OSHA/NIOSH	35 ppm (40 mg/m^3)
Ceiling	200 ppm (229 mg/m^3) (measured over a 5-min period; an instantaneous reading of 1500 ppm shall not be exceeded)

(*Quick Guide*, 1993)

The poisonous properties of carbon monoxide were utilized as far back as the time of Cicero (106–43 B.C.) where charcoal fumes were used by suicide attempters and in the execution of criminals (Shepherd, 1983). Ramazini, a 17th century physician, wrote of the miners who inhaled carbon monoxide vapors which "pervert and pollute the natural composition of the nervous fluid, and the result is palsy, torpor, and the maladies above mentioned" (cited in Shephard, 1983, pp. 3–4).

Carbon monoxide is a colorless, odorless, lighter than air (specific gravity=0.98) gas that is produced both as a natural oxidation product of methane, and as an incomplete combustion product from burning carbon-based fuels. It is the most abundant air pollutant, with an estimated 360–600 million tons produced annually from man-made sources of CO, including burning waste material, emissions from gasworks, garages and service stations, coke ovens and blast furnaces, and defective heaters and stoves. Tobacco smoke is also a significant source of CO in a closed environment (World Health Organization, 1979b).

Heavy cigarette smokers can raise their carbon monoxide hemoglobin (COHb) levels to levels as high as 18% (Winter & Miller, 1976). Methylene chloride paint stripper is an unusual source, since it is metabolized to CO, with a 3-h exposure in a ventilated room capable of producing COHb saturations of 8–16% (Stewart, 1975).

Epidemiology

Carbon monoxide exposure is one of the leading causes of the approximately 10,000 reported cases of poisoning each year in the United States. Of these, about 3800 deaths per year result from CO exposure, 40% of which are accidental and 60% the result of suicide (Smith & Brandon, 1970). In countries employing indoor coal fires (e.g., China), acute CO poisoning is a common and life-threatening problem (He et al., 1993).

Toxicological Effects

Carbon monoxide inhibits cellular respiration mechanisms. It displaces oxygen from hemoglobin and is about 250 times more "eager" to bind to hemoglobin. However, since the oxygen in carbon monoxide cannot be used by the body, a carbon monoxide-saturated hemoglobin starves the body of oxygen. In addition, carbon monoxide interacts with the hemoglobin molecule to further reduce its ability to give up any remaining oxygen, worsening oxygen deprivation. It has been suggested that hypoxia from hemoglobin binding is not sufficient, in itself, to produce demyelinating effects, but usually does so in the context of cerebral ischemia from systemic hypotension and acidosis (Ginsberg, 1985). Free radical formation leading to lipid peroxidation and direct toxicity from CO binding to intracellular proteins are two additional neuropathological mechanisms proposed to account for CO neurotoxicity (James, 1989; Penney, 1990).

Elimination

The half-life of carbon monoxide following exposure is approximately 250 to 320 min and about 15% of COHb is eliminated after 1 h (Ginsberg, 1980). Elimination under 100% oxygen is reduced to about 80 min and in a hyperbaric chamber at 3 atmospheres, the biological half-life of carbon monoxide is 23.5 min (Stewart, 1975). The exact relationship between CO uptake and elimination is more complex and depends on length and continuity of exposure (Stewart, 1975).

Carbon monoxide is especially toxic to the brain and the heart, "the most susceptible organs in the body to an oxygen deficit" (Grandstaff, 1974, p. 292). Severity of organ damage and symptom expression is generally considered to be largely dependent on the percentage of carbon monoxide bound to blood hemoglobin, although this is disputed by Myers, Mitchell, and Cowley (1983) who state that they have seen unconscious patients with COHb levels of under 1%, and "apparently well oriented patients" with COHb levels of 45% (p. 279).

TABLE 8.2. Carbon Monoxide Hemoglobin Levels and Severity of Symptoms

COHb	Symptoms
0–10%	No symptoms
10–20%	Mild headache or shortness of breath
20–30%	Throbbing headache
30–40%	Confusion, vomiting, dim vision, poor judgment
40–50%	Confusion, possible fainting, rapid heartbeat
50–60%	Unconsciousness, seizures, coma
60–70%	Coma, death
>70%	Rapidly fatal

As a general rule, however, exposure victims will exhibit symptoms as a function of how much carbon monoxide is in the blood (COHb) at the time of intoxication. Increasingly severe symptoms are expected with heavier exposures (see Table 8.2).

Neurological and Neuropsychological Effects

Short-Term and Low-Level Exposures

Short-term or low-level exposure to carbon monoxide does not appear to affect brain functions significantly (Laties & Merigan, 1979). Benignus, Vernon, Otto, Prah, and Benignus (1977) exposed subjects to as much as 200 ppm of carbon monoxide (a COHb of 12.62%; see Table 8.3) and found no effects on vigilant attention. Heavy smokers, whose COHb levels can raise to 18%, typically do not show significant, acute neuropsychological or psychiatric impairments as a result of smoking.

Very low levels of CO exposure may produce slight changes in light sensitivity as a function of mild hypoxia (e.g., Richalet, Duval-Arnould, Darnaud, Keromes, & Rutgers, 1988). Laboratory experiments with higher CO concentrations are inconsistent, with some studies showing small effects of 45 min exposure to CO (COHb of 4.5%) (Ramsey, 1972) while others that tested COHb levels as high as 34% were unable to show slowing of either arm, wrist, or finger movements (Fodor & Winneke, 1972). The subjects of Bender et al. (cited in Shephard, 1983) were said to show deterioration of performance on the Purdue Pegboard Test after 3 h exposure to 10 Pa CO (COHb = 7.2%) while Stewart et al. (1970) report no decrements. Shephard (1983) has suggested that the sensitivity of various tasks to CO exposure is related to task difficulty, as research subjects appear to be able to draw on "spare mental capacity" and maintain performance by increasing concentration (Shephard, 1983).

Minimal acute experimental effects were found in healthy subjects exposed to 100 ppm of CO for 2 h and required to perform single or dual information-processing tasks (Mihevic, Gliner, & Horvath, 1983). Subjects showed mild impairment in performing simultaneous tasks only when CO exposure was com-

TABLE 8.3. Relationship between Inhaled CO and COHb Level

ppm inhaled CO	COHb
1	0.5%
3	0.8
7	1.1
9	1.5
10	1.9
30	5.0
50	7.9
70	10.7
90	13.2
100	14.5
500	45.4
1,000	62.4
5,000	89.2
10,000	94.3

bined with secondary tasks of moderate difficulty. No decrements in motor performance were found.

It is generally considered that COHb levels below 20% do not produce significant nervous system damage in healthy individuals. However, patients with preexisting heart disease or brain damage may be deleteriously affected by much lower levels of exposure. (Winter & Miller, 1976). A review of the carbon monoxide clinical literature shows that low-level, short-term exposure to carbon monoxide is "probably not neurotoxic in normal, healthy men" (Harbin, Benignus, Muller, & Barton, 1988, p. 93). Stewart (1975) suggests that small COHb elevations (e.g., up to 5%) may alter visual thresholds, but more complex neuropsychological tasks are not significantly affected by COHb saturations below 10%. Benignus, Muller, and Malott (1990) reviewed the literature on human, experimental exposure and reported that "no study in which dose-effects functions involving COHb and a behavioral response has been replicated" (p. 112) and reiterate that human data show no convincing effect below 20% COHb. The same authors reviewed laboratory animal COHb studies, which showed nonsignificant differences in data values compared with controls at COHb levels below 20%. Above 20%, "effects began to be statistically significant in the individual reports and rapidly increased in magnitude as COHb was further increased. Eventually effects became large in magnitude, corresponding to as much as a 50% reduction in response rates" (Benignus et al., 1990, p. 116). They suggest that human data show a similar nonlinear positive relationship with exposure with "large and reliable" effects predicted to occur at COHb levels greater than 20%. The authors correctly note that this pattern of exposure data which holds for healthy young males at rest, may not apply to other populations. Without a more complete theory of CO/nervous system interactions, it is not clear whether negative or positive experimental results more accurately describe CO effects.

Severe Poisoning

Moderate cases of carbon monoxide poisoning may be misdiagnosed or unrecognized. In a prospective study of patients with neurologic complaints examined in the emergency room during winter, Heokerling, Leiken, Terzian, and Maturer (1990) concluded that "unrecognized carbon monoxide poisoning occurs in a small but important fraction of patients with wintertime neurologic illness . . ." (p. 29). In another study of inner-city patients (who often use alternative heating sources), 23.6% of patients examined with flulike symptoms had COHb levels above 10% (Dolan, Haltom, Barrows, Short, & Ferriell, 1987). General malaise, nausea, or cough related to CO intoxication may be misdiagnosed as viral or bacterial infection.

In more severe poisoning (i.e., patients who were rendered unconscious or those with COHb levels of 50% or above), carbon monoxide can cause a variety of neurological sequelae as to provoke Garland and Pearce's (1967) observation that CO poisoning is associated with almost every known neurological syndrome. For example, in severe acute CO poisoning (about half of which in a recent survey were attempted suicides), half of the total group displayed abnormal mental status. Depression was common, although this symptom probably existed before exposure. Other alterations in emotional function, including delirium, persisted up to 4 weeks. Permanent dementing process occurred in 4% of these patients (Ginsberg, 1980). Other chronic sequelae of severe exposure can include parkinsonism, persistent vegetative state, akinetic mutism, agnosia, apraxia, impairments in vision, amnesia, confabulation, and psychosis.

Fatal exposure to CO shows effects typical of hypoxic–ischemic lesions, including cerebral vessel congestion, edema, and petechial white matter hemorrhage. Cortical and hippocampal lesions are found in 50% of patients. MR imaging indicates lesions of the globus pallidus as well as bilateral lesions in the temporal, occipital, and posterior parietal lobes (Horowitz, Kaplan, & Sarpel, 1987).

Severe Acute Exposure

Patients with severe acute exposure may develop brain damage and severe mental impairments. Severe exposure may produce diffuse or frontal lobe brain damage, or damage to the basal ganglia. Cerebellar or basal ganglia damage may impair ability to move or walk smoothly. These patients may suffer general lowering of cognitive function, as well as specific memory, visuospatial impairments, and personality changes.

He *et al.* (1993) studied somatosensory (SEP), visual (VEP), and brain stem auditory evoked potentials (BAEP) in 88 adults with acute CO poisoning, one subgroup of which was studied immediately after poisoning. The other consisted of 35 patients with delayed encephalopathy after a period of pseudorecovery from 2 to 35 days. Long-latency SEP components (N32 and N60) were impaired in both groups, suggesting diffuse cortical dysfunction. SEP changes were correlated with severity of impairment. Six patients with an abnormal shorter-latency component N20 have subcortical lesions confirmed by CT. The II-IV interpeak latency of BAEP was prolonged in comatose patients compared with conscious patients. Almost eighty percent of patients with moderate to deep coma and impaired brain stem reflexes had abnormal BAEP, compared with 12.5% of patients with slight coma and intact brain stem reflexes. VEP were prolonged in 67.9% of patients classified as demented from delayed encephalopathy. VEP recovery was noted to parallel clinical improvement of mental functions. Abnormal BAEP predicted longer duration of unconsciousness and higher mortality. Consistently abnormal VEP P100 was associated with poor prognosis, as was VEP that became longer over repeated assessments.

Delayed Exposure Effects

CO intoxication is also more rarely associated with delayed neuropsychological and psychiatric sequelae. In certain patients, high-level acute exposure to carbon monoxide appears to cause delayed deterioration of brain function. Typically, the pattern consists of several days to weeks of apparent recovery, followed by acute-onset dementia, deterioration of neurological status, and sometimes permanent disability or death. Patients with delayed-onset sequelae showed a mixture of psychiatric and neurological symptoms, including "apathetic masklike facial expression, symptoms of dementia, such as amnesia and disorientation, urinary incontinence and hypokinesia" (Min, 1986, p. 82). Other frequent symptoms included irritability, distractibility, various types of apraxia and behavioral abnormalities, including silly smiles, frowns, or repetitive behaviors (Min, 1986).

Min (1986) examined the records of 738 patients admitted to hospital with accidental CO poisoning. Eleven percent of these patients (86) went on to develop delayed sequelae. Almost all of these patients were rendered unconscious by CO poisoning; only 1% of the original 738 patients developed delayed sequelae without first being rendered unconscious. This small group showed obvious neurological deficit including glabella sign, and grasp reflex. Approximately half of the patients in this study with delayed sequelae had abnormal EEGs, and Min's review of the literature concluded that more than 90% of such patients have abnormal EEGs (Min, 1986).

Neuropathological findings from delayed CO intoxication include extensive regions of white matter myelin injury, bilateral necrosis of the globus pallidus, and damage to the hippocampus and focal cortical areas (Zagami, Lethlean, & Mellick, 1993). Lesions of the globus pallidus are not inevitable; neither of two cases examined with MRI had this "characteristic" form of localized damage (Zagami *et al.*, 1993). The first patient showed abnormal MRI signal intensity in gray matter of the frontal lobes bilaterally, as well as in the left posterior parietal cortex. The second patient's MRI was characterized by extensive bilateral white matter lesions. The patients with delayed CO encephalopathy in the He *et al.* (1993) investigation showed similar neuropathology, including "extensive demyelination in subcortical white matter and softening of bilateral globus pallodum" (p. 225). SEP and VEP abnormalities were found, along with prolonged P100 latencies without configuration or amplitude changes.

When Min followed up 56 of the delayed impairment group, he found that 60% (34) recovered to "almost premorbid levels," 19% (11) showed slight weakness and memory impairment, 14% (8) had moderate memory impairment, 8% (5) did not show cognitive recovery from initial demented state, and 10% (6) died of infections.

There did not appear to be any predictors useful in determining who will develop delayed CO neurotoxicity, although the syndrome seems to occur more frequently in patients with histories of neurologically related illness (e.g., diabetes mellitus, hypertension) (Min, 1986).

Chronic, Low-Level Exposure

Chronic CO poisoning at lower exposure levels may produce several CNS effects, initially on auditory and visual vigilance (attention), and also in peripheral vision. For example, Johnson *et al.* (1974) found highway toll collectors (who also may have been exposed to lead) to perform more poorly on parallel processing tasks as a function of CO levels. Workers using propane-fueled forklifts or similar apparatus in enclosed areas are at risk for both acute and chronic CO exposure effects since a single propane-fueled forklift is capable of producing 60 liters of carbon monoxide per minute. At that rate, even in a 60,000 m^3 warehouse, it would take only 9 minutes to exceed EPA standards (Fawcett *et al.*, 1992). One report of workers so exposed noted symptoms of global headache, nausea, lightheadedness, and sometimes syncope or unconsciousness (Fawcett *et al.*, 1992).

Extended exposure to high levels of CO may affect other neuropsychological functions. In one such case, a 48-year-old woman ran a business for 6 years in a basement home office that also housed a furnace. After 3 years of headaches, depression, and self-reported mental confusion, her furnace was found to be producing 180 ppm of carbon monoxide, three times OSHA limits (Ryan, 1990). Exposure was without any history of unconsciousness, although the patient reported an episode of near syncope. Neuropsychological test results showed poor verbal memory and learning, impaired immediate and delayed visual recall, and poor retrieval under cued recall. Measures of perceptual speed were unremarkable. We have seen a case similar to Ryan's where the patient developed chronic flulike symptoms and significant impairments in memory. The patient was referred for neuropsychological testing to determine if cognition, behavior, or affect presents in a manner consistent with chronic carbon monoxide exposure.

Case Report: Low-Level Exposure to CO without Unconsciousness

Tests Administered

Wechsler Memory Scale subtests: Logical and Figural Memory, immediate and 30-minute delayed recall, Associate Learning Test

Halstead–Reitan Battery subtests: Finger Tapping Test, Trail-Making Tests A&B, Category Test, Tactual Performance Test, Sensory Perceptual Examination (Selected Items)
Wechsler Adult Intelligence Scale subtests: Digit Symbol Test, Block Design Test, Symbol Digit Test

Other Tests

Stroop Color-Word Test
Luria–Nebraska Pathognomonic Scale
Spatial Relations Test
Grooved Pegboard Test
Grip Strength Test
Shipley Institute of Living Scale
Beck Depression Inventory
Profile of Mood States
Personal Problems Checklist
Minnesota Multiphasic Personality Inventory

Clinical Interview. The patient appeared to be uncertain about dates and sequences of her illness, the following information having been obtained with the collaboration of her husband. The patient is a 65-year-old white female with 2 years of college and a professional occupation degree. She is married and has several children with her husband, who is also a professional.

The patient's physical complaints were said to have begun in 1980, initially presenting as gastroenteritis followed by migratory neuritis. The patient's principal symptoms at that time included right foot fatigue when driving and using the accelerator, as well as sharp sensations "like being stuck with a needle loud enough that you'd yell out" while in bed.

The patient's physical complaints increased over the next several years, and included frequent urinary tract infections, as well as episodes of pneumonia, depression, neuritis, and paresthesias. "My legs felt like lead; numb with feelings of cold and wet around the buttocks." Shooting pains increased during urinary tract infections.

By 1985, the patient was experiencing chronic infections and depression; "bad depressions like somebody pulled shades down in the room." Panic attacks, fear of water or drinking, and insomnia also began. The patient gained about 30 pounds. Symptoms of a hiatus hernia were discovered at this time but no other diagnosis was discovered to account for the patient's many symptoms. Other abnormalities developing during this time included spasms in back and stomach, and fatigue that required her to go to bed immediately after dinner. She reports that morning awakening was accompanied by loss of tone in her facial muscles so profound that "my face was hanging." The patient experienced bloody stools during this period but a GI examination was negative. Blackening of fingernails and toenails was also reported.

In 1986–1987 the patient began losing her balance in the shower, withdrew from social and noisy situations, and ceased gardening, a favorite activity. After an episode when she fell on the house stairs and blacked out, the patient became afraid of walking outside or unaccompanied. The patient also experienced the development of severe allergies during this time.

Deteriorated cognitive functioning was also reported during this period, especially complaints of memory dysfunction. "I would see a movie and I'd completely forget." The patient reports that she could read a magazine or newspaper article and forget it so completely that she would not recognize it on rereading. There were also episodes of apparent disorientation. "Once I came to a stop light and didn't remember what to do... I thought I was getting Alzheimer's... I'd have to check and recheck, I just couldn't seem to stay on top of things.... My mother told my sister there was something wrong with me."

To learn about her perceived condition, the patient became active in Alzheimer's organizations, even though she reports that there have been no dementing conditions in her long-lived family. Prior to 1980 she claims to have been in excellent health with no preexisting physical or emotional impairments.

The patient denies any history of emotional illness prior to 1980. Since 1980, she briefly saw a psychologist whom she did not find helpful, and also visited a psychiatrist who prescribed Ativan for her depressive symptoms. Premorbid use of alcohol was "two to three drinks per week and a little more during vacation." The patient denies having used alcohol since her urinary tract infections began.

The patient and her husband were unable to discover the cause of her increasingly impaired physical and mental condition. It was initially attributed to stress by both parties, since her brother had been diagnosed with cancer, and three other relatives had died within a year. "We thought all the problems were due to stress."

Other facts eventually led to another conclusion. In 1980, the patient and her husband replaced two old furnaces in their home, along with their hot water heater. After that time, they noticed that plants were constantly dying and silverware turning black almost immediately on removal from storage. The patient also noticed that she felt much improved after leaving the house to visit her mother for the weekend. However, she would fall ill again after coming home. The patient's husband became aware that his wife appeared to show her worst symptoms in only one part of the house. "Then I saw that the junction on the pipe to the water heater had opened up and I saw a carbon streak on the pipe from the hot water heater below the flue." At this point, her husband replaced the furnace and called a repairman. This first repairman did not notice a problem, but when the patient's husband had an industrial hygienist inspect the house, he noted that a pipe from the hot water heater was black and rusted. The patient's husband was told that the pipe had been installed incorrectly and that without a blower, carbon monoxide fumes would leave the pipe junction to be picked up in the furnace and blown throughout the house.

Immediately prior to the realization that a defectively installed hot water heater was discharging carbon monoxide into the house, the patient had become so disabled that she was finding it painful even to lie in bed. She reports slow recovery of physical health and mental function since that time. Residual impairments and distortion of vision continue to be noted by the patient with flashes of light, zigzags, and occasional loss of peripheral vision in the right eye.

Memory is said to be much improved; however, the patient reports that occasional problems remain attempting to remember written material. Because of her many allergies, the patient remains socially isolated and restricted in activity. "I don't place as many demands on myself. I can't go out because of perfumes and allergies." The patient occupies her time with cooking and household chores. She talks to friends by phone, tries to read, and listen to music.

Sleep is improving although the patient still wakes up several times a night. Appetite is good on a restricted rotation diet where the patient is allowed to eat at most two foods per meal. She is not allowed seasoning or sweets.

Affect is also said to be much improved with no recurrences of depression other than occasional "blue days" since replacement of the hot water heater and therapy for allergies. The patient reports that her fear of walking by herself is subsiding, as is her fear of eating.

Finally, since the heater and pipes were repaired, the patient has been able to grow plants once again in her home. The patient notes that flowering plants bought since Thanksgiving 1987 remain alive at the present time.

Current medications include vitamin C and thyroid replacement hormone 1.5 grains per day. The patient has apparently not yet received EEG, CT, visual evoked potential, or nerve conduction studies to examine nervous system correlates of her complaints.

Behavioral Observations. The patient arrived on time for testing, accompanied by her husband. She was neatly dressed and composed. The patient was oriented to time, place, and person. Speech was normal in volume, rate, and timbre. Affect was unremarkable, although the patient appeared to become mildly distressed at her performance in several of the more difficult neuropsychological tests. There was no evidence of fatigue or loss of motivation during an evaluation running longer than 5 hours. The patient was unable to provide a consistent history without the assistance of her husband. Therefore, both parties were interviewed together.

Neuropsychological Test Results
Intellectual, language, and general cognitive functioning. The patient's estimated level of intellectual ability is in the Superior range, a level consistent with this patient's college education and advanced professional training. Expressive and receptive speech were normal with no indications of aphasia, dysarthria, alexia, or agraphia.

Attentional processes were also unimpaired. The patient consistently showed above average to superior performance in tests that required continuous attention as well as rapid, efficient, and flexible cognitive processing.

Verbal abstract reasoning was above average and consistent with the patient's advanced education. Alternately, the patient showed mild levels of impairment on a task requiring nonverbal inductive reasoning, spatial memory, and problem solving.

Learning and memory. The patient's complaints of poor memory appear to apply to memory for nonverbal spatial information rather than verbal data. Using age-adjusted norms, the patient showed approximately normal verbal memory for complex story paragraphs, both for immediate recall and for surprise 30-minute delayed recall.

Learning of easy and difficult verbal paired-associates (e.g., "fruit–apple"; "crush–dark") was near perfect after only one learning trial. Subsequent trials were recalled perfectly.

In contrast, nonverbal memory showed mild to moderate impairment, suggesting that the patient is unable to retain spatial information with the same facility as verbal information.

Sensorimotor functions. The patient is right-handed and shows right hand dominance in speed, strength, and coordination. Motor speed of the dominant hand is excellent and unimpaired, while the left hand is mildly slowed in comparison. Motor coordination on a task requiring fine motor skills and spatial manipulation of materials shows mild to moderate bilateral slowing that appears to be a function of the task's spatial components rather than lack of motor speed *per se.*

The patient shows intact basic sensory abilities in the hands (e.g., touch, pressure, sharp–dull), with no indications of suppression. However, her ability to perceive and integrate complex stimuli through tactual perception alone is moderately to severely impaired (astereognosis). For example, the patient showed very poor performance bilaterally when attempting to identify numbers drawn on her fingertips. In a somewhat easier task, requiring the patient to identify common objects by touch alone, the patient performed adequately with her dominant hand, but failed to correctly identify half of the objects presented to her nondominant hand.

Perceptions of auditory stimuli were adequate for both ears with no indications of suppression when both ears were simultaneously stimulated.

Visuospatial functions. There is no indication of constructional apraxia; the patient is able to copy simple and complex geometric figures without error. Visual fields appeared intact on casual examination. In contrast, the patient's ability to spatially integrate complex tactually perceived materials is significantly worse than expectation. The patient had great difficulty with a task that required her to place wooden shapes in a formboard while blindfolded. She took longer than expected and appeared to experience great difficulty when attempting to match the shape of a block to its particular place on the board. In addition, her memory for block shapes was quite poor. In an immediate recall task, the patient was only able to remember two of the ten blocks she had just manipulated for 30 minutes.

Personality functioning. Results of the patient's personality evaluation suggest a profile consistent with great reserve about revealing oneself psychologically. The patient's personality profile is consistent with that of a very conventional and very feminine-interest-centered individual who denies or minimizes psychological difficulties and emphasizes an optimistic view of life. A person with this profile is unlikely to be very comfortable handling anger or self-assertion. Given the list of stressors cited by the patient in clinical interview, depression is probably greater than the patient is willing to admit on formal psychological tests. It is apparent from clinical history and interview that the patient experienced substantial psychological reaction to the stressors of illness, including social isolation, withdrawal from favored activities, and phobias. By history, depression was quite severe during the past several years and may have been the result of organic and reactive components.

Summary and Conclusions. The patient is a 65-year-old white female who was referred for neuropsychological evaluation as a consequence of deteriorating health presumptively tied to a faulty hot water heater installation and continuing carbon monoxide exposure over a period of 6-7 years. Current neuropsychological evaluation suggests superior premorbid intellectual functioning and unimpaired verbal memory. In contrast, the patient shows significant and moderate to severe deficits in tactual apprehension of complex stimuli (astereognosis), as well as impaired spatial memory and spatial reasoning. These deficits are clearly inconsistent with the patient's premorbid educational or employment level as well as her current estimated intellectual level. Deficits are also worse than might be expected when the patient's performance is compared with relevant age norms.

While there is no invariant pathognomonic syndrome associated with chronic carbon monoxide intoxication, the patient's increasing impairment of cognition and affect that only halted after fixing the heating system is certainly consistent with chronic carbon monoxide poisoning.

The effect of chronic carbon monoxide exposure would be to subject body and brain tissues to prolonged hypoxia. Carbon monoxide brain lesions found at autopsy resemble other types of hypoxic–ischemic lesions. Hippocampus, globus pallidus, and white matter lesions are also reported. Neuropsychological abnormalities produced in other carbon monoxide-exposed patients are similar to cortical dysfunctions experienced by this patient.

Thus, it is entirely possible that the patient's deficits in memory and complex spatial integration could have been caused by prolonged exposure to carbon monoxide. Her pattern of performance does not appear typical of common dementing syndromes; it is also worse in some abilities than age norms and premorbid functioning would suggest. Deficits are consistent with parietal lobe cortical dysfunction with possible extension into subcortical structures. The patient's poor performance on complex tactual stimuli also suggests that PNS involvement should be ruled out.

By the patient's self-report, she has experienced some recovery of function since replacement of their hot water heater installation. This provides some basis for optimism for continued recovery. However, since CO-exposed patients show quite variable recovery patterns, each case must be longitudinally assessed on an individual basis. Reevaluation after 1-2 years is suggested to track recovery of function.

Diagnosis (DSM-III-R). 294.80 Organic Mental Syndrome Not Otherwise Specified; 293.83 Organic Mood Syndrome—Depressed.

Recommendations.

1. Repeat neuropsychological tests in 1-2 years to reassess cognitive function.
2. Neurological workup to rule out alternative causes of behavioral impairment. Evaluation of PNS functions would be useful to rule out noncortical contributions to observed sensory abnormalities.
3. If social phobias and depressive behaviors persist, supportive counseling may be useful.

Follow-Up Testing. The patient was referred for follow-up neuropsychological testing to determine the course and severity of neuropsychological impairments noted on first testing one year ago and attributed to chronic carbon monoxide exposure.

Tests Administered

Wechsler Memory Scale subtests (Russell Modification):
Logical and Figural, Memory immediate and 30-minute delayed recall, Associate Learning Test
Halstead–Reitan Battery subtests:
Finger Tapping Test, Trail-Making Tests A&B, Category Test, Tactual Performance Test, Sensory Perceptual Examination (Selected Items), Grip Strength (Dynamometer)
Wechsler Adult Intelligence Scale subtests: Digit Symbol Test, Block Design Test

Other Tests

Stroop Color-Word Test
Luria–Nebraska Pathognomonic Scale
Spatial Relations Test
Grooved Pegboard Test
Shipley Institute of Living Scale
Beck Depression Inventory
Profile of Mood States
Minnesota Multiphasic Personality Inventory

Clinical Interview. In the clinical interview preparatory to her second neuropsychological exam, the patient described herself as "more alert and comprehending better ... now I can read the newspaper," but still having difficulty several days each week. Her current difficulties include some comprehension and memory problems when holding conversations with her husband.

The patient has almost completely ceased feeling disoriented while she is driving, although she had one recent episode last week, where she had to "stop and think 'what does a green light mean?' " The patient stated that her improvement has been "very gradual" and continual for the past several years.

The patient denied depression on second clinical interview, although she finds herself crying when she believes she is in the presence of a toxic chemical, an experience that suggests to the patient that "I don't have any control emotionally." The patient stated that she has difficulty viewing her tears as an emotional reaction to exposure because "I get sort of dulled." However, she stated that these episodes happen "infrequently ... not more than once a month, and only when I get into something really bad," for example, when her neighbors sprayed the lawn with pesticide.

Despite her emotional episodes, the patient described herself as having a positive attitude and credits that attitude with her improvement. The patient strongly believes that she has improved over time and since her last testing.

Sleep was described as good until recently, but has recently deteriorated, with the patient attributing this change to stress resulting from increased levels of activity. The patient has difficulty staying asleep, and reads for 1–3 hours before she is able to resume sleeping. Appetite is good and stable. The patient continues to have balance problems but hides it from others. "I always walk near the wall. . . . I'm careful, now I don't fall anymore."

The patient takes vitamins and minerals in a multivitamin supplement. She also takes megadoses of vitamin C. She does not use drugs or drink alcohol. The patient still reports occasional "lights" visible at the periphery of her vision, and she has seen an eye doctor "who says 'it's in the brain, not the eyes.' " The patient never received hyperbaric treatment during her period of carbon monoxide exposure, but takes pure oxygen several times a month.

Behavioral Observations: The patient is a white female with excellent hygiene, dress, and grooming.

The patient arrived on time for evaluation, accompanied by her husband. The patient was able to participate in clinical interview and neuropsychological testing, but had several interruptions, resulting from coughing, pain, or emotional upset. Toward the end of a test of complex spatial skills, for example, the patient experienced weakness and abdominal pain. The patient attributed the pain to the presence of an unknown toxin in the immediate environment. In her report of severe stomach pain, she stated that she was "hemorrhaging internally" whenever she experienced pain of that sort. This comment was verified by her husband, who verified that the patient was suffering from an intestinal disorder similar to Crohn's disease. Cooperation was generally good, although the patient expressed frustration about her inability to perform well on tests requiring integration of spatial information though tactile stimulation.

Neuropsychological Test Results

General cognitive functions. On a screening measure of IQ emphasizing vocabulary and verbal abstract reasoning, the patient scored in the middle of the Bright Normal range (IQ est.=124). While the patient's individual test scores were several points lower for this test on the second testing, correction for age eliminates any significant differences between the two scores. The patient's level of IQ is consistent with her educational background.

Attentional skills were excellent and unimpaired. The patient equaled and in some cases exceeded her previous test performance on tasks that required rapid, flexible visual scanning, attention in the presence of distraction, and efficient processing of repetitive information. For example, the patient was able to rapidly scan a page and draw connecting lines between alternating numbers and letters (e.g., A-1-B-2-C-3, etc.).

Abstract reasoning showed improved capabilities in the area of nonverbal inductive reasoning. The patient was able to determine spatial similarities and differences among sets of geometric figures (Category Test) at an umimpaired level of performance. This is an improvement over her prior test results, where performance was in the moderate range of impairment.

Sensorimotor functions. Fine motor speed is excellent and unimpaired bilaterally, an improvement over the patient's prior test results where nondominant hand speed showed mild impairment. In contrast, fine motor coordination has deteriorated since the patient's prior testing. The patient performed at a severely impaired level on a speeded pegboard test, with dominant hand speed being particularly poor and almost at the bottom of measurable norms.

The patient had difficulty discriminating single versus simultaneous tactile stimulation. She could not discriminate single or simultaneous points separated by 5 mm, an abnormal result. At 10 mm, the patient was able to accurately distinguish single versus simultaneous points in all fingers except the fifth finger of the dominant hand. The patient's left hand generally showed less consistent accuracy than the dominant right hand.

The patient showed very severe impairments in interpreting and integrating complex tactile perceptions (astereognosis). For example, she made several errors when asked to name a finger touched while blindfolded, and failed to discriminate large letters or numbers drawn on the back of her wrist. She was almost completely unable to accurately name numbers traced on her fingertips. Her performance was severely impaired and showed no improvement since her last evaluation 1½ years ago. The patient improved somewhat when asked to identify common objects by touch alone. She accurately named a key, eraser, and paper clip placed in her palm, but called a quarter "a nickel."

Gross motor strength was approximately unchanged from prior testing and unremarkable for the patient's age and sex.

Visuospatial functions. The patient showed above-average age-corrected scores on tests of visuospatial manipulation. For example, she was able to correctly assemble block patterns to match a template. In contrast, when the patient did not have the benefit of visual input, her performance became grossly impaired. On the Tactual Performance Test, which requires the placement of wooden shapes in a formboard while blindfolded, the patient showed severely impaired performance. The patient was unable to correctly apprehend shapes by touch alone and could not match the wooden forms with their proper indentations on the formboard. In addition, she was impaired in her ability to recall or localize those forms on paper after the task was completed. Performance on this test was worse than on the patient's previous assessment $1^{1}/_{2}$ years ago.

Although fields were not tested, the patient showed accurate visual perception in all tasks.

Memory and learning. The patient showed good and unimpaired verbal memory and learning. Recall of short story paragraphs remained approximately equal to prior test results. Recall remained good after a 30-minute delayed recall interval. Nonverbal memory for visually presented information was excellent and showed improvement since prior testing. There was no loss of visual information over a 30-minute delayed recall interval, an excellent result. Nonverbal memory for shapes apprehended by touch alone was moderately impaired.

Personality functioning. Compared with her previous testing, the patient's personality profile showed her to be slightly better able to reveal herself psychologically. Stress is still experienced in somatic ways, and the patient's profile is consistent with a general denial of psychologically based symptoms, with a preference for specific somatic concerns (e.g., headache, GI symptoms). This pattern is also slightly less pronounced than was evident in prior testing, although it continues to remain an important feature of current personality functioning. At times, the patient expressed symptoms that appeared to be quite unusual, as in her belief that an unknown toxin was affecting her behavior in session.

The patient continued to exhibit a profile consistent with traditionally feminine interests and attitudes, including conscientiousness, sensitivity, passivity, and submissiveness.

The patient's current personality profile has become less elevated relative to prior testing, in the direction of more normal attitudes and affects. The patient continues to deny depression or other psychologically based influences on her current health status. Just as in the former assessment, however, depression and psychologically based difficulties remain greater than the patient is willing to admit on formal psychological tests.

Summary and Recommendations. This is a 67-year-old white female referred for neuropsychological evaluation as part of a longitudinal follow-up of chronic carbon monoxide exposure.

The patient continues to experience sensorimotor and spatial integration deficits. There is bilateral loss of ability to discriminate one-versus two-point touch on the fingertips, some errors of fingertip localization, and grossly impaired ability to integrate complex information through touch alone. Fine motor speed was adequate but fine motor coordination that required spatial orienting of pegs before placement was severely impaired.

Other neuropsychological functions were largely intact, including memory, attention, and abstract reasoning. There was evidence for some neuropsychological improvement in these areas in the second assessment.

The patient's pattern of neuropsychological abnormality has become somewhat less diffuse than previously, and appears to be consistent with lesions of the right parietal cortex, although PNS involvement may also contribute to this patient's very poor performance on spatial and sensorimotor tasks. The lack of further deterioration in other neuropsychological areas is inconsistent with ongoing degenerative dementia, e.g., Alzheimer's disease, but consistent with a history of CO-induced neuropsychological deterioration with stabilization subsequent to correction of a faulty heating system.

Carbon monoxide exposure presentations vary widely among patients, but the patient's symptoms could certainly be consistent with unresolved sequelae to chronic, low-level CO exposure.

Diagnosis (DSM-III-R). 294.80 Organic Mental Disorder Not Otherwise Specified—Probable CO exposure etiology.

The frequency of such impairments in low-level chronic exposure to carbon monoxide is unknown at this time and awaits larger-scale clinical/epidemiological investigations.

There is animal evidence suggesting that repeated exposure to carbon monoxide may have cumulative or synergistic effects on neurochemical systems. Elovaara and Rantanen (1983) found that biochemical changes in brain function were "determined more by the number of poisonings than by the dose of a single exposure." The authors suggest that cumulative exposures to CO can result in irreparable brain-cell injury.

Neuropsychological Recovery

A case study of acute and chronic effects of carbon monoxide has been reported by Vincente (1980), who administered serial neuropsychological evaluations to a patient 4 and 13 months post-CO intoxication. Deficits were noted in cognitive efficiency and flexibility, rhythm, short-term verbal and visual memory, orientation, and complex learning. At 13 months posttrauma the patient displayed patchy areas of recovery, e.g., normal performance on Trails A and B, but continued having memory and spatial deficits. Bryer, Heck, and Reams (1988) reported a case of CO intoxication and coma who was administered the Wechsler Adult Intelligence Scale (WAIS) initially and 11 years after exposure. Overall IQ remained stable with particular deficits in visuo/spatial problem-solving (Object Assembly).

Recovery from CO intoxication is dependent on severity, chronicity, and number of repeated exposures. Systematic neuropsychological investigation is difficult because of variability among clinical cases, although Veil et al. (1970) suggest that most patients continue to show abnormalities 1 year postexposure.

Case Report: CO Intoxication with Unconsciousness

The following case is unfortunately rather typical of many neurotoxicity referrals in its complex and possibly confounding medical history. The patient had documented carbon monoxide intoxication but also a history of several closed-head injuries. The patient's low level of education further contributes to the difficulty of accurate assessment. However, despite such a checkered neurological history, she maintained an unbroken work history until her carbon monoxide exposure.

Tests Administered

Shipley Institute of Living Scale
Finger Tapping Test
Wechsler Memory Scale subtests: Logical and Figural Memory, immediate and delayed recall
Trail-Making Tests A & B
Stroop Color-Word Test
Digit Symbol Test
Luria–Nebraska Pathognomonic Scale
Spatial Relations Test
Grooved Pegboard Test
Minnesota Multiphasic Personality Inventory
Personal Problems Checklist for Adults
Clinical Interview

Background Information. The patient was a black female in her 40s with an eighth grade education. Her most recent job was that of a stockroom clerk/manager which she performed responsibly with good supervisory evaluations. She was referred for neuropsychological evaluation from the Occupational Medicine Clinic at Cook County Hospital to determine the cognitive and emotional sequelae of carbon monoxide intoxication induced by a defective space heater in her apartment.

Two days prior to her hospitalization, the patient stated that she awakened in the morning with stomach pains, but when she visited a physician that day, she was apparently sent home. The following day, the patient awakened in the morning reportedly unable to control the movement of her arms and legs. Ms. Y. then attempted to dress herself, enlisting the aid of her 12-year-old son, but their combined efforts were not successful and the patient gave up and went back to sleep. Ms. Y. remembered awakening at noon and returning to sleep.

She was discovered unconscious by her adult daughter at 5:30 P.M. that same day. Ms. Y. remained in coma for approximately 1 week, but has no personal recollection of how long she was unconscious.

The patient believed that her memory had deteriorated since the accident, and indeed, she had forgotten her initial testing appointment the week before. She also admitted to forgetting that she is cooking on the stove and frequently burns her food. Prior to CO exposure, accidental burning of food would occur at most two or three times a year. In contrast, the patient reportedly has burned her cooking "almost every day" since she was discharged from the hospital. The patient also stated that she forgets where she hides her money, with resulting permanent loss of $60 several months ago. In addition, the patient stated that she has been unable to adequately continue in her employment as a supply clerk, having returned to work for 3 days and finding she could no longer perform the functions of her job. She denied any periods of disorientation.

The patient's medical history included two closed-head injuries, the first when she was struck in the occipital region with a stick brandished by her fiancé. The second, more recent head injury occurred in the 1970s where the patient struck her head in an auto accident, cracking the windshield. The patient remained conscious in both cases, did not seek medical attention, and was able to return to work without incident. There was no reported history of seizures, high fevers, or other periods of unconsciousness. The patient denied alcohol or drug abuse.

Discussion. The patient's current test results suggested mild to moderate neuropsychological dysfunction consistent with carbon monoxide intoxication. The patient's most significant deficits involved both visual memory and her ability to organize, plan, self-monitor, and maintain attention. For example, the patient showed poor memory for abstract designs, and when required to match simple patterns to numbers, performed slowly and reversed many patterns. She also performed poorly and made several errors when attempting to draw lines between alternating letters and numbers (Trails B).

Some areas of poor performance may be more easily tied to the patient's relatively low education (eighth grade). For example, the patient did poorly on an IQ screening measure which suggested that her vocabulary and abstract abilities were well below average. However, the patient's reportedly good record of employment prior to her accident suggests that nonverbal memory and organizing abilities necessary for effective performance as a stockroom distributor are significantly decremented as a result of her recent exposure to carbon monoxide.

Personality evaluation suggested that the patient is coping poorly with the stress of her injury and subsequent unemployment. She reported feelings of depression, poor sleep, tension, social isolation, and low self-esteem. The patient attempted to handle current life stressors with inflexible and probably ineffective coping strategies. She may have had periods where her defenses deteriorated and she appeared disorganized and unable to cope. The patient's current personality functioning may be related to structural sequelae of carbon monoxide intoxication and coma, as depression and poor coping are commonly reported neuropsychological consequences of these types of cases.

Summary and Recommendations. This is a black female in her early 40s rendered unconscious for approximately 1 week as a result of carbon monoxide intoxication from a faulty space heater. The patient exhibited deficits in visual memory, nonverbal abstraction, planning, attention, and concentration, all of which are consistent with known effects of CO intoxication and coma. Lower than average intellectual screening results are probably more attributable to education than to CO exposure. The patient is coping poorly with social and psychological consequences of her disability; she is depressed, isolated, tense, and does not cope well emotionally.

The patient will most likely not be able to resume her employment in the near future. Cognitive rehabilitation of what appear to be right hemisphere deficits in nonverbal memory and abstraction may allow the patient to return to work in the future. Supportive psychotherapy to increase self-image and coping skills is also indicated; a referral to the patient's local mental health center would be desirable.

CASSAVISM

Found in populations that employ cassava root as a food staple, symptoms are similar to those reported for lathyrism. Etiology appears to be associated with the high cyanide content of the plant coupled with a low-protein diet that cannot supply sulfur required to detoxify the cyanide (Spencer *et al.*, 1993). Clinical manifestations include symmetrical spastic paraparesis, hyperreflexia, bilateral extensor plantar signs, and pyramidal signs in the upper extremities (Spencer *et al.*, 1993). Other symptoms include visual impairment with possible optic atrophy, dysarthria, hearing loss, and leg numbness. Walking impairments may be permanent.

CHLORINE

Conversion factor	1 ppm = mg/m^3
Odor threshold	0.08 ppm
IDLH	30 ppm
OSHA	0.5 ppm (1.5 mg/m^3)
STEL	1 ppm (3 mg/m^3)
NIOSH	TWA not given
Ceiling	0.5 ppm (1.45 mg/m^3) (15 min)
(*Quick Guide*, 1993)	

Chlorine gas exposure is relatively uncommon but has clear neurotoxic consequences. Several reports of chlorine-induced brain damage are complicated by fatal exposure levels (Adelson & Kaufman, 1971) or nonfatal exposure complicated by concurrent hypoxia and premorbid low intelligence (Auerbach & Hodnett, 1990). A case that addresses previous shortcomings was discussed by Holmes and Haben (1991) who report two nonfatal cases of chlorine poisoning without hypoxia. Symptoms in both cases included double vision, tremor, and numbness, as well as neuropsychological deficit. The first case involved a 30-year-old high school-educated male who had been attempting to add chlorine to a tank in an enclosed building. The tank exploded and released high concentrations of chlorine gas. The patient was in the building for about 30 seconds, left the building, and received medical treatment. Initial trauma was confined to breathing difficulty, with negative CT and MRI. Recovery was complicated by pneumonia. Initial neuropsychological examination at 4 months postinjury showed generalized decrements in IQ, memory, psychomotor speed, problem-solving, fine motor speed, expressive language, and visuospatial functions. Eight-month follow-up evaluation showed patchy recovery of memory and visuospatial functions, but poorer dominant hand motor speed. Similar findings occurred in an older male whose case was also assessed by the same authors. The mechanism of chlorine induced brain-damage is not known at this time.

CYANIDE

as HCN	
Conversion factor	1 ppm (1.12 mg/m^3)
Odor threshold	0.1–5.0 ppm
IDLH	50 ppm
OSHA/NIOSH	TWA not given
STEL	4.7 ppm (5 mg/m^3)
(*Quick Guide*, 1993)	

Cyanides occur naturally in certain foods, and are produced in industrial processes. In small quantities, as vitamin B_{12} (cyanocobalamin) it is essential to human health. Automobile exhaust is the largest source of cyanide in the air, with 11–14 mg/mile for cars without catalytic converters or with defective converters, and 1 mg/mile for cars equipped with fully functional converters. Cyanide is also released into the air from metallurgical industries, petroleum refineries, wastewater treatment sites, iron and steel factories, and organic chemical plants (ATSDR, 1989b). An estimated 143,720 workers may be exposed to cyanide in various compounds, including those employed in electroplating, metallurgy, pesticide application, firefighting, steel production, gas workers operation, and metal cleaning. Manufacture of pharmaceuticals, photoengraving and photography, cyanide production, tanneries, and blacksmithing are also potential sources of exposure (ATSDR, 1989b). Cyanide is inhaled from cigarette smoke with dose per cigarette ranging from 10 to 400 μg. The Agency for Toxic Substances and Disease Registry estimates that between 25 and 30% of the U.S. population daily inhales between 250 and 10,000 μg cyanide per day.

Exposure to 270 ppm of hydrogen cyanide is immediately fatal, with lower exposure of 110 ppm producing death in 30-60 min (see Table 8.4).

Levels as low as 18 ppm cause headaches, weakness, and nausea after several hours of exposure. Characteristic odor of bitter almond is first detected at about 0.6 ppm. Risk of cyanide toxicity increases when coupled with deficiencies of iodine, protein, B_{12}, or riboflavin.

Cyanide affects many target organ systems, including the cardiovascular, pulmonary, endocrine, and nervous system. The extraordinary toxicity of the cyanide ion results from its rapid reaction with the trivalent iron of cytochrome oxidase. The cytochrome oxidase–cyanide complex "renders cells unable to utilize oxygen, and they rapidly become hypoxic and die" (Politis et al., 1980, p. 617).

Neurological effects of nonlethal exposure to cyanide are typically noted in the range of 5 to 44 ppm (6 to 49 mg/m^3). Symptoms include deafness, visual impairments, and incoordination. Patients poisoned by cyanide have exhibited abnormal EEG, dilated pupils, hemiparesis, headache, fatigue, vertigo, tinnitus, "sensory obtusion," dyspnea, tremor, and loss of consciousness. Heavy cigarette smokers may show cyanide-induced "tobacco amblyopia," optic atrophy, Leber's hereditary optic atrophy, retrobulbar neuritis, pernicious anemia, and optic atrophy (ATSDR, 1989b).

Cyanide causes damage to cortical neurons, cerebellar cortex, hippocampus corpora striata, and substantia nigra. White matter damage in the corpus callosum may also occur (Norton, 1986). A suicide attempter who ingested between 975 and 1300 mg of KCN (390-519 mg CN) developed CNS depression and behavioral changes in the immediate aftermath, with parkinsonian symptoms following in subsequent months. When the patient died of a drug/alcohol overdose 19 months later, neuropathology was noted in the globus pallidus and putamen, a finding inconsistent with overdose and attributed to cyanide neurotoxicity.

Mechanisms of neurotoxicity, selective encephalopathic effects, and cyanide-produced neurotransmitter abnormalities are not well understood (ATSDR, 1989b). It has been suggested that focal encephalopathic effects may correspond to vascular distribution and vascular insufficiency; damage may be both cytotoxic

TABLE 8.4. Human Fatality Data: Hydrogen Cyanide in Air[a]

Response	Concentration	
	mg/liter	ppm
Immediately fatal	0.3	270
Estimated human LDC_{50} after 10 min	0.61	546
Fatal after 10 min	0.2	181
Fatal after $1/2$–1 hr or longer or dangerous to life	0.12–0.15	110–135

[a]Hartung (1981).

and anoxic. Workers who have been exposed to cyanide have experienced a variety of CNS and PNS effects, including dilated pupils, hemiparesis, headache, malaise, hemianopsia, abnormal EEG, and numbers or tingling in the hands. Fumigators working with HCN experienced tachycardia, vertigo, tinnitus, headache, epigastric burning, vomiting, weakness, tremor, "sensory obtusion," nystagmus, vertigo, equilibrium disturbances, and loss of consciousness (Carmelo, 1955). It has been suggested that tobacco amblyopia, Leber's hereditary optic atrophy, retrobulbar neuritis, complicating pernicious anemia, and optic atrophy may involve defective metabolism of cyanide to thiocyanate (ATSDR, 1989b).

Few neuropsychological studies of cyanide poisoning are available. In a woman who drank a cyanide-adulterated soft drink, neuropsychological evaluation 1 year postpoisoning showed impaired visuospatial memory, poor performance on visuospatial tasks, and impaired abstract reasoning (Carella et al., 1988). At 5 years postpoisoning, the patient presented with clinically parkinsonian features. She showed a diffusely irritative EEG, especially in left temporo-occipital regions. Visual evoked response showed decreased amplitude of P100. BAEP was delayed to the peak of VI wave. Neuropsychological test results were not reported in detail, but were characterized as stable, with continued impairments in visual and constructional ability, and abstract reasoning. MRI showed cerebellar and cerebral atrophy with "marked" ventricular dilation. Abnormally high signal intensity was noted in both cerebellar hemispheres, posterior subcortical white matter, both cerebral hemispheres, globus pallidus, posterior putamen, and the right parietal cortex.

CYCADS

Ingestion of starch made from the false sago palm *(Cycas circinalis)* has been held responsible for attack rates of ALS (motor neuron disease) with parkinsonism and dementia (ALS/P-D) in Guam and the Mariana Islands among the Chamorro people 50–100 times higher than the U.S. incidence. Flour prepared from the seed of *C. circinalis* contains neurotoxins, the excitotoxin B-*N*-methyl amino-L-alanine and cycasin, a potent neurotoxin (Spencer et al., 1993).

EXTREMELY LOW-FREQUENCY ELECTROMAGNETIC RADIATION (ELF)

There is considerable controversy over the health effects of magnetic fields associated with power lines and video display terminals (VDTs). The capability of electromagnetic fields to influence natural electric fields in the human body has been documented. For example, exposure to high-level magnetic fields produces "magnetophosphenes," a flickering light perceived by the retina. Several incidents of elevated cancer risk in children living near power lines have also heightened concern.

pain increased and cold sores began to develop. Her feet "felt like a truck ran over them." One of her arms became so weak she was unable to move it.

A daylong evaluation at a teaching hospital confirmed the diagnosis of EMS.

The patient denied any history of chronic preexisting illnesses. However, her family history is positive for rheumatoid arthritis and multiple CVAs in the father.

Current complaints include fatigue, memory problems, problems with intensive problem-solving, sustained attention, transient numbness in the right foot, arm, and leg, and occasional joint pain. There are reported to be red marks on the patient's legs, but these have not progressed. Raynaud-like symptoms began in the patient's hands, ears, and nose within the past year, but have lessened this year. Other current symptoms include dry skin, joint pain, frequent colds, easy bruising, muscle cramping, irregular menstrual periods, recurrent cystitis, canker sores, and dermatological abnormalities.

Currently, the patient takes about one or two coated aspirin per month, especially after sitting in a confined position or carrying heavy articles. The patient takes Anaprox for uterine cramps during menstruation, about 3 days per month. She began taking multivitamins approximately 1 month ago, along with 500 mg vitamin C each day.

There is no history of head injury or significant auto accident. The patient denied any periods of unconsciousness or syncope. There is no history of auditory or visual hallucinations, or delusions. The patient denied any history of suicidal thoughts, plans, or actions.

Prior to taking L-tryptophan, the patient had been an excellent athlete, engaging in strenuous outdoor exercise and both aerobic and anaerobic health club exercise. She had jogged up to 13 miles at a time during the 1980s, and had been able to bicycle for up to 50 miles, but now can sustain such effort for only 4–5 miles. Before taking L-tryptophan, she was able to perform strenuous aerobic exercises for 45–60 minutes. Her current limits are about 15–30 minutes of very low-level exercise.

Mental Health History. The patient had seen a psychologist twice before, once related to overwork, and again for a brief reactive depression. Symptoms remitted completely in each situation after brief treatment. Since developing EMS, the patient was seen in cognitive therapy and depression reactive to the loss of her previously active life-style. She was tried on various antidepressants, but discontinued them after side effects developed.

Academic History. The patient's grades in high school were "average" C's but she applied herself in college and graudated with high honors and Phi Beta Kappa. She went on to receive a master's degree.

Behavioral Observations and Mental Status. The patient arrived on time for testing. Dress was neat and hygiene was good. The patient was oriented to time, place, person, and situation and was able to sustain cooperation throughout a full day of testing with only a single intrusion of emotional disruption. This occurred when she was given her first memory test and tears welled up in her eyes when she began having difficulty. At this point, the patient quickly mastered her feelings and continued with no discernible effect on subsequent tests.

Neuropsychological Test Results
Intellectual, language, and general cognitive functioning. General intelligence, as estimated from a test of vocabulary and verbal concept formation, was at the juncture between Average and High Average, a result consistent with the patient's employment and education. Speech was normal in volume, rate, and timbre. There were no deficits in expressive or receptive speech, and no indication of dysarthria or aphasia. The patient's most notable impairment was in the area of rapid, sustained, visual attention. In one test, for example, the patient was required to press a button after seeing the numbers 1 and 9 presented consecutively on a computer panel. Performance on this test was severely impaired relative to individuals in her age group; performance was the bottom 1% of population norms. The patient remained impaired in a more complex version of the same task, where attention to the numbers 1–9 in a center panel was distracted by side panels showing randomly flashing numbers. The patient showed moderate impairments in attention, with performance falling in the bottom 10% of the population. In addition, there were many errors on this task involving nonresponse (omissions) rather than impulsive or late responses (commission), suggesting poor capacity to maintain rapid, flexible visual attention.

Other less difficult attentional tasks showed normal performance. The patient performed normally on tasks involving repetitive nondistracting visual stimuli, and sustained attention to verbal and nonverbal auditory stimuli.

Concept formation and inductive problem-solving were unremarkable and unimpaired. Two tests that have been linked to frontal lobe function were excellent and unimpaired. In the first (WCST), which required flexible set shifting, the patient performed so well that the task was discontinued early. The patient also showed high average performance in a test of speeded word production which also requires cognitive flexibility and prefrontal lobe function. Finally, the patient showed average skill in a task involving sustained inductive reasoning.

Sensorimotor functions. Fine motor speed was in the average range for both hands, with expected dominant hand preference. Fine motor coordination was mildly low in the dominant hand, but normal in the nondominant hand.

Memory and learning. Verbal memory for categorized word lists (e.g., tools, spices, fruits) showed mild decrement at first presentation, but improved to normal recall after several presentations. The patient showed mild decrements in immediate list recall, but used explicit cues to boost her performance to the average range. Twenty-minute delayed recall showed mild impairment for both free recall and cued recall. The patient's principal verbal memory difficulty involved a severe inability to correctly recall only previously heard items without intrusions.

The patient also performed within the normal range for recall and recognition of a more difficult list without explicit categories, by attempting mnemonic strategies to boost her recall.

Visual memory for figures was adequate when the patient was required to draw designs immediately and after surprise 30-minute recall. However, the patient's ability to discriminate more complex designs from one another was relatively poor, especially when sustained attention was required to distinguish new designs from old. The patient correctly identified these visually perceived designs items more poorly than 96% of the population. In addition, she made more incorrect visual memory discriminations than 70.5% of the population. The patient's ability to discriminate items correctly was in the impaired range.

Spatial memory for items perceived by touch alone was moderately impaired for both item shape and item location.

Visuospatial functions. The patient showed no evidence of constructional apraxia or other impairment of visuospatial information processing. The patient was able to identify common objects that were presented as scrambled puzzle pieces and could rapidly assemble block designs from a template.

Personality functioning. The patient denied having any unusual emotional reactions the week of her neuropsychological examination and portrayed herself as relaxed, good-natured, friendly, and cheerful. However, scores on a self-report depression inventory placed the patient in the Mild range of depression, with admissions of self-criticism, increased crying, irritability, loss of interest in other people, and increased fatigue.

Personality testing on a more extensive inventory suggested generalized elevations in most scales, most likely indicating acute disturbance and a large number of physical and reactive psychological difficulties. The patient's profile is consistent with a high degree of concern over physical malfunction, which is not unrealistic under the circumstances. In addition, the patient's scale scores indicate concern over the loss of control over mind and emotions, as well admission of chronically stressful and unrewarding life circumstances.

Among the stresses listed by the patient include job stress, fear of job failure because of high expectations, and difficulties restructuring her marriage in light of physical disabilities. These appear to be largely reactive to EMS disease onset.

Summary and Conclusions. This is a white female referred for neuropsychological evaluation to determine the nature and extent of neurological involvement of diagnosed EMS. Behaviorally, the patient reports a significant loss of physical stamina since being diagnosed with EMS. Although she exercises and attempts to increase her endurance, she experiences moderate to severe limitations on her endurance and capabilities relative to premorbid function.

Neuropsychologically, the patient shows severe deficits in a visual attention test that requires rapid visual decision-making over a sustained period. The test involved is commonly used to assess attention deficit difficulties, and results are consistent with patient report of difficulty handling rapidly presented, simultaneous task demands.

Memory was an additional area of significant dysfunction. The patient displayed mild impairments in verbal memory, which would probably have been worse had she not employed mnemonic strategies to boost her encoding and recall. Even so, she displayed abnormally high numbers of recall intrusions, suggesting an inability to discriminate among correct versus incorrect words. Visual memory and discrimination of finely detailed abstract drawings was also moderately to severely impaired, as was memory for information perceived through tactile sensation alone.

In contrast, the patient's overall intelligence, problem-solving capability, and abstract reasoning were unimpaired, compared with premorbid estimates. The patient's overall pattern of impairment suggests principal difficulties with attention and rapid transfer of information across the brain. Such difficulties may indicate diffuse white matter damage from EMS.

Since it is a recently discovered syndrome, there is no literature on the lifetime course of EMS. Therefore, the patient's course of disease or recovery is unknown. It is likely, however, that her current neuropsychological deficits will be long-lasting, if not permanent. Because the patient has been attempting self-rehabilitation since her diagnosis, her performance during this testing must be considered to be at her optimal level. Thus, it is likely that she would have been even more impaired, had she not been practicing her own form of cognitive rehabilitation strategy (e.g., use of mnemonics to aid memorization).

Although the patient has symptoms of depression and psychological stress, these were not considered sufficient to be clinically disruptive during the neuropsychological examination. Thus, current results are considered to be unaffected by emotional status. Degree of stress response does indicate the need for continued stress management psychotherapy. It is also recommended that the patient's work efficiency would be enhanced by limiting distractions and simultaneous interruptions of her work. In general, work efficiency will be maximized if the patient is able to structure her tasks to occur sequentially rather than simultaneously.

ETHYL CHLORIDE

Conversion factor	1 ppm = 2.68 mg/m^3
Odor threshold	4.2 ppm
IDLH	20,000 ppm
OSHA	1000 ppm (2600 mg/m^3)
NIOSH	Data not given
(*Quick Guide*, 1993)	

Ethyl chloride ("Ethyl Gaz," "Ethyl Four Star") was developed by Basil Valentine in the 17th century. It was used as a local anesthetic until a Swedish dentist discovered its general anesthetic properties by accidentally rendering a patient unconscious after spraying his gums for a dental extraction.

Ethyl chloride gas is volatile and capable of producing anesthesia within 1-4 min. Elimination through the lungs is slow because of high blood solubility. The gas causes vagal stimulation and may cause arrhythmias (Nordin, Rosenqvist, & Hollstedt, 1988). In the United States, it is employed by physical therapists and sports medicine clinicians to treat as a local anesthetic for muscle and tendon injury. In spray form, it is an abusable with neurological and psychological sequelae. Ethyl chloride abuse may be found in industry (Nordin et al., 1988), among homosexual males in place of amyl nitrate (Hersh, 1991), and in other circumstances (Hes, Cohn, & Strefler, 1979).

Short-term symptoms of confusion, hallucinations, ataxia, and short-term memory impairment have been noted. Nordin et al. (1988) report the case of a 52-year-old male who sniffed ethyl chloride periodically for 30 years, along with alcohol and barbiturate abuse. Prior to hospital admission, the patient had inhaled 100 ml per day of ethyl chloride. Toxicological screens for alcohol and other abused drugs were negative on admission. Euphoria, visual and auditory hallucinations, amnestic syndrome, and diffuse moderate neuropathy were noted, but had largely remitted after 3 weeks. A neuropsychological screening battery administered 6 weeks after hospital admission also failed to find impairments in memory or general intelligence.

Another case study involving acute use of 200-300 ml per day noted fine horizontal nystagmus, ataxia, and paranoid auditory hallucinations (Hes et al., 1979). Cerebellar signs, nystagmus, and finger tremor disappeared by the end of the third week. Intelligence testing was unremarkable but more formal neuropsychological testing was not performed.

ETHYLENE OXIDE

Conversion factor	1 ppm = 1.83 mg/m^3
Odor threshold	257-690 ppm
IDLH	800 ppm
OSHA	< 1 ppm; 5 ppm (15 min excursion)
NIOSH	< 0.1 ppm (0.18 mg/m^3)
Ceiling	5 ppm (9 mg/m^3) (10 min/day)
Odor threshold	257-690 ppm
(*Quick Guide*, 1993)	

Ethylene oxide (EtO) is a highly reactive gas primarily used to disinfect materials that would be damaged by heat sterilization but is also used to produce ethylene glycol, polyesters, and detergents. Because the gas readily penetrates cellophane and other wrappings, it is used by manufacturers of medical supplies to disinfect prepackaged materials. Because EtO is the preferred (and sometimes the only) way to sterilize these materials, exposure is widespread. From 75,000 to over 100,000 workers are exposed to EtO in the health care industries alone

(Gross, Haas, & Swift, 1979; Landrigan, Meinhardt, & Gordon, et al., 1984). Many more individuals may come into contact with the gas during other industrial processes involving ethylene glycol, including fumigation, as well as production of pigments, rocket propellants, and ethylene glycol (Gross et al., 1979).

Neurological disorder in the form of sensorimotor neuropathy can develop with 3–5 months of repeated exposure to several hundred parts per million of EtO. Sensorimotor neuropathy develops with a more chronic course with exposure of approximately 10 ppm. Ohnishi and Murai (1993) have reviewed the neurological literature on EtO in various cohorts of patients, and summarize the results as follows:

- 10/12 patients studied showed muscle weakness and decreased sensation in distal lower limbs.
- Ankle jerk was absent in 9/10 patients and decreased in 1/10.
- Nerve conduction was abnormal in 8/10.
- Three patients showed mild histological abnormalities including decreased density of large myelinated fibers and other evidence of mild axonal degeneration.
- CSF showed elevated protein in 2/6 patients.
- Needle EMG showed neurogenic changes in 8/11 patients.
- Stopping exposure allows sensorimotor improvement in a majority of cases.

Dermatological, conjunctival, and mucosal irritation are consequences of exposure to EtO. Asthma, pulmonary edema, nausea, and vomiting have also been noted (Deschamps et al., 1992; Finelli, Morgan, Yaar, & Granger, 1983). CNS and PNS sequelae have been reported, including headache, abnormal EEG, seizures, and generalized sensorimotor polyneuropathy. Neurological examinations of some patients show substantial recovery at 2 weeks postexposure, although there appears to be significant individual variation with some individuals recovering after 6 months but others failing to recover on 1 year follow-up (Gross et al., 1979; Finelli et al., 1983). Related compounds of propylene oxide and butylene oxide are also neurotoxic, although their neurotoxicity is not as great as EtO (Onishi & Murai, 1993).

Neuropsychological Effects

Both individual case studies and small group comparisons have found human neurotoxic sequelae from EtO exposure. For example, Crystal, Schaumburg, Grober, Fuld, & Lipton (1988) report neurological and cognitive impairments in a 29-year-old female with 10 years of EtO exposure. Symptoms began 7 years after initial exposure, and included headache, fatigue, dizziness, memory impairment, and loss of concentration. Symptoms were worse at the end of the workday but better on weekends. Time-weighted average EtO levels were 2.4 ppm, but levels near the sterilizing equipment were 4.2 ppm. General physical examination 3 months postexposure showed loss of lower limb vibration sense, but no other symptoms. One year postexposure, the patient was anxious, emotionally labile, and showed poor word memory and digit span. Sensation to cold was decreased to midthigh, deep tendon reflexes were 2+. Vibration thresholds were 3 standard deviations higher than the reference population mean. EEG showed mild background slowing, but MRI was normal.

Neuropsychological examination showed low average IQ that was inconsistent with demonstrated reading proficiency and an associate college degree. Cued verbal recall, visual memory, and attentional impairments were noted, along with perceptual and visuomotor slowing. Memory problems were characterized as a pattern of "intact learning with profound forgetting."

Several studies by Estrin and colleagues suggest significant neuropsychological impairment as a consequence of EtO exposure. Estrin et al. (1986) used a computerized neuropsychological battery to compare the performance of eight female hospital workers with chronic ethylene oxide exposure to controls matched for age, sex, and education. Ethylene oxide subjects were exposed for a mean of 11.6 years. Subjects were administered several neurological tests and a computerized neuropsychological battery that included the Continuous Performance Test, Simple Reaction Time, Digit Span Forward, Symbol Digit, Pattern Memory and Horizontal Addition and Hand–Eye Coordination. A significant dose–response relationship was found between scores on the Continuous Performance Test and years of exposure to ethylene oxide, which persisted after controlling for age. Exposed subjects also performed significantly more poorly on the Hand–Eye Coordination Test, with the Symbol Digit Test approaching significance ($p=0.06$). While no other test result achieved significance, the exposed group performed more poorly on all tests. Unfortunately the authors analyzed their data with multiple t tests and significance levels were not corrected for multiple comparisons.

The only neurological test that correlated with ethylene oxide exposure was a significant (age corrected) dose–response relationship between years of exposure and diminished sural nerve conduction velocity (Estrin et al., 1986).

A more recent study (Estrin, Bowler, Lash, & Becker, 1990) evaluated ten hospital workers chronically exposed to EtO and compared their performance with controls matched for age, sex, race, and education. EtO levels were not monitored during exposure, but independent air samples ranged from 15 ppm to 250 ppm. All subjects received a neuropsychological screening battery, the computerized Neurobehavioral Evaluation system (NES), P300, a two-electrode ERP signal, and several other neurological tests (i.e., reflex, pinprick, vibration).

Results of neurological examination showed significantly reduced ankle reflex reduction in the exposed groups. Nerve conduction data did not differentiate exposed patients from controls. P300 showed significantly lower amplitude in the exposed group.

Neuropsychological testing using Bonferroni-adjusted significance levels showed equivalent vocabulary skills between patients and controls, an expectable finding in an education-matched sample. Trails A was significantly worse in exposed patients, and borderline significant differences were found on a cancellation task and WMS Visual Reproduction. The computerized NES battery showed significantly poorer performance on finger tapping, a finding that was not verified with a conventional finger tapper. All exposed patients showed significantly elevated tension, anger, depression, fatigue, and confusion on the NES Profile of Mood States. MMPI scales Hypochondriasis, Depression, and Hysteria were clinically abnormal and significantly higher in exposed patients.

Problems in the study are the inclusion of neuropsychological data from subjects with histories of diabetes and polio, as well as unknown influence of depression and psychological factors on neuropsychological results. Nonetheless, this and previous studies (e.g., Crystal *et al.*, 1988) suggest the likelihood of neuropsychological and neurological impairments resulting from EtO exposure.

Case Report

Tests Administered

Finger Tapping Test
Wechsler Memory Scale subtests: Logical and Figural Memory, immediate
    and delayed recall, Digits Forward and Backward, Associate Learning Test
Trail-Making Tests A & B
WAIS-R Digit Symbol subtest Stroop Color-Word Test
Luria–Nebraska Pathognomonic Scale
Spatial Relations Test
Grooved Pegboard Test
Grip Strength Test
Shipley Institute of Living Scale
Profile of Mood States

Background Information. The patient was an African-American male in his late 30s referred for neuropsychological testing to evaluate cognitive and emotional sequelae of chronic ethylene oxide exposure. Mr. N. reported that he operated a sterilizer of plastic hospital materials for 5 years, ending in 1985 when he was switched to the operation of a UPS machine. The sterilizer used ethylene oxide gas as part of the sterilization process. Mr. N. stated that for the first 4 years of his exposure, the sterilizing machines had no exhaust systems, thereby allowing the gas to escape in his direction whenever he opened the door to remove the equipment.

Mr. N. first noted physical symptoms during his third year of employment; these included dry, damaged hair and grayish skin discoloration. He remembered wondering about the effects of sterilizer chemicals, especially as they appeared corrosive to his clothing and shoes. Subsequently, the patient developed "rough headaches." In addition, he claimed to have "passed out" about 10–12 times during his period of exposure for periods of up to a minute.

Mr. N. denied any history of head injury, high fevers, or seizure disorder. There was no current or past history of alcohol or drug abuse. The patient reported an episode of hypertension occurring in February of 1985, but stated that it had resolved in a subsequent evaluation. He is currently prescribed ergotamine for headache.

Discussion. Current neuropsychological test results suggest a pattern of deficits attributable to both ethylene oxide exposure and preexisting learning disability. The patient shows severe deficits in visual memory, learning new verbal material, and significant difficulties in sustaining attention and concentration; deficits probably not attributable to low education alone.

Other areas of poor performance are more likely attributable to preexisting learning disability, including very poor reading and spelling skills, and the manipulation of abstract verbal concepts. Mr. N. admitted to school difficulties that limited his achievement in school and which may have caused him to leave high school before graduation.

Personality evaluation in clinical interview did not suggest any remarkable difficulties. Mr. N. denied symptoms of depression and had no history of psychiatric hospitalization or use of mental health services. The patient admitted to feeling angry and abused by his employers who, he believes, withheld information that would have allowed him to make a knowledgeable decision about continuing to work with ethylene oxide. However, he showed no evidence of unusual suspicion or delusion concerning his exposure.

Summary and Recommendations. This is a male in his 30s referred for neuropsychological evaluation after a 5-year history of work-related ethylene oxide exposure. The patient showed educational deficits that by pattern and history appear to predate exposure. Other dysfunctions seem consistent with published literature on ethylene oxide neurotoxicity. CNS effects of ethylene oxide have also been reported and the patient's poor visual memory and inability to sustain attention and concentration may be CNS sequelae of exposure.

FOODS

Most types of food poisoning are the result of bacterial bloom or other contamination. Important exceptions are related to foods with excitotoxic potential. Animal studies have noted the extreme vulnerability of the infant and adolescent hypothalamus to glutamate. Infant mice treated with glutamate developed a neuroendocrine syndrome that included skeletal shortening, obesity, and reproductive failure (Zorumski & Olney, 1992). Processed and artificially sweetened foods are potential sources of various excitotoxins, including Glutamate (Glu) and Aspartate (Asp), and it is of concern that "immature humans are being fed multiple excitotoxins, including Asp contained in Aspertame (Nutrasweet®), Glu, Asp, and cysteine contained in protein hydrolysates" (p. 284). The authors warn that there may be no outward signs of hypothalamic destruction and that "immature humans, on the basis of species and age, may be exceedingly vulnerable to Glu-induced brain damage" (p. 284).

FOOD ADDITIVES

Commonly available flavor enhancers, sweetners, and other food additives are being discovered to have potentially neurotoxic properties. The mechanism by which neurons are damaged or destroyed is called *excitotoxicity*—the capacity to silently destroy brain neurons through overactivation. Processed and artificially sweetened foods are the main sources of glutamate and other excitotoxins, including Aspartate (Asp), found in Aspartame (Olney, 1994). Glutamate (Glu), one of the most commonly available food additives, is the most well-known example of a neurotoxicant which, in in young animals, "requires only a single feeding . . . for neurons to be rapidly and irreversibly destroyed" (Olney, 1994, p. 537). Animal studies have repeatedly found extreme vulnerability of the infant and adolescent hypothalamus to glutamate. Infant mice treated with glutamate developed a neuroendocrine syndrome that included skeletal shortening, obesity, and reproductive failure (Zorumski & Olney, 1992). There is apparently a very small margin of safety for children eating certain processed foods. Olney (1994) notes that a cup of soup containing 1300 mg of glutamate would provide a 10 kg child with about 130 mg/kg of glutamate, when the amount needed to destroy hypothalamic neurons in animals is only 2–3 times that total—a very thin margin of safety, given the fact that glutamates and other excitotoxins may appear in more than one food per meal. "Immature humans are being fed multiple excitotoxins, including Asp contained in Aspertame (Nutrasweet®); Glu, Asp, and cysteine contained in protein hydrolysates" (p. 284)." There may be no outward signs of hypothalamic destruction. Adults may not be immune from excitotoxic damage, since they maintain blood glutamate concentrations twenty times higher than the adult mice, upon whom much of the research has been done. Olney (1994) disputes food industry research on effects of glutamate on primates, since the animal subjects were anesthesized with a glutamate antagonist during the experiment and so were provided with protection against excitotoxic damage.

FORMALDEHYDE

Conversion factor	1 ppm = 1.25 mg/m^3
Odor threshold	0.027–0.9770 ppm
IDLH	30 ppm
OSHA	0.75 ppm
STEL	2 ppm
NIOSH	0.016 ppm
Ceiling	0.1 ppm (15 min)

(*Quick Guide*, 1993)

Formaldehyde is a highly irritating and potentially carcinogenic liquid or gas used in germicides, fungicides, embalming fluids, artificial fibers, and dyes

(Anger & Johnson, 1985). It is a by-product of cigarette smoke, auto and diesel exhaust and is gradually emitted from urea foam insulation, particle board, and other construction materials (Kilburn, Warshaw, et al., 1985). Formaldehyde is also employed in histology laboratories to fix animal and human tissue for pathological diagnosis (Kilburn, Seidman, & Warshaw, 1985). Occupational exposure to formaldehyde occurs in histology technicians, fiberglass batt makers, and other occupations where formaldehyde is produced. NIOSH estimates that 1.7 million workers are exposed to formaldehyde in their jobs (Anger & Johnson, 1985).

Nonoccupational exposure to formaldehyde occurs primarily in homes insulated with urea–formaldehyde insulation. Workers and tenants in these residences have reported symptoms of eye irritation, chest tightness, and neuropsychological symptoms, including memory loss, irritability, and headache (Kilburn, Seidman, & Warshaw, 1985). Dialysis patients are exposed to formaldehyde used as a disinfectant in capillary flow dialyses, where exposure during a single dialysis session may be as high as 126 mg (Chang & Gershwin, 1992). Formaldehyde is also one of the most ubiquitous molecules in the environment. Normal human tissue level ranges from 3 to 12 ng/g tissue and humans can detect the odor of formaldehyde as low as 0–5 ppm (Chang & Gershwin, 1992).

Formaldehyde is reported anecdotally to be associated with a variety of neurological complaints, including headache, dizziness, fatigue, paresthesias, and depression (Change & Gershwin, 1992).

Neuropsychological Effects

Relatively little information is available on the neuropsychological effects of formaldehyde exposure. Kilburn, Seidman, and Warshaw (1985b) surveyed neurobehavioral symptoms among histology technicians and formaldehyde batt makers. Reported symptoms included sleep disorder, altered sense of balance, loss of concentration, and memory deficit. Mood alterations were also present. Greater numbers of symptoms were reported by workers employed in higher exposure occupations.

Kilburn, Seidman, and Warshaw (1985a) examined female histology technicians who had daily exposure to formaldehyde but who were also exposed to xylene and toluene. Neuropsychological symptoms claimed by the technicians included recent memory loss, loss of concentration, dizziness, and lightheadedness. Reported symptoms of decreased concentration and recent memory were "three times as frequent" in a group exposed to formaldehyde for 4 hours per day compared with controls. The authors' analysis suggested that neuropsychological symptoms increased with duration of exposure to formaldehyde rather than to toluene or xylene.

Cripe and Dodrill (1988) examined 13 subjects with chronic formaldehyde exposure who had been removed from exposure for several months. Examination included an expanded Halstead–Reitan Battery and MMPI. Eight of twenty measures were significantly different from an age-, sex-, and education-matched control group; however, when patients with learning disability, neurological or cardiac disease were excluded, all differences disappeared. Only MMPI scales were significantly different between groups with formaldehyde exposure subjects showing elevations on Hs, D, and Hy scales; a profile similar to that of head-injured patients. Problems in the study include lack of information of alcohol or medication use in the control group, small numbers of subjects, and the formaldehyde subjects being personal injury litigants.

The largest study thus far of formaldehyde-exposed workers was reported by Kilburn and Warshaw (1992). They longitudinally surveyed 318 histology workers and procured neuropsychological data from 418 additional exposed workers from a single testing. Subjects were tested across a variety of neuropsychological, neurological (EMG blink), and neurophysiological (e.g., balance, color vision) tests. No direct air sampling was performed and the authors relied on 1983 NIOSH estimates, of ten representative histology labs, which showed formaldehyde levels to vary between 0.2 and 5 ppm, xylene between 8.9 and 12.6 ppm, and chloroform from 2 to 19.1 ppm. Results were largely negative, with "virtually identical results" across 4 years of longitudinally tested subjects, and thus no evidence of progressive impairment. Weaknesses of the study include the lack of direct on-site measurement of levels for subjects, and the omission of long- or intermediate-term memory recall testing, a more sensitive measure than immediate recall.

A more recent case study by Kilburn (1994) does indicate clear neuropsychological and electrophysiological abnormality caused by more profligate use of formaldehyde (Kilburn, 1994). Three of the patients were lab technicians who had used formaldehyde over many years to fix whole animal preparations. The other patients had been covered several times by formaldehyde and phenol from manufacturing spills. Two of the patients developed seizures—all four had clear neuropsychological impairments in choice reaction time, a culture fair IQ test, and block design and digit symbol tests. Psychomotor speed on the trailmaking tests were impaired in all subjects, as was fine motor coordination on a pegboard test. Verbal and visual short term memory were impaired in three out of four patients. All four patients were considered disabled.

JET FUELS

These lead-free fuels consist of aromatic, olefin, and saturated hydrocarbons, including small amounts of benzene and toluene (Knave et al., 1978). Acute neuropsychological effects on fuel workers include dizziness, headache, and fatigue. Chronic exposure appears to produce symptoms of "neuraesthenia, anxiety and/or mental depression." Simple reaction time showed an increasing linear trend over time that was significantly greater for exposed individuals than for controls (Knave et al., 1978, p. 29). Performance decrements relative to matched controls were also found in a modified version of the Bourdon–Wiersma Vigilance Test where subjects were required to draw lines over groups of four dots.

The authors characterized these results as effects on attention and sensorimotor speed. No differences in memory function or dexterity were found (Knave et al., 1978).

In a more recent study, Struwe, Knave, and Mindus (1983) found significantly increased use of medical service for emotional problems among jet fuel-exposed workers compared with matched controls. Of the most severely symptomatic subjects, 7 of 14 were judged to have mild organic brain syndrome. Unfortunately, while the authors listed the neuropsychological tests employed (i.e., "SRB, Benton, dots, digit symbol, blot and pins") (Struwe et al., 1983, p. 58), they did not provide data, criteria, or statistics related to the diagnosis.

HYDROGEN SULFIDE

Conversion factor	1 ppm = 1.42 mg/m^3
Odor threshold	0.001–0.13 ppm
IDLH	300 ppm
OSHA	10 ppm (14 mg/m^3)
STEL	15 ppm (21 mg/m^3)
NIOSH	Not given
Ceiling	10 ppm (15 mg/m^3) (10 min)
(*Quick Guide*, 1993)	

Hydrogen sulfide is an extremely toxic gas that in concentrations above 700 ppm can be quickly fatal. In toxicity and rapidity of action it is said to resemble hydrogen cyanide. Like many other neurotoxic materials, the dangerous nature of hydrogen sulfide exposure has been known for centuries. Descriptions of hydrogen sulfide poisoning similar to current reports have been available since 1785 (Yant, 1929), although little information was available in the United States until the 1920s (Poda, 1966). Principal sources of exposure include gypsum and sulfur mines, sewers, tanneries, and other confined areas where decomposition of organic matter occurs. Production, storage, and transport of high-sulfur petroleum, gas manufacture, and manufacture of chemicals, dyes, and pigments are sources of hydrogen sulfide. Pumping of manure, especially swine manure, poses high risk of exposure (Tvedt, Skyberg, Aaserud, Hobbesland, & Mathiesen, 1991).

The more frequently cited effects in human exposure are local respiratory disturbance and eye irritations. The characteristic "rotten egg" odor of hydrogen sulfide is detectable at concentrations of 0.02–0.13 ppm and becomes intense at 20–30 ppm. However, exposure to 100–150 ppm causes rapid fatigue or paralysis of the olfactory nerve and eliminates the perception of odor (Beauchamp, Bus, Popp, et al., 1983).

Exposure to levels below the threshold of metabolic clearance (e.g., 20 ppm) appears to be well tolerated with no cumulative effects. However, systemic effects can be rapid and severe if hydrogen sulfide is absorbed more rapidly into the blood than it can be oxidized.

Descriptive studies have generally found that overexposed workers recover rapidly when not actually rendered unconscious by hydrogen sulfide. Individuals rendered unconscious by hydrogen sulfide may develop encephalopathy and abnormal CT that is more than simply the effect of anoxia. Hydrogen sulfide exposure interferes with mitochondrial cytochrome oxidase and may therefore cause direct cellular toxicity. It has also been suggested that acute systemic hypotension must also occur to account for the severity of hydrogen sulfide lesions (Matsuo, Cummins, & Anderson, 1979).

The risk of long-term neurotoxic injury caused by H_2S exposure found little support in two earlier reports (Burnett, King, Grace, & Hall, 1977; Arnold, Dufresne, Alleyne, & Stuart, 1985). However, Tvedt *et al.*, (1991) state that these reports underestimate the risk because they include cases of very low exposure and did not follow patients longitudinally. Five of their six patients cohorts who had been unconscious for 5–20 min showed chronic neurological and neuropsychological impairments with memory and motor functions most affected. One worker was too demented to be tested, and another (Patient 1) who was probably exposed for less than 1 min lost his sense of smell (anosmia) for 3 years. Another patient (No. 6) had temporarily reduced hearing and an impaired audiogram. Findings bear out that very short duration exposure with unconsciousness (Patient 1) does not seem to produce permanent disability, although recovery is protracted. Medical and neuropsychological findings are summarized in Tables 8.5 and 8.6 and document the severity and extent of findings.

TABLE 8.5. Medical Findings in H_2S-Exposed Patients at Follow-up, and Duration of Unconsciousness[a]

Patient number	Age[b]	Years[c]	Main affected functions	CT			H_2S exposed unconscious (min)
				EEG	PEG	Test[d]	
1	46	5	(Smell, vision, memory)[e]	−[f]	−	−	<1
2	31	8	Dementia, motor function, vision	−			5–10
3	59	7	Memory, motor function (vision)	+[g]	+	+	~15
4	53	6	Motor function (memory, vision)	(+)		+	~10
5	30	10	Visual abilities, memory	+	−	+	<15
6	31	6	Motor function, vision, memory	+	+	+	15–20

[a]From Tvedt *et al.* (1991).
[b]Age at the time of the poisoning.
[c]Number of years between the poisoning and the last examination.
[d]Neuropsychological examination.
[e]Temporary and uncertain findings are given in parenthesis.
[f]Normal.
[g]Pathological finding.

TABLE 8.6. Neuropsychological Test Results of Five Patients with H_2S Poisoning

	Patient				
	1	3	4	5	6
Age when examined (years)	51	66	59	40	36
Years since the poisoning	5	7	$4\frac{1}{2}$	10	5
Similarities	11 (7)^r	4 (6)	13 (8)	12 (7)	17 (8)
Digit Span	9 (8)	9 (9)	7 (3)	7 (3)^h	10 (9)
Vocabulary (split half)^a	46 (8)^r	24 (5)	24 (4)	34 (5)	38 (5)
Digit Symbol	36 (8)^r	21 (7)	21 (6)	32 (5)	39 (7)
Picture Completion	16 (12)	8 (6)	7 (4)	5 (2)^g	15 (10)
Block Design	24 (7)	20 (7)	24 (7)	14 (2)^g	13 (12)
Benton Vis. Ret. Test, errors	4	7	8	6	3
Ten words, learning, five trials	12^{h,c}	7^g	9	10^h	12^h
Ten words, retention 1 h	11^{h,c}	3^g	7	9^h	11^h
Fifteen word pairs, learn. errors	19	67^{g,x}	12	27	42^h
Fifteen word pairs, ret. errors	9^h	—^{c,e}	6	11^g	8^h
Visual Gestalt, learn. errors	4^c	9	—	7^h	7^h
Visual Gestalt, ret. errors	5^r	11	—	13^g	8^h
Trail-Making, A	37	50	42	41	40
Trail-Making, B	129^h	166^h	170^h	230^g	73
Finger Tapping, dom. hand	50	41	17^g	49	36^g
Finger Tapping, nondom. hand	48	40	17^g	47	35^g
Pegboard, dom. hand	58	—^{d}	185^g	75	79
Pegboard, nondom. hand	70	150^g	155^g	83	83
Static steadiness. number, dom. hand	73^r	235^g	210^g	150	131
Static steadiness, number nondom. hand	44^r	222^g	>210^g	133	260^g
Reaction time, 10th percentile^f	21	20	—	20	19
Reaction time, median	23	22	—	24	22
Reaction time, 90th percentile	28	27	—	31	27

From Tvedt et al. (1991). Patient 2 was too demented to be tested. Norwegian age-adjusted scaled score (median 10) for the WAIS tests in parentheses.
^{a}The Vocabulary Norwegian scaled scores are probably too low.
^{b}Twelve instead of ten words, maximum number of words correct in five trials.
^{c}Stopped because of slow learning, and retention then could not be tested.
^{d}Patient 3 had lost a finger in a working accident.
^{e}Results at the first examination 1 year after the accident. Not retested.
^{f}Simple visual reaction time 5 min.
^{g}Test results that are definitely reduced compared with age and estimated premorbid function.
^{h}Test results that are probably reduced.

HYPERBARIC NITROGEN

First described in 1835, nitrogen narcosis occurs when inert nitrogen in scuba tanks enters the bloodstream and causes swelling of neural cells, interfering with nerve transmission. Experimental studies have indicated decreased speed of processing, manual dexterity, choice reaction time, memory and verbal fluency under intoxicated conditions (O'Reilly, 1974; Rostain, Lemaire, Gardette-

Chauffour, Doucet, & Naquet, 1983; Logue, Schmitt, et al., 1986). The time course and pressure necessary to induce neuropsychological sequelae of nitrogen are unclear, but one controlled double-blind study failed to demonstrate impaired performance with brief exposure to 4 atm (100 foot depth equivalent) (Gallway et al., 1990).

IONIZING RADIATION

Present in X rays, nuclear power generators, uranium, and other natural and artificial sources, ionizing radiation can be fatal at high doses, and produce delayed cancer or genetic alterations at subfatal exposure levels. Human developmental neurotoxic effects of ionizing radiation show the highest risk for severe mental retardation to occur with exposure 8–15 weeks after fertilization, the time when the "greatest proliferation of neurons and their migration to the cerebral cortex occurs." The frequency of retardation is related in a linear model to the radiation dose received by fetal tissue (Schull, Norton, & Jensh, 1990, p. 257). Intelligence testing on survivors of Hiroshima/Nagasaki obtained 10 years after the bombing suggested no effect on individuals exposed within 0–7 weeks or after 25 weeks postfertilization. Again, individuals exposed between 8 and 15 weeks were most severely affected, with individuals from the 16- to 25-week group damaged to a lesser extent. Similar patterns were seen in scholastic performance. Exposure levels of 1.0 Gy and above were associated with severe neurotoxic consequences, including mental retardation (Schull et al., 1990).

Uncertainty and stress about the possibility of being exposed to radiation may itself induce psychiatric disorder. House, Sax, Rumack, and Holness (1992) report a case of three workers who were exposed to high levels of ionizing radiation in a nuclear power plant accident. None of the workers developed symptoms of radiation sickness. One worker developed anxiety symptoms that seemed to be triggered by discussion of the exposure. Another worker with a past history of recurrent anxiety began experiencing daily headache, nausea, apathy, poor concentration, and sleep onset difficulty. He experienced vivid nightmares of his exposure and was diagnosed as having posttraumatic stress disorder (PTSD). Symptoms persisted for 6 months and gradually diminished. The third worker showed only minor anxiety symptoms, after being told that his dose was the lowest of the three.

LATHYRUS SATIVUS (NEUROLATHYRISM)

Lathyrus sativus is a chickling pea, whose flour when eaten as a staple for 2 months or more causes a severe neurotoxic syndrome. When eaten as a major component of the diet, or as a famine food when other crops fail, the result is a form of motor-neuron disease with sudden onset of aching around the waist, calf muscle rigidity, leg weakness, paresthesias, and coarse upper extremity movement

(Jahan & Ahmad, 1993; Zorumski & Olney, 1992). The disease is not accompanied by dementia.

Neuropathological studies show degeneration of spinal long tracts, especially corticospinal tracts, severe loss of Betz cells in the motor cortex, especially in the precentral sulcus, and paracentral lobule (Spencer, 1990; Zorumski & Olney, 1992). The disease is endemic to parts of India, Bangladesh, Ethiopia, India, China, and Nepal, with an astounding attack rate of up to 2.5% of the regional population.

The active neurotoxin is an excitotoxin determined through primate studies to be β-N-oxalyl-L-α,β-diaminopropionic acid (ODAP or BOAA). The disorder appears to be preventable with a small amount of vitamin C added daily to the diet, but it is unlikely that populations that depend on this flour have access to vitamin C-containing foods or vitamins (Jahan & Ahmad, 1993). Soaking seeds overnight in lime water and boiling them for 25 minutes removed all ODAP (Jahan & Ahmad, 1993).

OZONE

Conversion factor	1 ppm = 2.00 mg/m^3
Odor threshold	0.0076–0.036 ppm
IDLH	10 ppm
OSHA	0.1 ppm (0.2 mg/m^3)
STEL	0.3 ppm (0.6 mg/m^3)
NIOSH	TWA not given
Ceiling	0.1 ppm (0.2 mg/m^3)

(*Quick Guide*, 1993)

Considered to be a ubiquitous urban pollutant, more than half of the U.S. population lives in areas that exceed the National Ambient Air Quality standard (Weschler, Shields, & Naik, 1989). Indoor concentrations may be significantly related to outdoor levels (Weschler *et al.,* 1989). Ozone exposure produces throat irritations, lung damage, respiratory changes, and in severe exposure cases, fatal pulmonary edema (Menzel, 1984; Nasr, 1971). Humans and animals exposed to ozone concentrations as low as 0.26 ppm for as little as 2 h develop shallow breathing, caused by reflex bronchoconstriction. Breathing usually returns to normal within 30 min after exposure ends (Menzel, 1984).

Symptoms of acute ozone poisoning include severe headache, dizziness, burning sensation in the eyes and throat, and choking sensation. Lasting symptoms after acute exposure may include generalized complaints that could be misdiagnosed as functionally based, including weakness, fatigability, weight loss, and shortness of breath.

Psychological effects of acute (2 h) exposure to 2 ppm of ozone were chronicled by Griswold, Chambers, and Motley (1956), with the first author of the study voluntarily exposing himself to ozone in an enclosed fumigation chamber.

The exposed author described "a very marked effect in coordination and articulation and expression of oral thoughts during the last half hour [of exposure] and until retiring [that evening]." Other symptoms included chest pain, loss of appetite, hypogeusia, poor sleep, and a "definite feeling of extreme tiredness for about two weeks."

PENTABORANE

Conversion factor	1 ppm = 2.62 mg/m^3
Odor threshold	0.97 ppm
IDLH	3 ppm
OSHA/NIOSH	0.005 ppm (0.01 mg/m^3)
STEL	0.015 ppm (0.03 (mg/m^3)
(*Quick Guide*, 1993)	

Pentaborane is a volatile liquid boron hydride used in the manufacture of semiconductors. It is selectively neurotoxic with toxicity said to be equal to or greater than hydrocyanide (Roush, 1959). Mechanism of CNS toxicity is thought to be caused by depletion of monoamine neurotransmitters (Merritt, Schultz, & Wykes, 1964).

The neuropsychological effects of pentaborane intoxication were addressed by Hart *et al.* (1984) who studied a group of rescue squad personnel who were exposed to pentaborane. Patients showed impairments in sustained attention using Rennick's digit vigilance task, verbal memory from the logical memory subtest of the WMS, and visual memory (BVRT) and visuospatial skills (WAIS-R Block Design). Percentage of patients impaired on these tests varied from 36 to 57% based on published norms. Higher ventricular brain ratios were noted in five of the patients, but EEGs were normal. Degree of exposure was statistically related to reported symptoms and neuropsychological measures of attention (Silverman *et al.*, 1985). Seven of the patients were additionally diagnosed with PTSD although it is not clear from the report the degree to which PTSD overlapped with positive neuropsychological or VBR finding.

PHENOL

Conversion factor	1 ppm = 3.91 mg/m^3
Odor threshold	0.06 ppm
IDLH	250 ppm
OSHA/NIOSH	5 ppm (19 mg/m^3)
Ceiling (NIOSH)	15.6 ppm (60 mg/m^3) (15 min; skin)
(*Quick Guide*, 1993)	

Phenol is a colorless or white solid in pure form, but is usually used as a liquid. It is most widely used in industries that make plastics but is also used to make caprolactam (for nylon and other fibers) and bisphenol A (for epoxies and resins) (ATSDR, 1989).

Phenol is a component of many analgesic or antiseptic over-the-counter medications including ointments, ear and nose drops, cold sore lotions, mouthwashes, toothache drops, analgesic rubs, throat lozenges, and antiseptic lotions. It retains medical applications as an antiseptic and an anesthetic. Phenol is a component of 15% of hazardous waste sites on the U.S. National Priorities list (ATSDR, 1989).

There do not appear to be any group studies detailing neurological or neuropsychological impairments from phenol exposure, although there are several case reports in the literature. For example, Merliss (1972) reports a case of muscle pain and weakness in a patient exposed to phenol, cresol, and xylenol. Neudorfer and Wolpert (1976) report a variety of psychometric and psychiatric abnormalities in a 30-year-old engineer that they attributed to phenol intoxication. Animal studies indicate hindlimb paralysis in guinea pigs exposed to 26–52 ppm for 41 days, but rabbits and mice exposed to similar amounts of phenol for 88 and 74 days showed no neurological impairments. Rats continuously exposed to 26 ppm showed muscle tremor, twitching, and disturbances of rhythm and posture after 3–5 days of exposure and difficulty walking on a tilted plane after 15 days; a degree of impairment characterized as severe (Dalin & Kristofferson, 1974).

In rats, the lowest observable adverse effect level is 26 ppm. Lethal human oral dose is estimated at 140 mg/kg (5000 ppm in water). Dermal application of phenol may be fatal. In one report, death occurred after 25% of the patient's body was exposed to liquid phenol (Lewin & Cleary, 1982). No human studies of neurological effects of phenol exposure have entered the literature. Phenol is absorbed readily and excreted almost completely in the urine.

POLYCHLORINATED BIPHENYLS

Arochler 1242
    Conversion factor   1 ppm = 10.72 mg/m^3 (approx.)
    IDLH   10 mg/m^3
    42% chlorine: OSHA   1 mg/m^3 (skin)
    NIOSH   0.001 mg/m^3
54% chlorine (Arochlor 1254)
    Conversion factor   1 ppm = 13.55 mg/m^3 (approx.)
    IDLH   5 mg/m^3
    OSHA   0.5 mg/m^3 (skin)
    NIOSH   0.001 mg/m^3
(*Quick Guide*, 1993)

Polychlorinated biphenyls (PCBs) were introduced during the 1930s in high-voltage electrical and electronic applications, and in the production of paint, plastics, and carbonless copying paper. When introduced, they were thought of as an improvement over mineral oils, since they did not explode or catch fire (Rogan & Gladen, 1992), but today they are "counted among the most toxic chemicals," do not degrade in the environment, and exhibit increasing concentration in body tissues along the food chain (Reggiani & Bruppacher, 1985, p. 225). Half lives in the human body range from 1.5 to 5.7 years, depending on the compound and serum concentration (Taylor & Lawrence, 1992). Because of these factors, PCBs are found in background levels throughout the developed world. Major sources of high-level contamination include industrial exposures and consumption of fatty fish caught in contaminated water (e.g., Lake Michigan) (Jacobson, Jacobson, & Humphrey, 1990).

There have been two major episodes of mass poisoning with PCBs, the first during 1968 in western Japan. About 1600 people were exposed to a mixture of PCBs and other polychlorinated compounds when they ingested rice oil contaminated with these substances (Reggiani & Bruppacher, 1985).

The second incident was similar to the first. It occurred in 1978 in Taiwan when approximately 2000 people were poisoned with PCB-contaminated cooking oil. PCBs had been used as a heat conductor in the cooking oil processing plant, and a pipe containing PCBs had leaked into the cooking oil (Chia & Chu, 1984). This unfortunate epidemic has given researchers most of the currently available information about the neurotoxic properties of PCBs.

PCB exposure has been found to cause a variety of symptoms, the most obvious being severe acnelike eruptions. Early stage exposure symptoms include eye discharge, disturbance of vision, fatigability, malaise, and nonspecific symptoms such as cough, poor appetite, sore limbs, pruritus, and abnormal menstruation (Lu & Wong, 1984).

There is some dispute as to whether PCB itself is neurotoxic or whether mass exposures with demonstrated neurotoxicity were caused by the presence of more toxic polychlorinated dibenzofurans (PCDFs) (James et al., 1993). PCB workers tend to consistently report subjective neurological symptoms, including headache, nervousness, fatigue, loss of appetite, memory impairment, and paresthesias, but standard neurological examination does not validate these complaints. Large-scale neuropsychological testing to more sensitively search for these abnormalities has apparently not been performed; however, Peper et al. (1993) report results of a small study in which blood levels of both dioxins and furan were significantly correlated with picture completion, a recognition score, Stroop, Trails A, visual memory, and paired associates.

Developmental Effects

Several studies have examined the neonatal effects of PCB exposure and have noted that Japanese children exposed to PCBs were hypotonic, apathetic, and dull (Harada, 1976). Japanese children born to PCB-exposed mothers were intellectually impaired, and showed a variety of developmental delays in areas of language, speech, and psychomotor functions with clinical impressions of delay of up to 10% of exposed children, versus 3% of controls (Rogan & Gladen, 1992). Children born to exposed U.S. mothers were also hypoactive and hypotonic at birth (Rogan et al., 1987). In utero-PCB-exposed Taiwanese children, however, were characterized as hyperactive (Hsu et al., 1988).

Jacobson et al. (1990) evaluated two cohorts of children including those of Lake Michigan sportfish-eating mothers, and those with long-term familial exposure to PCBs or PBBs from contaminated farm products. Children were given the McCarthy Scales of Children's Abilities, reaction time tests, and had their weights and measurements taken. Maternal alcohol and cigarette use, medical history, and intelligence estimates were obtained during home visits, as were mothers' assessments of their child's personality and social functioning. Children were also independently rated by examiners on activity and behavior. After confounding variables were removed from a regression analysis, prenatal PCB exposure showed a dose-dependent association with lower weight and activity level at age 4 years. "Thirty-one percent of the children with levels of 9 ng/ml or more were rated in the bottom tenth percentile for the sample as a whole, and none were in the top percentile" (Jacobson et al., 1990, p. 323). The authors indicate that weight deficit was associated with prenatal but not postnatal exposure, and that children failed to catch up by age 4. Possible explanations proposed by the authors include toxicity to neuroendocrine or CNS growth regulatory mechanisms, greater vulnerability of fetal cells, incomplete blood–brain barrier development, and/or undeveloped drug metabolization mechanisms in the fetus. Prenatal PCB exposure is also associated with 5- to 8-point poorer performance on the Bayley Scales of Infant Development at 6, 12, 18, and 24 months (Rogan & Gladen, 1991).

Neurological Manifestations

Chia and Chu (1984) examined 35 consecutive dermatological admissions to their hospital and report that 31 (88.6%) had had one or more neurological complaints. Headaches and dizziness were found in about one-third of the patients. Paresthesias in fingers and toes were reported in 65.7% of patients. Hypo- or hyperesthesia were found in 37% of the cases. Deep tendon reflexes were characterized as absent or sluggish in 17% of patients, and another one-third reported back and limb pain.

EEG examinations of six cases (17%) were abnormal in individuals with no prior seizure history. Mean nerve conduction velocity was significantly slower in the exposed group compared with a control group. The authors found evidence of peripheral neuropathy in 28.6% of their patients. Despite the finding of abnormal EEG in one-fifth of the sample, and headache or dizziness in two-fifths, the authors suggested that their evidence did not support CNS dysfunction. Unfortunately, neuropsychological measures were not applied to test that hypothesis. Further, there do not appear to be any neuropsychological studies of PCB poisoning in the literature at this time. Available evidence strongly suggests the need for such a project, and it can be hoped that neuropsychologists who see PCB-exposed patients will attempt relevant research.

There is some debate in the recent literature about the neurotoxicity of "pure" PCBs, and some authors have suggested that neurotoxicity observed in the Yusho and Yu-Cheng mass poisoning patients was actually related to the presence of PCDFs (James et al., 1993). Rogan and Gladen (1992) find few studies of neurotoxicity among workers exposed to such "clean" PCBs.

PROPANE

Conversion factor	1 ppm = 1.83 mg/m^3
Odor threshold	12,225–20,000 ppm
IDLH	20,000 ppm
OSHA/NIOSH	1000 ppm (1800 mg/m^3)
(*Quick Guide*, 1993)	

Propane is an alkane gas at room temperature that is usually stored in canisters under pressure, where it exists as a liquid. There are few case reports available on exposure to this gas but recreational abuse carries a significant risk of sudden death. It appears to produce hypoxemia and may induce cardiac arrhythmia. At high concentrations, propane is anesthetic, which increases the risk of involuntary asphyxiation. One case of chronic (6 months), daily propane inhalation has been reported, where the subject experienced euphoria, ataxia, and lightheadedness, without loss of consciousness (Wheeler, Rozycki, & Smith, 1992). The individual reported severe headaches on the morning following abuse, but mental status exam and neurological exam were normal. Neuropsychological testing was not performed.

SILICON AND SILICONE

Silicon forms about 28% of the earth's crust by weight. It is found naturally in the human skeleton, and in the environment as an oxide or a silicate. Silicon is a principal component of glass, clay, and microelectronic devices. Silicone processes pure silicon through a variety of steps into a high-molecular-weight polymer. Of late, there has been considerable concern related to undesirable side effects caused by silicone breast augmentation implants. Case reports and anecdotal accounts have suggested the existence of silicone-mediated autoimmune diseases, including scleroderma, Reynaud's syndrome, systemic sclerosis, systemic lupus erythematosus (SLE), and Sjögren's syndrome, but their occurrence is thought to be rare (e.g., 39 to 115 cases per million for development of SLE) (Yoshida et al., 1993). Yoshida et al. (1994) have noted that there is substantial similarity between reactions to the "oily substance" released from silicone breast implants and autoimmune dysfunction from adulterated cooking oil (toxic oil syndrome), and adulterated L-tryptophan (eosinophilia myalgia syndrome). Several of these disorders have clear neuropsychological sequelae (e.g., EMS, SLE), but the connection between silicone exposure and immunologic or neuroimmunologic abnormalities is inconsistent and uncommon, and symptom etiology has not been verified as of this writing. No neuropsychological studies of silicone-mediated immune disorders have entered the literature.

TOXIC OIL SYNDROME

Like EMS, toxic oil syndrome (TOS) is an example of a food or food contaminant form of neurotoxic disorder. TOS was an autoimmune disease brought about in Spain when approximately 20,000 people consumed adulterated cooking oil. The oil was from rapeseed and denatured with aniline, being intended for industrial use, but sold illegally door-to-door as an inexpensive olive oil (Yoshida et al., 1994). Oleic anilide is the toxic agent that most likely caused the syndrome, which produces symptoms very similar to EMS. Early stage symptoms included eosinophilia, edema, myalgia, neuropathy, malaise, and occasionally death from respiratory failure. One percent of earlier phase patients had clear CNS abnormalities. Later stages saw the development of sicca and Reynaud's syndromes, with death caused by thromboembolisms and ischemic colitis, secondary infections, and hemorrhage (Yoshida et al., 1994). Both TOS and EMS are associated with elevations in tryptophan metabolism, but oxidative damage is the most widely accepted etiology of TOS symptoms. An elaborate model of TOS immunotoxicity has been proposed by Yoshida et al. (1994).

9

Psychosomatic Disorders

Psychosomatic and psychological disorders are implicit diagnostic rule-outs in any evaluation of neurotoxic exposure. These maladies neither presume nor eliminate the possibility of neurotoxic nervous system damage in that neurotoxic injury to limbic system and emotional substrates may indicate organic mood disorder. However, current understanding of neurotoxic brain damage indicates that severe emotional dysfunction in the absence of neuropsychological impairment is not credible from the standpoint of neurotoxic pathology, and therefore must be considered of functional origin. This clinical supposition may change, of course, if a mechanism is uncovered to explain selective damage to affective subcortical structures. Regardless, differential diagnosis of functional disorder versus toxicant-induced brain lesion has significant implications for treatment and recovery. Neuropsychological impairments suggest the need for environmental exposure management, further medical screening, and perhaps cognitive rehabilitation. An examination that indicates functional impairments directs interventions into psychiatric or stress management modalities.

MULTIPLE CHEMICAL SENSITIVITY

A curious interaction between psyche and soma has come to the attention of health care workers in the phenomenon of *multiple chemical sensitivity* (MCS). Also known as *ecological* or *ecologic illness, twentieth century syndrome, total allergy syndrome,* or other terms (see Terr, 1989), MCS is a diagnostic disorder of uncertain etiology, but usually characterized by the following symptoms (after Cullen, 1987):

1. Acquired in relation to documented environmental exposures, insults, or illnesses
2. Involve more than one organ system
3. Symptoms recur and abate in response to predictable stimuli
4. Symptoms elicited by chemicals of diverse structure and toxicologic action
5. Symptoms elicited by demonstrable, albeit ultra-low-level exposures
6. Exposures far below levels known to cause damage elicit symptoms
7. No single test explains symptoms

Symptoms may occasionally appear quite bizarre, as in the case of an MCS patient interviewed for an article in *The New York Times Magazine,* who complained of the symptoms induced by the interviewer's polyester pants.

MCS is characterized by remission or dissipation of symptoms when exposure is terminated. Symptom profile resembles an acute response to a stimulus, rather than chronic Type 2A, 2B, or 3 solvent encephalopathy; disorders that are characterized by impairments independent of exposure (Fiedler, Maccia, & Kipen, 1992). Clinically, MCS symptoms tend to overlap with those of somatoform disorder, mass psychogenic poisoning, and posttraumatic stress disorder, suggesting at least a partial psychosomatic component to the disorder. In many cases, there are no observable physical correlates of the disease other than patient report. In Terr's (1986) review of 50 litigating cases, 31 had no physical or laboratory abnormalities.

Despite recent public interest and debate, symptoms resembling MCS are not new. The novel *Madame Bovary* contains a graphic description of an individual who may have been the earliest "chemically sensitive" patient. The condition known as "neurasthenia" traced to Glenard in 1986, has many similarities with MCS, including fatigue, mental dullness, headache, abdominal pain, and poor appetite. The disorders are both largely subjective in character (Gots, 1993).

Patient Profile

Patients volunteering for MCS studies are typically female Caucasians with good education and above-average socioeconomic status (e.g., Black, Rathe, & Goldstein, 1990). However, it is not clear whether this profile is confounded by self-selection biases and high cost of treatments (Mooser, 1987). Most MCS patients have an extensive history of health care with less than satisfactory results, and a "profound aversion to psychiatry" (Gots, Hamosh, Flamm, & Carr, 1993). Such patients are convinced that they are being damaged by the environment, perhaps as the result of an unusual disease process that has not yet been discovered. They have discovered the "truth" about their symptomatology, as the following excerpt of a letter to the author from a patient claiming MCS should make clear:

> I have a passion to fully realize what's going on in my body and mind.... Obviously my problems are psychological (your field) in nature, however, they are very much secondary to the environment–brain–body connection. Chemical-based products are omnipresent and unavoidable and they have a toxic effect on me.... Who is more of an expert about how a little understood disease erodes the functioning of the body and mind? A doctor who has read a stack of subjectively written articles via black and white facts, who additionally sees nothing under a microscope, *or* a patient who experiences, lives with the process of intermittent dysfunction, who also reads subjective articles, yet additionally feels with firsthand knowledge, the *truth* written in some theoretical study.... I have seen doctors and others who feel *I am* terribly objective and intelligent about my problems and rational about my concern of my future medical needs and the realistic potential of further dementia.... I am not so naive any longer to expect a "cure" but I do sincerely hope that by using me for research, there might

be a better understanding of what has gone wrong.... I am desperate for hope that my suffering could possibly assist in some manner with advancement of medical science.... There must be a *reason*, mind you, a *divine purpose* for my life to continue on since this chemical exposure.

This patient's world view is consistent with the hypothesis of Gots (1993) that, because of the general decline of trust in "expert" conventional medicine coupled with the public's embrace of a participatory, holistic health ideal, each individual becomes "a personal laboratory, 'understanding' his or her own body, [and] 'knowing' how various factors or substances influenced it" (p. 11).

Such patients often develop a life-style, observed by Brodsky (1983) and Black *et al.* (1990), that is heavily involved in activities, literature, friendships, and support groups that validate their symptoms. Many are in treatment with so-called "clinical ecologists" who prescribe a variety of vitamins, "immune boosters," and other nostrums to relieve symptoms. Recommendations often include removal of all offending products from the immediate environment, resulting in expensive home renovation, building of "clean rooms" or "safe rooms," usually surrounded in ceramic tile or glass, and sometimes, moving to a supposedly less polluted area. The majority have been instructed to wear oxygen or charcoal masks when they cannot avoid coming into contact with a noxious substance. Following these recommendations often has a secondary effect of isolating patients from friends, family, and employers.

Litigation often accompanies the attempt to be validated as a "chemically sensitive" individual. In one cohort of 50 MCS patients, 40 were involved in workers' compensation litigation and 6 were parties to other types of litigation (Terr, in Gots *et al.*, 1993). This author has seen many patients apply for Social Security disability payments for MCS inability to work.

Etiology of MCS: Multiple Diagnostic Possibilities

The unusual symptom constellation experienced by MCS patients eludes easy explanation. MCS syndrome has been explained as a disorder of immune regulation, biochemical imbalance, nutritional deficiency, and psychological/psychiatric abnormalities.

Immune Regulation

One explanation proposed for MCS is that it produces vasculitis of autoimmune etiology; causing environmental toxins to induce antibodies for blood vessel antigens (Terr, 1989; Rea, 1977). Another autoimmune hypothesis was proposed by McGovern, Lazaroni, Hicks, *et al.* (1983) to involve T-suppressor cell toxicity. Support for immunontoxicity of selected substances has entered the literatures; e.g., subjects exposed to the pesticide chlorpyrifos were found to have developed high rates of atopy and antibiotic sensitivities, and show several other markers of immunologic abnormalities (Thrasher, Madison, & Broughton,

TABLE 9.1. Indoor Air Pollutants and Immunologic Effects[a]

Substance	Immunologic effect
Smoke, dust, volatile or gaseous chemicals	Inflammation secondary to irritant and adjuvant reactions
Biological antigens (bacteria, molds, plants, mites, insects, pets); industrial chemicals (beryllium, isocyanates)	Inflammation from hypersensitivity reactions
Vinyl chloride	Autoimmune reactions
Benzene, trinitrotoluene	Immunosuppression
Benzene	Neoplastic proliferation
Dimethylsulfide, trimethylamine	Conditioned neuroendocrine effects on immune functions (in animal models)

[a]Vogt (1991).

1993). Table 9.1 lists known immunologic effects of several other toxins found in occupational settings and as indoor air pollutants (Vogt, 1991).

Biochemical

Some support for biochemical abnormality in MCS patients comes from Rea *et al.*, who found elevated organic pollutants and reduced erythrocyte chromium. However, other studies have failed to find abnormalities of biochemistry in MCS patients. Another explanation was proposed by Levine and Reinhardt (1983) who view MCS as resulting from failure of human antioxidant mechanisms, leaving free radicals produced by environmental toxins to cause systemic damage.

Anecdotal family history of chemically sensitive patients is said to be positive for allergy, thyroid dysfunction, collagen vascular disease, and other disorders (Ross, 1992). The possibility that MCS is a form of autoimmune disease has been raised frequently (e.g., by Ross, 1992; Levin & Byers, 1992; Kippen *et al.*, 1992). At present, however, there is no way to differentiate biochemical or autoimmune abnormalities said to occur as a result of MCS from alternative and less obscure causation, such as chronic stress, depression, anxiety, or other factors. Thus, it is not clear whether (1) this physiologically stressed population more easily experiences MCS because of increased physiological vulnerability, (2) MCS is responsible in some way for a subset of complaints, or (3) preexisting diagnoses cause the complaints attributable to MCS.

Psychological

Psychological explanations fall into five categories: (1) those that classify MCS patients as having a psychological disorder with physical symptom expression in the form of a hysterical or somatoform disorder (Kahn & Letz, 1989; Schottenfeld, 1987), (2) those resulting from severe psychologically traumatic threat of poisoning (posttraumatic stress disorder), (3) those that appear to be related to conditioning and generalization of autonomic and/or neurotoxic abnormalities, (4) MCS is an *iatrogenic* psychological disorder produced by physician/clinical ecologist "treatment" (Terr, in Gots *et al.,* 1993), and (5) MCS patients represent a mixed group of psychiatric disorders with no underlying factor, except "that some person in the medical community has told them they have an environmental illness and has fostered that belief" (Black, in Gots *et al.,* 1993, p. 73).

Schottenfeld (1987) supports a common opinion in the psychiatric community that MCS-related disorders fall under the DSM-III-R categorization of Somatoform Disorders. He also suggests that MCS patients may be a heterogeneous group, including subjects who (1) show unusual sensitivities that are not allergic or immune system-based, (2) somaticizers who exaggerate normal body symptoms, (3) somaticizers who amplify irritant symptoms, (4) primary psychiatric diagnoses, and (5) psychiatric diagnosis reactive to and exacerbated by MCS (Schottenfeld, 1987). Support for a high incidence of psychiatric disorders was found in several studies, including Stewart and Raskin (1985) who reported that all 18 patients referred to a university psychiatric clinic for MCS symptoms were found to have somatoform, psychotic, or personality disorder. Brodsky (1983) described 8 MCS patients as having long histories of somatization, undocumentable illnesses, and high health care utilization. This is further supported by Black *et al.* (1990) who found that lifetime prevalence of major mental disorders, especially major depression, anxiety disorders, and somatoform disorders, was significantly greater in MCS patients compared with age- and sex-matched controls. Anecdotally, the authors note that many of their patients would be diagnosed under the DSM-III category of hypochondriasis, but this disorder was not evaluated in their structured interview.

The argument that such patients are misdiagnosed, or that psychiatric diagnosis was *secondary* to MCS, has been addressed by Simon, Katon, and Sparks (1990), who administered structured interviews and self-report measures to 37 symptomatic aerospace plastics workers filing workmen's compensation claims for MCS-related symptoms. New composite plastics had been introduced into the workplace, with measurable, but subtoxic, levels of phenol, formaldehyde, and methyl ethyl ketone. Symptomatic workers complained of headache, fatigue, dizziness, nausea, shortness of breath, and cognitive impairments.The strongest predictor of environmental illness development was a history of psychiatric morbidity that predated chemical exposure, as recorded on the Diagnostic Interview Schedule (DIS). Just over half of the 13 subjects with diagnosed MCS had such histories, while only 1 of the 23 symptomatic, non-MCS patients had such history. The MCS group was also more likely to be classified as having somatoform disorder. None of the subjects met the criteria for posttraumatic stress disorder (PTSD). Like the study by Black *et al.* (1990), results are specific to small numbers of patients. Further, workers came from a single factory and were litigating exposure effects. Nevertheless, the authors' argument for similarity between MCS and other somatically influenced disorders including fibromyalgia, chronic fatigue syndrome, and chronic pelvic pain is intriguing and deserves further study.

A further argument for the psychosomatic and/or stress-related nature of MCS is that similar symptoms are triggered by stimuli that differ across cultures. In Sweden, for example, thousands of employees have claimed that they are "allergic" to electricity, and present many symptoms similar to MCS, even though no controlled study has uncovered a direct relationship between symptoms and electrostatic or magnetic fields (Berg, Arnetz, Lidén, Eneroth, & Kallner, 1992).

Fiedler *et al.* (1992) performed medical, neuropsychological, and psychiatric screening on 11 patients who met Cullen's criteria for MCS. Previous medical and psychiatric history was deemed noncontributory, although screening of psychiatric factors was minimal, using only patient report on a questionnaire. Four patients had history of asthma. While several subjects had isolated immunological finds outside the range of normal, no significant or consistent immunologic dysfunction was detected. Four subjects met SCID-III-R criteria for depression. Mean MMPI profile appeared similar to a "conversion V" result, with primary elevations on scales 1 (hypochondriasis) and 3 (hysteria). Six of eight female subjects presented profiles "characteristic of a somatoform disorder" (p. 534). Scores on a measure of psychosocial adjustment were comparable to those experienced by depressed subjects and alcoholic subjects. The group as a whole did not appear to have significant neuropsychological impairments, although four subjects had impaired CVLT findings. CVLT findings were characterized by impaired initial learning rather than impaired recall (a pattern also seen in depressed patients). The patient with history of as well as concurrent asthma was also impaired on several other tests (Digit Span, Digit Symbol, Visual Reproduction I), but there is no information given on medications or other acute concomitants of asthma that might impair performance on these tests.

The strongest, and seemingly conclusive study proving that the majority of claimed MCS patients have some form of primary psychological and/or somatoform disorder rather than a physiological or immunologic hypersensitivity disease is that of Staudenmayer, Selner, and Buhr (1993) who subjected claimed MCS patients (including those whose diagnoses were "confirmed" by clinical ecologists) to controlled trials of chemical provocation in a specially built exposure chamber of porcelainized steel, glass, and aluminum. The chamber provided air filtered with a HEPA filter, activated charcoal, and a Purafil bed, which removed more than 90% of volatile organic compounds and atmospheric oxidants, and more than 99% of all particles from filtered air.

Patients were exposed to chemical challenges using chemicals selected by each patient as capable of eliciting symptoms, at concentrations and durations reported to induce symptoms (from 15 min to 2 h). Distinctive odors of certain substances were masked by a tolerated masking agent, e.g., peppermint, cinnamon, or anise.

Initial exposure consisted of single-blind control challenges of clean air and a masking agent. Subjects who did not react to this control condition entered into a double-blind condition, where they were presented with a random series of chemical and sham challenges. A challenge was deemed to cause a positive reaction on the occurrence of either (1) objective physiological reaction (e.g., hives, blood pressure drop), (2) symptom self-rating by the patient, or (3) the presentation of a postchallenge increase in symptom self-rating.

Results were analyzed with a signal detection paradigm that analyzed *sensitivity* (percentage of positive reactions to real chemical agents), *specificity* (true negative responses to sham challenges), and *efficiency* (percentages of correct identifications of active and sham challenges). MCS patients showed only 33.3% sensitivity, 64.7% specificity, and most tellingly 52.4% efficiency, that is, their reactions occurred at a pure chance level. No patient was ever discovered to have a general chemical sensitivity, although Dr. Staudenmayer has validated specific reactions to individual chemicals, e.g., TDI asthma (Gots *et al.*, 1993).

At a subsequent conference on MCS, Dr. Staudenmayer reported that of 20 patients who were presented with their negative challenge results and who subsequently underwent psychotherapy, 75% "recognized they had psychological problems . . . , addressed these problems and recognized that their projection on to the environment was erroneous and they gave up their belief. Twenty-five percent would not give up the belief" (Gots *et al.*, 1993, p. 76).

MCS patients probably include a heterogeneous group of psychological disorders including somatoform disorders along with more conventional character pathology. We have seen a number of MCS patients who fit the diagnostic category of "Borderline" personality disorder. These patients can usually be distinguished by marginal work and social adjustment, with near-dissociative experiences of emotional dyscontrol. One such patient in our experience showed intense interpersonal rage at a significant other, followed by a *post hoc* search for the presumed toxicant that caused the episode, including leaves burning one block away, cooking odors, or perfumes. When there was an absence of a readily identifiable odorant, the patient concluded that toxic chemical exposure occurred below olfactory threshold.

These patients are typically quite refractory toward suggestions of psychological or interpersonal difficulties, insisting that their bodies are undergoing a toxic assault for which they require diagnosis and elaborate detoxification treatments. Patients in our clinical experience have readily volunteered for any physical diagnostic procedure with the potential to validate their symptoms. At the same time, they tend to be refractory to psychiatric explanations for symptoms, even when possible stress–immune system interactions are included in an explanation.

A third subcategory of MCS patients appear to suffer the effects of a conditioned aversion syndrome. Bolla-Wilson, Wilson, and Bleecker (1988) describe several case studies where headache, dizziness, memory difficulty, and limb pain were experienced in response to common environmental odorants such as hair spray, perfume, gasoline, and cigarette smoke. Noting that MCS does not develop with odorless neurotoxins, the authors suggest that MCS symptoms develop as part of a Pavlovian conditioning paradigm. MCS develops when a chemical odor becomes a conditioned stimulus for the unconditioned stimulus (toxin) after being paired with it. Subsequently, symptoms are produced by the odor alone.

Generalization may occur to new odors and response intensity covaries with odor similarity to the original conditioned odor. In addition, Cone and Shusterman (1991) note the role of acute overexposure in producing conditioned behavioral sensitization. They cite several cases of recurrent panic or hyperventilation in patients who were able to tolerate certain odors until they were overexposed. Subsequent exposure to odors triggered symptoms.

While conditioning effects proposed by Bolla-Wilson *et al.* explain psychiatric symptom production, conditioning may also provide an explanation for allergic and immune dysfunction claimed by a subset of MCS patients. Evidence was recently obtained indicating that the immune system is also subject to conditioning. MacQueen *et al.* (1989) paired mucosal mast cell-stimulating egg albumin/*Bordetella pertussis* injections with a flashing light. They found that actual histamine release could be trigged by the CS alone, suggesting that "true" allergic reactions could be produced by situations favorable to classical conditioning.

Finally, even non-MCS individuals appear to believe that certain odors will negatively affect their level of health. For example, Knasko (1993) found that when subjects were exposed to aversive odors, they believed their performance levels on simple addition and attention tasks suffered, and that the odor exposure affected their health and mood. In reality, subjects did not differ from controls on any index of health, mood, or performance. This suggests that MCS patients may have an unrelated psychiatric overlay to a normal human reaction to aversive odor perception. Since exposure to mucosal irritants increases general levels of stress and taxes premorbid coping mechanisms (Staudenmayer *et al.*, 1993), it is not unreasonable to assume that certain patients with low stress threshold (or high levels of concomitant stress) may be pushed "over the edge" by the chronic low-level, but additive stress of an aversive odor.

Treatments

Optimal treatment would address the multifactorial influences of MCS. Removing the patient from clinically toxic exposure is obvious and prudent. However, it would be difficult or impossible to sequester patients away from newsprint, wood, paint, and other outgassing materials, although even this has been attempted, with individuals banding together and moving to relatively pristine environments (Belkin, 1990). In our experience, many patients with claimed MCS do not easily accept the possibility of multiple causation for treatments, preferring to view themselves as "canaries in a coal mine," i.e., that they are more sensitive and vulnerable than the average patients to a demonstrably toxic environment. It is usual for them to insist on a unifactorial relationship between symptoms and toxic exposure; that their symptoms are the direct result of toxic exposure, without psychological intervening variables. This viewpoint is unfortunately fueled by treaters who share and amplify their fears. Therefore, while biofeedback and hypnosis have shown some efficacy (e.g., Podell, 1983;

Bolla-Wilson, Wilson, & Bleeker, 1988), convincing patients to undergo psychological treatments is often difficult or impossible.

SICK BUILDING SYNDROME

Sick building syndrome (also called "tight building syndrome") is characterized by mucosal irritation of the eyes, nose, throat, and lower airway, dermatological reactions, fatigue, headache, nausea, and other nonspecific somatic and psychological symptoms. The symptom constellation is distinguished from *building-related illness,* which includes medical diseases related to poor indoor air quality, such as hypersensitivity pneumonitis, humidifier fever, asthma, and legionellosis (Welch, 1991). Symptoms worsen during the workday and may diminish after workers leave the building. The disorder is prevalent, with a Danish study finding office work-related mucosal irritation estimated at 44% for women and 25% for men [versus 21% women and 12% men in the general population (Valbjorn *et al.,* 1986, cited in Skov *et al.,* 1989)]. Morrow (1992) estimates that from 800,000 to 1.2 million commercial buildings in the United States evoke at least some sick building related symptoms, the impact of which is sufficiently severe as to cause an estimated 500,000 lost workdays each year.

There has been an international effort to determine the etiology of sick building syndrome, culminating in several internationally attended conferences on indoor air quality. One of the more recent was held in Toronto in 1990 with 542 scheduled presentations (Lebowitz & Walkinshaw, 1992). Like MCS, sick building syndrome has been variously viewed as being of chemical, biological, psychosocial, or physical origin (Skov *et al.,* 1989). Early studies suggested associations among observed symptoms and formaldehyde levels (Main *et al.,* 1983; Olsen & Dossing, 1982), total hydrocarbon concentrations (Skov *et al.,* 1989), smoking (Skov *et al.,* 1987), interpersonal cooperation and stress. Some forms of sick building syndrome may be related to a combination of building design inadequacies and odorant perception. Cone and Shusterman (1991) found such a problem in an elementary school, where a combination of fireproofing odorant and poor air ventilation rates produced symptoms of headache, dizziness, abdominal pain, cough, runny nose, and itchy eyes. Other recent studies conclude that multifactorial explanations may best explain complaints. For example, Skov *et al.* (1989) found that individual factors (e.g., females under 40 with history of allergies or migraines), type of work (handling of carbon paper, photocopying and VDT work), and psychosocial factors related to job dissatisfaction were all related to experience of symptom production.

Norback, Torgen, and Edling (1990) found similar results, with a mixture of psychosocial factors, personal habits, and chemical exposure associated with symptoms. Nonspecific hyperreactivity, sick leave related to airway illness, current smoking history, exposure to static electricity, and total indoor hydrocarbon concentrations were all related to symptoms. Bauer *et al.* (1992) showed that patients in sick buildings had greater distrust of authority, defensiveness, anxiety, and confusion, but psychological inventories alone did not distinguish patients with sick building symptoms from nonsymptomatic workers.

Little literature exists concerning whether neuropsychological deficits accompany sick building syndrome. For example, Otto, Mulhare, Rose, *et al.*, (*Annals of the New York Academy of Sciences,* 1992) attempted to produce an experimental analogue of sick building syndrome by exposing 66 healthy male subjects to a mixture of volatile organic solvents for 2.75 h. Subjectively experienced symptoms of discomfort were recorded, and objective neuropsychological data were collected on the Neurobehavioral Evaluation System (NES). Though the authors hypothesize that vigilance and attention tasks would be most likely to prove sensitive to low-level volatile hydrocarbon emissions, based on the known anesthetic/depressant effects of solvents, no neuropsychological abnormalities accompanied complaints of strong odor, impaired air quality, increased headache, and increased sleepiness. Unfortunately, as the authors point out, their experimental population "is perhaps the least likely subset of the general population to be affected by chemical exposure." A redesign of the experiment to include self-described "chemically sensitive" individuals, and repeated exposure designs could provide more complete information. However, the possibility of better controlled research could be hampered by difficulty producing double-blind designs with olfactory stimulants, and in finding self-identified chemically sensitive patients willing to undergo an aversive procedure.

The need for testing of subpopulations of chemically sensitive individuals is underscored by two recent studies. In the first, Koren, Graham, and Devlin (1992) exposed 14 subjects to a mixture of volatile organic compounds followed by nasal lavage to monitor neutrophil (PMN) evidence of inflammatory response. A subset (4/14) of subjects who were otherwise "normal" with no history of allergies showed a significant increase in PMNs immediately after exposure and 18 h later. The second study by Berg *et al.* (1992) analyzed blood samples of 47 employees with video-display-related skin problems. Persons with skin complaints showed significant elevated stress hormones prolactin and thyroxine compared with controls during work, but not during a leisure day.

It is likely from the above data that sick building syndrome will prove multifactorial in nature, although full-blown MCS still appears to best fit one or more psychiatric, rather than organic disorders.

MASS PSYCHOGENIC ILLNESS

Mass psychogenic illness is usually characterized by a rapid spread of physical and psychosomatic symptoms in the context of a real or imagined environmental trigger (e.g., a chemical odor; an "invisible gas"). While the substance is believed to be harmful, industrial hygiene evaluation usually fails to discover any substance in sufficient quantity to produce neuropsychological or psychiatric complaints.

Behavioral contagion may affect entire communities, as occurred in a residential area of Memphis, Tennessee, where residents complained to their physicians of various maladies after false rumors surfaced that the town was built on a toxic waste dump (Schwartz, 1985). Symptoms of mass psychogenic illness are similar to those found in sick building syndrome, e.g., throat irritation, dryness or irritation of the eyes, nose, mouth, or throat, dizziness, anxiety, or fatigue.

The model individual likely to be affected has been characterized as a highly stressed female who is dissatisified with her job. Hall and Johnson (1989) showed that 33% of the variance in the characterization of a mass psychogenic illness outbreak was explained by five factors: work intensity, mental strain, home and work problems, education, and gender, in decreasing order. Ryan and Morrow (1992) suggest that individuals prone to MPI have premorbid psychological vulnerability and chronically high levels of stress. Pennebaker (1982) hypothesizes that MPI patients have elevated levels of autonomic arousal from tension, anxiety, and stress. Morrow and Ryan contribute a form of biopsychosocial model, speculating that both building characteristics and psychological make-up contribute to the development of MPI. There is little evidence that Sick Building Syndrome results from MPI (Morrow, 1992).

WORKPLACE STRESS

A World Health Organization report indicates that approximately one half of the working population are unhappy with their jobs (Levi, 1989). A 1985 report estimates that 11 million workers report health-damaging levels of mental stress while at work. In 1988, 21% of all Social Security Administration Disability allowances were for mental disorders (*Social Security Bulletin*, 1989). The effects of stress are capable of directly and indirectly interfering with neuropsychological function. Direct effects include depression and mental disorders. Indirect effects of stress include hypertension and maladaptive behavior, including substance abuse or chemical dependence. Over 5 billion doses of Valium and other minor tranquilizers are prescribed yearly.

It has been suggested that high psychosocial stress and low input into the work process may be physically and emotionally dangerous (*Social Security Bulletin*, 1989). Work stress may behaviorally manifest as affective disturbance or maladaptive behavior or lifestyle, including chemical dependency and/or alcohol abuse. Neuropsychological methods may be a main source of differential diagnosis of these syndromes, since the nonspecific complaints of work stress and/or substance abuse are often similar to those from neurotoxic exposure.

Steps that can be taken to evaluate work stressors are an essential inclusion in an occupational neuropsychological work-up. In addition to standard personality inventories, specified questions about work adjustment are necessary. Sauter *et al.* (1990) specified several avenues of questioning that can increase knowledge of psychosocial risk factors in work stress:

1. Personal control or discretion over workload and work pace: Research indicates emotional distress results from lack of worker participation in these factors.
2. Consistency of work schedule: Shift work and night work are more stressful than consistent work scheduling.
3. Role ambiguity: Ambiguous role or conflictual position in the workplace has been linked to hypertension and increased heart rate.
4. Security: Fear of forced retirement, low job security.
5. Interpersonal relationships.
6. Job content: Narrow, repetitive jobs are more stressful.
7. Psychological factors: The meaning of job attributed by the individual, e.g., one person might find the job of a food server as interesting, another highly stressful and demeaning because of differing expectations.

Individuals with traumatic stress responses often complain of neuropsychological impairments, principally in memory and attention. While clinical cases must be considered on an individual basis, present evidence suggests that there are few, if any, neuropsychological correlates of stress when comparing groups of PTSD and anxiety disorder patients against controls (Zalewski & Thompson, 1994).

Neuropsychological complaints may also be interwoven with serious PTSD and it is sometimes difficult to partial out which effects are of direct toxic etiology and which are reactive to the fact or the worry about poisoning. Unlike the robust relationship between MCS and prior psychopathology, there is no good evidence to support a relationship between PTSD and elevated levels of premorbid mental illness (Silverman *et al.*, 1985).

POSTTRAUMATIC STRESS DISORDER

Previously identified in war veterans, PTSD is a diagnosis made with increasing frequency in the workplace, in individuals who have survived harrowing experiences. Immediate or latent threats to life or physical integrity are stressors capable of producing PTSD in the workplace.

Psychiatric criteria for PTSD are described in Table 9.2; however, workplace-related PTSD may differ somewhat from strict criteria. Neurotoxic exposure or exposure to potentially damaging, albeit nonlethal substances may produce PTSD symptoms. High levels of sustained stress as a result of uncertain outcome is almost certainly causal, as was the case with three nuclear power plant workers who were exposed to high, but not damaging levels of radiation (House *et al.*, 1992). One of the workers with a past history of recurrent anxiety began experiencing daily headache, nausea, apathy, poor concentration, and sleep onset difficulty. He experienced vivid nightmares of his exposure and was diagnosed as having PTSD. Symptoms persisted for 6 months and gradually diminished.

patient shows abnormal neuropsychological test results ... therefore ... patient shows neurotoxic brain damage." Diagnosis must be both inductive and deductive; evidence must point to exposure as the primary etiology, while alternative explanations are systematically ruled out. Patient self-report, by itself, is probably the least valid clue to toxic exposure, and self-report neuropsychological questionnaires may be unable to distinguish between neurotoxic exposure profiles and patterns indicative of high levels of personal distress (Dunn et al., 1993). There are few universal pathognomonic effects resulting from neurotoxic exposure, which makes it imperative to determine that symptoms are more likely than not related to exposure effects. For causation to be inferred, several criteria must inhere, as follows.

A. *Documented exposure.* Perhaps the most crucial piece of information; all subsequent diagnostic pronouncements will depend on the adequacy and accuracy of exposure data. What constitutes toxic "exposure" will vary considerably from substance to substance; the examiner cannot simply conclude exposure from patient symptoms but must make a comprehensive effort to determine whether significant exposure took place. This requires answers to the following questions.

B. *Are symptoms credible from a toxicological perspective?* Based on the toxic properties of the substance in question, is it likely that exposure was damaging? For example, the toxicological and clinically observed effects of toluene exposure indicate that brief, low-level, open-air exposure is unlikely to produce neurotoxic damage. Alternatively, serious neurotoxic brain damage, if not fatality, may occur from large, single exposures to such substances as carbon disulfide, carbon monoxide, or cyanide.

C. *Do symptoms make sense with respect to brain–behavior relationships?* Neuropsychological test data must make sense with respect to both internal consistency of deficits obtained as well as ecological validity when comparing test behavior to real-world capabilities. Within a test battery, the patient with complaints of severe attentional deficit should not complete the PASAT without error but fail Speech Sounds Perception and Seashore Rhythm. The patient who struggles to produce ten taps in a 10-second interval on the Finger Oscillation Test should not be able to wiggle his hand rapidly in the air nor keep his regular weekly tennis date at the health club.

D. *Are test symptoms ecologically valid?* There should be a clear and convincing relationship between degree of impairment on test performance and degree of impairment in everyday life. In their review of the ecological validity of neuropsychological test results, McCue & Goldstein (1990) have shown a substantial relationship between test performance and everyday capabilities in a brain-injured population. As a form of brain injury, neurotoxic exposure should likewise be expected to be related to everyday capabilities. However, the compensable nature of a litigable injury claim makes it imperative that the clinician ascertain that test performance does indeed reflect real-world capabilities. This author contested the ecological validity of a set of test results where the claimant's grossly impaired neuropsychological test performance on tasks involving attention and concentration conflicted with ability to perform supervised flying lessons in a small aircraft. In a similar case, neurotoxic attentional abnormalities were alleged, but the claimant's abnormal test results made no sense in the context of his continued squash playing at the local health club. Another litigant claimed severe attention, concentration, and motor control deficits that forbade him from working. In the meantime, he passed the time at home, assembling plastic models as a hobby (a hobby that obviously required concentration and fine motor control).

Secondary gain in the form of large damage awards is an obvious influence on patient behavior in these types of cases. Only careful correlation of patient history and clinical observation with test results will allow the neuropsychologist to differentiate between valid psychometric indications for brain damage, and elaborate exaggerations in the service of secondary gain.

E. *Is there a unique relationship of exposure to symptom?* If symptoms developed *prior* to neurotoxic exposure, they cannot be of neurotoxic etiology, no matter how similar and "textbook" they may appear. The patient who claims intellectual deterioration after solvent exposure, but who had long-standing difficulties with academic performance (see Matarazzo, 1990), obviously cannot have academic deficits attributed to neurotoxic insult.

F. *There are no other illnesses, drugs, stresses, or other psychological diagnoses that might account for symptoms equally well.* Neurotoxic injuries do not have the same external validation criteria that diagnoses of cancer or heart disease have; unless there is moderate to severe damage, there are no abnormal cells or blocked blood vessels to observe. Symptoms produced are not unique to toxic exposure, but may be found in unrelated disease states, psychological illnesses, drug or alcohol abuse, or prescription drug use, among others.

G. *There is no evidence of malingering or a pattern that suggests conscious motivation or deliberate manipulation of performance to exaggerate impairments.* Patterns on the MMPI, wildly inconsistent performance within or across similar tests, and exaggerated performance on forced-choice tests of malingering (e.g., Hiscock & Hiscock, 1989) may all provide evidence of less than credible cooperation and performance.

Neuropsychologists who participate in the forensic process must be prepared to subject their oral or written opinions to intensive scrutiny. Every recorded opinion will not only be gone over line by line, but the neuropsychologist can be repeatedly questioned about those opinions in future depositions or trials for the rest of his or her professional life. The only defensible posture by the neuropsychological toxicologist is that of the research–clinician who understands that opinions must be justifiable by scientific, clinical, neuropsychological, and historical data. In the following example, this author was asked by a defense attorney to critique the likelihood of alleged neurotoxic exposure to hydrogen sulfide in light of a plaintiff's history and claimed symptoms.

Attorney at Law
Chicago, Illinois

Dear Attorney:

In a review of the book, *Hydrogen Sulfide,* published by University Park Press, 1979, I will compare the effects listed in the book with the symptoms reported by the plaintiff in this case. I will concentrate on human exposure, and most especially on the neuropsychological effects of H_2S exposure. The pages listed are from the book, and the italicized comments are the contrasting plaintiff complaints.

p. 25. "Inflammation of the eyes and mucus membranes are observed in sewer and tunnel workers who had been exposed chronically to hydrogen sulfide"

The plaintiff failed to report either eye or mucous membrane irritation when he described his alleged "exposure."

p. 47. "The characteristic olfactory response to hydrogen sulfide is an important aspect of its toxicology. Its typical 'rotten-egg' odor is detectable by olfaction at very low concentrations." Approximately odor threshold is about .1–.2 ppm. Offensive odor is 3–5 ppm. Threshold of serious eye injury is 50–100 ppm, and olfactory paralysis is 150–250 ppm.

The plaintiff failed to report smelling this rotten-egg odor, therefore, there was no significant exposure under 5ppm. One cannot argue that the plaintiff was exposed to enough H_2S to paralyze his sense of smell, since that level would also cause "serious eye injury" and possibly other immediate effects. The plaintiff never had either an eye injury or any other immediate effects.

p. 47. "Hydrogen sulfide is an irritant gas. Its direct action on tissues induces local inflammation of the moist membranes of the eye and respiratory tract.... Subacute effects are more often related to exposures of several hours. Eye irritation becomes noticeable in several minutes and respiratory tract irritation without coughing may occur in approximately 30 minutes."

The plaintiff reports that he walked onto the dock several times to say "hello" to a another employee. He did not experience immediate eye irritation, but thought he felt symptoms 4 days later.

p. 48. "The term 'acute hydrogen sulfide intoxication' has been used most often to describe episodes of systemic poisoning characterized by rapid onset and predominance of signs and symptoms of nervous system involvement."

The plaintiff failed to report any symptoms of nervous system problems in his conversation with his employer one month after his alleged "exposure." Therefore, there was no acute hydrogen sulfide intoxication as defined by the Subcommittee on Hydrogen Sulfide.

p. 48. "In 'subacute intoxication,' signs and symptoms of eye and respiratory tract irritation prevail . . . subacute effects are more often related to exposures of several hours."

The plaintiff failed to report any of those symptoms while "walking around the dock." In addition, there is no indication that he stayed on the dock for several hours. The plaintiff also has smoked a pack of cigarettes each day since he was a teenager. There is no indication that he had ever experienced any difficulties more than would be expected from his pack-per-day habit, e.g., bronchitis. In fact, medical evaluation of his lung function showed that the plaintiff's lung volume had actually improved after his alleged exposure. Therefore, there was no subacute intoxication syndrome as defined by the Subcommittee on Hydrogen Sulfide.

Let us review the human exposure cases cited in *Hydrogen Sulfide* where there is information about postexposure symptoms. The reader will quickly see that the plaintiff's symptoms do not fit those of typical hydrogen sulfide exposure cases, either during alleged exposure or afterwards.

Ahlborg (1951): Studied Swedish shale oil workers exposed to H_2S. 58 workers **unconscious,** 14 hospitalized but no deaths. Only 25 percent of those developed symptoms. Most (80) percent had a history of repeated poisoning and symptoms which developed immediately or soon after acute poisoning.

Kaipainen (1955). Farmer **collapses unconscious** from hydrogen sulfide exposure and is resuscitated two hours later. EKG suggests heart attack, but patient recovered after 1.5 months with only "slight, persistent dizziness."

Hurwitz and Taylor (1954). Sewer worker descends into manhole to investigate "the foulest smell I have ever come across." **Became unconscious,** developed convulsions, artificial respiration needed. Slow recovery and evidence of heart attack. Neurological disabilities and pain on exertion.

Kemper (1966) reports the case of a refinery worker who had been **unconscious** for several minutes after exposure to diethanolamine charged with hydrogen sulfide. The patient survived intermittent respiratory failure cardiovascular collapse, convulsions, and other symptoms. Had mild depression and lassitude for several months, along with amnesia for the day of the accident.

Milby (1962) reports a case where workers disposed of a cylinder of pressurized H_2S by shooting it open; they were then overcome by the vapor. Two of the men became **unconscious,** but were treated with oxygen and **recovered completely within a few days.**

Kleinfeld et al. (1964) report a case where 12 plant employees were overcome when the chemical manufactured by the plant reacted with sewer water to form hydrogen sulfide. Of the 12 severely intoxicated individuals 2 died and all the rest recovered without sequelae. Forty others became ill but did not lose consciousness. **All the survivors recovered without sequelae.**

The same authors report that a 60-year-old man who was exposed to hydrogen sulfide "developed severe respiratory distress but did not lose consciousness." In the hospital, it was discovered that he had pulmonary edema. **He recovered completely within three days.**

Poda (1966) discussed the case of 42 workers who were rendered unconscious at a heavy-water production plant that used hydrogen sulfide in the production process. "Workers who inhaled sufficient gas to cause staggering or loss of consciousness often developed a syndrome characterized by nervousness, dry cough, nausea, headache and insomnia." Nonetheless, the only serious case was in a mechanic found unconscious without pulse or respiration. He recovered after 3 days and suffered no sequelae.

The authors of *Hydrogen Sulfide* report that most cases in the medical literature are similar to these cited "in which serious poisoning with unconsciousness preceded the appearance of sequelae."

Thus, by the human case studies cited in *Hydrogen Sulfide,* where there was information about recovery exposure and duration of neurological problems, most workers **recovered completely, even when rendered unconscious.** There is not a single case that remotely resembles the plaintiff's in the case literature on hydrogen sulfide. Every single case that produced significant long-term H_2S cognitive impairment also showed instantaneous effects in the exposed individual. There is not a single instance similar to the plaintiff's, who was able to operate a boat, take detailed notes about sailing conditions **1 week after** his imagined exposure, and lucidly discuss his "symptoms" on the phone with his employer **1 month after** his supposed exposure.

Even now, the plaintiff's periodic physical complaints alternate with his self-report of being able to play basketball! Patients with nervous system damage of the sort the plaintiff is claiming do not recover enough to play a fast, complicated ball game only to reassert symptoms afterward. Therefore, review of the book *Hydrogen Sulfide* and the plaintiff's present behavior show us that his symptoms have nothing to do with hydrogen sulfide.

How should we understand the plaintiff's symptoms?

1. A review of medical records shows that the plaintiff has always complained of pains and physical problems, long before he believes he was exposed to hydrogen sulfide. Thus, his physical complaints are nothing new.
2. The plaintiff's claim that threats of violence against his employer are due to organic effects of toxic exposure ignores his long premorbid history of violent and impulsive behavior. In the past, this plaintiff kicked one man in the testicles and kicked another in the face after knocking him down. The plaintiff himself was knocked unconscious with a metal pipe in another fight. Thus, his threats of violence, now directed against his employer, are the way he has always behaved.

3. The plaintiff has admitted to drinking 15–21 drinks per week. This is more than enough to worsen impulsive behavior in an already unstable individual.
4. The plaintiff clearly tried to fake some of the test results on his neuropsychological examination in an attempt to look impaired. A detailed look at that report shows his symptoms are not real or are a product of the psychiatric medication he was taking.

It should be clear, therefore, that (1) the plaintiff did not have a real toxic exposure, (2) he does not have symptoms consistent with those in the literature, (3) that his preexisting personality and physical complaints are all similar to current complaints, and (4) he does not have a believable pattern of either impairment or recovery. Therefore, in every important way, the plaintiff's symptom development does not resemble hydrogen sulfide exposure.

Please feel free to contact me if you have any further questions about this case.

Sincerely,
David E. Hartman, Ph.D.

Neuropsychological expertise in neurotoxic exposure cases is particularly valuable when providing objective evidence of behavioral changes in the context of what is known about neurotoxic exposure effects. Psychologists unused to controversy may be uncomfortable about criticizing the reasoning or data of an expert hired by the opposing side, but too this may be required. Again, a clear, data-driven explanation is a much more effective persuader of judges and juries than inveighing against colleagues on a more personal level.

APPLYING RESEARCH DATA TO CLINICAL CASES

Data from studies of individual or group exposure cases must be carefully scrutinized not only for applicability to the individual case, but also for methodological error or possible bias in interpretation. Studies "refuting" industrial neurotoxicity that are funded by the industry or the makers of the potentially injurious substance should be viewed with suspicion. A more subtle form of bias was seen in one study funded by industry that found neuropsychological deficits in the affected population, but labeled them as "subclinical" because they were not "validated" by psychiatric observation.

Needleman (1993) enumerates several statistical biases found in lead intoxication studies that have applicability more generally to review of neurotoxicity studies. These include:

- *Arbitrary worship of the p value of 0.05.* Significance levels of 0.05 or less have become the mark of "causality" when the actual meaning of this α level is simply that there is one possibility in 20 that results were due to chance. This "sanctification" of a particular α level may have more to do

FORENSIC AND PRIVATE PRACTICE ISSUES

with the fact that "Fisher was given permission to reprint only two lines from the table of P values" than to any intrinsic cutoff for causality at 0.05.
- *Overcontrol of "confounds."* Control of all possible "socioheredity" factors may actually take away variance properly attributable to the particular neurotoxic exposure. For example, removing variance for diagnosis of hyperactivity, developmental delay, or mental retardation in childhood lead studies may be taking away the influence of data from subjects who were affected by lead severely enough to merit other lead-related diagnoses.
- *Ignoring the possibility of multigeneration toxic exposures.* For example, poor parenting ability or low parent IQ may be the result of parents living or working under the same neurotoxic conditions as their forebears, a possibility that should be investigated before removing parenting or similar factors from the statistical model.
- *Accepting the null hypothesis from studies with inadequate power.* Power analyses are rarely published, and findings that have accepted the null hypothesis may be using small data sets, or overanalyzing the data. Needleman gives an example of a study with 48 subjects that used a multiple regression analysis with 17 covariates—producing a power "between zero and .30" to reject the null hypothesis (p. 164).
- *Underestimating the epidemiological significance of a true, but "small" effect.* In childhood lead studies, an IQ difference of 6 points "predicts a four-fold increase in the proportion of significantly impaired children" (p. 165).
- *Demands for proof of causality*—the unrealistic expectation that all variables are properly controlled and accounted for. Epidemiological studies can be reanalyzed using infinite combinations of variables and yet bring the reviewer no closer to a secure definition of "causality." The purpose of group studies is to accrue data to which causality is inferred. Causality, therefore, is an inference made from data, it is not *produced* by data analysis.
- *Evaluating isolated studies.* When multiple studies are available, simple "balloting" or tallying of positive or negative studies does not constitute an adequate review of data. Metaanalysis or other quantitative review techniques are available and are preferred.

References

Aaserud, O., Hommeren, O. J., Tvedt, B., Nakstad, P., Mowe, G., Efskind, J., Russell, D., Jorgensen, E. B., Nyberg-Hansen, R., Rootwelt, J., & Gjerstad, L. (1990). Carbon disulfide exposure and neurotoxic sequelae among viscose rayon workers. *American Journal of Industrial Medicine, 18*, 25-37.

A bad drug's benefit. (1985, December 9). *Newsweek*, p. 84.

Abel, E. L. (1976a). *The scientific study of marihuana*. Chicago: Nelson-Hall.

Abel, E. L. (1976b). Marijuana and memory. In E. L. Abel (Ed.), *The scientific study of marihuana* (pp. 113-116). Chicago: Nelson-Hall.

Abel, E. L. (1976c). Retrieval of information after use of marihuana. In E. L. Abel (Ed.), *The scientific study of marihuana* (pp. 121-124). Chicago: Nelson-Hall.

Abel, E. L., & Sokol, R. J. (1987). Incidence of fetal alcohol syndrome and economic impact of FAS-related anomalies. *Drug and Alcohol Dependence, 19(1)*, 51-70.

Abraham, H. (1982). A chronic impairment of colour vision in users of LSD. *British Journal of Psychiatry, 140*, 518-520.

Acker, C. (1986). Neuropsychological deficits in alcoholics: The relative contributions of gender and drinking history. *British Journal of Addiction, 81*, 395-403.

Acker, W., Aps, E. J., Majumdar, S. R., Shaw, G. K., & Thomson, A. D. (1982). The relationship between brain and liver damage in chronic alcoholic patients. *Journal of Neurology, Neurosurgery and Psychiatry, 45*, 984-987.

Adams, J., Vorhees, C. V., & Middaugh, L. D. (1990). Developmental neurotoxicity of anticonvulsants: Human and animal evidence on phenytoin. *Neurotoxicology and Teratology, 12*, 203-214.

Adams, K. M., Grant, I., & Reed, R. (1980). Neuropsychology in alcoholic men in their late thirties: One year follow-up. *American Journal of Psychiatry 137*, 928-931.

Adams, K. M., Gilman, S., Koeppe, R. A., Kluin, K. J., Brunberg, J. A., Dede, D., Berent, S., & Kroll, P. D. (1994). Neuropsychological deficits are correlated with frontal hypometabolism in positron emission tomography studies of older alcoholic patients. *Alcoholism: Clinical and Experimental Research, 17*, 205-210.

Adelson, L., & Kaufman, J. (1971). Fatal chlorine poisoning: Report of two cases with clinicopathologic correlation. *American Journal of Clinical Pathology, 56*, 430-442.

Agarwal, S. B. (1993). A clinical, biochemical, neurobehavioral, and sociopsychological study of 190 patients admitted to hospital as a result of acute organophosphorus poisoning. *Environmental Research, 62*, 63-70.

Agency for Toxic Substances and Disease Registry. (1988). *The nature and extent of lead poisoning in children in the United States: A report to Congress*. Atlanta, GA: Centers for Disease Control.

Agency for Toxic Substances and Disease Registry. (1989). *Toxicological profile for chlordane*. Washington, DC: U.S. Department of Health and Human Services.

Agency for Toxic Substances and Disease Registry. (1989). *Toxicological profile for cyanide*. Washington, DC: U.S. Department of Health and Human Services.

Agency for Toxic Substances and Disease Registry. (1989). *Toxicological profile for phenol*. Washington, DC: U.S. Department of Health and Human Services.

Agency for Toxic Substances and Disease Registry. (1990). *Toxicological profile for styrene*. Washington, DC: U.S. Department of Health and Human Services.

Agency for Toxic Substances and Disease Registry. (1990). *Toxicological profile for tetrachloroethylene*. Washington, DC: U.S. Department of Health and Human Services.

Agency for Toxic Substances and Disease Registry. (1991). *Antimony*. Washington, DC: U.S. Department of Health and Human Services.

Aghajanian, G. K. (1994). Serotonin and the action of LSD in the brain. *Psychiatric Annals, 24*, 137-141.

Agnew, J., Schwartz, B. S., Bolla, K. I., Ford, D. P., & Bleecker, M. L. (1991). Comparison of computerized and examiner-administered neurobehavioral testing techniques. *Journal of Occupational Medicine, 33 (11), 1156-1162.*

Agrawal, A. K., Srivastava, S. P., & Seth, P. K. (1982). Effect of styrene on dopamine receptors. *Bulletin of Environmental Contamination and Toxicology, 29,* (4) 400-403.

Aiello, I., Sau, G. F., Pluliga, M. V., Lentinu, M. E., Muzzu, S., Posadinu, D., & Traccis, S. (1992). Evoked potentials in patients with non-neurological Wilson's disease. *Journal of Neurology, 239*, 65-68.

Albers, J. W. (1990). Standardized neurological testing in neurotoxicology studies. In B. Johnson (Ed.), *Advances in neurobehavioral toxicology: Applications in environmental and occupational health* (pp. 151-164). Chelsea, MI: Lewis Publishers.

Aldridge, W. N., & Johnson, M. K. (1971). Side effects of organophosphorus compounds: Delayed neurotoxicity. *Bulletin of the World Health Organization, 44,* 59-63.

Alexeeff, G. V., & Kilgore, W. W. (1983). Methyl bromide. *Residue Reviews, 88,* 153.

Al-Hachim, G. M., & Al-Baker, A. (1973). Effects on chlordane on conditioned avoidance response, brain seizure threshold and open-field performance of prenatally-treated mice. *British Journal of Pharmacology, 49,* 311-315.

Allen, N., Mendell, J. R., Billmaier, D. J., Fontaine, R. E., & O'Neill, J. (1975). Toxic polyneuropathy due to methyl *n*-butyl ketone. An industrial outbreak. *Archives of Neurology, 32,* 209-218.

Allison, W. M., & Jerrom, D. W. A. (1984). Glue sniffing: A pilot study of the cognitive effects of long-term use. *International Journal of the Addictions, 19,* 453-458.

Almirall-Hernàndez, P., Mayor-Rios, J., del Castillo-Martin, N., Rodrigues-Notario, R., & Romàn-Hernàndez, J. (1987). *Manual de Recomendaciones para la Evaluación Psicológica en Trabajadores Expuestos a Sustancias Neurotóxicas.* Havana, Cuba: Minesterio de Salud Publica, Instituto del Trabajo, Departmento de Psicologicá.

Alterman, A. I., & Tarter, R. E. (1985). Assessing the influence of confounding subject variables in neuropsychological research in alcoholism and related disorders. *International Journal of Neuroscience, 26,* (1-2) 75-84.

Alterman, A. I., Tarter, R. E., Petrarulo, E. W., & Baughman, T. G. (1984). Evidence for impersistence in young male alcoholics. *Alcoholism: Clinical and Experimental Research, 8,* 448-450.

Altman, L. K. (1991, February 19). Storing wine in crystal decanters may pose lead hazard. *The New York Times,* p. B6.

Altura, B. M. (1984). Introduction and overview. *Alcohol, 1,* 321-323.

Altura, B. M., & Altura, B. T. (1984). Alcohol, the cerebral circulation and strokes. *Alcohol, 1,* 325-331.

Altura, B. M., Altura, B. T., & Gebrewold, A. (1983). Alcohol-induced spasms of cerebral blood vessels: Relation to cerebrovascular accidents and sudden death, *Science, 220,* 331-333.

American Conference of Governmental Industrial Hygienists. (1982). *Documentation of the threshold limit values.* Cincinnati, OH: ACGIH Publications Office 1980-1982.

American Conference of Governmental Industrial Hygienists. (1984). *Threshold limit values for chemical substances and physical agents in the work environment with intended changes for 1983-1984.* Cincinnati, OH: ACGIH Publications Office 1983-1984.

American Psychiatric Association. (1980). *Diagnostic and Statistical Manual of Mental Disorders DSM-III.* Washington, DC: Author.

American Psychiatric Association. (1994). *Diagnostic and Statistical Manual of Mental Disorders DMS-IV.* Washington, DC: Author.

REFERENCES

Amler, R. W., & Lybarger, J. A. (1993). Research program for neurotoxic disorders and other adverse health outcomes at hazardous chemical sites in the United States of America. *Environmental Research, 61,* 279-284.
Amr, M., Allam, M., Osmaan, A. L., El-Batanouni, M., Samra, G., & Halim, Z. (1993). Neurobehavioral changes among workers in some chemical industries in Egypt. *Environmental Research, 63,* 295-300.
Andersen, I., Lundqvist, G. R., Molhave, L., Pedersen, O. F., Proctor, D. F., Vaeth, M., & Wyon, D. P. (1983). Human response to controlled levels of toluene in six-hour exposures. *Scandinavian Journal of Work, Environment and Health, 9,* (5) 405-418.
Anderson, A. (1982, July). Neurotoxic follies. *Psychology Today,* pp. 30-42.
Andrews, L. S., & Snyder, R. (1986). Toxic effects of solvents and vapors. In C. D. Klaasen, M. O. Amdur, & J. Doull (Eds.), *Toxicology,* third edition. New York: Macmillan.
Anger, W. K. (1984). Neurobehavioral testing of chemicals: Impact on recommended standards. *Neurobehavioral Toxicology and Teratology, 6,* 147-153.
Anger, W. K. (1985). Neurobehavioral tests used in NIOSH-supported worksite studies, 1973-1983. *Neurobehavioral Toxicology and Teratology, 7,* 359-368.
Anger, W. K. (1990). Human neurobehavioral toxicology testing. In R. W. Russell, R. E. Flattau, & A. M. Pope (Eds.), *Behavioral measures of neurotoxicity: Report of a symposium* (pp. 69-85). Washington, DC: National Academy Press.
Anger, W. K., & Cassitto, M. G. (1993). Individual-administered human behavioral test batteries to identify neurotoxic chemicals. *Environmental Research, 61,* 93-106.
Anger, W. K., & Johnson, B. L. (1985). Chemicals affecting behavior. In J. L. O'Donoghue (Ed.), *Neurotoxicity of industrial and commercial chemicals* (Vol. I, pp. 52-148). Boca Raton: CRC Press.
Anger, W. K., Moody, L., Burg, J., Brightwell, W. S., Taylor, B. J., Russo, J. M., Dickerson, N., Setzer, J. V., Johnson, B. L., & Hicks, K. (1986). Neurobehavioral evaluation of soil and structural fumigators using methyl bromide and sulfuryl fluoride. *NeuroToxicology, 7,* 137-156.
Anger, W. K., Cassitto, M. G., Liang, Y.-X., Amador, R., Hooisma, J., Chrislip, D. W., Mergler, D., Kiefer, M., Hötnagl, J., Fournier, L., Dudek, B., & Zsögön, E. (1993). Comparison of performance from three continents on the WHO-recommended neurobehavioral core test battery. *Environmental Research, 62,* 125-147.
Angotzi, G., Camerino, D., Carboncini, F., Cassitto, M. G., Ceccarelli, F., Cioni, R., Paradiso, C., & Sartorelli, E. (1982). Neurobehavioral follow-up study of mercury exposure. In R. Gilioli, M. G. Cassitto, & V. Foa (Eds.), *Neurobehavioral methods in occupational health* (pp. 247-253). New York: Pergamon Press.
Annau, Z. (1981). The neurobehavioral toxicity of trichlorethylene. *Neurobehavioral Toxicology and Teratology, 3,* 417-424.
Annest, J. L., Pirkle, J. L., Makuc, D., Neese, J. W., Bayse, D. D., & Kovar, M. G. (1983, June 9). Chronological trend in blood lead levels between 1976 and 1980. *New England Journal of Medicine, 308,* 1373-1377.
Antti-Poika, M. (1982a). Prognosis of symptoms in patients with diagnosed chronic organic intoxication. *International Archives of Occupational and Environmental Health, 51,* 81-89.
Antti-Poika, M. (1982b). Overall prognosis of patients with diagnosed chronic organic solvent intoxication. *International Archives of Occupational and Environmental Health, 51,* 127-138.
Araki, S., & Honma, T. (1976). Relationships between lead absorption and peripheral nerve conduction velocities in lead workers. *Scandinavian Journal of Work Environment and Health, 4,* 25.
Araki, S., & Murata, K. (1993). Determination of evoked potentials in occupational and environmental medicine: A review. *Environmental Research, 63,* 133-147.
Araki, S., Abe, A., Ushio, K., & Fujino, M. (1971). A case of skin atrophy, neurogenic muscular and anxiety reaction following long exposure to styrene. *Japanese Journal of Industrial Health, 13,* 427-431.
Aratani, J., Suzuki, H., & Hashimoto, K. (1993). Measurement of vibratory perception threshold in workers exposed to organic solvents. *Environmental Research, 61,* 357-361.

Arcia, E., & Otto, D. A. (1992). Reliability of selected tests from the Neurobehavioral Evaluation System. *Neurotoxicology and Teratology, 14*, 103-110.

Ardila, A., Rosselli, M., & Strumwasser, S. (1991). Neuropsychological deficits in chronic cocaine abusers. *International Journal of Neuroscience, 57(1-2)*, 73-79.

Arfini, G., Mutti, A., Vescovi, P., Ferroni, C., Ferrari, M., Giaroli, C., Passeri, M., & Franchini, I. (1987). Impaired dopaminergic modulation of pituitary in workers occupationally exposed to styrene: Further evidence from PRL response to TRH stimulation. *Journal of Occupational Medicine, 29*, 826-830.

Arlien-Søberg, P. (1985). Chronic effects of organic solvents on the central nervous system and diagnostic criteria. In *Chronic Effects of Organic Solvents on the Central Nervous System and Diagnostic Criteria* (pp. 197-218). Copenhagen: World Health Organization.

Arlien-Søberg, P. (1992). *Solvent neurotoxicity.* Boca Raton, FL: CRC Press.

Arlien-Søberg, P., Bruhn, P., Gyldensted, C., & Melgaard, B. (1979). Chronic painters' syndrome: Toxic encephalopathy in house painters. *Acta Neurologica Scandinavica, 60*, 149-156.

Arlien-Søberg, P., Zilstorff, K., Grandjean, B., & Pedersen, L. (1981). Vestibular dysfunction in occupational chronic solvent intoxication. *Clinical Otolaryngology, 6*, 285-290.

Arlien-Søberg, P., Henriksen, L., Gade, A., Gyldensted, C., & Paulson, O. B. (1982). Cerebral blood flow in chronic toxic encephalopathy in house painters exposed to organic solvents. *Acta Neurologica Scandinavica, 66*, 34-41.

Arlien-Søberg, P., Hansen, L., Ladefoged, O., & Simonsen, L. (1992). Report on a conference on organic solvents and the nervous system. *Neurotoxicology and Teratology, 14*, 81-82.

Arnold, I., Dufresne, R., Alleyne, B., & Stuart, P. (1985). Health implication of occupational exposures to hydrogen sulfide. *Journal of Occupational Medicine, 27*, 373-376.

Arria, A. M., Tarter, R. E., Warty, V., & Van Thiel, D. H. (1990). Vitamin E deficiency and psychomotor dysfunction in adults with primary biliary cirrhosis. *American Journal of Clinical Nutrition, 52(2)*, 383-390.

Asada, Y., Shin, K., Sumiyoshi, A., Ishikawa, M., & Nakamura, N. (1988). Disseminated necrotizing encephalopathy induced by methotrexate therapy alone. *Acta Pathologica Japan, 38*, 1305-1312.

Astley, S. J., & Little, R. E. (1990). Maternal marijuana use during lactation and infant development at one year. *Neurotoxicology and Teratology, 12*, 161-168.

Astrand, I. (1975). Uptake of solvents in the blood and tissues of man. A review. *Scandinavian Journal of Work Environment and Health, 1*, 199-218.

Astrup, P. (1972). Some physiological and pathological effects of moderate carbon monoxide exposure. *British Medical Journal, 4(838)*, 447-452.

Atchison, W. J. (1989). Effects of neurotoxicants on synaptic transmission: Lessons learned from electrophysiological studies. *Neurotoxicology and Teratology, 10*, 393-416.

Aub, J. C., Fairhall, L. T., Minot, A. S., & Reznikoff, P. (1926). *Lead poisoning* (Medicine Monographs 7). Baltimore: Williams & Wilkins.

Auerbach, V., & Hodnett, C. (1990). Neuropsychological follow-up in a case of severe chlorine gas poisoning. *Neuropsychology, 4*, 105-112.

Austin, L. S., Waid, L. R., Kurent, J. E., Barton, S. D., Heyes, M. P., & Silver, R. M. (1991). *Neuropsychiatric sequelae of eosinophilia myalgia syndrome secondary to l-tryptophan ingestion.* Paper presented at the 1991 annual meeting of the National Academy of Neuropsychology, Dallas, TX.

Axelson, O., Hane, M., & Hogstedt, C. (1976). A case-referent study on neuropsychiatric disorders among workers exposed to solvents. *Scandinavian Journal of Work, Environment and Health, 2(1)*, 14-20.

Bach, M. J., Arbit, J., & Bruce, D. L. (1974). Trace anesthetic effect on viligance. In C. Xintaras, B. L. Johnson, & I. deGroot (Eds.), *Behavioral toxicology: Early detection of occupational hazards* (HEW Publication No. (NIOSH) 74-126, pp. 41-50). Washington, DC: U.S. Government Printing Office.

Baker, E. L. (1983a). Neurologic and behavioral disorders. In B. S. Levy & D. H. Wegman (Eds.), *Occupational health: Recognizing and preventing work-related disease* (pp. 317-330). Boston: Little, Brown.

Baker, E. L. (1983b). Neurological disorders. In W. N. Rom (Ed.), *Environmental and occupational medicine* (pp. 313–327). Boston: Little, Brown.
Baker, E. L., & Fine, L. J. (1986). Solvent neurotoxicity: The current evidence. *Journal of Occupational Medicine, 28,* 126–129.
Baker, E. L., & Letz, R. (1986). Neurobehavioral testing in monitoring hazardous workplace exposures. *Journal of Occupational Medicine, 28,* 987–990.
Baker, E. L., & Seppäläinen, A. M. (1986). Human aspects of solvent neurobehavioral effects. In J. Cranmer & L. Golberg (Eds.), Proceedings of the workshop on neurobehavioral effects of solvents. *NeuroToxicology, 7,* 43–56.
Baker, E. L., Folland, D. S., Taylor, T. A., Frank, M., Peterson, W., Lovejoy, G., Cox, D., Housworth, J., & Landrigan, P. J. (1977). Lead poisoning in children of lead workers: Home contamination with industrial dust. *New England Journal of Medicine, 296(5),* 260–261.
Baker, E. L., Feldman, R. G., White, R. F., & Harley, J. P. (1983). The role of occupational lead exposure in the genesis of psychaitric and behavioral disturbances. *Acta Psychiatrica Scandinavica, 67*(Suppl. 303), 38–48.
Baker, E. L., Feldman, R. G., White, R. F., Harley, J. P., Dinse, G. E., & Berkey, C. S. (1983). Monitoring neurotoxins in industry: Development of a neurobehavioral test battery. *Journal of Occupational Medicine, 25(2),* 125–130.
Baker, E. L., Letz, R. E., Fidler, A. T., Shalat, S., Plantamura, D., and Lyndon, M. (1985). A computer-based neurobehavioral evaluation system for occupational and environmental epidemiology: Methodology and validation studies. *Neurobehavioral Toxicology and Teratology, 7,* 369–378.
Baker, E. L., Letz, R. E., and Fidler, A. T. (1985). A computer-administered neurobehavioral evaluation system for occupational and environmental epidemiology. *Journal of Occupational Medicine, 27,* 206–212.
Baker, E. L., Smith, T. J., & Landrigan, P. J. (1985). The neurotoxicity of industrial solvents: A review of the literature. *American Journal of Industrial Medicine, 8,* 207–217.
Baker, E. L., White, R. F., Pothier, L. J., Berkey, C. S., Dinse, G. E., Travers, P. H., Harley, J. P., & Feldman, R. G. (1985). Occupational lead neurotoxicity: Improvement in behavioural effects after reduction of exposure. *British Journal of Industrial Medicine, 42(8),* 507–516.
Baker, E. L., Letz, R. E., Eisen, E. A., Pothierl, J., Plantamura, D. L., Larson, M., & Wolford, R. (1988). Neurobehavioral effects of solvents in construction painters. *Journal of Occupational Medicine, 30(2),* 116–123.
Baker, R., & Woodrow, S. (1984). The clean light image of the electronics industry: Miracle or mirage? In W. Chavkin (Ed.), *Double exposure: Women's health hazards on the job and at home* (pp. 21–36). New York: Monthly Review Press.
Baker, S. J., Chrzan, G. J., Park, C. N., & Saunders, J. H. (1986). Behavioral effects of 0% and 0.05% blood alcohol in male volunteers. *Neurobehavioral Toxicology and Teratology, 8,* 77–81.
Baldridge, E. B., & Bessen, H. A. (1990). Phencyclidine. *Emergency Medicine Clinics of North America, 8,* 541–549.
Baloh, R., Sturm, R., Green, R., & Gleser, G. (1975). Neuropsychological effects of chronic asymptomatic increased lead absorption. *Archives of Neurology, 32,* 326–330.
Balster, R. L. (1989). Toxic materials in the aerospace industry. Testimony before the United States Senate Subcommittee on Toxic Substances, Environmental Oversight and Research and Development, Committee on Environmental & Public Works, March 6, 1989.
Balter, M. B., Manheimer, D. I., Mellinger, D. G., & Uhlenhuth, E. H. (1984). A cross-national comparison of anti-anxiety/sedative drug use. *Current Medical Research Opinion, 8*(Suppl. 4), 5–18.
Bank, W. J. (1980). Thallium. In P. S. Spencer & H. H. Schaumburg (Eds.), *Experimental and clinical neurotoxicology* (pp. 570–577). Baltimore: Williams & Wilkins.
Bardodej, Z., & Bardodejova, E. (1970). Biotransformation of ethylbenzene styrene and alpha-methylstyrene in man. *American Industrial Hygiene Association Journal, 31,* 206–209.
Barlow, D. H., & Hersen, M. (1984). *Single case experimental designs.* New York: Pergamon Press.
Barnes, G. E. (1979). Solvent abuse: A review. *International Journal of the Addictions, 14,* 126.

Barnes, J. M. (1961). Psychiatric sequelae of chronic exposure to organophosphorus insecticides. *Lancet, 2,* 102–103.

Barry, P. S. I. (1975). A comparison of concentrations of lead in human tissues. *British Journal of Industrial Medicine, 32,* 119–139.

Bartlett, E. J., Brodie, J. D., Simkowitz, P., Dewey, S. L., Rusinek, H., Wolf, A. P., Fowler, J. S., Volkow, N. D., Smith, G., Wolkin, A., & Cancro, R. (1994). Effects of haloperidol challenge on regional cerebral glucose utilization in normal human subjects. *American Journal of Psychiatry, 151,* 681–686.

Bass, M. (1970). Sudden sniffing death. *Journal of the American Medical Association, 212,* 2075–2079.

Bauer, L. O. (1993). Motoric signs of CNS dysfunction associated with alcohol and cocaine withdrawal. *Psychiatry Research, 47,* 69–77.

Bauer, R. M., Greve, K. W., Besch, E. L., Schramke, C. J., Crouch, J., Hicks, A., Ware, M. R., & Lyles, W. B. (1992). The role of psychological factors in the report of building-related symptoms in sick building syndrome. *Journal of Consulting & Clinical Psychology, 60(2),* 213–219.

Baxter, L. R., Schwartz, J. M., Phelps, M. E., Mazziotta, J. C., Barrio, J., Warson, R. A., Engel, J., Guze, B. H., Selin, C., & Sumida, R. (1988). Localization of neurochemical effects of cocaine and other stimulants in the human brain. *Journal of Clinical Psychiatry, 49*(Suppl.), 23–26.

Bear, D., Rosenbaum, J., & Norman, R. (1986). Aggression in cat and human precipitated by a cholinesterase inhibitor. *Psychosomatics, 27,* 53–56.

Beaubrun, M. H., & Knight, F. (1973). Psychiatric assessment of 30 chronic users of cannabis and 30 matched controls. *American Journal of Psychiatry, 130(3),* 309–311.

Beck, A. T., Ward, L. H., Mendelson, M., Mock, J., & Erbaugh, J. (1961). An inventory for measuring depression. *Archives of General Psychiatry, 4,* 561–571.

Becker, C. E., & Lash, A. (1990). Detecting subtle human CNS dysfunction: Challenge for toxicologists in the 1990's. *Clinical Toxicology, 28,* vii–xi.

Beckett, W. S., Moore, J. L., Keogh, J. P., & Bleecker, M. L. (1986). Acute encephalopathy due to occupational exposure to arsenic. *British Journal of Industrial Medicine, 43,* 66–67.

Becking, G. C., Boyes, W. K., Damstra, T., & MacPhail, R. C. (1993). Assessing the neurotoxic potential of chemicals—A multidisciplinary approach. *Environmental Research, 61,* 164–175.

Beckmann, J., & Mergler, D. (1985). Symptomatology questionnaires. In P. Grandjean (Ed.), *Neurobehavioral methods in occupational and environmental health: Symposium report* (pp. 26–28). Copenhagen: World Health Organization.

Begleiter, H., Porjesz, B., & Tenner, M. (1980). Neuroradiological and neuropsychological evidence of brain deficits in chronic alcoholics. *Acta Psychiatrica Scandinavica, 62*(Suppl. 286), 313.

Behse, F., & Carlson, F. (1978). Histology and ultrastructure of alterations in neuropathy. *Muscle and Nerve, 1,* 368.

Belcher, H. M. E., & Wallace, P. M. (1991). Neurodevelopmental evaluation of children with intrauterine cocaine exposure. *Pediatric Research, 29,* 7A.

Belkin, L. (1990, December 2). Seekers of clean living head for Texas hills. *The New York Times,* pp. 1, 21.

Bell, A. J., Cole, A., Eccleston, D., & Ferrier, I. N. (1993). Lithium neurotoxicity at normal therapeutic levels. *British Journal of Psychiatry, 162,* 689–692.

Bellinger, D. C., & Stiles, K. M. (1993). Epidemiologic approaches to assessing the developmental toxicity of lead. *NeuroToxicology, 14,* 151–160.

Bellinger, D. C., Needleman, H. L., Leviton, A., Waternaux, C., Rabinowitz, M. B., & Nichols, M. L. (1984). Early sensory motor development and pre-natal exposure to lead. *Neurobehavioral Toxicology and Teratology, 6,* 387–402.

Bellinger, D., Leviton, A., Needleman, H. L., Waternaux, C., & Rabinowitz, M. (1986). Low level lead exposure and infant development in the first year. *Neurobehavioral Toxicology and Teratology, 8,* 151–161.

Bellinger, D., Leviton, A., Rabinowitz, M., Allred, E., Needleman, H., & Schoenbaum, S. (1991). Weight gain and maturity in fetuses exposed to low levels of lead. *Environmental Research, 54,* 151–158.

Bellinger, D., Hu, H., Titlebaum, L., & Needleman, H. L. (1994a). Attentional correlates of dentin and bone lead levels in adolescents. *Archives of Environmental Health, 49,* 98–105.

Bellinger, D., Leviton, A., Allred, E., & Rabinowitz, M. (1994b). Pre- and postnatal lead exposure and behavior problems in school-aged children. *Environmental Research, 66,* 12-30.
Benetou-Marantidou, A., Nakou, S., & Micheloyannis, J. (1988). Neurobehavioral estimation of children with life-long-increased lead exposure. *Archives of Environmental Health, 43,* 392-395.
Benignus, V. A. (1981). Neurobehavioral effects of toluene: A review. *Neurobehavioral Toxicology and Teratology, 3,* 407-415.
Benignus, V. A., Vernon, A., Otto, D. A., Prah, J. D., & Benignus, G. (1977). Lack of effects of carbon monoxide on human vigilance. *Perceptual and Motor Skills 45* (3, Pt. 1), 1007-1014.
Benignus, V. A., Muller, K. E., & Malott, C. M. (1990). Dose-effects functions for carboxyhemoglobin and behavior. *Neurotoxicology and Teratology, 12,* 111-118.
Benitz, W. E., & Tatro, D. S. (1984). Disulfiram intoxication in a child. *Journal of Pediatrics, 105,* 487-489.
Benson, M. D., & Price, J. (1985). Cerebellar calcification and lead. *Journal of Neurology, Neurosurgery, and Psychiatry, 48,* 814-818.
Benton, A. J. (1985). Some problems associated with neuropsychological assessment. *Bulletin of Clinical Neuroscience, 50,* 11-15.
Berg, M., Arnetz, B. B., Lidén, S., Eneroth, P., & Kallner, A. (1992). Techno-stress and VDU skin symptoms. *Journal of Occupational Medicine, 34,* 698-701.
Berg, R. A., & Kelafant, G. A. (1993). Neuropsychological sequelae of chronic organic solvent exposure [Abstract]. *Archives of Clinical Neuropsychology, 8,* 213-214.
Berglund, M. (1981). Cerebral blood flow in chronic alcoholics. *Alcoholism: Clinical and Experimental Research, 5,* 295-303.
Berglund, M., & Ingvar, D. H. (1976). Cerebral blood flow and its regional distribution in alcoholism and in Korsakoff's psychosis. *Journal of Studies on Alcohol, 37,* 586-597.
Berglund, M., Bliding, G., Bliding, A., & Risberg, J. (1980). Reversibility of cerebral dysfunction in alcoholism during the first seven weeks of abstinence—A regional cerebral blood flow study. *Acta Psychiatrica Scandinavica, 62*(Suppl. 286), 119-127.
Bergman, H. (1984). The Halstead-Reitan neuropsychological test battery. *Scandinavian Journal of Work Environment and Health, 10*(Suppl. 1), 30-32.
Bergman, H., Norlin, B., Borg, S., & Fyro, B. (1975). Field dependence in relation to alcohol consumption: A co-twin control study. *Perceptual and Motor Skills, 41(3),* 855-859.
Bergman, H., Borg, S., Hindmarsh, T., Idestrom, C.-M., & Mutzell, S. (1980a). Computed tomography of the brain and neuropsychological assessment of male alcoholic patients and a random sample from the general male population. *Acta Psychiatrica Scandinavica, 62*(Suppl. 286), 47-56.
Bergman, H., Borg, S., Hindmarsh, T., Idestrom, C.-M., & Mutzell, S. (1980b). Computed tomography of the brain, clinical examination and neuropsychological assessment of a random sample of men from the general population. *Acta Psychiatrica Scandinavica, 62*(Suppl. 286), 77-88.
Bergman, H., Borg, S., & Holm, L. (1980c). Neuropsychological impairment and exclusive abuse of sedatives or hypnotics. *American Journal of Psychiatry, 137,* 215-217.
Bergman, H., Borg, S., Hindmarsh, T., Ideström, C.-M., & Mützell, S. (1980d). Computerized tomography of the brain and neuropsychological assessment of alcoholic patients. In H. Begleiter (Ed.) *Biological effects of alcohol* (pp. 771-786). New York: Plenum Press.
Bergman, H., Borg, S., Engelbrektson, K., & Vikander, B. (1989). Dependence on sedative-hypnotics: Neuropsychological impairment, field dependence and clinical course in a 5-year follow-up study. *British Journal of Addiction, 84(5),* 547-553.
Berman, S. M., Whipple, S. C., Fitch, R. J., & Noble, E. P. (1993). P3 in young boys as a predictor of adolescent substance abuse. *Alcohol, 10,* 69-76.
Berndt, D. J., Berndt, S. M., & Kaiser, C. F. (1984). Multidimensional assessment of depression. *Journal of Personality Assessment, 48,* 489-494.
Berry, G. J. (1976). *Neuropsychological assessment of solvent inhalers. Final report to the National Institute on Drug Abuse.* Washington, DC: U.S. Government Printing Office.
Berry, G. J., Heaton, R. K., & Kirby, M. W. (1977). Neuropsychological deficits of chronic inhalant abusers. In B. H. Rumack & A. R. Temple (Eds.), *Management of the poisoned patient* (p. 931). Princeton: Science Press.
Berry, J., van Gorp, W. G., Herzberg, D. S., Hinkin, C., Boone, K., Steinman, L., & Wilkins, J. N.

(1993). Neuropsychological deficits in abstinent cocaine abusers: Preliminary findings after two weeks of abstinence. *Drug and Alcohol Dependence, 32,* 231–237.

Bertoni, J., & Sprenkle, P. (1988). Lead acutely reduces glucose utilization in the rat brain especially in higher auditory centers. *NeuroToxicology, 9,* 235–242.

Bhattacharya, A., Shukla, R., Dietrich, K. N., Miller, J., Bagchee, A., Bornschein, R. L., Cox, C., & Mitchell, T. (1993). Functional implications of postural disequilibrium due to lead exposure. *Neurotoxicology, 14(2–3),* 179–189.

Bidstrup, P. L. (1961). (Letter to the Editor). *Lancet, 2,* 103.

Bigler, E. D. (1984). *Diagnostic clinical neuropsychology.* Austin: University of Texas Press.

Billmaier, D., Yee, H. T., Allen, N., Craft, B., Williams, N., Epstein, S., & Fontaine, R. (1974). Peripheral neuropathy in a coated fabrics plant. *Journal of Occupational Medicine, 16,* 665–671.

Binder, L. M., & Pankratz, L. (1987). Neuropsychological evidence of a factitious memory complaint. *Journal of Clinical and Experimental Neuropsychology, 9,* 167–171.

Bingham, E. (1974). Worker exposure to metals: Metals seminar keynote address. In C. Xintaras, B. L. Johnson, & I. deGroot (Eds.), *Behavioral toxicology: Early detection of occupational hazards* (HEW Publication No. (NIOSH) 74-126, pp. 199–206). Washington, DC: U.S. Government Printing Office.

Bishop, C. M. (1992). A case of methyl bromide poisoning. *Occupational Medicine, 42,* 106–108.

Biskind, M. S., & Mobbs, R. F. (1972). Psychiatric manifestations from insecticide exposure. *Journal of the American Medical Association, 220,* 1248.

Bjornaes, S., & Naalsund, L. U. (1988). Biochemical changes in different brain areas after toluene inhalation. *Toxicology, 49(2–3),* 367–374.

Black, B. (1990). Matching evidence about clustered health events with toxic tort law requirements. *American Journal of Epidemiology, 132(1),* S79–S86.

Black, D. W., Rathe, A., & Goldstein, R. B. (1990). Environmental illness: A controlled study of 26 subjects with "20th century disease." *Journal of the American Medical Association, 246,* 3166–3170.

Blake, D. R., Winyard, P., Lunec, J., Williams, A., Good, P. A., Crewes, S. J., Gutteridge, J. M. C., Rowley, D., Halliwell, B., Cornish, A., & Hider, R. C. (1985). Cerebral and ocular toxicity induced by desferrioxamine. *Quarterly Journal of Medicine,* new series, 56, 345–355.

Bleecker, M. L. (1984). Clinical neurotoxicology: Detection of neurobehavioral and neurological impairments occurring in the workplace and the environment. *Archives of Environmental Health, 39,* 213–218.

Bleecker, M. L., & Bolla-Wilson, K. (1985). Neuropsychological impairment following inorganic arsenic exposure. Unmasking a memory disorder. *Neurobehavioral methods in occupational health* (Document 3, pp. 172–176). Copenhagen: World Health Organization.

Bleecker, M. L., Agnew, J., Keogh, J. P., & Stetson, D. S. (1982). Neurobehavioral evaluation in workers following a brief exposure to lead. In R. Gilioli, M. G. Cassitto, & V. Foa (Eds.), *Neurobehavioral methods in occupational health* (pp. 255–262). New York: Pergamon Press.

Bleecker, M. L., Bolla, K. I., Agnew, J., Schwartz, B. S., & Ford, D. P. (1991). Dose-related subclinical neuropsychological effects of chronic exposure to low levels of organic solvents. *American Journal of Industrial Medicine, 19,* 715–728.

Block, R. I., & Berchou, R. (1984). Alprazolam and lorazepam effects on memory acquisition and retrieval processes. *Pharmacology, Biochemistry and Behavior, 20,* 233–241.

Block, R. I., & Ghoneim, M. M. (1993). Effects of chronic marijuana use on human cognition. *Psychopharmacology, 110,* 219–228.

Blum, K., Noble, E. P., Sheridan, P. J., Montgomery, A., Ritchie, T., Ozkaragoz, T., Fitch, R. J., Wood, R., Finley, O., & Sadlack, F. (1993). Genetic predisposition in alcoholism: Associations of the D_2 dopamine receptor Taq1 B1 RFLP with severe alcoholics. *Alcohol, 10,* 59–67.

Boeckx, R. L. (1979). The clinical chemistry of lead poisoning: New approaches to an old problem. Special review article. *Clinical Proceedings, CHNMC, 35,* 216–231.

Boja, J. W., Cline, E. J., Carroll, F. I., Lewin, A. H., Philip, A., Dannals, R., Wong, D., Scheffel, U., & Kuhar, M. (1992). High potency cocaine analogs: Neurochemical, imaging, and behavioral studies. *Annals of the New York Academy of Sciences, 654,* 282–291.

Bolla, K. I. (1991). Neuropsychological assessment for detecting adverse effects of volatile organic compounds on the central nervous system. *Environmental Health Perspectives, 95*, 93-98.
Bolla, K. I. (1994). Agent Orange: Neuropsychological effects. In The Institute of Medicine (Ed.), *Veterans and Agent Orange* (pp. 640-671). Washington, D.C.: The Institute of Medicine.
Bolla, K. I., Briefel, G., Spector, D., Schwartz, B., Wieler, L., Herron, J., & Gimenez, L. (1992). Neurocognitive effects of aluminum. *Archives of Neurology, 49*, 1021-1026.
Bolla-Wilson, K. (1986). *Neurasthenia vs. depression. Common presentation of CNS toxicity in the workplace.* Paper presented at the American Academy of Neurology, New Orleans, April 1986, Annual Course 108, Occupational and Environmental Neurology.
Bolla-Wilson, K., Wilson, R. J., & Bleecker, M. L. (1988). Conditioning of physical symptoms after neurotoxic exposure. *Journal of Occupational Medicine, 30*, 684-686.
Bolter, J. F., & Hannon, R. (1986). Lateralized cerebral dysfunction in early and late stage alcoholics. *Journal of Studies on Alcohol, 47*, 213-218.
Bolter, J. F., Stanczik, D. F., & Long, C. J. (1983). Neuropsychological consequences of acute, high level, gasoline inhalation. *Clinical Neuropsychology, 5*, 47.
Bornstein, R. A., McLean, D. R., & Ho, K. (1985). Neuropsychological and electrophysiological examination of a patient with Wilson's disease. *International Journal of Neuroscience, 26*, 239-247.
Bornstein, R. A., Watson, G. D., & Kaplan, M. J. (1985). Effects of flurazepam and triazolam on neuropsychological performance. *Perceptual and Motor Skills, 60*, 47-52.
Bornstein, R. A., Watson, G. D., & Pawluk, L. K. (1985). Effects of chronic benzodiazepine administration on neuropsychological performance. *Clinical Neuropharmacology, 8*, 357-361.
Bos, P. M. J., Gerrit, deM., & Bragt, P. C. (1991). Critical review of the toxicity of methyl-*n*-butyl ketone: Risk from occupational exposure. *American Journal of Industrial Medicine, 20*, 175-194.
Bowers, M. B., Goodman, E., & Sim, V. M. (1964). Some behavioral changes in man following anticholinesterase administration. *Journal of Nervous and Mental Disease, 138*, 383.
Bowler, R. M., Mergler, D., Huel, G., Harrison, R., & Cone, J. (1991). Neuropsychological impairment among former microelectronics workers. *NeuroToxicology, 12*, 87-104.
Bowman, M., & Pihl, R. O. (1973). Cannabis: Psychological effects of chronic heavy use. *Psychopharmacologia, 29*, 159-170.
Boxer, P. A. (1985). Occupational mass psychogenic illness. *Journal of Occupational Medicine, 27*, 867-872.
Boyde, T. A., Ernhart, C. B., Green, T. H., Sokol, R. J., et al. (1991). Prenatal alcohol exposure and sustained attention in the preschool years. *Neurotoxicology and Teratology, 13(1)*, 49-55.
Bozarth, M. A., & Wise, R. A. (1985). Toxicity associated with long-term intravenous heroin and cocaine self-administration in the rat. *Journal of the American Medical Association, 254*, 81-83.
Bracy, O. L. (1984). Using computers in neuropsychology. In M. D. Schwartz (Ed.), *Using computers in clinical practice* (pp. 257-268). New York: Haworth Press.
Braff, D. L., & Saccuzzo, D. P. (1982). Effect of antipsychotic medication on speed of information processing in schizophrenic patients. *American Journal of Psychiatry, 139*, 1127-1130.
Branconnier, R. J. (1985). Dementia in human populations exposed to neurotoxic agents: A portable microcomputerized dementia screening battery. *Neurobehavioral Toxicology and Teratology, 7*, 379-386.
Brandt, J., & Butters, N. (1986). The alcoholic Wernicke-Korsakoff syndrome and its relationship to neuropsychological functioning. In I. Grant & K. M. Adams (Eds.), *Neuropsychological assessment of neuropsychiatric disorders* (pp. 441-477). London: Oxford University Press.
Brandt, J., & Provost, D. G. (1985). On the dissimilar effects of alcohol and aging on the perception of cognitive failings. *Alcohol, 2*, 633-635.
Brandt, J., Butters, N., Ryan, C., & Bayog, R. (1983). Cognitive loss and recovery in long-term alcohol abusers. *Archives of General Psychiatry, 40*, 435-442.
Braun, C. M., & Richer, M. (1993). A comparison of functional indexes, derived from screening tests, of chronic alcoholic neurotoxicity in the cerebral cortex, retina and peripheral nervous system. *Journal of Studies on Alcohol, 54(1)*, 11-16.

Bravaccio, F., Ammendola, A., Barruffo, L., & Carlomagno, S. (1981). H-Reflex behavior in glue (*n*-hexane) neuropathy. *Clinical Toxicology, 18,* 1369–1375.

Breggin, P. R. (1983). *Psychiatric drugs: Hazards to the brain.* Berlin: Springer.

Brewer, C. (1984). How effective is the standard dose of disulfiram? A review of the alcohol-disulfiram reaction in practice. *British Journal of Psychiatry, 144,* 200–202.

Briving, C., Jacobson, I., Hamberger, A., Kjellstrand, P., Haglid, K. G., & Rosengren, L. E. (1986). Chronic effects of perchloroethylene and trichloroethylene on the gerbil brain amino acids and glutathione. *Neurotoxicologist, 7,* 101–108.

Brizer, D. A., & Manning, D. W. (1982). Delirium induced by poisoning with anticholinergic agents. *American Journal of Psychiatry, 139,* 1343–1344.

Broadhurst, A. D. (1980). The effect of propranolol on human psychomotor performance. *Aviation, Space, and Environmental Medicine, 51,* 176–179.

Brodsky, C. M. (1983). "Allergic to everything": A medical subculture. *Psychosomatics, 24,* 731–742.

Brooker, A. E., Wiens, A. N., & Wiens, D. A. (1984). Impaired brain functions due to diazepam and meprobamate abuse in a 53-year-old male. *Journal of Nervous & Mental Disease, 172(8),* 498–501.

Brown, R. T., Madan-Swain, A., Pais, R., Lambert, R. G., Baldwin, K., Casey, R., Frank, N., Sexson, S. B., & Ragab, A. (1992). Cognitive status of children treated with central nervous system prophylactic chemotherapy for acute lymphocytic leukemia. *Archives of Clinical Neuropsychology, 7,* 481–497.

Bruce, D. (1985). On the origin of the term "neuropsychology." *Neuropsychologia, 23,* 813–814.

Bruce, D. L., & Bach, M. J. (1975). Psychologic studies of human performance as affected by traces of enflurane and nitrous oxide. *Anesthesiology, 42,* 194.

Bruce, D. L., & Bach, M. J. (1976). Effects of trace anesthetic gases on behavioral performance of volunteers. *British Journal of Anaesthesiology, 48,* 871–876.

Brugnone, F., Maranelli, G., Zotti, S., Zanella, I., De Paris, P., Caroldi, S., & Betta, A. (1992). Blood concentration of carbon disulphide in "normal" subjects and in alcoholic subjects treated with disulfiram. *British Journal of Industrial Medicine, 49,* 658–663.

Bruhn, P., Arlien-Søberg, P., Gyldensted, C., & Christensen, E. L. (1981). Prognosis in chronic toxic encephalopathy: A two year follow up study in 26 house painters with occupational encephalopathy. *Acta Neurologica Scandinavica, 64,* 259–272.

Bryer, J. B., Heck, E. T., & Reams, S. H. (1988). Neuropsychological sequelae of carbon monoxide toxicity at eleven-year follow-up. *The Clinical Neuropsychologist, 2,* 221–227.

Buchsbaum, M. S., Wu, J., Haier, R., Hazlett, E., Ball, R., Katz, M., Sokolski, K., Lagunas-Solar, M., & Langer, D. (1987). Positron emission tomography assessment of effects of benzodiazepines of regional glucose metabolic rate in patients with anxiety disorder. *Life Sciences, 40 (25),* 2393–2400.

Buckholtz, N. S., & Panem, S. (1986). Regulation and evolving science: Neurobehavioral toxicology. *Neurobehavioral Toxicology and Teratology, 8,* 89–96.

Buge, A., Supino-Viterbo, V., Rancurel, G., & Pontes, C. (1981). Epileptic phenomena in bismuth toxic encephalopathy. *Journal of Neurology, Neurosurgery, and Psychiatry, 44,* 621–627.

Bukowski, J. A., Sargent, E. V., & Pena, B. M. (1992). Evaluation of the utility of a standard history questionnaire in assessing the neurological effects of solvents. *American Journal of Industrial Medicine, 22,* 337–345.

Burbacher, T. M., Rodier, P. M., & Wiess, B. (1990). Methylmercury developmental neurotoxicity: A comparison of effects in humans and animals. *Neurotoxicology and Teratology, 12,* 191–202.

Burnett, W. W., King, E. G., Grace, M., & Hall, W. F. (1977). Hydrogen sulfide poisoning: Review of 5 years' experience. *Canadian Medical Association Journal, 117,* 1277–1280.

Burns, R. S., Lerner, S. E., Corrado, R., James, S. H., & Schnoll, S. H. (1975). Phencyclidine—States of acute intoxication and fatalities. *Western Journal of Medicine, 123(5),* 345–349.

Burns, T. G., Cantor, D. S., & Holder, G. S. (1994). Neurofunctional correlates of long-term arsenic exposure in humans. Paper presented at the 1994 annual meeting of the National Academy of Neuropsychology, Orlando, Florida.

Butters, N. (1985). Alcoholic Korsakoff's syndrome: Some unresolved issues concerning etiology, neuropathology and cognitive deficits. *Journal of Clinical and Experimental Neuropsychology, 7,* 181–210.

Butters, N., & Cermak, L. S. (1976). Neuropsychological studies of alcoholic Korsakoff patients. In G. Goldstein & C. Neuringer (Eds.), *Empirical studies of alcoholism* (pp. 153-195). Cambridge, MA: Ballinger.

Byers, R. K., & Lord, E. E. (1943). Late effects of lead poisoning on mental development. *American Journal of Diseases of Children, 66,* 471-494.

Byrne, J. M., Camfield, P. R., Clark-Tovesnard, M., & Hondas, B. J. (1987). Effects of phenobarbital on early intellectual and behavioral development: A concordant twin case study. *Journal of Clinical and Experimental Neuropsychology, 9,* 393-398.

Byrne, R. (1982). *The 637 best things anybody ever said.* New York: Fawcett Crest.

Cabrera, M. A. E. (1990). Neurotoxicology in Mexico and its relation to the general and work environment. In B. L. Johnson (Ed.), *Advances in neurobehavioral toxicology* (pp. 25-28). Chelsea, MI: Lewis Publishers.

Cala, L. A., & Mastaglia, F. L. (1981). Computerized tomography in chronic alcoholics. *Alcoholism: Clinical and Experimental Research, 81,* 283-294.

Cala, L. A., Jones, B., Wiley, B., & Mastaglia, F. L. (1980). A computerized axial tomography (CAT) study of alcohol induced cerebral atrophy: In conjunction with other correlates. *Acta Psychiatrica Scandinavica, 62*(Suppl. 286), 31-40.

Calabrese, E. J. (1985). *Toxic susceptibility: Male/female differences.* New York: Wiley.

Calabrese, E. J. (1986). Sex differences in susceptibility to toxic industrial chemicals. *British Journal of Industrial Medicine, 43,* 577-579.

Call, D. W. (1973). A study of Halon 1301 (CBrF3) toxicity under simulated flight conditions. *Aerospace Medicine, 44,* 202-204.

Callender, T. J., Morrow, L., Subramanian, K., Duhon, D., & Ristovy, M. (1993). Three-dimensional brain metabolic imaging in patients with toxic encephalopathy. *Environmental Research, 60,* 295-319.

Calne, D. B. (1991). Neurotoxins and degeneration in the central nervous system. *NeuroToxicology, 12(3),* 335-339.

Camfield, P. R., Camfield, C. S., Dooley, J. M., Gordon, K., Jollymore, S., & Weaver, D. F. (1992). Aspartame exacerbates EEG spike-wave discharge in children with generalized absence epilepsy. *Neurology, 42,* 1000-1003.

Campagna, K. D. (1986). Drug information forum: What are designer drugs? *U.S. Pharmacist, 11(5),* 16-17.

Campbell, D. D., Evans, M., Thomson, J. L., & Williams, M. J. (1971). Cerebral atrophy in young cannabis smokers. *Lancet, 2,* 1219-1224.

Campbell, D. D., Lockey, J. E., Petajan, J., Gunter, B. J., & Rom, W. N. (1986). Health effects among refrigeration repair workers exposed to fluorocarbons. *British Journal of Industrial Medicine, 43,* 107-111.

Caprio, R. J., Margulis, H. L., & Joselow, M. M. (1974). Lead absorption in children and its relation to urban traffic densities. *Archives of Environmental Health, 28,* 195-197.

Carel, R. S., Belmaker, I., Potashnik, G., Levine, M., & Blau, R. (1992). Delayed health sequelae of accidental exposure to bromine gas. *Journal of Toxicology and Environmental Health, 36,* 273-277.

Carella, F., Grassi, M. P., Savoiardo, M., Contri, P., Rapuzzi, B., & Mangoni, A. (1988). Dystonic-parkinsonian syndrome after cyanide poisoning: Clinical and MRI findings. *Journal of Neurology, Neurosurgery and Psychiatry, 51,* 1345-1348.

Carlen, P. L., & Wilkinson, D. A. (1980). Alcoholic brain damage and reversible deficits. *Acta Psychiatrica Scandinavica, 62*(Suppl. 286), 103-118.

Carlen, P. L., Wortzman, G., Holgate, R. C., Wilkinson, D. A., & Rankin, J. G. (1978). Reversible cerebral atrophy in recently abstinent chronic alcoholics measured by computed tomography scans. *Science, 200,* 1076-1078.

Carlen, P. L., Wilkinson, D. A., Wortzman, G., Holgate, R., Cordingley, J., Lee, M. A., Huszar, L., Moddel, G., Singh, R., Kiraly, L., & Rankin, J. G. (1981). Cerebral atrophy and functional deficits in alcoholics without clinically apparent liver disease. *Neurology, 31,* 377-385.

Carlin, P. L., Penn, R. D., Fornazzari, L., Bennett, J., Wilkinson, D. A., & Wortzman, G. (1986).

Computerized tomographic scan assessment of alcoholic brain damage and its potential reversibility. *Alcoholism: Clinical and Experimental Research, 10,* 226–232.

Carlin, A. S. (1986). Neuropsychological consequences of drug abuse. In I. Grant & K. M. Adams (Eds.), *Neuropsychological assessment of neuropsychiatric disorders* (pp. 478–497). London: Oxford University Press.

Carlin, A. S., & Trupin, E. (1977). The effects of long-term chronic cannabis use on neuropsychological functioning. *International Journal of the Addictions, 12,* 617–624.

Carlin, A. S., Grant, K., Adams, K. M., & Reed, R. (1979). Is phencyclidine (PCP) abuse associated with organic brain impairment? *American Journal of Drug and Alcohol Abuse, 6,* 273–281.

Carmelo, S. (1955). New contributions to the study of subacute-chronic hydrocyanic acid intoxication in men. *Rassegna di Medicina Industriale, 24,* 254–271.

Carroll, J. B. (1980). Individual difference relations in psychometric and experimental cognitive tasks. In L. L. Thurstone Psychometric Laboratory Report No. 163. Chapel Hill, NC: The University of North Carolina. (NTIS No. AD-A086 057)

Carter, W. E., & Doughty, E. (1976). Social and cultural aspects of cannabis use in Costa Rica. *Annals of the New York Academy of Science, 282,* 2–16.

Cassitto, M. G. (1983). Current behavioral techniques. In R. Gilioli, M. G. Cassitto, & V. Foa (Eds.), *Neurobehavioral methods in occupational health* (pp. 27–38). New York: Pergamon Press.

Cassitto, M. G., Gilioli, R., & Camerino, D. (1989). Experiences with the Milan Automated Neurobehavioral System (MANS). *Neurotoxicology and Teratology, 11,* 571–574.

Cassitto, M. G., Camerino, D., Hänninen, H., & Anger, W. K. (1990). International collaboration to evaluate the WHO Neurobehavioral Core Test Battery. In B. L. Johnson (Ed.), *Advances in neurobehavioral toxicology: Applications in environmental and occupational health* (pp. 203–223) Chelsea, MI: Lewis Publishers.

Cassitto, M. G., Camerino, D., Imbriani, M., Contardi, T., Masera, L., & Gilioli, R. (1993). Carbon disulfide and the central nervous system: A 15-year neurobehavioral surveillance of an exposed population. *Environmental Research, 63,* 252–263.

Casswell, S., & Marks, D. F. (1973a). Cannabis and temporal disintegration in experienced and naive subjects. *Science, 179,* 803–805.

Casswell, S., & Marks, D. F. (1973b). Cannabis-induced impairment of performance on a divided attention task. *Nature, 241,* 60–61.

Castleman, M. B. I., & Ziem, G. E. (1988). Corporate influence on threshold limit values. *American Journal of Industrial Medicine, 13,* 531–559.

Caston, J. C., Roufs, J. B., Applebaum, M. L., Tracy, K., Ringel, R., Kooistra, C., Bass, G., Tiller, W. H., & Long, J. (1992). Treatment of refractory eopsinophilia–myalgia syndrome associated with the ingestion of L-tryptophan-containing products with divalproex sodium: Case reports. *Advances in Therapy, 9,* 310–331.

Cavalleri, A., Trimarchi, F., Minoia, C., & Gallo, G. (1982). Quantitative measurement of visual field in lead exposed workers. In R. Gilioli, M. G. Cassitto, & V. Foa (Eds.), *Neurobehavioral methods in occupational health* (pp. 263–269). New York: Pergamon Press.

Cavanaugh, J. B. (1983). Some clinical and neuropathological correlations in four solvent intoxications. In N. Cherry & H. A. Waldron (Eds.), *The neuropsychological effects of solvent exposure* (p. 722). Havant, Hampshire: Colt Foundation.

Cavanaugh, J. B. (1985). Mechanisms of organic solvent toxicity: Morphological changes. In *Chronic effects of organic solvents on the central nervous system and diagnostic criteria* (pp. 110–135). Copenhagen: World Health Organization.

Cawte, J. (1985). Psychiatric sequelae of manganese exposure in the adult, foetal and neonatal nervous systems. *Australian and New Zealand Journal of Psychiatry, 19,* 211–217.

Centers for Disease Control. (1983). NIOSH recommendations for occupational health standards. *Morbidity and Mortality Weekly Report, 32,* Suppl. Atlanta: U.S. Department of Health and Human Services, 15–245.

Centers for Disease Control. (1985, January). *Preventing lead poisoning in young children* (Document No. 992230). U.S. Department of Health and Human Services.

Centers for Disease Control. (1991). *Preventing lead poisoning in young children: A statement by*

REFERENCES

the Centers for Disease Control. Atlanta: U.S. Department of Health and Human Services, Public Health Service.
Cezard, C., Demarquilly, C., Boniface, M., & Haguenoer, J. M. (1992). Influence of the degree of exposure to lead on relations between alcohol consumption and the biological indices of lead exposure: Epidemiological study in a lead acid battery factory. *British Journal of Industrial Medicine, 49(9),* 645–647.
Chaffin, D. B., & Miller, J. M. (1974). Behavioral and neurological evaluation of workers exposed to inorganic mercury. In C. Xintaras, B. L. Johnson, & I. deGroot (Eds.), *Behavioral toxicology: Early detection of occupational hazards* (HEW Publication No. (NIOSH) 74-126, pp. 213–239. Washington, DC: U.S. Government Printing Office.
Chan, T. Y. K. (1995). Anticholinergic poisoning due to Chinese herbal medicines. *Veterinary and Human Toxicology, 37,* 156–157.
Chandra, S. V. (1983). Psychiatric illness due to manganese poisoning. *Acta Psychiatrica Scandinavica, 67*(Suppl. 303), 49–54.
Chang, C. C., & Gershwin, M. E. (1992). Perspectives on formaldehyde toxicity: Separating fact from fantasy. *Regulatory Toxicology and Pharmacology, 16,* 150–160.
Chang, L. W. (1980). Mercury. In P. S. Spencer & H. H. Schaumburg (Eds.), *Experimental and clinical neurotoxicology* (pp. 508–526). Baltimore: Williams & Wilkins.
Chang, L. W. (1982). Pathogenetic mechanisms of the neurotoxicity of methylmercury. In K. N. Prasad & A. Vernadakis (Eds.), *Mechanisms of actions of neurotoxic substances* (pp. 51–66). New York: Raven Press.
Chang, Y.-C. (1990). Patients with n-hexane induced polyneuropathy: A clinical follow up. *British Journal of Industrial Medicine, 47,* 485–489.
Chang, Y.-C. (1991). An electrophysiological follow up of patients with n-hexane polyneuropathy. *British Journal of Industrial Medicine, 48,* 12–17.
Chapman, L. J., Sauter, S. L., Henning, R. A., Dodson, V. N., Reddan, W. G., & Matthews, C. G. (1990). Differences in frequency of finger tremor in otherwise asymptomatic mercury workers. *British Journal of Industrial Medicine, 47,* 838–843.
Charness, M. E. (1994). Brain lesions in alcoholics. *Alcoholism: Clinical and Experimental Research, 17,* 2–11.
Chasnoff, I. J., Burns, W. J., Hatcher, R. P., & Burns, K. A. (1983). Phencyclidine: Effects on the fetus and neonate. *Developmental Pharmacology and Therapeutics, 6,* 404–408.
Chasnoff, I. J., Burns, K. A., Burns, W. J., & Schnoll, S. H. (1986). Prenatal drug exposure: Effects on neonatal and infant growth and development. *Neurotoxicology and Teratology, 8,* 357–362.
Chasnoff, I. J., Griffith, D. R., MacGregor, S., Dirkes, S., & Burns, K. A. (1989). Temporal patterns of cocaine use in pregnancy: Perinatal outcome. *Journal of the American Medical Association, 26,* 1731–1744.
Checkoway, H., Costa, L. G., Camp, J., Coccini, T., Daniell, W. E., & Bills, R. L. (1992). Peripheral markers of neurochemical function among workers exposed to styrene. *British Journal of Industrial Medicine, 49,* 560–565.
Chemical Regulation Reporter. (1986, March 14). Health hazards, p. 1598.
Cherry, N. (1993). Neurobehavioural effects of solvents: The role of alcohol. *Environmental Research, 62(1),* 155–158.
Cherry, N., & Gautrin, D. (1990). Neurotoxic effects of styrene: Further evidence. *British Journal of Industrial Medicine, 47,* 29–37.
Cherry, N., & Waldron, H. A. (Eds.). (1983). *The neuropsychological effects of solvent exposure.* Havant, Hampshire: Colt Foundation.
Cherry, N., Rodgers, B., Venables, H., Waldron, H. A., & Wells, G. G. (1981). Acute behavioral effects of styrene exposure: A further analysis. *British Journal of Industrial Medicine, 38,* 346–350.
Cherry, N., Venables, H., & Waldron, H. A., (1981). *A test battery to measure the behavioral effects of neurotoxic substances.* London: TUC Centenary Institute of Occupational Health.
Cherry, N., Venables, H., & Waldron, H. A. (1983). The acute behavioral effects of solvent exposure. *Journal of the Society of Occupational Medicine, 33,* 13–18.

Cherry, N., Venables, H., & Waldron, H. A. (1984). Description of the tests in the London School of Hygiene test. *Scandinavian Journal of Work Environment and Health, 10*(suppl 1), 18–19.

Cherry, N., Hutchins, H., Pace, T., & Waldron, H. A. (1985). Neurobehavioral effects of repeated occupational exposure to toluene and paint solvents. *British Journal of Industrial Medicine, 42,* 291–300.

Cherry, N. M., Labrèche, F. P., & McDonald, J. C. (1992). Organic brain damage and occupational solvent exposure. *British Journal of Industrial Medicine, 49,* 776–781.

Chia, L.-G., & Chu, F.-L. (1984). Neurological studies on polychlorinated biphenyl (PCB)-poisoned patients. In M. Kuratsune & R. Shapiro (Eds.), *PCB poisoning in Japan and Taiwan* (pp. 117–126). New York: Liss.

Chia, L.-G., & Chu, F.-L. (1985). A clinical and electrophysiological study of patients with polychlorinated biphenyl poisoning. *Journal of Neurology, Neurosurgery, and Psychiatry, 48,* 894–901.

Chia, S. E., Ong, C. N., Phoon, W. H., Tan, K. T., & Jeyaratnam, J. (1993). Neurobehavioral effects on workers in a video tape manufacturing factory in Singapore. *NeuroToxicology, 14,* 51–56.

Chisolm, J. J. (1992). BAL, EDTA, DMSA and DMPS in the treatment of lead poisoning in children. *Clinical Toxicology, 30,* 493–505.

Chmielewski, C., & Golden, C. (1980). Alcoholism and brain damage: An investigation using the Luria–Nebraska Neuropsychological Battery. *International Journal of Neuroscience, 10,* 99–105.

Christensen, A. (1975). *Luria's neuropsychological investigation.* New York: Spectrum.

Christensen, A. (1984). Neuropsychological investigation with Luria's methods. *Scandinavian Journal of Work, Environment and Health, 10*(Suppl. 1), 33–34.

Christensen, E., Moller, J. E., & Faurbye, A. (1970). Neuropathological investigation of 28 brains from patients with dyskinesia. *Acta Psychiatrica Scandinavica, 46(1),* 14–23.

Christodoulou, G. N., Kokkovi, A., Lykouras, E. P., Stefanis, O. N., & Papadimitriou, G. N. (1981). Effects of lithium on memory. *American Journal of Psychiatry, 138(6),* 847–848.

Clark, D. C., Pisani, V. D., Aagesen, C. A., Sellers, D., & Fawcett, J. (1984). Primary affective disorder, drug abuse, and neuropsychological impairment in sober alcoholics. *Alcoholism: Clinical and Experimental Research, 8,* 399–404.

Clark, G. (1971). Organophosphate insecticides and behavior, a review. *Aerospace Medicine, 42,* 735–740.

Classen, W., & Laux, G. (1988). Sensorimotor and cognitive performance of schizophrenic inpatients treated with haloperidol, flupenthixol, or clozapine. *Pharmacopsychiatry, 21,* 295–297.

Claudio, L. (1992). An analysis of the U.S. Environmental Protection Agency neurotoxicity testing guidelines. *Regulatory Toxicology and Pharmacology, 16,* 202–212.

Cocores, J. A., Miller, N. S., Pottash, A. C., & Gold, M. S. (1988). Sexual dysfunction in abusers of cocaine and alcohol. *American Journal of Drug & Alcohol Abuse, 14(2),* 169–173.

Cohen, C., & Frank, A. L. (1994). Liver disease following occupational exposure to 1,1,1 trichloroethane: A case report. *American Journal of Industrial Medicine, 26,* 237–241.

Cohen, N., Modai, D., Golik, A., Pik, A., Weissgarten, J., Sigler, E., & Averbukh, Z. (1986). An esoteric occupational hazard for lead poisoning. *Clinical Toxicology, 24,* 59–67.

Cohen, S. (1978). An international perspective on solvents and aerosols. In *Solvents, adhesives and aerosols* (pp. 71–79). Addiction Research Foundation of Ontario.

Cohen, S. (1980). Adolescence and drug abuse: Biomedical consequences. *National Institute of Drug Abuse Research Monograph Series, 38,* 104–112.

Cohen, S. (1984). Cocaine: Acute medical and psychiatric complications. *Psychiatric Annals, 14,* 747–749.

Cohen, S., & Edwards, A. (1969). LSD and organic brain impairment. *Drug Dependence, 2,* 14.

Cohen, W. J. (1974). Lithium carbonate, haloperidol, and irreversible brain damage. *Journal of the American Medical Association, 230,* 1283–1287.

Cohr, K.-H. (1985). Definition and practical limitation of the concept organic solvents. In Joint WHO/Nordic Council of Ministers Working Group (Eds.), *Chronic effects of organic solvents on the central nervous system and diagnostic criteria* (Document 5). Copenhagen: World Health Organization, Regional Office for Europe.

Cohr, K.-H., & Stokholm, J. (1979). Toluene. A toxicologic review. *Scandinavian Journal of Work, Environment and Health, 5(2)*, 71-90.
Coles, C. D., Platzman, K. A., Smith, I., James, M. E., & Falek, A. (1992). Effects of cocaine and alcohol use in pregnancy on neonatal growth and neurobehavioral status. *Neurotoxicology and Teratology, 14*, 23-33.
Colrain, I. M., Taylor, J., McLean, S., Butter, R., Wise, G., & Montgomery, I. (1993). Dose dependent effects of alcohol on visual evoked potentials. *Psychopharmacology, 112*, 383-388.
Colvin, M., Myers, J., Nell, V., Rees, D., & Cronje, R. (1993). A cross-sectional survey of neurobehavioral effects of chronic solvent exposure on workers in a paint manufacturing plant. *Environmental Research, 63*, 122-132.
Cone, J. E., & Shusterman, D. (1991). Health effects of indoor odorants. *Environmental Health Perspectives, 95*, 53-59.
Conry, J. (1990). Neuropsychological deficits in fetal alcohol syndrome and fetal alcohol effects. *Alcoholism: Clinical and Experimental Research, 14*, 650-655.
Cook, T. L., Smith, B. A., Starkweather, J. A., Winter, P. M., & Eger, E. S. (1978). Behavioral effects of trace and subanesthetic halothane and nitrous oxide in man. *Anesthesiology, 49*, 419-424.
Cooper, J. R., Bloom, F. E., & Roth, R. H. (1982). *The biochemical basis of neuropharmacology.* London: Oxford University Press.
Coremans, P., Lambrecht, G., Schepens, P., Vanwelden, J., & Verhaegen, H. (1994). Anticholinergic intoxication with commercially available thorn apple tea. *Journal of Clinical Toxicology, 32*, 589-592.
Cornelius, J. R., Soloff, P. H., & Reynolds, C. F. (1984). Paranoia, homicidal behavior, and seizures associated with phenylpropanolamine. *American Journal of Psychiatry, 141*, 120-121.
Cote, L. & Crutcher, M. D. (1985). Motor functions of the basal ganglia and diseases of transmitter metabolism. In E. R. Kondel & J. H. Schwartz (Eds.). *Principles of neural science* (pp. 523-535). Amsterdam: Elsevier.
Cotton, H. A. (1912). Comparative psychological studies of the mental capacity in cases of dementia praecox and alcoholic insanity. *Nervous and Mental Disease Monographs, Series 9*, 123-154.
Cousens, P., Waters, B., Said, J., & Stevens, M. (1988). Cognitive effects of cranial irradiation in leukaemia: A survey and meta-analysis. *Journal of Child Psychology & Psychiatry & Allied Disciplines, 29(6)*, 839-852.
Coye, M. J. (1985). The health effects of agricultural productions: I. The health of agricultural workers. *Journal of Public Health Policy, 6*, 349-370.
Coye, M. J., Barnett, P. G., Midtling, J. E., Velasco, A. R., Romero, P., Clements, C. L., O'Malley, M. A., Tobin, M. W., & Lowry, L. (1986). Clinical confirmation of organophosphate poisoning of agricultural workers. *American Journal of Industrial Medicine, 10*, 399-409.
Cranmer, J. M., & Golberg, L. (Eds.). (1986). Proceedings of the workshop on neurobehavioral effects of solvents. *NeuroToxicology, 7*, 195.
Crapper, D. R., & De Boni, U. (1980). Aluminum. In P. S. Spencer & H. H. Schaumburg (Eds.), *Experimental and clinical neurotoxicology* (pp. 326-335). Baltimore: Williams & Wilkins.
Crider, R. A., & Rouse, B. A. (1988). Inhalant overview. *NIDA Research Monographs, 85*, 1-7.
Cripe, L. I., & Dodrill, C. B. (1988). Neuropsychological test performances with chronic low level formaldehyde exposure. *The Clinical Neuropsychologist, 2*, 41-48.
Crystal, H. A., Schaumburg, H. H., Grober, E., Fuld, P. A., & Lipton, R. B. (1988). Cognitive impairment and sensory loss associated with chronic low-level ethylene oxide exposure. *Neurology, 38(4)*, 567-569.
Cullen, M. R. (1987). The worker with multiple chemical sensitivities: An overview. In M. R. Cullen (Ed.), *Workers with multiple chemical sensitivities* (pp. 655-662). Philadelphia: Hanley & Belfus.
Culver, C., & King, F. (1974). Neuropsychological assessment of undergraduate marijuana and LSD users. *Archives of General Psychiatry, 31*, 707-711.
Curley, A., & Garrettson, L. K. (1969). Acute chlordane poisoning. *Archives of Environmental Health, 18*, 211-215.
Current Intelligence Bulletin 48. (1987). *Organic solvent neurotoxicity* (DHHS Publication No. 87-

104). Public Health Services, Centers for Disease Control, National Institute for Occupational Safety and Health.

Curtis, M. F., & Keller, L. W. (Co-chairman). (1986). Exposure issues in the evaluation of solvent effects. *NeuroToxicology, 7,* 524.

Cynn, V. E. H. (1992). Persistence and problem-solving skills in young male alcoholics. *Journal of Studies on Alcohol, 53,* 57–62.

Dackis, C. A., & Gold, M. S. (1985). New concepts in cocaine addiction: The dopamine depletion hypothesis. *Neuroscience and Biobehavioral Review, 9,* 469–477.

Daigle, R. D. (1990). Anabolic steroids. *Journal of Psychoactive Drugs, 22,* 77–80.

Daigle, R. D., Clark, H. W., & Landry, M. J. (1988). A primer on neurotransmitters and cocaine. *Journal of Psychoactive Drugs, 20,* 283–295.

Dalin, N. M., & Kristofferson, R. (1974). Physiological effects of a sublethal concentration of inhaled phenol on the rat. *Annals Zoologica Fenn, 11,* 193–199.

Daniell, W., Barnhart, S., Demers, P., Costa, L. G., Eaton, D. L., Miller, M., & Rosenstock, L. (1992). Neuropsychological performance among agricultural pesticide applicators. *Environmental Research, 59(1),* 217–228.

Darley, C. F., & Tinklenberg, J. R. (1974). Marijuana and memory. In L. L. Miller (Ed.), *Marijuana: Effects on human behavior.* New York: Academic Press.

Darley, C. F., Tinklenberg, J. R., Roth, W. T., & Atkinson, R. C. (1974). The nature of storage deficits and state dependent retrieval under marijuana. *Psychopharmacologia, 37,* 139–149.

Darley, C. F., Tinklenberg, J. R., Roth, W. T., Hollister, L. E., & Atkinson, R. C. (1973). Influence of marijuana on storage and retrieval processes in memory. *Memory and Cognition, 1,* 196–200.

Darley, C. F., Tinklenberg, J. R., Roth, W. T., Vernon, S., & Koppell, B. S. (1977). Marijuana effects on long-term memory assessment and retrieval. *Psychopharmacologia, 52,* 239–241.

David, O. J., Grad, G., McGann, B., and Kolton, A. (1982). Mental retardation and "nontoxic" lead levels. *American Journal of Psychiatry, 139,* 806–809.

David, O. J., Hoffman, S. P., Clark, J., Grad, G., & Swerd, J. (1983). The relationship of hyperactivity to moderately elevated lead levels. *Archives of Environmental Health, 38,* 341–346.

Davies, J. E. (1990). Neurotoxic concerns of human pesticide exposures. *American Journal of Industrial Medicine, 18,* 327–331.

Davis, J. M., Elias, R. W., & Grant, L. D. (1993). Current issues in human lead exposure and regulation of lead. *NeuroToxicology, 14,* 15–28.

Dean, A., Pugh, J., Embrey, K., Cain, J., Lane, L., Brackin, B., & Thompson, F. E. (1984). Organophosphate insecticide poisoning among siblings—Mississippi. *Morbidity and Mortality Weekly Report, 33,* 592–594.

Del Campo, A. F. M., McMurray, R. G., Besser, G. M., & Grossman, A. (1992). Effect of 12-hour infusion of naloxone on mood and cognition in normal male volunteers. *Biological Psychiatry, 32,* 344–353.

Delpech, A. L. D. (1856). *Mémoire sur les accidents que développe cher les ouvriers en cautchoue l'inhalation du sulfure de carbone en vapeur. Lu à l'Académie de Médecine dans la séance du 15 janvier 1856.* Paris: Labé.

Dempsey, G. M., & Meltzer, H. L. (1977). Lithium toxicity: A review. In L. Roizin, H. Shiraki, & N. Grcevic (Eds.), *Neurotoxicology* (pp. 171–183). New York: Raven Press.

Dening, T. R. (1985). Psychiatric aspects of Wilson's disease. *British Journal of Psychiatry, 147,* 677–682.

de Obaldia, R., Leber, W. R., & Parsons, O. A. (1981). Assessment of neuropsychological functions in chronic alcoholics using a standardized version of Luria's Neuropsychological Technique. *International Journal of Neuroscience, 14(1–2),* 85–93.

Deptula, D., & Pomara, N. (1990). Effects of antidepressants on human performance: A review. *Journal of Clinical Psychopharmacology, 10,* 105–111.

DeRenzi, E., & Vignolo, L. A. (1962). The Token Test: A sensitive test to detect receptive disturbances in aphasics. *Brain, 85,* 665–678.

Derban, L. K. (1974). Outbreak of food poisoning due to alkyl-mercury fungicide on southern Ghana state farm. *Archives of Environmental Health, 28(1),* 49–52.

REFERENCES

Deschamps, D., Rosenberg, N., Soler, P., Maillard, G., Fournier, E., Salson, D., & Gervais, P. (1992). Persistent asthma after accidental exposure to ethylene oxide. *British Journal of Industrial Medicine, 49,* 523–525.
de Wit, H., & Wise, R. A. (1977). Blockade of cocaine reinforcement in rats with the dopamine blocker pimoxide but not with the noradrenergic blockers phentolalamine or phenoxybenzamine. *Canadian Journal of Psychology, 31,* 195–203.
de Wit, H., Dudish, S., & Ambre, J. (1993). Subjective and behavioral effects of diazepam depend on its rate of onset. *Psychopharmacology, 112,* 324–330.
Dial, J. (1992). The interaction of alcohol and cocaine: A review. *Psychobiology, 20,* 179–184.
Diaz-Mayans, J., Laborda, J., & Nunez, A. (1986). Hexavalent chromium effects on motor activity and some metabolic aspects of Wistar albino rats. *Comparative Biochemistry and Physiology, 83,* 191–195.
Dick, R. B., Setzer, J. V., Taylor, B. J., & Shukla, R. (1989). Neurobehavioral effects of short duration exposures on acetone and methyl ethyl ketone. *British Journal of Industrial Medicine, 46,* 111–121.
Deitrich, K. M., Succop, P. A., Berger, O. G., Hammond, P. B., & Bornschein, R. L. (1991). Lead exposure and the cognitive development of urban preschool children: The Cincinnati Lead Study Cohort at age 4 years. *Neurotoxicology and Teratology, 13,* 203–211.
Dietrich, K. M., Succop, P. A., Berger, O. G., & Keith, R. W. (1992). Lead exposure and the central auditory processing abilities and cognitive development of urban children: The Cincinnati Lead Study Cohort at age 5 years. *Neurotoxicology and Teratology, 14,* 51–56.
Deitrich, K. M., Berger, O. G., Succop, P. A., Hammond, P. B., & Bornschein, R. L. (1993). The developmental consequences of low to moderate prenatal and postnatal lead exposure: Intellectual attainment in the Cincinnati Lead Study Cohort following school entry. *Neurotoxicology and Teratology, 15,* 37–44.
Dille, J. R., & Smith, P. W. (1964). Central nervous system effects of chronic exposure to organophosphate insecticides. *Aerospace Medicine, 35,* 475–478.
Dixit, R. R., Husain, H., Mukhtar, H., & Seth, P. (1981). Effect of diethyl maleate on acrylamide-induced neuropathy in rats. *Toxicology Letters, 6,* 417–421.
Dolan, M. C., Haltom, T. L., Barrows, G. H., Short, C. S., & Ferriell, K. M. (1987). Carboxyhemoglobin levels in patients with flu-like symptoms. *Annals of Emergency Medicine, 16,* 782–786.
Dolcourt, J. L., Hamrick, H. J., O'Tuama, L. A., Wooten, J., & Baker, E. L. (1978). Increased lead burden in children of battery workers: Asymptomatic exposure resulting from contaminated work clothing. *Pediatrics, 62,* 563–566.
Domino, E. F., & Luby, E. D. (1981). Abnormal mental states induced by phencyclidine as a model of schizophrenia. In E. F. Domino (Ed.), *PCP (phencyclidine): Historical and current perspectives* (pp. 401–418). Ann Arbor: NPP Books.
Donovan, D. M., Kivlahan, D. R., & Walker, D. (1984). Clinical limitations of predicting treatment outcome among alcoholics. *Alcoholism: Clinical and Experimental Research, 8,* 470–475.
Dornbush, R. L. (1974). Marijuana and memory: Effects of smoking on storage. *Transactions of the New York Academy of Sciences, 36,* 94–100.
Dornbush, R. L., Fink, M., & Freedman, A. M. (1976). Marijuana, memory and perception. In E. L. Abel (Ed.), *The scientific study of marijuana* (pp. 133–140). Chicago: Nelson-Hall.
Doty, R. L., Deems, D. A., Frye, R. E., Pelberg, R., & Shapiro, A. (1988). Olfactory sensitivity, nasal resistance, and autonomic function in patients with multiple chemical sensitivities. *Archives of Otolaryngology—Head & Neck Surgery, 114,* 1422–1427.
Drake, A. I., Hannay, H. J., & Gam, J. (1990). Effects of chronic alcoholism on hemispheric functioning: An examination of gender differences for cognitive and dichotic listening tasks. *Journal of Clinical and Experimental Neuropsychology, 12,* 781–797.
Drejer, K., Theilgaard, A., Teasdale, T. W., Schulsinger, F., & Goodwin, D. W. (1985). A prospective study of young men at high risk for alcoholism: Neuropsychological assessment. *Alcoholism: Clinical and Experimental Research, 9,* 498–502.
Driscoll, C. D., Streissguth, A. P., & Riley, E. P. (1990). Prenatal alcohol exposure: Comparability of effects in humans and animal models. *Neurotoxicology and Teratology, 12,* 231–237.

DuBois, K. P. (1971). The toxicity of organophosphorus compounds to mammals. *Bulletin of the World Health Organization, 44*, 233–240.

Duckett, S. (1986). Abnormal deposits of chromium in the pathological human brain. *Journal of Neurology, Neurosurgery, and Psychiatry, 49*, 296–301.

Dudek, B. (1985). The effect of perchloroethylene on mental functions. In *Neurobehavioral methods in occupational and environmental health* (pp. 141–146). Copenhagen: World Health Organization, Regional Office for Europe.

Dudek, B., & Bazylewicz-Walczak, B. (1993). Adaptation of the WHO NCTB for use in Poland for detection of effects of exposure to neurotoxic agents. *Environmental Research, 61(2)*, 349–356.

Dudley, A. W., Jr., Chang, L. W., Dudley, M. A., Bowman, R. E., & Katz, J. (1977). Review of effects of chronic exposure to low levels of halothane. In L. Roizin, H. Shiraki, & N. Grcevic (Eds.), *Neurotoxicology* (pp. 137–146). New York: Raven Press.

Duffy, F. H., & Burchfiel, J. L. (1980). Long term effects of the organophosphate sarin on EEGs in monkeys and humans. *NeuroToxicology, 1*, 667–689.

Duffy, F. H., Burchfiel, J. L., Bartels, P. H., Gaon, M., & Sim, V. M. (1979). Long-term effects of an organophosphate upon the human electroencephalogram. *Toxicology and Applied Pharmacology, 47*, 161–176.

Dunn, J. T., Brown, R. S., Lees-Haley, P. R., & English, L. T. (1993). *Neurotoxic and neuropsychologic symptom base rates: A comparison of three groups.* Paper presented at the 13th annual conference of the National Academy of Neuropsychology, Phoenix, Arizona, October 28–30.

Durwen, H. F., Hufnagel, A., & Elger, C. E. (1992). Anticonvulsant drugs affect particular steps of verbal memory processing—An evaluation of 13 patients with intractable complex partial seizures of left temporal lobe origin. *Neuropsychologia, 30*, 623–631.

Echeverria, D., Fine, L., Langolf, G., Schort, A., & Sampaio, C. (1989). Acute neurobehavioral effects of toluene. *British Journal of Industrial Medicine, 46*, 483–495.

Echeverria, D., Fine, L., Langolf, G., Schort, A., & Sampaio, C. (1991). Acute behavioral comparisons of toluene and ethanol in human subjects. *British Journal of Industrial Medicine, 48*, 750–761.

Eckardt, M. J., & Martin, P. R. (1986). Clinical assessment of cognition in alcoholism. *Alcoholism: Clinical and Experimental Research, 10*, 123–127.

Eckardt, M. J., Ryback, R. S., & Pautler, C. P. (1980). Neuropsychological deficits in alcoholic men in their mid thirties. *American Journal of Psychiatry, 137*, 932–936.

Eckerman, D. A., Carroll, J. B., Foree, C. M., Gullion, M., Lansman, E. R., Long, E. R., Waller, M. B., & Wallsten, T. S. (1985). An approach of brief testing for neurotoxicity. *Neurobehavioral Toxicology and Teratology, 7*, 387–394.

Ecobichon, D., & Joy, R. (1982). *Pesticides and neurological diseases.* Boca Raton, FL: CRC Press.

Ecobichon, D., & Joy, R. M. (1994). *Pesticides and neurological diseases*, 2nd ed. Boca Raton, FL: CRC Press.

Edling, C. (1980). Anesthetic gases as an occupational hazard—A review. *Scandinavian Journal of Work Environment and Health, 6*, 85–93.

Edling, C. (1985). Nervous systems symptoms and signs associated with long-term organic solvent exposure. In *Chronic effects of organic solvents on the central nervous system and diagnostic criteria* (pp. 149–155). Copenhagen: World Health Organization.

Edling, C., & Ekberg, K. (1985). No acute behavioral effects of exposure to styrene: A safe level of exposure? *British Journal of Industrial Medicine, 42*, 301–304.

Edling, C., Ekberg, K., Ahlborg, G., Alexander, R., Barregard, L., Ekenvall, L., Nilsson, L., & Svensson, B. G. (1990). Long term follow up of workers exposed to solvents. *British Journal of Industrial Medicine, 47*, 75–82.

Edling, C., Lindberg, A., & Ulfberg, J. (1993). Occupational exposure to organic solvents as a cause of sleep apnoea. *British Journal of Industrial Medicine, 50*, 276–279.

Egeland, G. M., Burkhart, G. A., Schnorr, T. M., Hornung, R. W., Fajen, J. M., & Lee, S. T. (1992). Effects of exposure to carbon disulphide on low density lipoprotein cholesterol concentration and diastolic blood pressure. *British Journal of Industrial Medicine, 49*, 287–293.

Ehrenberg, R. L., Vogt, R. L., Smith, A. B., Brondum, J., Brightwell, W. S., Hudson, P. J., McManus, K. P., Hannon, W. H., & Phipps, F. C. (1991). Effects of elemental mercury exposure at a thermometer plant. *American Journal of Industrial Medicine, 19,* 495-507.

Eisdorfer, C., Nowlin, J., & Wilkie, F. (1970). Improvement of learning in the aged by modification of autonomic nervous system activity. *Science, 170,* 1327-1329.

Eisen, L. N., Field, T. M., Bandstra, E. S., Roberts, J. P., Morrow, C., & Larson, S. K. (1991). Prenatal cocaine effects on neonatal stress behavior and performance on the Brazelton Scale. *Pediatrics, 83(3),* 477-480.

Ekberg, K., Barregard, L., Hagberg, S., & Sallsten, S. (1986). Chronic and acute effects of solvents on central nervous system functions in floorlayers. *British Journal of Industrial Medicine, 43,* 101-106.

Ellis, E. J. (1987). *Cerebral functional asymmetries in aging and chronic alcoholism.* Unpublished dissertation, Boston University Medical School, Boston.

Ellis, E. J., & Oscar-Berman, M. (1984). Effects of aging and alcoholism on recognition on dichotically presented stimuli. *The International Neuropsychological Society Bulletin, 13,* 14.

Elofsson, S., Gamberale, F., Hindmarsh, T., Iregren, A., Isaksson, A., Johnsson, I., Knave, B., Lydahl, E., Mindus, P., Persson, H. E., Philipson, B., Steby, M., Struwe, G., Soderman, E., Wennberg, A., & Widen, L. (1980). Exposure to organic solvents: A cross-sectional epidemiologic investigation on occupationally exposed car and industrial spray painters with special reference to the nervous system. *Scandinavian Journal of Work Environment and Health, 6,* 239-273.

Engstrom, J., Bjurstrom, R., Astrand, I., & Ovrum, P. (1978). Uptake, distribution and elimination of styrene in man. *Scandinavian Journal of Work Environment and Health, 4,* 315-323.

EPA. (1985). *Health assessment document for tetrachloroethylene (perchloroethylene)—Final report* (EPA 600/8-82/006F). Washington, DC: Author.

Erkkila, J., Armstrong, R., Riihimaki, V., Chettle, D. R., Paakkari, A., Scott, M., Somervaille, L., Starck, J., Kock, B., & Aitio, A. (1992). In vivo measurements of lead in bone at four anatomical sites: Long term occupational and consequent endogenous exposure. *British Journal of Industrial Medicine, 49(9),* 631-644.

Ernhart, C. B. (1992). A critical review of low-level prenatal lead exposure in the human: 2. Effects on the developing child. *Reproductive Toxicology, 6,* 21-40.

Ernhart, C. B., Morrow-Tlucak, M., Marler, M. R., & Wolf, A. W. (1987). Low level lead exposure in the prenatal and early preschool periods: Early preschool development. *Neurotoxicology and Teratology, 9,* 259-270.

Errico, A. L., Parsons, O. A., Kling, O. R., & King, A. C. (1992). Investigation of the role of sex hormones in alcoholics' visuospatial deficits. *Neuropsychologia, 30(5),* 417-426.

Escobar, A., & Aruffo, C. (1980). Chronic thinner intoxication: Clinico-pathologic report of a human case. *Journal of Neurology, Neurosurgery and Psychiatry, 43,* 986-994.

Eskelinen, L., Luisto, M., Tenkanen, L., & Mattei, O. (1986). Neuropsychological methods in the differentiation of organic solvent intoxication from certain neurological conditions. *Journal of Clinical and Experimental Neuropsychology, 8,* 239-256.

Estrin, W. J., Cavalieri, S. A., Wald, P., Becker, C. E., Jones, J. R., & Cone, J. E. (1986). *Evidence of neurologic dysfunction related to chronic ethylene oxide exposure.* Unpublished manuscript. Available from the first author, Department of Neurology, 4M71, San Francisco General Hospital, 1001 Potrero Avenue, San Francisco, CA 94110.

Estrin, W. J., Bowler, R. M., Lash, A., & Becker, C. E. (1990). Neurotoxicological evaluation of hospital sterilizer workers exposed to ethylene oxide. *Clinical Toxicology, 28,* 1-20.

Eustis, S. L., Haber, S. B., Drew, R. T., & Yang, R. S. H. (1988). Toxicology and pathology of methyl bromide in F344 rats and B6C3F1 mice following repeated inhalation exposure. *Fundamental and Applied Toxicology, 11,* 594-610.

Evans, R. W., & Gualtieri, C. T. (1985). Carbamazepine: A neuropsychological and psychiatric profile. *Clinical Neuropharmacology, 8,* 221-241.

Fabian, M. S., Jenkins, R. L., & Parsons, O. A. (1981). Gender, alcoholism, and neuropsychological functioning. *Journal of Consulting and Clinical Pathology, 49,* 138-140.

Fairchild, E. J. (1974). Welcoming remarks from NIOSH. In C. Xintarus, B. L. Johnson, & I. deGroot (Eds.), *Behavioral toxicology: Early detection of occupational hazards* (Hew Publication No. (NIOSH) 74-126, pp. 3-4. Washington, DC: U.S. Government Printing Office.

Fauman, M. A., & Fauman, B. J. (1981). Chronic phencyclidine (PCP) abuse: A psychiatric perspective. In E. F. Domino (Ed.), *PCP (phencyclidine): historical and current perspectives* (pp. 419-436). Ann Arbor: NPP Books.

Fawcett, T. A., Moon, R. E., Fracica, P. J., Mebane, G. Y., Theil, D. R., & Piantadosi, C. A. (1992). Warehouse workers' headache. Carbon monoxide poisoning from propane-fueled forklifts. *Journal of Occupational Medicine, 34(1)*, 12-15.

Fein, G. G., Schwartz, P. M., Jacobson, S. W., & Jacobson, J. L. (1983). Environmental toxins and behavioral development. A new role for psychological research. *American Psychologist, 38(11)*, 1188-1197.

Feldman, R. G. (1982a). Central and peripheral nervous system effects of metals: A survey. *Acta Neurologica Scandinavica, 66*(Suppl. 92), 143-166.

Feldman, R. G. (1982b). Neurological manifestations of mercury intoxication. *Acta Neurologica Scandinavica, 66*(Suppl. 92), 201-209.

Feldman, R. G., Niles, C. A., Kelly-Hayes, M., Sax, D., Dixon, W., Thompson, D. J., & Landau, E. (1979). Peripheral neuropathy in arsenic smelter workers. *Neurology, 29*, 939-944.

Feldman, R. G., Ricks, N. L., & Baker, E. L. (1980). Neuropsychological effects of industrial toxins: A review. *American Journal of Industrial Medicine, 1*, 211-227.

Feldman, R. G., White, R. F., Currie, J. N., Travers, P. H., & Lessell, S. (1985). Long-term follow-up after single toxic exposure to trichlorethylene. *American Journal of Industrial Medicine, 8*, 119-126.

Fergusson, D. M., Horwood, L. J., & Lynskey, M. T. (1993). Early dentine levels and subsequent cognitive and behavioural development. *Journal of Child Psychology & Psychiatry & Applied Disciplines, 34*, 215-227.

Fernandez, P. G. (1976). Alpha methyldopa and forgetfulness. *Annals of Internal Medicine, 85*, 128.

Fernberger, S. W. (1933), Shepherd Ivory Franz, 1874-1933. *The Psychological Bulletin, 30*, 741-742.

Ferraro, D. P. (1980). Acute effects of marijuana on human memory and cognition. *National Institute of Drug Abuse Research Monograph Series, 31*, 98-119.

Ferroni, C., Selis, L., Mutti, A., Folli, D., Bergamaschi, E., & Franchini, I. (1992). Neurobehavioral and neuroendocrine effects of occupational exposure to perchloroethylene. *NeuroToxicology, 13*, 243-248.

Fetzer, J., Kader, G., & Danahy, S. (1981). Lithium encephalopathy: A clinical, psychiatric, and EEG evaluation. *American Journal of Psychiatry, 138*, 1622-1623.

Fiedler, N., Maccia, C., & Kipen, H. (1992). Evaluation of chemically sensitive patients. *Journal of Occupational Medicine, 5*, 529-605.

Fields, S., & Fullerton, J. (1975). Influence of heroin addiction on neuropsychological functioning. *Journal of Consulting and Clinical Psychology, 43*, 114.

Finelli, P. F., Morgan, T. F., Yaar, I., & Granger, C. V. (1983). Ethylene oxide-induced polyneuropathy. *Archives of Neurology, 40*, 419-421.

Firth, J. B., & Stuckey, R. E. (1945). Decomposition of trilene in closed circuit anesthesia. *Lancet, 1*, 814.

Fischbein, A., Wallace, J., Sassa, S., Kappas, A., Butts, G., Rohl, A., & Kaul, B. (1992). Lead poisoning from art restoration and pottery work: Unusual exposures source and household risk. *Journal of Environmental Pathology, Toxicology, and Oncology, 11*, 7-11.

Fischman, M. W. (1984). The behavioral pharmacology of cocaine in humans. In J. Grabowski (Ed.), *Cocaine: Pharmacology, effects, and treatment of abuse* (NIDA Research Monograph 50, pp. 72-91). Washington, DC: Department of Health and Human Services.

Fisher, D. G., Sweet, J. J., & Pfaelzer-Smith, E. A. (1986). The influence of depression on repeated neuropsychological testing. *The International Journal of Clinical Neuropsychology, 8*, 14-18.

Fisher, K. (1985, December). Measuring effects of toxic chemicals: A growing role for psychology. *APA Monitor*, 13-14.

Fittro, K. P., Bolla, K. I., Heller, J. R., & Meyd, C. J. (1992). The Milan Automated Neurobehavioral System. Age, sex and education differences. *Journal of Occupational Medicine, 34,* 918–922.
Fitzhugh, L. C., Fitzhugh, K. B., & Reitan, R. M. (1960). Adaptive abilities and intellectual functioning in hospitalized alcoholics. *Quarterly Journal of Studies on Alcohol, 21,* 414–423.
Fitzhugh, L. C., Fitzhugh, K. B., & Reitan, R. M. (1965). Adaptive abilities and intellectual functioning in alcoholics: Further considerations. *Quarterly Journal of Studies on Alcohol, 26,* 402–411.
Fleming, A. J. (1982). The toxicity of antimony trioxide. Sponsored by E. I. Du Pont de Nemours and Co., Wilmington, DE.
Flodin, U., Edling, C., & Axelsson, O. (1984). Clinical studies of psychoorganic syndromes among workers with exposure to solvents. *American Journal of Industrial Medicine, 5,* 287–295.
Flodin, U., Ekberg, K., & Anderson, L. (1989). Neuropsychiatric effects of low exposure to styrene. *British Journal of Industrial Medicine, 46,* 805–808.
Foa, V. (1982). Evaluation, limits and perspectives of neurobehavioral toxicology in occupational health. In R. Gilioli, M. G. Cassitto, & V. Foa (Eds.), *Neurobehavioral methods in occupational health* (pp. 173–176). New York: Pergamon Press.
Fodor, G. G., & Winneke, G. (1972). Effect of low CO concentrations on resistance to monotony and on psychomotor capacity. *Staub-Reinhalt Luft, 32,* 46–54.
Foo, S. C., Jeyaratnam, J., & Koh, D. (1990). Chronic neurobehavioral effects of toluene. *British Journal of Industrial Medicine, 47,* 480–484.
Foo, S. C., Ngim, C. H., Salleh, I., Jeyaratnam, J., & Boey, K. W. (1993). Neurobehavioral effects in occupational chemical exposure. *Environmental Research, 60,* 267–273.
Fornazzari, L., Wilkinson, D. A., Kapur, B. M., & Carlen, P. L. (1983). Cerebellar, cortical and functional impairment in toluene abusers. *Acta Neurologica Scandinavica, 67,* 319–329.
Fortemps, E., Amand, G., Bomboir, A., Lauwerys, R., & Laterre, E. C. (1978). Trimethyltin poisoning: Report of two cases. *International Archives of Occupational and Environmental Health, 41,* 16.
Foster, K. R., Bernstein, D. E., & Huber, P. W. (1993). Science and the toxic tort. *Science, 261,* 1509, 1614.
Fotopoulos, S. S., Cook, M. R., Gerkovich, M. M., Koontz, E., Graham, C., & Cohen, H. H. D. (1984). In: *Contractors review meeting.* St. Louis, MO: Department of Energy, Electric Power Research Institute and New York State Power Lines Project.
Franchini, I., Ferri, F., Lommi, G., Lotta, S., Lucertini, S., & Mutti, A. (1982). Brain evoked potentials in solvent exposure. In R. Gilioli, M. G. Cassitto, & V. Foa (Eds.), *Neurobehavioral methods in occupational health* (pp. 197–204). New York: Pergamon Press.
Frantik, E., Hornychova, M., & Horvath, M. (1994). Relative acute neurotoxicity of solvents: Isoeffective air concentrations of 48 compounds evaluated in rats and mice. *Environmental Research, 66,* 173–185.
Freed, D. M., & Kandel, E. (1988). Long-term occupational exposure and the diagnosis of dementia. *NeuroToxicology, 9(3),* 391–400.
Freimer, M. L., Glass, J. D., Chaudhry, V., Tyor, W. R., Cornblath, D. R., Griffin, J. W., & Kunci, R. W. (1992). Chronic demyelinating polyneuropathy associated with eosinophilia–myalgia syndrome. *Journal of Neurology, Neurosurgery and Psychiatry, 55,* 352–358.
Freud, S. (1985). Project for a scientific psychology. In J. Strachey (Ed.), *The standard edition of the complete psychological works of Sigmund Freud* (Vol. I, pp. 283–346) (reprinted in 1975). London: Hogarth Press.
Freud, S. (Letter to Wilhelm Fleiss, 26 October 1896). In J. Masson (Ed.), *The complete letters of Sigmund Freud* (p. 201). Cambridge: Belknap Press, 1985.
Frezza, M., Dipadova, C., Pozzato, G., Terpin, M., Baraona, E., & Lieber, C. S. (1990). High blood alcohol levels in women: The role of decreased gastric alcohol dehydrogenase activity and first-pass metabolism. *New England Journal of Medicine, 322,* 95–99.
Fried, P. A. (1980). Marihuana use by pregnant women: Neurobehavioral effects in neonates. *Drug and Alcohol Dependence, 6,* 415–425.
Fried, P. A., & Watkinson, B. (1988). 12- and 24-month neurobehavioural follow-up of children

prenatally exposed to marihuana, cigarettes and alcohol. *Neurotoxicology and Teratology, 10(4),* 305–313.
Friedman, M. J., Culver, C. M., & Ferrell, R. B. (1977). On the safety of long-term treatment with lithium. *American Journal of Psychiatry, 134,* 1123–1126.
Froines, J. R., Baron, S., Wegman, D. H., & O'Rourke, S. (1990). Characterization of the airborne concentrations of lead in U.S. industry. *American Journal of Industrial Medicine, 18,* 1–17.
Frontali, N., Amantini, M. C., Spagnolo, A., Guarcini, A. M., & Saltari, M. C. (1981). Experimental neurotoxicity and urinary metabolites of the C5–C7 aliphatic hydrocarbons used as glue solvents in shoe manufacture. *Clinical Toxicology, 18,* 1357–1367.
Fujimura, Y., Araki, S., Murata, K., Yokoyama, K., & Handa, S. (1993). Assessment of the distribution of nerve conduction velocities in alcoholics. *Environmental Research, 61,* 317–322.
Fujita, A. J., Shibata, H., Amani, K., Kato, Y., Itomi, E., Suzuki, T., Nakazawa, T., & Takahashi, T. (1961). Clinical observations of three cases of acrylamide poisoning. *Nippon IJO Shimpo, 1869,* 37–40.
Fullerton, P. M. (1966). Chronic peripheral neuropathy produced by lead poisoning in guinea pigs. *Journal of Neuropathology and Experimental Neurology, 25,* 214.
Fung, Y. K., & Molvar, M. P. (1992). Toxicity of mercury from dental environment and from amalgam restorations. *Clinical Toxicology, 30,* 49–61.
Fung, Y. K., Molvar, M. P., Strom, A., Schneider, N. R., & Carlson, M. P. (1991). In vivo mercury and methyl mercury levels in patients at different intervals after amalgam restorations. *General Dentistry,* March–April, 89–92.
Gade, A., Mortensen, H., & Udesen, H. (1984). The pattern of intellectual impairment in toxic encephalopathy, or a study of the pattern, degree, and validity of intellectual impairment in toxic encephalopathy. Presented at the International Conference on Organic Solvent Toxicity, 15–17 October, Stockholm, Sweden.
Gade, A., Mortensen, H., Udesen, H., & Bruhn, P. (1985). On the importance of control data and background variables in the evaluation of neuropsychological aspects of brain functioning. *Neurobehavioral methods in occupational and environmental health* (Document 3, pp. 91–96). Copenhagen: World Health Organization.
Gallant, D. M. (1985). Alcoholism: The most common psychiatric illness. *Alcoholism: Clinical and Experimental Research, 9,* 297.
Gallant, D. M. (1986). Hypertension and alcohol consumption. *Alcoholism: Clinical and Experimental Research, 10,* 358.
Gallway, R. A., Millington, J. T., Van Gorp, W. G., Mirsky, A. F., Wolcott, C. L., & Wilmeth, J. B. (1990). *Neuropsychological consequences of hyperbaric nitrogen narcosis.* Presented at the 13th European Conference of the International Neuropsychological Society, Innsbruck, Austria, July 6.
Gamberale, F. (1985). Use of behavioral performance tests in the assessment of solvent toxicity. *Scandinavian Journal of Work Environmental and Health, 11*(Suppl. 1), 65–74.
Gamberale, F. (1990). Physiological and psychological effects of exposure to extremely low-frequency electric and magnetic fields on humans. *Scandinavian Journal of Work Environment and Health, 16*(Suppl. 1), 51–54.
Gamberale, F., & Hultengren, M. (1972). Toluene exposure: II. Psychophysiological functions. *Work Environment and Health, 9,* 131–139.
Gamberale, F., & Hultengren, M. (1973). Methylchloroform exposure. II. Psychophysiological functions. *Work Environment and Health, 10,* 82–92.
Gamberale, F., & Kjellberg, A. (1982). Behavioral performance assessment as a biological control of occupational exposure to neurotoxic substances. In R. Gilioli, M. G. Cassitto, & V. Foa (Eds.), *Neurobehavioral methods in occupational health* (pp. 111–121). New York: Pergamon Press.
Gamberale, F., Annwall, G., & Hultengren, M. (1978). Exposure to xylene and ethylbenzene: III. Effects on central nervous system functions. *Scandinavian Journal of Work Environment and Health, 4,* 204–211.

Gamberale, F., Anselm, B., Eneroth, P., Lindh, T., & Wennberg, A. (1989). Acute effects of ELF electromagnetic fields: A field study of linesmen working with 400 kV power lines. *British Journal of Industrial Medicine, 46*, 729-737.

Gamberale, F., Iregren, A., & Kjellberg, A. (1990). Computerized performance testing in neurotoxicology: Who, what, how, and where to? The SPES Example. In R. W. Russell, P. E. Flattau, & A. M. Pope (Eds.), *Behavioral measures of neurotoxicity* (pp. 359-394). Washington, DC: National Academy Press.

Garland, H., & Pearce, J. (1967). Neurological complications of carbon monoxide poisoning. *Quarterly Journal of Medicine, 144*, 445-455.

Garland, T. O., & Patterson, M. W. H. (1967). Six cases of acrylamide poisoning. *British Journal of Medicine, 4*, 134-138.

Garnier, R., Rosenberg, N., Puissant, J. M., Chauvet, J. P., & Efthymiou, M. L. (1989). Tetrahydrofuran poisoning after occupational exposure. *British Journal of Industrial Medicine, 46(9)*, 677-678.

Garrettson, L. K. (1984). Direct and indirect chemical exposure in children. *Clinics in Laboratory Medicine, 4*, 469-473.

Gawin, F. H., & Kleber, H. D. (1986). Abstinence symptomatology and psychiatric diagnosis in cocaine abusers. *Archives of General Psychiatry, 43*, 107-113.

Gay, D. D., Cox, R. D., & Reinhardt, J. W. (1979). Chewing releases mercury from filling. *Lancet, 1*, 985-986.

Gebhardt, C. A., Naeser, M. A., & Butters, N. (1984). Computerized measure of CT scans of alcoholics: Thalamic region related to memory. *Alcohol, 1(2)*, 133-140.

Gerhardsson, L., Attewell, R., Chettle, D. R., Englyst, V., Lundström, N. G., Nordberg, G. F., Nyhlin, H., Scott, M. C., & Todd, A. C. (1993). In vivo measurements of lead in bone in long-term exposed lead smelter workers. *Archives of Environmental Health, 48(3)*, 147-156.

Gershon, S., & Shaw, F. H. (1961). Psychiatric sequelae of chronic exposure to organophosphorus insecticides. *Lancet, 1*, 1371-1374.

Geschwind, N., & Behan, P. (1982). Left handedness: Association with immune disease, migraine and developmental learning disorder. *Proceedings of the National Academy of Sciences, USA, 79*, 5097-5100.

Ghadirian, A. M., Engelsmann, F., & Ananth, J. (1983). Memory functions during lithium therapy. *Journal of Clinical Psychopharmacology, 3*, 313-315.

Ghosh, S. K. (1976). Methyldopa and forgetfulness. *Lancet, 1*, 202-203.

Giannini, A. J. (1994). Inward the mind's I: Description, diagnosis, and treatment of acute and delayed LSD hallucinations. *Psychiatric Annals, 24*, 134-136.

Gibson, J. L. (1904). *A plea for painted railings and painted walls of rooms as the source of lead poisoning amongst Queensland children.* Sydney: Australas.

Gillham, R. A., Williams, N., Wiedmann, K., Butler, F., Larkin, J. G., & Brodie, M. J. (1988). Concentration-effect relationships with carbamazepine and its epoxide on psychomotor and cognitive function in eplieptic patients. *Journal of Neurology, Neurosurgery and Psychiatry, 51(7)*, 929-933.

Gimenez, E. R., Rodriguez, C. A., & Capone, L. (1990). Potential human exposure resulting from the chemical industry in Argentina. In B. L. Johnson (Ed.), *Advances in neurobehavioral toxicology* (pp. 59-74). Chelsea, MI: Lewis Publishers.

Ginsberg, M. (1980). Carbon monoxide. In P. S. Spencer & H. H. Schaumberg (Eds.), *Experimental and clinical neurotoxicology* (pp. 374-394). Baltimore: Williams & Wilkins.

Ginsberg, M. (1985). Carbon monoxide intoxication: Clinical features, neuropathology and mechanisms of injury. *Clinical Toxicology, 23*, 281-288.

Giovacchini, R. P. (1985). Abusing the volatile organic chemicals. *Regulatory Toxicology and Pharmacology, 5*, 18-37.

Girkem, W., Krebs, F. A., & Muller-Oerlinghausen, B. (1975). Effects of lithium on electromyographic recordings in man. *Pharmacopsychiatry, 10*, 79-82.

Gittelman, R., & Eskenazi, B. (1983). Lead and hyperactivity revisited. An investigation of nondisadvantaged children. *Archives of General Psychiatry, 40*, 827-833.

Glenn, S. W., & Parsons, O. A. (1992). Neuropsychological efficiency measures in male and female alcoholics. *Journal of Studies on Alcohol, 53,* 546-552.

Glenn, S. W., Sinha, R., & Parsons, O. (1993). Electrophysiological indices predict resumption of drinking in sober alcoholics. *Alcohol, 10,* 89-95.

Goetz, C. G. (1985). *Neurotoxins in clinical practice.* New York: SP Medical and Scientific Books.

Gold, J. H. (1969). Chronic perchloroethylene poisoning. *Canadian Psychiatric Association Journal, 14,* 627-630.

Gold, M. S. (1994). The epidemiology, attitudes and pharmacology of LSD use in the 1990's. *Psychiatric Annals, 24,* 124-126.

Gold, M. S., Verebey, K. (1984). The psychopharmacology of cocaine. *Psychiatric Annals, 14,* 714-723.

Gold, M. S., Washton, A. M., & Dackis, C. A. (1985). Cocaine abuse: Neurochemistry, phenomenology, and treatment. In N. J. Kozel & E. H. Adams (Eds.), *Cocaine use in America: Epidemiologic and clinical perspectives* (NIDA Research Monograph 61). Rockville, MD: NIDA.

Goldberg, T. E., Greenberg, R. D., Griffin, S. J., Gold, J. M., Kleinman, J. E., Pickar, D., Schultz, S. C., & Weinberger, D. R. (1993). The effect of clozapine on cognition and psychiatric symptoms in patients with schizophrenia. *British Journal of Psychiatry, 162,* 43-48.

Golden, C. J. (1978). *Diagnosis and rehabilitation in clinical neuropsychology.* Springfield, IL: Charles C. Thomas.

Golden, C. J., Hammeke, T. A., & Purisch, A. D. (1980). *The Luria-Nebraska Neuropsychological Battery: Manual.* Los Angeles: Western Psychological Services.

Golden, C. J., Kane, R., Sweet, J., Moses, J. A., Cardellino, J. P., Templeton, R., Vicente, P., & Graber, B. (1981). Relationship of the Luria-Nebraska to the Halstead-Reitan Neuropsychological Battery. *Journal of Consulting and Clinical Psychology, 49,* 410-417.

Goldings, A. S., & Stewart, M. (1982). Organic lead encephalopathy: Behavioral change and movement disorder following gasoline inhalation. *Journal of Clinical Psychiatry, 43,* 70-72.

Goldman, L. R., Mengle, D., Epstein, D. M., Fredson, D., Kelley, K., & Jackson, R. J. (1987). Acute symtpoms in persons residing near a field treated with the soil fumigants methyl bromide and chloropicrin. *The Western Journal of Medicine, 147,* 95-98.

Goldman, M. S. (1983). Cognitive impairment in chronic alcoholics. Some cause for optimism. *American Psychologist, 38(10),* 1045-1054.

Goldman, M. S. (1986). Neuropsychological recovery in alcoholics: Endogenous and exogenous processes. *Alcoholism: Clinical and Experimental Research, 10,* 136-144.

Goldman, M. S., Williams, D. L., & Klisz, D. K. (1983). Recoverability of psychological functioning following alcohol abuse: Prolonged visual-spatial dysfunction in older alcoholics. *Journal of Consulting and Clinical Psychology, 51,* 370-378.

Goldman-Rakic, P. S. (1987). Circuitry of the frontal association cortex and its relevance to dementia. *Archives of Gerontology and Geriatrics, 6(3),* 299-309.

Goldsmith, J. R. (1970). Contribution of motor vehicle exhaust, industry, and cigarette smoking to community carbon monoxide exposures. *Annals of the New York Academy of Sciences, 174(1), 122-134.*

Goldstein, G., & Shelly, C. (1980). Neuropsychological investigation of brain lesion localization in alcoholism. In H. Begleiter (Ed.), *Biological effects of alcoholism* (pp. 731-743). New York: Plenum Press.

Goldwater, L. J. (1972). *Mercury: A history of quicksilver.* Baltimore: York Press.

Golombok, S., Moodley, P., & Lader, M. (1988). Cognitive impairment in long-term benzodiazepine users. *Psychological Medicine, 18,* 365-374.

Goodwin, T. M. (1988). Toluene abuse and renal tubular acidosis in pregnancy. *Obstetrics and Gynecology, 71,* 715-718.

Goronzy, J. J., & Weyand, C. M. (1990). Eosinophilia, myopathy, and neuropathy in a patient with repeated use of L-tryptophan. *Klinische Wochenschrift, 68(14),* 735-738.

Gossel, T. A., & Bricker, J. D. (1984). *Principles of clinical toxicology.* New York: Raven Press.

Gots, R. E. (1993). Medical hypothesis and medical practice: Autointoxication and multiple chemical sensitivity. *Regulatory Toxicology and Pharmacology, 18,* 2-12.

Gots, R. E., Hamosh, T. D., Flamm, W. G., & Carr, C. J. (1993). Multiple chemical sensitivities: A symposium on the state of the science. *Regulatory Toxicology and Pharmacology, 18*, 61-78.
Goyer, R. A. (1986). Toxic effects of metals. In C. D. Klaassen, M. O. Amdur, & J. Doull (Eds.), *Casarett and Doull's toxicology* (3rd ed., pp. 582-635). New York: Macmillan Co.
Grabski, D. A. (1961). Toluene sniffing producing cerebellar degeneration. *American Journal of Psychiatry, 118*, 461-462.
Graeven, D. B., Sharp, J. G., & Glatt, S. (1981). Acute effects of phencyclidine (PCP) on chronic and recreational users. *American Journal of Drug and Alcohol Abuse, 8*, 39-50.
Graham, C., Cohen, H. D., Cook, M. R., Gerkovich, M. M., Phelps, J. W., & Fotopoulos, S. S. (1984). Human exposure to 60 Hz fields: Effects on performance, physiology and subjective state. In *Contractors review meeting*. St. Louis, MO: Department of Energy, Electric Power Research Institute and New York State Power Lines Project.
Grandjean, E. (1960). Trichloroethylene effects on animal behavior. *Archives of Environmental Health, 1*, 106-108.
Grandjean, P. (1983). Behavioral toxicity of heavy metals. In P. Zbinden, V. Cuomo, G. Racagni, & B. Weiss (Eds.), *Application of behavioral pharmacology in toxicology (pp. 331-340)*. New York: Raven Press.
Grandjean, P. (Ed.). (1985). *Neurobehavioral methods in occupational and environmental health: Symposium Report Number 6*. Copenhagen: World Health Organization, Regional Office for Europe.
Grandjean, P. (1993). International perspectives of lead exposure and lead toxicity. *NeuroToxicology, 14*, 9-14.
Grandjean, P., & Nielsen, T. (1979). Organolead compounds: Environmental health aspects (review). *Residue Reviews, 72*, 97-148.
Grandjean, P., & Jorgensen, P. J. (1990). Retention of lead and cadmium in prehistoric and modern human teeth. *Environmental Research, 53*, 6-15.
Grandjean, P., & Weihe, P. (1993). Neurobehavioral effects of intrauterine mercury exposure: Potential sources of bias. *Environmental Research, 61*, 176-183.
Grandjean, P., Arnvig, E., & Beckmann, J. (1978). Psychological dysfunctions in lead-exposed workers. *Scandinavian Journal of Work Environment and Health, 4*, 295-303.
Grandjean, P., Weihe, P., Jorgensen, P. J., Clarkson, T., Cernichiari, E., & Videro, T. (1992). Impact of maternal seafood diet on featl exposure to mercury, selenium, and lead. *Archives of Environmental Health, 47(3)*, 185-195.
Grandstaff, N. (1974). Carbon monoxide and human functions. In C. Xintaras, B. L. Johnson, & I. deGroot (Eds.), *Behavioral toxicology: Early detection of occupational hazards* (HEW Publication No. (NIOSH) 74-126, pp. 292-305). Washington, DC: U.S. Government Printing Office.
Grant, D. A., & Berg, E. A. (1948). A behavioral analysis of degree of reinforcement and ease of shifting to new responses in Weigl-type card sorting problem. *Journal of Experimental Psychology, 38*, 404-411.
Grant, I. (1987). Alcohol and the brain: Neuropsychological correlates. *Journal of Consulting and Clinical Psychology, 55*, 310-324.
Grant, I., & Mohns, L. (1975). Chronic cerebral effects of alcohol and drug abuse. *International Journal of the Addictions, 10(5)*, 883-920.
Grant, I., & Reed, R. (1985). Neuropsychology of alcohol and drug abuse. In A. I. Alterman (Ed.), *Substance abuse and psychopathology* (pp. 289-339). New York: Plenum Press.
Grant, I., Rochford, J., Fleming, T., & Stunkard, H. (1973). A neuropsychological assessment of the effects of moderate marihuana use. *Journal of Nervous and Mental Disease, 156*, 278-280.
Grant, I., Adams, K. M., Carlin, A. S. Rennick, P. M., Judd, L. L., Schooff, K., & Reed, R. (1978). Neuropsychological effects of polydrug abuse. In D. R. Wesson, A. S. Carlin, K. M. Adams, & G. Beschner (Eds.), *Polydrug abuse: The results of national collaborative study* (pp. 223-261). New York: Academic Press.
Grant, I., Adams, K. M., Reed, R., & Carlin, A. (1979a). Neuropsychological function in young alcoholics and polydrug abusers. *Journal of Clincal Neuropsychology, 1*, 39-47.

Grant, I., Adams, K., & Reed, R. (1979b). Normal neuropsychological abilities of alcoholic men in their late thirties. *American Journal of Psychiatry, 136(10),* 1263–1269.

Grant, I., Adams, K. M., & Reed, R. (1984). Aging, abstinence, and medical risk factors in the prediction of neuropsychologic deficit among long-term alcoholics. *Archives of General Psychiatry, 41,* 710–718.

Grcevic, N., Jadro-Santel, D., & Jukic, S. (1977). Cerebral changes in paraquat poisoning. In L. Roizin, H. Shiraki, & N. Grcevic (Eds.), *Neurotoxicology* (pp. 469–484). New York: Raven Press.

Greenberg, B. D., Moore, P. A., Letz, R., & Baker, E. L. (1985). Computerized assessment of human neurotoxicity: Sensitivity to nitrous oxide exposure. *Clinical Pharmacology and Therapeutics, 38,* 656–660.

Greenberg, J. (1971). The neurological effects of methyl bromide poisoning. *Industrial Medicine, 40,* 27–79.

Greenblatt, D. J., & Shader, R. I. (1974). *Benzodiazepines in clinical practice.* New York: Raven Press.

Greenfield, B. M., Moyer, J. W., & Sibit, R. R. (1991). The eosinophilia–myalgia syndrome and the brain. *Annals of Internal Medicine, 115,* 159–160.

Gregersen, P. (1988). Neurotoxic effects of organic solvents in exposed workers: Two controlled follow-up studies after 5.5 and 10.6 years. *American Journal of Industrial Medicine, 14,* 681–701.

Gregersen, P., & Stigsby, B. (1981). Reaction time of industrial workers exposed to organic solvents: Relationship to degree of exposure and psychological performance. *American Journal of Industrial Medicine, 2,* 313–321.

Gregersen, P., Angelso, B., Neilsen, T. E., Norgaard, B., & Uldal, C. (1984). Neurotoxic effects of organic solvents in exposed workers: An occupational, neuropsychological, and neurological investigation. *American Journal of Industrial Medicine, 5,* 201–225.

Griffen, J. W. (1981). Hexacarbon neurotoxicity. *Neurobehavioral Toxicology and Teratology, 3,* 437–444.

Griffiths, A. N., Tedeschi, G., & Richens, A. (1986). The effects of temazepam and nitrazepam on several measures of human performance. *Acta Psychiatrica Scandinavica, 74*(Suppl. 332), 119–186.

Griffiths, R. R., Bigelow, G. E., Stitzer, M. L., & McLeod, D. R. (1983). Behavioral effects of drugs of abuse. In G. Zbinden, V. Cuomo, G. Racagni, & B. Weiss (Eds.), *Applications of behavioral pharmacology* (pp. 367–382). New York: Raven Press.

Grigg, J. R. (1988). Nitrous oxide mood disorder. *Journal of Psychoactive Drugs, 20,* 449–450.

Grigsby, J., Rosenberg, N., Dreisbach, J., Busenbark, D., & Grisby, P. (1993). Chronic toluene abuse produces neurologic and neuropsychological deficits [Abstract]. *Archives of Clinical Neuropsychology, 8,* 229–230.

Griswold, S. S., Chambers, L. A., & Motley, H. L. (1956). Report of a case of exposure to high ozone concentrations for two hours. *A.M.A. Archives of Industrial Health, 15,* 108–110.

Gross, J. A., Haas, M. L., & Swift, T. R. (1979). Ethylene oxide neurotoxicity: Report of four cases and review of the literature. *Neurology, 29,* 978–983.

Gualtieri, T. (1990). The neuropharmacology of inadvertent drug effects in patients with traumatic brain injuries. *Journal of Head Trauma Rehabilitation, 5,* 32–40.

Guerit, J. M., Meulders, M., Amand, G., Roels, H. A., Buchet, J. P., Lauwerys, R., Bruaux, P., Claeys-Thoreau, F., Ducoffre, G., & Lafontaine, A. (1981). Lead neurotoxicity in clinically asymptomatic children living in the vicinity of an ore smelter. *Clinical Toxicology, 18,* 1257–1267.

Guirguis, S. S., Pelmear, P. L., Roy, M. L., & Wong, L. (1990). Health effects associated with exposure to anaesthetic gases in Ontario hospital personnel. *British Journal of Industrial Medicine, 47,* 490–497.

Gun, R. T., Grysorewicz, C., & Nettelbeck, J. (1978). Choice reaction time in workers using trichloroethylene. *Medical Journal of Australia, 1,* 535–546.

Gunnarsson, L. G., Bodin, L., Söderfeldt, B., & Axelson, O. (1992). A case-control study of motor neurone disease: Its relation to heritability, and occupational exposures, particularly to solvents. *British Journal of Industrial Medicine, 49(11),* 791–798.

Gurling, H. M., Curtis, D., & Murray, R. M. (1991). Psychological deficit from excessive alcohol consumption: Evidence from a co-twin study. *British Journal of Addiction, 86(2),* 151–155.

Gustafson, C., & Tagesson, C. (1985). Influence of organic solvent mixtures on biological membranes. *British Journal of Industrial Medicine, 42,* 591-595.
Guthrie, A. (1980). The first year after treatment: Factors affecting time course of reversibility of memory and learnig deficits in alcoholism. In H. Begleiter (Ed.), *Biological effects of alcohol* (pp. 757-770). New York: Plenum Press.
Hake, C. L., & Stewart, R. D. (1977). Human exposure to tetrachloroethylene: Inhalation and skin contact. *Environmental Health Prospectives, 21,* 231-238.
Halikas, J. A., Weller, R. A., Morse, C. L., & Hoffmann, R. G. (1985). A longitudinal study of marijuana effects. *International Journal of the Addictions, 20,* 701-711.
Hall, R. C. W., & Zisook, S. (1981). Paradoxical reactions to benzodiazepines. *British Journal of Clinical Pharmacology, 11*(Suppl.), 99s-104s.
Hallenbeck, W. H., & Cunningham-Burns, K. M. (1985). *Pesticides and human health.* Berlin: Springer-Verlag.
Halonen, P., Halonen, J.-P., Lang, H. A., & Karskela, V. (1986). Vibratory perception thresholds in shipyard workers exposed to solvents. *Acta Neurologica Scandinavica, 73,* 651-655.
Halstead, W. C. (1947). *Brain and intelligence.* Chicago: University of Chicago Press.
Hamada, R., Yoshida, Y., Nomoto, M., Osame, M., Igata, A., Mishima, I., & Kuwano, A. (1993). Computed tomography in fetal methylmercury poisoning. *Clinical Toxicology, 31,* 101-106.
Hamilton, A. (1985). Forty years in the poisonous trades. *American Journal of Industrial Medicine, 7,* 3-18.
Hamilton, A., & Hardy, H. L. (1974). *Industrial toxicology.* Acton, MA: Publishing Sciences Group.
Hanakago, R. (1979). Severe polyneuritis showed amyotrophy following gold therapy for bronchial asthma. In L. Manzo (Ed.), *Advances in neurotoxicology* (pp. 391-396). New York: Pergamon Press.
Hane, M., & Ekberg, K. (1984). Current research in behavioral toxicology. *Scandinavian Journal of Work Environment and Health,* Suppl. 1, 8-9.
Hane, M., Axelson, O., Blume, J., Hogstedt, C., Sundell, L., & Ydreborg, B. (1977). Psychological function changes among house painters. *Scandinavian Journal of Work Environment and Health, 3,* 91-99.
Hane, M., Hogstedt, C., & Sundell, L. (1980). Neuropsychiatric symptoms among solvent workers— A questionnaire for screening. *Lakartidningen, 77,* 437-439.
Hanin, I., Krigman, M. R., & Mailman, R. B. (1984). Central neurotransmitter effects of organotin compounds: Trials, tribulations and observations. *NeuroToxicology, 5,* 267-278.
Hänninen, H. (1971). Psychological picture of manifest and latent carbon disulphide poisoning. *British Journal of Industrial Medicine, 28,* 374-381.
Hänninen, H. (1982a). Behavioral effects of occupational exposure to mercury and lead. *Acta Neurologica Scandinavica, 66*(Suppl. 92), 167-175.
Hänninen, H. (1982b). Psychological test batteries: New trends and development. In R. Gilioli, M. G. Cassitts, & V. Foa (Eds.), *Neurobehavioral methods in occupational health* (pp. 123-130). New York: Pergamon Press.
Hänninen, H. (1983). Psychological test batteries: New trends and developments. In R. Gilioli, M. G. Cassitto, & V. Foa (Eds.), *Neurobehavioral methods in occupational health* (pp. 123-130). New York: Pergamon Press.
Hänninen, H. (1990). Methods in behavioral toxicology: Current test batteries and need for development. In R. W. Russel, P. E. Flattau, & A. M. Pope (Eds.), *Behavioral measures of neurotoxicity: Report of a symposium* (pp. 39-55). Washington, DC: National Academy Press.
Hänninen, H., & Lindstrom, K. (1979). *Behavioral test battery for toxic psychological studies.* Helsinki: Institute of Occupational Health.
Hänninen, H., Eskelinen, L., Husman, K., & Nurminen, M. (1976). Behavioral effects of long-term exposure to a mixture of organic solvents. *Scandinavian Journal of Work Environment and Health, 4,* 240-255.
Hänninen, H., Hernberg, S., Mantere, P., Vesanto, R., & Jalkanen, M. (1978). Psychological performance of subjects with low exposure to lead. *Journal of Occupational Medicine, 20,* 683-689.

Hänninen, H., Nurminen, M., Tolonen, M., & Martelin, T. (1978). Psychological tests as indicators of excessive exposure to carbon disulfide. *Scandinavian Journal of Psychology, 19*, 163-174.

Hänninen, H., Mantere, P., Hernberg, S., Seppäläinen, A. M., & Kock, B. (1979). Subjective symptoms to low-level exposure to lead. *NeuroToxicology, 1*, 333-347.

Hänninen, H., Antti-Poiki, M., Juntunen, J., & Koskenvuo, M. (1991). Exposure to organic solvents and neuropsychological dysfunction: A study on monozygotic twins. *British Journal of Industrial Medicine, 48*, 18-25.

Hannon, R., Butler, C. P., Day, C. L., Khan, S. A., Quittoriano, L. A., Butler, A. M., & Meredith, L. (1985). Alcohol use and cognitive functioning in men and women college students. In M. Galanter (Ed.), *Recent developments in alcoholism* (Vol. 3, pp. 241-252). New York: Plenum Press.

Hansen, O. N., Trillingsgaard, A., Beese, I., Lyngbye, T., & Grandjean, P. (1985). Neurobehavioral methods in assessment of children with low-level lead exposure. In *Neurobehavioral methods in occupational and environmental health* (Document 3, pp. 183-187). Copenhagen: World Health Organization.

Hansen, W. W., Helweg-Larsen, S., & Trojaborg, W. (1989). Long-term neurotoxicity with cisplatin, vinblastine, and bleomycin for metastatic germ cell cancer. *Journal of Clinical Oncology, 7*, 1457-1461.

Hanson, J. W., & Smith, D. W. (1975). The fetal hydantoin syndrome. *Journal of Pediatrics, 87*, 285-290.

Harada, M. (1976). Intrauterine poisoning: Clinical and epidemiological studies and significance of the problem. *Bulletin of the Institute of Const. Med. Jumamoto University 25* (Suppl.), 1-69.

Harbin, T. J., Benignus, V. A., Muller, K. E., & Barton, C. N. (1988). The effects of low-level carbon monoxide exposure upon cortical potentials in young and elderly men. *Neurotoxicology and Teratology, 10*, 93-100.

Harkonen, H., Lindstrom, K., Seppäläinen, A. M., Sisko, A., & Hernberg, S. (1978). Exposure response relationship between styrene exposure and central nervous system functions. *Scandinavian Journal of Work Environment and Health, 4*, 53-59.

Harper, C. G. (1979). Wernicke's encephalopathy: A more common disease than realized. A neuropathological study of 51 cases. *Journal of Neurology, Neurosurgery, and Psychiatry, 42*, 226-231.

Harper, C. G. (1983). The incidence of Wernicke's encephalopathy in Australia—A neuropathological study of 131 cases. *Journal of Neurology, Neurosurgery, and Psychiatry, 46*, 593-598.

Harper, C. G., & Blumberg, P. C. (1982). Brain weights in alcoholics. *Journal of Neurology, Neurosurgery, and Psychiatry, 45*, 838-840.

Harper, C., & Kril, J. (1985). Brain atrophy in chronic alcoholic patients: A quantitative pathological study. *Journal of Neurology, Neurosurgery, and Psychiatry, 48* 211-217.

Harper, C. G., Giles, M., & Finlay-Jones, R. (1986). Clinical signs in the Wernicke-Korsakoff complex: A retrospective analysis of 131 cases diagnosed at necropsy. *Journal of Neurology, Neurosurgery, and Psychiatry, 49*, 341-345.

Hart, R. P., Silverman, J. J., Garrettson, L. K., Schulz, C., & Hamer, R. M. (1984). Neuropsychological function following mild exposure to pentaborane. *American Journal of Industrial Medicine, 6(1)*, 37-44.

Hartitzsch, von B., Hoenich, N. A., Leigh, R. J., Wilkinson, R., Frost, T. H., Weddel, A., & Posen, G. A. (1972). Permanent neurological sequelae despite haemodialysis for lithium intoxication. *British Medical Journal, 4*, 757-759.

Hartlage, L., & Knowles, E. (1986). *Reaction time correlates of occasional substance abuse.* Presented at the 1986 annual meeting of the National Academy of Neuropsychologists, Las Vegas, Nevada, October 27-29.

Hartman, D. E. (1986a). On the use of clinical psychology software: Practical, legal and ethical concerns. *Professional Psychology: Research and Practice, 17*, 462-465.

Hartman, D. E. (1986b). Artificial intelligence or artificial psychologist?: Conceptual issues in clinical microcomputer use. *Professional Psychology: Research and Practice, 17*, 528-534.

Hartman, D. E. (1987). Neuropsychological toxicology: Identification and assessment of neurotoxic syndromes. *Archives of Clinical Neuropsychology, 2*, 45-65.

REFERENCES

Hartman, D. E. (1988a). Neuropsychology and the neurochemical lesion. *NeuroToxicology, 9(3),* 401-404.
Hartman, D. E. (1988b). *Neuropsychological toxicology: Identification and assessment of human neurotoxic syndromes.* Oxford, England: Pergamon Press.
Hartman, D. E. (1991). Reply to Reitan: Unexamined assumptions and the history of neuropsychology. *Archives of Clinical Neuropsychology, 6,* 147-166.
Hartman, D. E., Sweet, J. J., & Elvart, A. (1985). Neuropsychological effects of Wernicke's encephalopathy as a consequence of hyperemesis gravidarum: A case study. *International Journal of Clinical Neuropsychology, 7,* 204-207.
Hartman, D. E., Hessl, S., & Tarcher, A. (1992). Neurobehavioral evaluation of environmental toxin exposure: Rationale and review. In A. Tarcher (Ed.), *Principles and practice of environmental medicine* (pp. 241-262). New York: Plenum Press.
Hartung, R. (1981). Cyanides and nitriles. In *Patty's industrial hygiene and toxicology* (3rd ed., Vol. 10, pp. 4845-4900). New York: Wiley.
Hawkins, K. A., Schwartz-Thompson, J., & Kahane, A. (1994). Abuse of formaldehyde-laced marijuana may cause dysmnesia. *Journal of Neuropsychiatry, 2,* 67.
Hayes, W. L. (1971). Studies on exposure during the use of anticholinesterase pesticides. *Bulletin of the World Health Organization, 44,* 277-288.
Hayes, W., Durham, I., & Cueto, C. (1956). The effects of known repeated oral doses of chlorophenothane (DDT) in man. *Journal of the American Medical Association, 162,* 890-897.
He, F., Zhang, S., Wang, H., Li, G., Zhang, Z., Li, F., Dong, X., & Hu, F. (1989). Neurological and electroneuromyographic assessment of the adverse effects of acrylamide on occupationally exposed workers. *Scandinavian Journal of Work Environment and Health, 15,* 135-139.
He, F., Liu, X., Yang, S., Zhang, S., Xu, G., Fang, G., & Pan, X. (1993). Evaluation of brain function in acute carbon monoxide poisoning with multimodality evoked potentials. *Environmental Research, 60(2),* 213-226.
Heath, M. J. (1986). Solvent abuse using bromochlorodifluoromethane from a fire extinguisher. *Medical Science Law, 26,* 23-24.
Heaton, R. K., & Crowley, T. J. (1981). Effects of psychiatric disorders and their somatic treatments on neuropsychological test results. In S. B. Filskov & T. J. Boll (Eds.), *Handbook of clinical neuropsychology* (pp. 481-525). New York: Wiley.
Heckerling, P. S., Leiken, J. B., Ferzian, C. G., & Maturen, A. (1990). Acute carbon monoxide poisoning in patients with neurologic illness. *Clinical Toxicology, 28,* 29-44.
Heimann, H., Reed, C. F., & Witt, P. N. (1968). Some observations suggesting preservation of skilled motor acts despite drug-induced stress. *Psychopharmacologica, 13,* 287-298.
Henderson, T. W. (1990). Legal aspects of disease clusters: Toxic tort litigation: Medical and scientific principles in causation. *American Journal of Epidemiology, 132* (Suppl. 1), S69-S78.
Hendler, N., Cimini, C., Terrence, M. A., & Long, D. (1980). A comparison of cognitive impairment due to benzodiazepines and to narcotics. *American Journal of Psychiatry, 137,* 828-830.
Herkenham, M., Lynn, A. B., Little, M. B., Johnson, R. M., Melvin, L. S., de Costa, B. N., & Rice, K. C. (1974). Cannabinoid receptor localization in the brain. *Proceedings of the National Academy of Sciences of the United States of America, 87,* 1932-1936.
Herkenham, M., Lynn, A. B., Little, M. D., Johnson, M. R., Melvin, L. S., de Costa, B. R., & Rice, K. C. (1990). Cannabinoid receptor localization in brain. *Proceedings of the National Academy of Sciences of the United States of America, 87(5),* 1932-1936.
Hermle, L., Spitzer, M., Borchardt, D., Kovar, K.-A., & Gouzoulis, E. (1993). Psychological effects of MDE in normal subjects. *Neuropsychopharmacology, 8,* 171-176.
Hernberg, S. (1980). Evaluation of epidemiologic studies in assessing the long-term effects of occupational noxious agents. *Scandinavian Journal of Work Environment and Health, 6,* 163-169.
Hernberg, S. (1983). Does solvent poisoning exist? In N. Cherry & H. A. Waldron (Eds.), *The neuropsychological effects of solvent exposure* (pp. 63-72). Havant, Hampshire: Colt Foundation.
Herning, R. I., Glover, B. J., Koeppl, B., Phillips, R. L., & London, E. D. (1994). Cocaine-induced increases in EEG alpha and beta activity: Evidence for reduced cortical processing. *Neuropsychopharmacology, 11,* 1-9.

Herschman, Z. J., Silverstein, J., Blumberg, G., & Lehrfield, A. (1991). Central nervous system toxicity from nebulized atropine sulfate. *Clinical Toxicology, 29*, 272-277.
Hersh, J. H., Podruch, P. E., Rogers, G., & Weisskopk, B. (1985). Toluene embryopathy. *Journal of Pediatrics, 106(6),* 922-927.
Hersh, R. (1991). Abuse of ethyl chloride. *American Journal of Psychiatry, 148,* 270-271.
Hertzler, D. R. (1990). Neurotoxic behavioral effects of Lake Ontario salmon diets in rats. *Neurotoxicology and Teratology, 12,* 139-143.
Herzstein, J., & Cullen, M. R. (1990). Methyl bromide intoxication in four field-workers during removal of soil fumigation sheets. *American Journal of Industrial Medicine, 17,* 321-326.
Hes, J. P., Cohn, D. F., & Strefler, M. (1979). Ethyl chloride sniffing and cerebellar dysfunction. *Israeli Annals of Psychiatry, 17,* 122-125.
Hessl, S., & Frumkin, H. (1990). *Beyond neglect: The problem of occupational disease in the United States.* Chicago: Workplace Institute.
Higgins, S. T., Rush, C. E., Hughes, J. E., Bickel, W. K., Lynn, M., & Capeless, M. A. (1992). Effects of cocaine and alcohol alone and in combination on human learning and performance. *Journal of Experimental Analysis of Behavior, 58,* 87-105.
Hill, S. Y. (1980). Comprehensive assessment of brain dysfunction in alcoholic individuals. *Acta Psychiatrica Scandinavica, 62*(Suppl. 286), 57-75.
Hill, S. Y., & Steinhauer, S. R. (1993). Assessment of prepubertal and postpubertal boys and girls at risk for developing alcoholism with P300 from a visual discrimination task. *Journal of Studies on Alcohol, 54,* 350-358.
Hillbom, M., & Kaste, M. (1983). Ethanol intoxication: A risk factor for ischemic brain infarction. *Stroke, 14,* 694-695.
Hilton, M. E. (1989). How many alcoholics are there in the United States? *British Journal of Addiction, 84,* 459-460.
Hindmarch, I. (1980). Psychomotor function and psychoactive drugs. *British Journal of Clinical Pharmacology, 10,* 189-209.
Hindmarch, I., & Gudgeon, A. C. (1980). The effects of clobazam and lorazepam on aspects of psychomotor performance and car handling ability. *British Journal of Clinical Pharmacology, 10,* 145-150.
Hindmarch, I., Haller, J., Sherwood, N., & Kerr, J. S. (1990). Comparison of five anxiolytic benzodiazepines on measures of psychomotor performance and sleep. *Neuropsychology, 24,* 84-89.
Hipolito, R. N. (1980). Xylene poisoning in laboratory workers: Case reports and discussion. *Laboratory Medicine, 11,* 593-595.
Hirano, A., & Llena, J. F. (1980). The central nervous system as a target in toxic-metabolic states. In P. S. Spencer & H. H. Schaumberg (Eds.), *Experimental and clinical neurotoxicology* (pp. 24-34). Baltimore: Williams & Wilkins.
Hirata, M., & Kosaka, H. (1993). Effects of lead exposure on neurophysiological parameters. *Environmental Research, 63(1),* 60-69.
Hirschberg, A., & Lerman, Y. (1984). Clinical problems in organophosphorus insecticide poisoning: The use of a computerized information system. *Fundamentals of Applied Toxicology, 4,* S209-S214.
Hiscock, M., & Hiscock, C. K. (1989). Refining the forced-choice method for the detection of malingering. *Journal of Clinical and Experimental Neuropsychology, 11(6),* 967-974.
Ho, I. K. (1988). Interactions in neurotoxicology. *Neurotoxicology, 9,* 151-152.
Hoch, A. (1904). A review of some psychological and psysiological experiments done in connection with the study of mental diseases. *Psychological Bulletin of New York, i,* 241-257.
Hodgson, M. J., Block, G. D., & Parkinson, D. K. (1986). Organophosphate poisoning in office workers. *Journal of Occupational Medicine, 28,* 434-437.
Hoehn-Saric, R., Harris, G. J., Pearlson, G. D., Cox, C., Machline, S. R., & Camargo, E. E. (1991). A fluoxetine-induced frontal lobe syndrome in an obsessive compulsive patient. *Journal of Clinical Psychiatry, 52,* 131-133.
Hoff, A. L., Ollo, C., & Helms, P. M. (1986). *Reduced memory function in the elderly as a result of antidepressant medication: Preliminary findings.* Paper presented at the International Neuropsychological Society, European Conference.

Hoffman, P., & Tabakoff, B. (1980). Receptor and neurotransmitter changes produced by chronic alcohol ingestion. In L. Manzo (Ed.), *Advances in neurotoxicology* (pp. 107-116). New York: Pergamon Press.
Hoffman, R. E., Stehr-Green, P. A,. Webb, K. B., Evans, G., Knutsen, A. P., Schramm, W. F., Staake, J. L., Gibson, B. B., & Steinberg, K. K. (1986). Health effects of long-term exposure to 2,3,7,8-tetrachlorodibenzo-p-dioxin. *Journal of the American Medical Association, 255*, 2031-2038.
Hogstedt, C., Hane, M,. & Axelson, O. (1980). Diagnostic and health care aspects of workers exposed to solvents. In C. Zenz (Ed.), *Developments in occupational medicine* (pp. 249-258). Chicago: Year Book Medical Publishers.
Hogstedt, C., Hane, M., Agrell, A., & Bodlin, L. (1983). Neuropsychological test results and symptoms among workers with well defined long-term exposure to lead. *British Journal of Industrial Medicine, 40*, 99-105.
Hojer, J., Malmlund, H.-O., & Berg, A. (1993). Clinical features in 28 consecutive cases of laboratory confirmed massive poisoning with carbamazepine alone. *Clinical Toxicology, 31*, 449-458.
Holmberg, P. C. (1979). Central-nervous-system defects in children born to mothers exposed to organic solvents during pregnancy. *Lancet, 2(8135)*, 177-179.
Holmes, J. A., & Haban, G. F. (1991). *Neuropsychological deficits secondary to chlorine gas exposure.* Paper presented at the annual meeting of the National Academy of Neuropsychology, Dallas.
Holmes, J. H., & Gaon, M. D. (1956). Observations on acute and multiple exposures to anticholinesterase agents. *Transactions of the American Clinical Climatology Associations, 68*, 86-101.
Holmes, T. H., & Rahe, R. H. (1967). The Social Readjustment Rating Scale. *Journal of Psychosomatic Research, 11(2)*, 213-218.
Holmes, T. M., Buffler, P. A., Holguin, A. H., & Hsi, B. P. (1986). A mortality study of employees at a synthetic rubber manufacturing plant. *American Journal of Industrial Medicine, 9*, 355-362.
Hooisma, J., Emmen, H. H., Kulig, B. M., Muijser, H., & Poortvlient, D. (1990). Factor analysis of tests from the Neurobehavioral Evaluation System and the WHO Neurobehavioral Cove Test Battery. In B. L. Johnson (Ed.), *Advances in neurobehavioral toxicology: Applications in environmental and occupational health* (pp. 245-255). Chelsea, MI: Lewis Publishers.
Hormes, J. T., Filley, C. M., & Rosenberg, N. L. (1986). Neurologic sequelae of chronic solvent vapor abuse. *Neurology, 36*, 698-702.
Horowitz, A., Kaplan, R., & Sarpel, G. (1987). Carbon monoxide toxicity: MR imaging in the brain. *Radiology, 162*, 787-788.
Horton, A. M., & Wedding, D. (1984). *Clinical and behavioral neuropsychology.* New York: Praeger.
Horton, A. M., & Hartlage, L. C. (1993). Unexpected catastrophic consequences of l-tryptophan ingestion. *Archives of Clinical Neuropsychology, 8*, 234.
Houck, P., Nebel, D., & Milham, S. Jr. (1992). Organic solvent encephalopathy: An old hazard revisited. *American Journal of Industrial Medicine, 22(1)*, 109-115.
House, R. A., Sax, S. E., Rumack, E. R., & Holness, L. (1992). Medical management of three workers following a radiation exposure incident. *American Journal of Industrial Medicine, 22*, 249-257.
House, R. A., Liss, G. M., & Wills, M. C. (1994). Peripheral sensory neuropathy associated with 1,1,1-trichloroethane. *Archives of Environmental Health, 49*, 196-199.
Householder, J., Hatcher, R., Burns, W., & Chasnoff, I. (1982). Infants born to narcotic-addicted mothers. *Psychological Bulletin, 92(2)*, 453-468.
Howard, J., Kropenski, V., & Tyler, R. (1986). The long-term effects on neurodevelopment in infants exposed prenatally to PCP. In D. H. Clouet (Ed.), Phencyclidine: An update (pp. 237-251). Rockville, MD: NIDA Research Monograph 64.
Howard, M. L., Hogan, T. P., & Wright M. W. (1975). The effects of drugs on psychiatric patients' performance on the Halstead-Reitan neuropsychological test battery. *Journal of Nervous and Mental Disease, 161(3)*, 166-171.
Hsu, C., Chen, T., Soong, W., Sue, S., Liu, C., Tsung, C., Lin, S., Chang, S., & Liao, S. (1988). A 6-year followup study of intellectual and behavioral development of Yu-cheng children: Cross-sectional findings of the first field work. *Chinese Psychiatry, 2*, 26-40.

Huang, J., Kato, K., Shibata, E., Asaeda, N., & Takeuchi, Y. (1993). Nerve-specific marker proteins as indicators of organic solvent neurotoxicity. *Environmental Research, 63,* 82-87.

Huel, G., Tubert, P., Frery, N., Moreau, T., & Dreyfus, J. (1992). Joint effect of gestational age and maternal lead exposure on psychomotor development of the child at six years. *NeuroToxicology, 13,* 249-254.

Hughes, J. T. (1988). Brain damage due to paraquat poisoning: A fatal case with neuropathological examination of the brain. *NeuroToxicology, 9(2),* 243-248.

Hunter, J., Urbanowicz, M. A., Yule, W., & Lansdown, R. (1985). Automated testing of reaction time and its association with lead in children. *International Archives of Occupational and Environmental Health, 57,* 27-34.

Hunting, K. L., Matanoski, G. M., Larson, M., & Wolford, R. (1991). Solvent exposure and the risk of slips, trips, and falls among painters. *American Journal of Industrial Medicine, 20(13),* 353-370.

Husman, K. (1980). Symptoms of car painters with long-term exposure to a mixture of organic solvents. *Scandinavian Journal of Work Environment and Health, 6,* 19-32.

Husman, K., & Karli, P. (1980). Clinical neurological findings among car painters exposed to a mixture of organic solvents. *Scandinavian Journal of Work Environment and Health, 6,* 33-39.

Hutchings, D. E. (1990). Issues of risk assessment: Lessons from the use and abuse of drugs during pregnancy. *Neurotoxicology and Teratology, 12,* 183-189.

Iregren, A. (1990). Psychological test performance in foundry workers exposed to low levels of manganese. *Neurotoxicology and Teratology, 12,* 673-675.

Iregren, A., Tesarz, M., & Wigaeus-Hjelm, E. (1993). Human experimental MIBK exposure: Effects on heart rate, performance, and symptoms. *Environmental Research 63(1), 101-108.*

Irons, R., & Rose, P. (1985). Naval biodynamics laboratory computerized cognitive testing. *Neurobehavioral Toxicology and Teratology, 7,* 395-397.

Izmerov, N., & Tarasova, L. (1993). Occupational diseases developed as a result of severely injured nervous system: Acute and chronic neurotic effects. *Environmental Research, 62,* 172-177.

Jacobson, J. L., & Jacobson, S. W. (1988). New methodologies for assessing the effects of prenatal toxic exposure on cognitive functioning in humans. In M. Evans (Ed.), *Toxic contaminants and ecosystem health: A Great Lakes focus.* New York: Wiley.

Jacobson, J. L., Jacobson, S. W., & Humphrey, H. E. B. (1990). Effects of exposure to PCB's and related compounds on growth and activity in children. *Neurotoxicology and Teratology, 12,* 319-326.

Jacobson, J. L., Jacobson, S. W., Sokol, R. J., Martier, R. J., Martier, S. S. Ager, J. W., & Kaplan-Estrin, M. G. (1994). Teratogenic effects of alcohol on infant development. *Alcoholism: Clinical and Experimental Research, 17,* 174-183.

Jahan, K., & Ahmad, K. (1993). Studies on neurolathyrism. *Environmental Research, 60,* 259-266.

James, P. B. (1989). Hyperbaric oxygen in the treatment of carbon monoxide and smoke inhalation: A review. *Intensive Care World, 6,* 135-138.

James, R. C. (1985). Neurotoxicity: Toxic effects in the nervous system. In P. L. Williams & J. L. Burson (Eds.), *Industrial toxicology: Safety and health applications in the workplace* (pp. 123-127). New York: Van Reinhold.

James, R. C., Busch, H., Tamburro, C. H., Roberts, S. M., Schell, J. D., & Harbison, R. D. (1993). Polychlorinated biphenyl exposure and human disease. *Journal of Occupational Medicine, 35,* 136-148.

Janowsky, D. S., Meacham, M. P., Blaine, J. D., Schoor, M., & Bozzetti, L. P. (1976). Marijuana effects on simulated flying ability. *American Journal of Psychiatry, 133,* 384-388.

Jarkman, S., Skoog, K.-O., & Nilsson, S. (1985). The c-wave of the electroretinogram and the standing potential of the eye as highly sensitive measures of effects of low doses of trichloroethylene, methylchloroform, and halothane. *Documenta Ophthalmologica, 60,* 375-382.

Jarvis, M. J. (1993). Does caffeine enhance absolute levels of cognitive performance? *Psychopharmacology, 110,* 45-52.

Jellinger, K. (1977). Neuropathologic findings after neuroleptic long-term therapy. In L. Roizin, H. Shiraki, & N. Grcevic (Eds.), *Neurotoxicology.* New York: Raven Press.

Jeyaratnam, J. (1985). Health problems of pesticide usage in the third world. *British Journal of Industrial Medicine, 42,* 505-506.

Jeyaratnam, J. (1993). Occupational health issues in developing countries. *Environmental Research, 60*, 207-212.
Johannesson, G., Berglund, M., & Ingvar, D. H. (1982). Reduction of blood flow in cerebral white matter in alcoholics related to hepatic function: A CBF and EEG study. *Acta Neurologica Scandinavica, 65*, 190-202.
Johnson, B. L. (Ed.). (1990). *Advances in neurobehavioral toxicology: Applications in environmental and occupational health.* Chelsea, MI: Lewis Publishers.
Johnson, B. L., & Anger, W. K. (1983). Behavioral toxicology. In W. N. Rom (Ed.), *Environmental and occupational medicine* (pp. 329-350). Boston: Little, Brown.
Johnson, B. L., Cohen, A., Struble, R., Selzer, J. V., Anger, W. K., Gutnik, B. D., McDonough, T., & Houser, P. (1974). Field evaluation of carbon monoxide exposed toll collectors. In C. Xintaras, B. L. Johnson, & I. deGroot (Eds.), *Behavioral toxicology: Early detection of occupational hazards* (HEW Publication No. (NIOSH) 74-126, pp. 306-328.). Washington, DC: U.S. Government Printing Office.
Johnson, B. L., Boyd, J., Burg, J. R., Lee, S. T., Xintaras, C., & Albright, B. E. (1983). Effects on the peripheral nervous system of workers' exposure to carbon disulfide. *Neurotoxicology, 4*, 53-66.
Johnson, B. P., Meredith, T. J., & Vale, J. A. (1983). Cerebellar dysfunction after acute carbon tetrachloride poisoning. *Lancet, 2*, 968.
Johnson, J. L., Adinoff, B., Bisserbe, J.-C., Martin, P. R., Rio, D., Rohrbaugh, J. W., Zubovic, E., & Eckardt, M. J. (1986). Assessment of alcoholism-related organic brain syndromes with positron emission tomography. *Alcoholism: Clinical and Experimental Research, 10*, 237-240.
Johnson, L. D., Bachman, J. G., & O'Malley, P. M. (1979). *Drugs and the nation's high school students: Five year trends, 1979 highlights.* Rockville, MD: National Institute on Drug Abuse.
Jokstad, A., Thomassen, Y., Bye, E., Clench-Aas, J., & Aaseth, J. (1992). Dental amalgam and mercury. *Pharmacology and Toxicology, 70*, 308-313.
Jones, B. (1971). Verbal and spatial intelligence in short and long term alcoholics. *Journal of Nervous and Mental Disease, 153*, 292-297.
Jones, B., & Parsons, O. (1971). Impaired abstracting ability in chronic alcoholics. *Archives of General Psychiatry, 24*, 71-75.
Jones, B., & Parsons, O. (1972). Specific vs. generalized deficits of abstracting ability in chronic alcoholics. *Archives of General Psychiatry, 26*, 380-384.
Jones, B. M., & Vega, A. (1982). Cognitive performance measured on the ascending limb of the blood alcohol curve. *Psychopharmacologia, 23*, 99-114.
Jones, R. T. (1980). Human effects: An overview. In R. C. Peterson (Ed.), *Marijuana research findings 1980* (NIDA Research Monograph 31, pp. 54-80). Washington, DC: Department of Health and Human Services.
Jones, R. T. (1984). The pharmacology of cocaine. In J. Grabowski (Ed.), *Cocaine: Pharmacology, effects, and treatment of abuse* (NIDA Research Monograph 50, pp. 34-53). Washington, DC: Department of Health and Human Services.
Jordan, C. M., Whitman, R. D., Harbut, M., & Tanner, B. (1993). Neuropsychological sequelae of hard metal disease. *Archives of Clinical Neuropsychology, 8*, 309-326.
Jorgensen, N. K., & Cohr, K.-H. (1981). n-Hexane and its toxicologic effects: A review. *Scandinavian Journal of Work Environment and Health, 7*, 157-168.
Judd, L. L., Hubbard, B., Janowsky, D. S., Huey, L. Y., & Takanashi, K. (1977). The effects of lithium carbonate on the cognitive functions of normal subjects. *Archives of General Psychiatry, 34*, 355-357.
Juntunen, J. (1982a). Alcoholism in occupational neurology: Diagnostic difficulties with special reference to the neurological syndromes caused by exposure to organic solvents. *Acta Neurologica Scandinavica, 66*, 89-108.
Juntunen, J. (1982b). Neurological examination and assessment of the syndromes caused by exposure to neurotoxic agents. In R. Gilioli, M. G. Cassitto, & V. Foa (Eds.), *Neurobehavioral methods in occupational health* (p. 310). New York: Pergamon Press.
Juntunen, J., Hupli, V., Hernberg, S., & Luisto, M. (1980). Neurological picture of organic solvent poisoning in industry. *International Archives of Occupational and Environmental Health, 46*, 219-231.

Jusic, A. (1974). Anticholinesterase pesticides of organophosphorus type: Electromyographic, neurological and psychological studies in occupationally exposed workers. In C. Xintaras, B. L. Johnson, & I. deGroot (Eds.), *Behavioral toxicology: Early detection of occupational hazards* (HEW Publication No. (NIOSH) 74-126, pp. 182-190). Washington, DC: U.S. Government Printing Office.

Kahn, E., & Letz, G. (1989). Clinical ecology: Environmental medicine or unsubstantiated theory? *Annals of Internal Medicine, 111,* 104-105.

Kaij, L. (1960). *Alcoholism in twins: Studies on the etiology and sequels of abuse of alcohol.* Stockholm: Almqvist & Wiksell.

Kain, Z. N., Kain, T. S., & Scarpelli, E. M. (1992). Cocaine exposure *in utero:* Perinatal development and neonatal manifestations—review. *Clinical Toxicology, 30(4),* 607-636.

Kajiyama, K., Doi, R., Sawada, J., Hashimoto, K., Hazama, T., Nakata, S., Hirata, M., Yoshida, T., & Miyajima, K. (1993). Significance of subclinical entrapment of nerves in lead neuropathy. *Environmental Research, 60,* 248-252.

Kaku, D. A., & Lowenstein, D. H. (1990). Emergence of recreational drug abuse as a major risk factor for stroke in young adults. *Annals of Internal Medicine, 113(11),* 821-827.

Kaloyanova, F. P., & El Batawi, M. A. (1991). *Human toxicology of pesticides.* Boca Raton, FL: CRC Press.

Kamb, M. L., Murphy, J. J., Jones, J. L., Caston, J. C., Nederlog, K., Horney, L. F., Swygert, L. A., Falk, H., & Kilbourne, E. M. (1992). *Journal of the American Medical Association, 267,* 77-82.

Kang-Yum, E., & Oransky, S. H. (1992). Chinese patient medicine as a potential source of mercury poisoning. *Veterinary and Human Toxicology, 34,* 235-237.

Kane, J., Honigfeld, G., Singer, J., & Meltzer, H. (1988). Clozapine for the treatment-resistant schizophrenic. A double-blind comparison with chlorpromazine. *Archives of General Psychiatry, 45(9),* 789-796.

Kanluen, S., & Gottlieb, C. A. (1991). A clinical pathologic study of four adult cases of acute mercury inhalation toxicity. *Archives of Pathology and Laboratory Medicine, 115,* 56-60.

Kaplan, J. G., Kessler, J., Rosenberg, N., Pack, D., & Schaumburg, H. H. (1993). Sensory neuropathy associated with Dursban (chlorpyrifos) exposure. *Neurology, 43(11),* 2193-2196.

Kardel, T., & Stigsby, B. (1975). Period amplitude analysis of the electroencephalogram correlated with liver function in patients with cirrhosis of the liver. *Electroencephalography and Clinical Neurophysiology, 38,* 605-609.

Kardel, T., Zander Olsen, P., Stigsby, B., & Tonneson, K. (1972). Hepatic encephalopathy evaluated by autonomic period analysis of the electroencephalogram during lactulose treatment. *Acta Medica Scandinavica, 192,* 493-498.

Katbamna, B., Metz, D. A., Adelman, C. L., & Thodi, C. (1993). Auditory-evoked responses in chronic alcohol and drug abusers. *Biological Psychiatry, 33,* 750-752.

Katz, D. L., & Pope, H. G. (1990). Anabolic-androgenic steroid-induced mental status changes. In G. C. Lin & L. Erinoff (Eds.), *Anabolic steroid abuse* (NIDA Research Monograph 102, pp. 215-223). Rockville, MD: Department of Health and Human Services.

Katz, G. V. (1985). Metals and metalloids other than mercury and lead. In J. L. O'Donoghue (Ed.), *Neurotoxicity of industrial and commercial chemicals* (Vol. I, pp. 171-191). Boca Raton: CRC Press.

Kelafant, G. A., Kasarskis, E. J., Horstman, S. W., Cohen, C., & Frank, A. L. (1993). Arsenic poisoning in central Kentucky: A case report. *American Journal of Industrial Medicine, 24,* 723-726.

Kelafant, G. A., Berg, R. A., & Schleenbaker, R. (1994). Toxic encephalopathy due to exposure to 1,1,1 trichloroethane. *American Journal of Industrial Medicine, 25,* 439-446.

Kemppainen, R., Juntunen, J., & Hillbom, M. (1982). Drinking habits and peripheral alcoholic neuropathy. *Acta Neurologica Scandinavica, 65(1),* 11-18.

Kendrick, D. C., & Post, F. (1967). Differences in cognitive status between healthy, psychiatrically ill, and diffusely brain-damaged elderly subjects. *British Journal of Psychiatry, 113,* 75-81.

Kennedy, R. S., Wilkes, R. L., Dunlap, W. P., & Kuntz, L. A. (1987). Development of an automated performance test system for environmental and behavioral toxicology studies. *Perceptual and Motor Skills, 65,* 947-962.

Kessler, R. M., Parker, E. S., Clark, C. M., Martin, P. R., George, D. T., Weingartner, H., Sokoloff, L., Ebert, M. H., & Mishkin, M. (1984). Regional cerebral glucose metabolism in patients with alcoholic Korsakoff's syndrome. *Society of Neuroscience Abstracts, 10,* 541.

Key, M. M., Henschel, A. F., Butler, J., Ligo, R. N., Tabershaw, I. R., & Ede, L. (Eds.). (1977). *Occupation diseases: A guide to their recognition* (DHEW Publication No. (NIOSH) 77-181). Washington, DC: U.S. Government Printing Office.

Khalil-Manesh, F., Gonick, H. C., Weiler, E. W. J., Prins, B., Weber, M. A., Purdy, R., & Ren, Q. (1994). Effect of chelation treatment with dimercaptosuccinic acid (DMSA) on lead-related blood pressure changes. *Environmental Research, 65,* 86-99.

Khalsa, J. E. *Epidemiology of maternal drug abuse and its health consequences: Recent findings.* Washington, DC: National Institute on Drug Abuse, in press.

Khanna, V. K., Husain, R., & Seth, P. K. (1988). Low protein diet modifies acrylamide neurotoxicity. *Toxicology, 49,* 395-401.

Kiesswetter, E., Blaszkewicz, M., Vangala, R. R., & Seeber, A. (1994). Acute exposure to acetone in a factory and ratings of well-being. *NeuroToxicology, 15,* 597-602.

Kilbourne, E. M. (1992). Eosinophilia myalgia syndrome: Coming to grips with a new illness. *Epidemiologic Reviews, 14,* 16-36.

Kilburn, K. (1994). Neurobehavioral impairment and seizures from formaldehyde. *Archives of Environmental Health, 49,* 37-44.

Kilburn, K. H., & Warshaw, R. H. (1992). Neurobehavioral effects of formaldehyde and solvents on histology technicians: Repeated testing across time. *Environmental Research, 58,* 134-146.

Kilburn, K. H., Seidman, B. C., & Warshaw, R. (1985). Neurobehavioral and respiratory symptoms of formaldehyde and xylene exposure in histology technicians. *Archives of Environmental Health, 40,* 229-233.

Kilburn, K. H., Warshaw, R., Boylen, C. T., Johnson, S.-J. S., Seidman, B., Sinclair, R., & Takaro, T. (1985). Pulmonary and neurobehavioral effects of formaldehyde exposure. *Archives of Environmental Health, 40,* 254-260.

King, M. (1983). Long term neuropsychological effects of solvent abuse. In N. Cherry & H. A. Waldron (Eds.), *The neuropsychological effects of solvent exposure* (pp. 75-84). Havant, Hampshire: Colt Foundation.

Kinjo, Y., Higashi, H., Nakano, A., Sakamoto, M., & Sakai, R. (1993). Profile of subjective complaints and activities of daily living among current patients with Minamata disease after 3 decades. *Environmental Research, 63,* 241-251.

Kippen, H., Fiedler, N., Maccia, C., Yurkow, E., Todaro, J., & Laskin, D. (1992). Immunologic evaluation of chemically sensitive patients. *Toxicology and Industrial Health, 8,* 125-136.

Kishi, R., Doi, R., Fukuchi, Y., Satoh, H., Satoh, T., Ono, A., Moriwaka, F., Tashiro, J., Takahata, N., & The Mercury Workers Study Group (1993). Subjective symptoms and neurobehavioral performances of ex-mercury miners at an average of 18 years after the cessation of chronic exposure to mercury vapor. *Environmental Research, 62,* 289-302.

Kishi, R., Harabuchi, I., Katakura, Y., Ikeda, T., & Miyake, H. (1993). Neurobehavioral effects of chronic occupational exposure to organic solvents among Japanese industrial painters. *Environmental Research, 62,* 303-313.

Kjellstrand, P., Holmquist, B., Kanje, M., Alm, P., Romare, S., Jonsson, I., Mansson, L., & Bjerkemo, M. (1984). Perchloroethylene: Effects on body and organ weights and plasma butyrylcholinesterase activity. *Acta Pharmacologia et Toxicologica, 54(5),* 414-424.

Kjellstrom, T., Kennedy, P., Wallis, S., & Mantell, C. (1986). Physical and mental development of children with prenatal exposure to mercury from fish. Stage 1: Preliminary tests at age 4. Solna, Sweden: National Swedish Environmental Protection Board, Report 3080.

Klaassen, C. D., Amdur, M. O., & Doull, J. (Eds.). (1986). *Casarett and Doull's toxicology.* New York: Macmillan Co.

Klaucke, D. N., Johansen, M., & Vogt, R. L. (1982). An outbreak of xylene intoxication in a hospital. *American Journal of Industrial Medicine, 3(2),* 173-178.

Kleinknecht, R. A., & Donaldson, D. (1975). A review of the effects of diazepam on cognitive and psychomotor performance. *Journal of Nervous and Mental Disease, 161,* 399-411.

Klisz, D., & Parsons, O. A. (1977). Hypothesis testing in younger and older alcoholics. *Journal of Studies on Alcohol, 38*, 1718–1729.

Knasko, S. C. (1993). Performance, mood, and health during exposure to intermittent odors. *Archives of Environmental Health, 48*, 305–308.

Knave, B., Olson, B. A., Elofsson, S., Gamberale, F., Isaksson, A., Mindus, P., Persson, H. E., Struwe, G., Wennberg, A., & Westerholm, P. (1978). Long-term exposure to jet fuel. II. *Scandinavian Journal of Work, Environment and Health, 4(1), 19–45.*

Kocsis, J. H., Shaw, E. D., Stokes, P. E., Wilner, P., Elliot, A. S., Sikes, C., Myers, B., Manevitz, A., & Parides, M. (1993). Neuropsychologic effects of lithium discontinuation. *Journal of Clinical Psychopharmacology, 13(4)*, 268–275.

Kolb, L. C. (1987). A neuropsychological hypothesis explaining posttraumatic stress disorders. *American Journal of Psychiatry, 144(8)*, 989–995.

Konietzko, H., Elster, J., Bencsath, A., Drysch, K., & Weichard, H. (1975). Psychomotor responses under standardized trichloroethylene load. *Archives of Toxicology, 33*, 129–139.

Konietzko, H., Keilbach, J., & Drysch, K. (1980). Cumulative effects of daily toluene exposure. *International Archives of Occupational and Environmental Health, 46(1)*, 53–58.

Koren, H. L., Graham, D. E., & Devlin, R. B. (1992). Exposure of humans to a volatile organic mixture. III. Inflammatory response. *Archives of Environmental Health, 47*, 39–44.

Korgeski, G. P., & Leon, G. (1983). Correlates of self-reported and objectively determined exposure to Agent Orange. *American Journal of Psychiatry, 140*, 1443–1449.

Korman, M., Matthews, R. W., & Lovitt, R. (1981). Neuropsychological effects of abuse of inhalants. *Perceptual and Motor Skills, 53*, 547–553.

Korsak, R. J., & Sato, M. M. (1977). Effects of chronic organophosphate pesticide exposure on the central nervous system. *Clinical Toxicology, 11*, 83–95.

Kratz, A. E., Holliday, S. L., & Donahoe, C. P. (1994). The neuropsychological effects of isopropyl alcohol intoxication: A case of surreptitious abuse. Paper presented at the 1994 annual meeting of the National Academy of Neuropsychology, Orlando, Florida.

Krigman, M. R., Bouldin, T. W., & Mushak, P. (1980). Lead. In P. S. Spencer & H. H. Schaumburg (Eds.), *Experimental and clinical neurotoxicology* (pp. 490–507). Baltimore: Williams & Wilkins.

Kroft, C., & Cole, J. O. (1992). Adverse behavioral effects of psychostimulants. In J. M. Kane & J. E. Lieberman (Eds.), *Adverse effects of psychotropic drugs* (pp. 153–162). New York: The Guilford Press.

Kroll-Smith, J. S., & Couch, S. (1991). As if exposure to toxins were not enough: The social and cultural system as a secondary stressor. *Environmental Health Perspectives, 95*, 61–66.

Kruger, G., Weinhardt, F., & Hoyer, S. (1979). Brain energy metabolism and blood flow in bismuth encephalopathy. In L. Manzo (Ed.), *Advances in neurotoxicology* (pp. 63–68). New York: Pergamon Press.

Kuitunen, T., Aranko, K., Nuotto, E., Lindbohm, R., Mattila, M. J., Korte, T., & Seppala, T. (1994). Article title? *Alcohol, Drugs and Driving, 10*, 135–146.

Kyrklund, T. (1992). The use of experimental studies to reveal suspected neurotoxic chemicals as occupational hazards: Acute and chronic exposures to organic solvents. *American Journal of Industrial Medicine, 21(1)*, 15–24.

Kyrklund, T., Alling, C., Kjellstrand, P., & Haglid, K. G. (1984). Chronic effects of perchloroethylene on the composition of lipid and acyl groups in the cerebral cortex and hippocampus of the gerbil. *Toxicology Letters, 22*, 343–349.

Kyrklund, T., Kjellstrand, P., & Haglid, K. (1987). Brain lipid changes in rats exposed to xylene and toluene. *Toxicology, 5*, 123–133.

Laatsch, L., Hartman, D., & Stone, J. (1994). Transcallosal intraventricular tumor excision, alcohol abuse, and amnestic syndrome: Case study. *Journal of Neurology, Neurosurgery, and Psychiatry, 57(6)*, 766–767.

Labrèche, F. P., Cherry, N. M., & McDonald, J. C. (1992). Psychiatric disorders and occupational exposure to solvents. *British Journal of Industrial Medicine, 49*, 820–825.

Landauer, A. A., Pocock, D. A., & Prott, F. W. (1979). Effects of atenolol and propranolol on human performance and subjective feelings. *Psychopharmacology, 60*, 211–215.

Landrigan, P. J. (1983). Toxic exposures and psychiatric disease—Lessons from the epidemiology of cancer. *Acta Psychiatrica Scandinavica, 67*(Suppl. 303), 615.
Landrigan, P. J., & Curran, A. (1992). Lead—A ubiquitous hazard. *Environmental Research, 59,* 279–289.
Landrigan, P. J., Whitworth, R. H., Baloh, R. W., Staehling, N. W., Barthel, W. F., & Rosenblum, B. F. (1975). Neuropsychological dysfunction in children with chronic low-level lead absorption. *Lancet, 1,* 708–712.
Landrigan, P. J., Kreiss, K., Xintaras, C., Feldman, R. G., & Heath, C. W., Jr. (1980). Clinical epidemiology of occupational neurotoxic disease. *Neurobehavioral Toxicology, 2,* 43–48.
Landrigan, P. J., Meinhardt, T. J., Gordon, J., Lipscomb, J. A., Burg, J. R., Mazzuckelli, L. F., Lewis, T. R., & Lemen, R. A. (1984). Ethylene oxide: An overview of toxicologic and epidemiologic research (review). *American Journal of Industrial Medicine, 6(2),* 103–115.
Lang, L. (1993). Are pesticides a problem? *Environmental Health Perspectives, 101,* 578–583.
Langolf, G. D., Chaffin, D. B., Henderson, R., & Whittle, H. P. (1978). Evaluation of workers exposed to elemental mercury using quantitative tests of tremor and neuromuscular functions. *American Industrial Hygiene Association Journal, 39,* 976–984.
Langston, J. W. (1988). Aging, neurotoxins, and neurodegenerative disease. In R. D. Terry (Ed.), *Aging and the brain* (pp. 149–164). New York: Raven Press.
Langston, J. W., & Irwin, I. (1986). MPTP current concepts and controversies. *Clinical Neuropharmacology, 9,* 485–507.
Langworth, S., Almkvist, O., Soderman, E., & Wikstrom, B.-O. (1992). Effects of occupational exposure to mercury vapors on the central nervous system. *British Journal of Industrial Medicine, 49,* 545–555.
Larsby, B., Tham, R., Eriksson, B., Hyden, D., Odkvist, L., Liedgren, C., & Bunnfors, I. (1986). Effects of trichloroethylene on the human vestibulo-oculomotor system. *Acta Otolaryngolica (Stockholm), 101,* 193–199.
Lash, A. A., Becker, C. E., So, Y., & Shore, M. (1991). Neurotoxic effects of methylene chloride: Are they long lasting in humans? *British Journal of Industrial Medicine, 48(6),* 418–426.
Laties, V. G., & Merigan, W. H. (1979). Behavioral effects of carbon monoxide on animals and man. *Annual Review of Pharmacology & Toxicology, 19,* 357–392.
Laursen, P. (1990). A computer-aided technique for testing cognitive functions. *Acta Neurologica Scandinavica, Supplement, 131,* 1–108.
Law, D., Lash, A. A., Bowler, R., Estrin, W., & Becker, C. E. (1990). Evaluation of the construct validity of examiner-administered and computer-administered neuropsychological tests. In B. L. Johnson (Ed.), *Advances in neurobehavioral toxicology: Applications in environmental and occupational health* (pp. 263–271). Chelsea, MI: Lewis Publishers.
Leavitt, J., Webb, P., Struve, F., Straumanis, J., Norris, G., Nixon, F., Fitz-Gerald, M. J., & Patrick, G. (1991). *Neuropsychological effects associated with chronic marijuana abuse.* Presented at the 11th Annual Meeting of the National Academy of Neuropsychology, November 4, Dallas, Tx.
Lebowitz, M. D., & Walkinshaw, D. S. (1992). Indoor air '90: Health effects associated with indoor air contaminants. *Archives of Environmental Health, 47,* 6–7.
Lee, G. P., & DiClimente, C. C. (1985). Age of onset versus duration of problem drinking on the alcohol use inventory. *Journal of Studies on Alcohol, 46(5),* 398–402.
Lee, S.-H., & Lee, S. H. (1993). A study on the neurobehavioral effects of occupational exposure to organic solvents in Korean workers. *Environmental Research, 60,* 227–232.
Legg, J. F., & Stiff, M. P. (1976). Drug-related test patterns of depressed patients. *Psychopharmacology, 50(2),* 205–210.
Leggett, R. W. (1993). An age-specific kinetic model of lead metabolism in humans. *Environmental Health Perspectives, 101(7),* 598–616.
Leira, H. L., Myhr, G., Nilsen, G., & Dale, L. G. (1992). Cerebral magnetic resonance imaging and cerebral computerized tomography for patients with solvent-induced encephalopathy. *Scandinavian Journal of Work Environment and Health, 18,* 68–70.
Leirer, V. O., Yesavage, J. A., & Morrow, D. G. (1989). Marijuana, aging task difficulty effects on pilot performance. *Aviation, Space & Environmental Medicine, 60,* 1145–1152.

Le Quesne, P. M. (1979). Neurological disorders due to toxic occupational hazards. *Practitioner, 223,* 40–47.
Le Quesne, P. M. (1982). Metal-induced diseases of the nervous system. *British Journal of Hospital Medicine,* 534–537.
Le Quesne, P. M. (1984). Neuropathy due to drugs. In P. J. Dyck, P. K. Thomas, E. H. Lambert, & R. Bunge (Eds.), *Peripheral neuropathy* (Vol. II, pp. 2162–2179). Philadelphia: Saunders.
Lerman, Y., Hirshberg, A., & Shteger, Z. (1984). Organophosphate and carbamate pesticide poisoning: The usefulness of a computerized clinical information system. *American Journal of Industrial Medicine, 6,* 17–26.
Letz, R. (1993). Covariates of computerized neurobehavior test performance epidemiologic investigations. *Environmental Research, 61,* 124–132.
Letz, R. (1990). The Neurobehavioral Evaluation System: An international effort. In B. L. Johnson (Ed.), *Advances in neurobehavioral toxicology: Applications in environmental and occupational health* (pp. 189–201). Chelsea, MI: Lewis Publishers.
Letz, R., & Singer, R. (1985). Neuropsychological tests. In P. Grandjean (Ed.), *Neurobehavioral methods in occupational and environmental health* (Document 6, pp. 17–18). Copenhagen: World Health Organization.
Letz, R., Mahoney, F. C., Hershman, D. L., Woskie, S., & Smith, T. J. (1990). Neurobehavioral effects of acute styrene exposure in fiberglass boatbuilders. *Neurotoxicology and Teratology, 12,* 665–668.
Levin, A. S., & Byers, V. S. (1992). Multiple chemical sensitivities: A practicing clinician's point of view—Clinical and immunologic research findings. *Toxicology and Industrial Health, 8,* 95–110.
Levin, E. D., Schantz, S. K., & Bowman, R. E. (1992). Use of the lesion model for examining toxicant effects on cognitive behavior. *Neurotoxicology and Teratology, 14,* 131–141.
Levin, H. S. (1991). Pioneers in research on the behavioral sequelae of head injury. *Journal of Clinical and Experimental Neuropsychology, 13,* 133–154.
Levin, H. S. (1974). Behavioral effects of occupational exposure to organophosphorate pesticides. In C. Xintaras, B. L. Johnson, & I. deGroot (Eds.), *Behavioral toxicology: Early detection of occupational hazards* (HEW Publication No. (NIOSH) 74-126, pp. 154–163). Washington, DC: U.S. Government Printing Office.
Levin, H. S., & Rodnitzky, R. L. (1976). Behavioral aspects of organophosphorous pesticides in man. *Clinical Toxicology, 9,* 391.
Levin, H. S., Rodnitzky, R. L., & Mick, D. L. (1976). Anxiety associated with exposure to organophosphate compounds. *Archives of General Psychiatry, 33,* 225–228.
Levine, S. A., & Reinhardt, J. H. (1983). Biochemical-induced pathology initiated by free radicals, oxidant chemicals, and therapeutic drugs in the etiology of chemical hypersensitivity disease. *Orthomolecular Psychiatry, 12,* 166–183.
Levinson, J. L. (1985). Neuroleptic malignant syndrome. *American Journal of Psychiatry, 142,* 1137–1144.
Lewi, Z., & Bar-Khayim, Y. (1964). Food-poisoning from barium carbonate. *Lancet, 2,* 342–343.
Lewin, J. F., & Cleary, W. T. (1982). An accidental death caused by the absorption of phenol through skin: A case report. *Forensic Science International, 19(2),* 177–179.
Lewis, J. E., & Hordan, R. B. (1986). Neuropsychological assessment of phencyclidine abusers. In D. H. Clouet (Ed.), *Phencyclidine: An update* (NIDA Research Monograph 64, pp. 190–208). Washington, DC: Department of Health and Human Services.
Lezak, M. D. (1984). Neuropsychological assessment in behavioral toxicology—Developing techniques and interpretive issues. *Scandinavian Journal of Work Environment and Health, 10*(Suppl. 1), 25–29.
Liang, Y., Chen, Z., Sun, R., Fang, Y., & Yu, J. (1990). Application of the WHO Neurobehavioral Core Test and other neurobehavioral screening methods. In B. L. Johnson (Ed.), *Advances in neurobehavioral toxicology: Applications in environmental and occupational health* (pp. 225–243). Chelsea, MI: Lewis Publishers.
Liang, Y., Sun, R., Sun, Y., Chen, Z., & Lin, L. (1993). Psychological effects of low exposure to mercury vapor: Application of a computer administered neurobehavioral evaluation system. *Environmental Research, 60,* 320–327.

Lilis, R., Fischbein, A., Diamond, S., Anderson, H. A., Selikoff, I. J., Blumberg, W. E., & Eisinger, J. (1977). Lead effects among secondary lead smelter workers with blood lead levels below 80 µg/100 ml. *Archives of Environmental Health, 32(6),* 256–266.
Liljequist, R., Linnoila, M., & Mattila, M. J. (1974). Effect of two weeks' treatment with chlorimipramine and nortriptyline, alone or in combination with alcohol, on learning and memory. *Psychopharmacologica, 39(2),* 181–186.
Lin, J.-L., & Lim, P.-S. (1993). Massive ingestion of elemental mercury. *Clinical Toxicology, 31,* 487–492.
Lindberg, N., Basch, K. E., & Lindberg, E. (1982). Psychotherapeutic examination of patients with suspected chronic solvent intoxication. An overview. *Psychotherapy and Psychosomatics, 37,* 36–63.
Lindboe, C. F., & Løberg, E. M. (1988). The frequency of brain lesions in alcoholics. *Journal of the Neurological Sciences, 88(1–3),* 107–113.
Lindbohm, M.-L., Taskinen, H., Kyyrönen, Sallmén, Anttila, A., & Hemminki, K. (1992).
Linden, C. H. (1990). Volatile substances of abuse. *Emergency Medicine Clinics of North America, 8,* 559–578.
Lindstrom, J., & Wickstrom, G. (1983). Psychological function changes among maintenance house painters exposed to low levels of organic solvent mixtures. *Acta Psychiatrica Scandinavica, 67*(Suppl. 303), 81–91.
Lindstrom, K. (1980). Changes in psychological performance of solvent-poisoned and solvent-exposed workers. *American Journal of Industrial Medicine, 1,* 69–84.
Lindstrom, K. (1981). Behavioral changes after long-term exposure to organic solvents and their mixtures. *Scandinavian Journal of Work Environment and Health, 7*(Suppl. 4), 48–53.
Lindstrom, K. (1982). Behavioral effects of long-term exposure to solvents. *Acta Neurologica Scandinavica, 66*(Suppl. 92), 131–141.
Lindstrom, K. (1984). The Rorschach test in behavioral toxicology. *Scandinavian Journal of Work Environment and Health, 10*(Suppl. 1), 20–23.
Lindstrom, K., Harkonen, H., & Hernberg, S. (1976). Disturbances in psychological functions of workers occupationally exposed to styrene. *Scandinavian Journal of Work Environment and Health, 3,* 129–139.
Lindstrom, K., Antti-Poika, M., Tolla, S., & Hyytianinen, A. (1982). Psychological prognosis of diagnosed organic solvent intoxication. *Neurobehavioral Toxicology and Teratology, 4,* 581–588.
Lindstrom, K., Riihimaki, H., & Hänninen, K. (1984). Occupational solvent exposure and neuropsychiatric disorders. *Scandinavian Journal of Work Environment and Health, 10,* 321–323.
Linz, D. H., DeGarmo, P. L., Morton, W. E., Weins, A. N., Coull, B. M., & Maricle, R. A. (1986). Organic solvent-induced encephalopathy in industrial painters. *Journal of Occupational Medicine, 28,* 119–125.
Linz, D. H., Barrett, E. T. Jr., Pflaumer, J. E., & Keith, R. E. (1992). Neuropsychologic and postural sway improvement after Ca(++)-EDTA chelation for mild lead intoxication. *Journal of Occupational Medicine, 34(6),* 638–641.
Lishman, W. A. (1978). *Organic psychiatry: The psychological consequences of cerebral disorder.* Oxford: Blackwell Scientific Publications.
Liss, G. M. (1988). Peripheral neuropathy in two workers exposed to 1,1,1-trichloroethane. *Journal of the American Medical Association, 260,* 2217.
Liss, G. M., & House, R. A. (1995). Toxic encephalopathy due to 1,1,1-trichloroethane exposure. *American Journal of Industrial Medicine, 27,* 445–446.
Lister, R. G. (1985). The amnesic action of benzodiazepines in man. *Neuroscience and Biobehavioral Review, 9,* 87–94.
Livingston, J. M., & Jones, C. R. (1981). Living area contamination by chlordane used for termite treatment. *Bulletin of Environmental Contamination Toxicology, 27,* 406–411.
Loberg, T. (1978). Neuropsychological deficits in alcoholics: Lack of personality (MMPI) correlates. In H. Begleiter (Ed.), *Biological effects of alcoholism* (pp. 797–808). New York: Plenum Press.
Loberg, T. (1986). Neuropsychological findings in the early and middle phases of alcoholism. In I. Grant & K. M. Adams (Eds.), *Neuropsychological assessment of neuropsychiatric disorders* (pp. 441–477). London: Oxford University Press.

Lodge, A. (1976). Developmental findings with infants born to mothers on methadone maintenance: A preliminary report. In G. Beschner & R. Brotman (Eds.), *Symposium on comprehensive health care for addicted families and their children* (NIDA Services Research Report). Washington, DC: U.S. Government Printing Office.

Lograno, D. E., Matteo, F., Trabucchi, M., Govoni, S., Cagiano, R., Lacomba, C., & Cuomo, V. (1993). Effects of chronic ethanol intake at a low dose on the rat brain dopaminergic system. *Alcohol, 10(1)*, 45-49.

Logue, P. E., Schmitt, F. A., Rogers, H. E., & Strong, G. B. (1986). Cognitive and emotional changes during a simulated 686-m deep dive. *Undersea Biomedical Research, 13(2)*, 225-235.

LoMonaco, M., Milone, M., Batocchi, A. P., Padua, L., Restuccia, D., & Tonali, P. (1992). Cisplatin neuropathy: Clinical course and neurophysiological findings. *Journal of Neurology, 239*, 199-204.

London, E. D., Cascella, N. G., Wong, D. F., Phillips, R. L., Dannais, R. F., Links, J. M., Herning, R., Grayson, R., Jaffe, J. H., & Wagner, H. N. (1990). Cocaine-induced reduction of glucose utilization in human brain. *Archives of General Psychiatry, 47*, 567-574.

London, W. P. (1985). Treatment outcome of left-handed versus right-handed alcoholic men. *Alcoholism: Clinical and Experimental Research, 9*, 503-504.

Long, J. A., & McLachlan, J. F. C. (1974). Abstract reasoning and perceptual-motor efficiency in alcoholics: Impairment and reversibility. *Quarterly Journal of Studies on Alcohol, 35*, 1220-1229.

Longstreth, W. T., Rosenstock, L., & Heyer, N. J. (1985). Potroom palsy? Neurologic disorder in three aluminum smelter workers. *Archives of Internal Medicine, 145*, 1972-1975.

Lopez-Ibor, J. J., Soria, J., Canas, F., & Rodriguez-Gamazo, M. (1985). Psychological aspects of the toxic oil syndrome catastrophe. *British Journal of Psychiatry, 147*, 352-365.

Lorenz, H., Omlor, A., Walter, G., Haab, A., Steigerwald, F., & Buchter, A. (1990). Nachweis von hirnschädigungen durch tetrachlorethen. *Zbl. Arbeitsmed, 40*, 355-364.

Lorscheider, F. L., and Vimy, M. J. (1990). Mercury from dental amalgam. *Lancet, 336*, 1578-1579.

Lovinger, D. M. (1994). Excitotoxicity and alcohol-related brain damage. *Alcoholism: Clinical and Experimental Research, 17*, 19-27.

Lu, Y.-C., & Wong, P.-N. (1984). Dermatological, medical and laboratory findings of patients in Taiwan and their treatments. In M. Kuratsune & R. Shapiro (Eds.), *PCB poisoning in Japan and Taiwan* (pp. 81-116). New York: Liss.

Lubit, R., & Russett, B. (1984). The effects of drugs on decision-making. *Journal of Conflict Resolution, 28*, 85-102.

Lucki, I., Rickels, K., & Geller, A. M. (1986). Chronic use of benzodiazepines and psychomotor and cognitive test performance. *Psychopharmacology, 88*, 426-433.

Lund, A. K., & Wolfe, A. C. (1991). Changes in the incidence of alcohol-impaired driving in the United States, 1973-1986. *Journal of Studies on Alcohol, 52*, 293-301.

Lund, Y., Nissen, M., & Rafaelsen, O. J. (1982). Long-term lithium treatment and psychological functions. *Acta Psychiatrica Scandinavica, 65*, 233-244.

Luria, A. (1973). *The working brain*. New York: Basic Books.

Lusins, J., Zimberg, S., Smokler, H., & Gurley, K. (1980). Alcoholism and cerebral atrophy: A study of 50 patients with CT scan and psychologic testing. *Alcohol Clinical Experiment Research, 4(4)*, 406-411.

MacFarland, H. N. (1986). Toxicology of solvents. *Journal of the American Industrial Hygiene Association, 47*, 704-707.

Mackay, C. J., Campbell, L., Samuel, A. M., Alderman, K. J., Idzikowski, C., Wilson, H. K., & Gompertz, D. (1987). Behavioral changes during exposure to 1,1,1-trichlorothane: Time-course and relationship to blood solvent levels. *American Journal of Industrial Medicine, 11*, 223-239.

MacQueen, G., Marshall, J., Perdue, M., Shepard, S., & Bienenstock, J. (1989, January). Pavlovian conditioning of rat mucosal mast cells to secrete rat mast cell protease II. *Science, 243*, 83-85.

Maddox, V. H. (1981). The historical development of phencyclidine. In E. F. Domino (Ed.), *PCP (phencyclidine): Historical and current perspectives* (p. 18). Ann Arbor: NPP Books.

Mahaffey, K. R., Annest, J. L., Roberts, J., & Murphy, R. S. (1982). National estimates of blood lead levels: United States 1976-1980: Association with selected demographic and socioeconomic factors. *New England Journal of Medicine, 307*, 573-579.

REFERENCES

Maizlish, N. A., Langolf, G. D., Whitehead, L. W., Fine, L. J., Albers, J. W., Goldberg, J., & Smith, P. (1985). Behavioural evaluation of workers exposed to mixtures of organic solvents. *British Journal of Industrial Medicine, 42*, 579–590.

Mallov, J. S. (1976). MBK neuropathy among spray painters. *Journal of the American Medical Association, 235*, 1455–1457.

Malm, G., & Lying-Tunell, U. (1980). Cerebellar dysfunction related to toluene sniffing. *Acta Neurologica Scandinavica, 62*, 188–190.

Mantere, P., Hänninen, H., Hernberg, S., & Luukkonen, R. (1984). A prospective followup study on psychological effects in workers exposed to low levels of organic solvent mixtures. *Scandinavian Journal of Work Environment and Health, 10*, 43–50.

Marcopulos, B. A., & Graves, R. E. (1990). Antidepressant effect on memory in depressed older persons. *Journal of Clinical and Experimental Neuropsychology, 12*, 655–663.

Marecek, J., Shapiro, I. M., Burke, A., Katz, S. H., & Hediger, M. L. (1983). Low-level lead exposure in childhood influences neuropsychological performance. *Archives of Environmental Health, 38*, 355–359.

Marks, G., & Beatty, W. K. (1975). *The precious metals of medicine.* New York: Charles Scribner's Sons.

Marlowe, M., Stellern, J., Errera, J., & Moon, C. (1985). Main and interaction effects of metal pollutants on visual-motor performance. *Archives of Environmental Health, 40*, 221–224.

Maroni, M., Bulgheroni, C., Cassitto, M. G., Merluzzi, F., Gilioli, R., & Foa, V. (1977). A clinical, neuropsychological and behavioral study of female workers exposed to 1,1,1-trichloroethane. Scandinavian Journal of Work Environment and Health, 3, 16–22.

Martin, P. R., McCool, B. A., & Singleton, C. K. (1994). Genetic sensitivity to thiamine deficiency and development of alcoholic organic brain disease. *Alcoholism: Clinical and Experimental Research, 17*, 31–37.

Martin, R. W., Duffy, J., Engel, A. G., Lie, J. T., Bowles, C. A., Moyer, T. P., & Gleich, G. J. (1990). The clinical spectrum of the eosinophilia–myalgia syndrome associated with L-tryptophan ingestion: Clinical features in 20 patients and aspects of pathophysiology. *Annals of Internal Medicine, 113(2)*, 124–134.

Maskin, A. (1989). Cancerphobia: An emerging theory of compensable damages. *Journal of Occupational Medicine, 31*, 427–431.

Massagli, T. L. (1991). Neurobehavioral effects of phenytoin, carbamazepine, and valproic acid: Implications for use in traumatic brain injury. *Archives of Physical Medicine Rehabilitation, 72*, 219–226.

Massioui, F. E., Lille, F., Lesevre, N., Hazemann, P., Garnier, R., & Dally, S. (1990). Sensory and cognitive event related potentials in workers chronically exposed to solvents. *Clinical Toxicology, 28*, 203–219.

Masson, J. M. (1985). *The complete letters of Sigmund Freud to Wilhelm Fliess, 1987–1904.* Cambridge, MA: The Belknap Press.

Matarazzo, J. D. (1972). *Wechsler's measurement and appraisal of adult intelligence.* Baltimore: MD: Williams & Wilkins.

Matarazzo, J. D. (1985). Clinical psychological test interpretations by computer: Hardware outpaces software. *Computers in Human Behavior, 1*, 235–253.

Matarazzo, J. D. (1990). Psychological assessment versus psychological testing: Validation from Binet to the school, clinic and courtroom. *American Psychologist, 45*, 999–1017.

Mathew, B., Kapp, E., & Jones, R. T. (1989). Commercial butane abuse: A disturbing case. *British Journal of Addiction, 84*, 563–564.

Mathew, R. J., & Wilson, W. H. (1991a). Evaluation of the effects of diazepam and an experimental anti-anxiety drug on regional cerebral blood flow. *Psychiatry Research: Neuroimaging, 40*, 125–134.

Mathew, R. J., & Wilson, W. H. (1991b). Substance abuse and cerebral blood flow. *Archives of General Psychiatry, 148*, 292–305.

Mathew, R. J., Margolin, R. A., & Kessler, R. M. (1985). Cerebral function, blood flow and metabolism: A new vista in psychiatric research. *Integrative Psychiatry, 3*, 214–215.

Mathew, R. J., Wilson, W. H., & Tant, S. R. (1989). Regional cerebral blood flow changes associated with amyl nitrite inhalation. *British Journal of Addiction, 84,* 293-299.

Mathur, A. K., Chandra, S. V., & Tandon, S. K. (1977). Comparative toxicity of trivalent and hexavalent chromium to rabbits. III. Morphological changes in some organs. *Toxicology, 8,* 53-61.

Matikainen, E., & Juntunen, J. (1985). Examination of the peripheral autonomic nervous system in occupational neurology. In *Neurobehavioral methods in occupational and environmental health* (Document 3, pp. 57-60). Copenhagen: World Health Organization.

Matikainen, E., Forsman-Grönholm, L., Pfäffli, P., & Juntunen, J. (1993). Nervous system effects of occupational exposure to styrene: A clinical and neurophysiological study. *Environmental Research, 61,* 84-92.

Matsumoto, T., Fukaya, Y., Yoshitomi, S., Arafuka, M., Kubo, N., & Ohno, Y. (1993). Relations between lead exposure and peripheral neuromuscular functions of lead-exposed workers— Results of tapping test. *Environmental Research, 61,* 299-307.

Matsuo, F., Cummins, J. W., & Anderson, R. E. (1979). Neurological sequelae of massive hydrogen sulfide inhalation (letter). *Archives of Neurology, 36(7),* 451-452.

Matsushita, T., Goshima, E., Miyakki, H., Maeda, K., Takeuchi, Y., & Inoue, T. (1969). Experimental studies for determining the MAC value of acetone. 2. Biological reactions in the "six-day exposure" to acetone. *Sangyo Igaku, 11,* 507-515.

Maurissen, J. P. J., & Weiss, B. (1980). Vibration sensitivity as an index of somatosensory function. In P. S. Spencer & H. H. Schaumburg (Eds.), *Experimental and clinical neurotoxicology* (pp. 767-774). Baltimore: Williams & Wilkins.

Mayeno, A. N., Lin, F., Foote, C. S., Loegering, D. A., Ames, M. M., Hedberg, C. W., & Gleich, G. J. (1990). Characterization of "peak E," a novel amino acid associated with eosinophiliamyalgia syndrome. *Science, 250(4988),* 1707-1708.

Mayersdorf, A., & Israeli, A. (1974). Toxic effects of chlorinated hydrocarbon insecticides on the human electroencephalogram. *Archives of Environmental Health, 28(3),* 159-163.

McCaffrey, R. J., Ortega, A., Orsillo, S. M., Haase, R. F., & McCoy, G. C. (1992). Neuropsychological and physical side effects of metoprolol in essential hypertensives. *Neuropsychology, 6,* 225-238.

McCann, U. D., & Ricaurte, G. A. (1993). Reinforcing subjective effects of (+/-) 3,4-methylenedioxymethamphetamine ("ecstasy") may be separable from its neurotoxic actions: Clinical evidence. *Journal of Clinical Psychopharmacology, 13(3),* 214-217.

McCarron, M. M., Schultze, B. W., Thompson, G. A., Conder, M. C., & Goetz, W. A. (1981). Acute phencyclidine intoxication: Clinical patterns, complications, and treatment. *Annals of Emergency Medicine, 10(6),* 290-297.

McCrady, B. S., & Smith, D. E. (1986). Implications of cognitive impairment for the treatment of alcoholism. *Alcoholism: Clinical and Experimental Research, 10,* 145-149.

McDowell, D. M., & Kleber, H. B. (1994). MDMA: Its history and pharmacology. *Psychiatric Annals, 24,* 127-130.

McEvoy, J. P., & Freter, S. (1989). The dose-response relationship for memory impairment by anticholinergic drugs. *Biological Psychiatry, 30,* 135-138.

McGlaughlin, A. I. G., Kazantzis, G., King, E., Teare, D., Porter, R. J., & Owens, R. (1962). Pulmonary fibrosis and encephalopathy associated with the inhalation of aluminum dust. *British Journal of Industrial Medicine, 19,* 253.

McGlothlin, W., Arnold, D., & Freedman, D. (1969). Organicity measures following repeated LSD ingestion. *Archives of General Psychiatry, 21,* 704-709.

McGovern, J. J. Jr., Lazaroni, J. A., Hicks, M. F., Adler, J. C., & Cleary, P. (1983). Food and chemical sensitivity: Clinical and immunologic correlates. *Archives of Otolaryngology, 109(5),* 292-297.

McKhann, C. F. (1926). Lead poisoning in children: Therapy. *American Journal of Diseases of Children, 32,* 386-392.

McLachlan, D. R. C., & De Boni, U. (1980). Aluminum in human brain disease—An overview. *NeuroToxicology, 1,* 316.

McNair, D. M., Lorr, M., & Droppleman, L. F. (1971). *EITS manual—Profile of mood states.* San Diego: Educational and Industrial Testing Service.

McNally, W. D. (1925). Two deaths from the administration of barium salts. *Journal of the American Medical Association, 84,* 1805-1807.

Mearns, J., Dunn, J., & Lees-Haley, P. R. (1994). Psychological effects of organophosphate pesticides: A review and call for research by psychologists. *Journal of Clinical Psychology, 50*, 286–294.
Meibach, R. C., Glick, S. D., Cox, R., & Maayani, S. (1979). Localization of phencyclidine-induced changes in brain energy metabolism. *Nature, 282*, 625–626.
Melamed, J. I., & Bleiberg, J. (1986). *Neuropsychological deficits in freebase cocaine abusers after cessation of use.* Paper presented at the 93rd annual convention of the American Psychological Association, Washington, DC, August 23.
Melgaard, B., Arlien-Søberg, P., & Brulin, P. *Chronic toxic encephalopathy in styrene exposed workers.* Unpublished manuscript, Department of Neurology, Rigshospitalet, Copenhagen, Denmark
Melgaard, B., Danielsen, U. T., Sorensen, H., & Ahlgren, P. (1986). The severity of alcoholism and its relation to intellectual impairment and cerebral atrophy. *British Journal of Addiction, 81*, 77–80.
Melges, F. T., Tinklenberg, J. R., Hollister, L. E., & Gillepsie, H. K. (1970a). Marihuana and temporal disintegration. *Science, 168*, 1118–1120.
Melges, F. T., Tinklenberg, J. R., Hollister, L. E., & Gillespie, H. K. (1970b). Temporal disintegration and depersonalization during marihuana intoxication. *Archives of General Psychiatry, 23*, 204–210.
Mendhiratta, S. S., Wig, N. N., & Verma, S. K. (1978). Some psychological correlates of long-term heavy cannabis users. *British Journal of Psychiatry, 132*, 482–486.
Mendhiratta, S. S., Varma, V. K., Dang, R., Malhotra, A. K., Das, K., & Nehra, R. (1988). Cannabis and cognitive functions: A re-evaluation study. *British Journal of Addiction, 83*, 749–753.
Mennear, J. H. (1993). Carbon monoxide and cardiovascular disease: An analysis of the weight of evidence. *Regulatory Toxicology and Pharmacology, 17*, 77–84.
Menzel, D. B. (1984). Ozone: An overview of its toxicity in man and animals. *Journal of Toxicology and Environmental Health, 13*, 183–204.
Mergler, D., & Blain, L. (1987). Assessing color vision loss among solvent-exposed workers. *American Journal of Industrial Medicine, 12(2)*, 195–203.
Mergler, D., Belanger, S., De Grosbois, S., & Vachon, N. (1988a). Chromal focus of acquired chromatic discrimination loss and solvent exposure among printshop workers. *Toxicology, 49(2–3)*, 341–348.
Mergler, D., Blain, L., Lemaire, J., & Lalande, P. (1988b). Colour vision impairment and alcohol consumption. *Neurotoxicology and Teratology, 10*, 255–260.
Mergler, D., Bowler, R., & Cone, J. (1990). Color vision loss among disabled workers with neuropsychological impairment. *Neurotoxicology and Teratology, 12*, 669–672.
Mergler, D., Huel, G., Bowler, R., Frenette, B., & Cone, J. (1991). Visual dysfunction among former microelectronics assembly workers. *Archives of Environmental Health, 46(6)*, 326–334.
Mergler, D., Huel, G., Bowler, R., Iregren, A., Bélanger, S., Baldwin, M., Tardif, R., Smargiassi, A., & Martin L. (1994). Nervous system dysfunction among workers with long-term exposure to manganese. *Environmental Research, 64*, 151–180.
Merliss, R. R. (1972). Phenol marasmus. *Journal of Occupational Medicine, 14(1)*, 55–56.
Merritt, J. H., Schultz, E. J., & Wykes, A. A. (1964). Effect of decaborane on the norepinephrine content of the rat brain. *Biochemistry and Pharmacology, 13*, 1364–1366.
Messiha, F. S. (1993). Fluoxetine: Adverse effects and drug–drug interactions. *Clinical Toxicology, 31*, 603–630.
Messite, J., & Bond, M. B. (1980). Occupational health considerations for women at work. In C. Zenz (Ed.), *Developments in occupational medicine* (pp. 43–57). Chicago: Year Book Medical Publishers.
Metcalf, D. R., & Holmes, J. H. (1969). EEG, psychological and neurological alteration in humans with organophosphorous exposure. *Annals of the New York Academy of Science, 160*, 357.
Meulenbelt, J., deGroot, G., & Savelkoul, T. J. F. (1990). Two cases of acute toluene intoxication. *British Journal of Industrial Medicine, 47*, 417–420.
Michaelis, E. K. (1990). Fetal alcohol exposure: Cellular toxicity and molecular events involved in toxicity. *Alcoholism: Clinical and Experimental Research, 14*, 819–826.
Midtling, J. E., Barnett, P. G., Coye, M. G., Velasco, A. R., Romero, P., Clements, C. L., O'Malley,

M. A., Tobin, M. W., Rose, T. G., & Monosson, I. H. (1985). Clinical management of field worker organophosphate poisoning. *Western Journal of Medicine, 142,* 514-518.

Mihevic, P. M., Gliner, J. A., & Horvath, S. M. (1983). Carbon monoxide exposure and information processing during perceptual-motor performance. *International Archives of Occupational and Environmental Health, 51,* 355-363.

Mikkelsen, S. (1980). A cohort study of disability pension and death among painters with special regard to disabling presenile dementia as an occupational disease. *Scandinavian Journal of Social Medicine,* Suppl. 16, 34-43.

Mikkelsen, S. (1986). Minority view on the classification of clinical cases. *NeuroToxicology, 7(4),* 55.

Mikkelsen, S., Browne, E., Jorgensen, M., & Gyldensted, C. (1985). Association of symptoms of dementia with neuropsychological diagnosis of dementia and cerebral atrophy. In *Chronic effects of organic solvents on the central nervous system and diagnostic criteria* (pp. 166-184). Copenhagen: World Health Organization.

Mikkelsen, S., Jørgensen, M., Brown, E., & Gyldensted, C. (1988). Mixed solvent exposure and organic brain damage. A study of painters. *Acta Neurologica Scandinavica, 78*(Suppl. 118), 1-143.

Mikuriya, T. H., & Aldrich, M. R. (1988). Cannabis 1988: Old drug, new dangers, the potency question. *Journal of Psychoactive Drugs, 20,* 47-55.

Milanovic, L., Spilich, G., Vucinic, G., Knezevic, S., Ribaric, B., & Mubrin, Z. (1990). Effects of occupational exposure to organic solvents upon cognitive performance. *Neurotoxicology and Teratology, 12(6),* 657-660.

Miller, J. M., Chaffin, D. B., & Smith, R. G. (1975). Subclinical psychomotor and neuromuscular changes in workers exposed to inorganic mercury. *American Industrial Hygiene Association Journal, 36,* 725-733.

Miller, L. (1985). Neuropsychological assessment of substance abusers: Review and recommendations. *Journal of Substance Abuse Treatment, 2,* 517.

Miller, L. L., & Branconnier, R. J. (1983). Cannabis: Effects on memory and the cholinergic limbic system. *Psychological Bulletin, 93,* 441-456.

Miller, L., Drew, W. G., & Kiplinger, G. C. (1976). Effects of marijuana on recall of narrative material and Stroop colorword performance. In E. L. Abel (Ed.), *The scientific study of marijuana* (pp. 117-120). Chicago: Nelson-Hall.

Miller, L. L., Cornett, T. L., Brightwell, D. R., McFarland, D. J., Drew, W. G., & Wikler, A. (1977). Marijuana: Effects on storage and retrieval of prose material. *Psychopharmacology, 51,* 311-316.

Miller, L. L., Cornett, T. L., & Nallan, G. (1978). Marijuana: Effect on non-verbal free recall as a function of field dependence. *Psychopharmacology, 58,* 297-301.

Miller, L. L., Cotnett, T. L., & Wikler, A. (1979). Marijuana: Effects on pulse rate, subjective estimates of intoxication and multiple measures of memory. *Life Sciences, 25,* 1325-1330.

Miller, N. S., & Gold, M. S. (1990). Benzodiazepines: Tolerance, dependence, abuse and addiction. *Journal of Psychoactive Drugs, 22,* 23-33.

Miller, N. S., & Gold, M. S. (1994). LSD and Ecstasy: Pharmacology, phenomenology and treatment. *Psychiatric Annals, 24,* 131-133.

Min, S. K. (1986). A brain syndrome associated with delayed neuropsychiatric sequelae following acute carbon monoxide intoxication. *Acta Psychiatrica Scandinavica, 73,* 80-86.

Minocha, A., Barth, J. T., Roberson, D. G., Herold, D. A., & Spyker, D. A. (1985). Impairment of cognitive and psychomotor function by ethanol in social drinkers. *Veterinary and Human Toxicology, 27,* 533-536.

Misra, U. K., Nag, D., & Murti, C. R. (1984). A study of cognitive functions in DDT sprayers. *Industrial Health, 22(3),* 199-206.

Mitchell, M. C. (1985). Alcohol-induced impairment of central nervous system function: Behavioral skills involved in driving. *Journal of Studies on Alcohol, 10,* 109-116.

Mittenberg, W., & Motta, S. (1993). Effects of chronic cocaine abuse on memory and learning. *Archives of Clinical Neuropsychology, 8,* 477-484.

Modestin, J., Toffler, G., & Drescher, J. P. (1992). Neuroleptic malignant syndrome: Results of a prospective study. *Psychiatry Research, 44,* 251-256.

Moen, B. E., Kyvik, K. R., Engelsen, B. A., & Riise, T. (1990). Cerebrospinal fluid proteins and

free amino acids in patients with solvent induced chronic toxic encephalopathy and healthy controls. *British Journal of Industrial Medicine, 47,* 277-280.
Money, J., Alexander, D., & Walker, H. T. (1965). *Manual for a standardized road-map test of direction sense.* Baltimore: Johns Hopkins University Press.
Montoya-Cabrera, M. A. (1990). Neurotoxicology in Mexico and its relation to the general and work environment. In B. L. Johnson (Ed.), *Advances in neurobehavioral toxicology: Applications in environmental and occupational health* (pp. 35-57). Chelsea, MI: Lewis Publishers.
Moodley, P., Golombok, S., Shine, P., & Lader, M. (1993). Computed axial brain tomograms in long-term benzodiazepine users. *Psychiatry Research, 48,* 135-144.
Mooser, S. B. (1987). The epidemiology of multiple chemical sensitivities (MCS) (review). (1987). *Occupational Medicine: State of the Art Reviews, 2(4),* 663-668.
Moreau, A., Jones, B. D., & Banno, V. (1986). Chronic central anticholinergic toxicity in manic depressive illness mimicking dementia. *Canadian Journal of Psychiatry, 31(4),* 339-341.
Morgan, B. B., Jr., & Repko, J. D. (1974). Evaluation of behavioral functions in workers exposed to lead. In C. Xintaras, B. L. Johnson, & I. deGroot (Eds.), *Behavioral toxicology: Early detection of occupational hazards.* (HEW Publication No. (NIOSH) 74-126, pp. 248-265). Washington, DC: U.S. Government Printing Office.
Morgan, D. P. (1982). Pesticide toxicology. In A. T. Yu (Ed.), *Survey of contemporary toxicology* (p. 136). New York: Wiley.
Morgan, D. P. (1989). *Recognition and management of pesticide poisoning.* Washington, DC: U.S. Government Printing Office, No. 540/9-88-001.
Morley, R., Eccleston, D. W., Douglas, C. P., Greville, W. E. J., Scott, D. J., & Anderson, J. (1970). Xylene-poisoning—A report on one fatal case and two cases of recovery after prolonged unconsciousness. *British Medical Journal, 3,* 442-443.
Morris, R. A., & Sonderegger, T. B. (1984). Legal applications and implications for neurotoxin research of the developing organism. *Neurobehavioral Toxicology and Teratology, 6,* 303-306.
Morrow, L. A. (1992). Sick building syndrome and related workplace disorders. *Otolaryngology— Head and Neck Surgery, 106,* 649-654.
Morrow, L. A., Ryan, C. M., Goldstein, G., & Hodgson, M. J. (1989). A distinct pattern of personality disturbance following exposure to mixtures of organic solvents. *Journal of Occupational Medicine, 31,* 743-746.
Morrow, L. A., Callender, T., Lottenberg, S., Buchsbaum, M. S., Hodgson, M. J., & Robin, N. (1990). PET and neurobehavioral evidence of tetrabromoethane encephalopathy. *Journal of Neuropsychiatry and Clinical Neurosciences, 2,* 431-435.
Morrow, L. A., Ryan, C. M., Hodgson, M. J., & Robin, N. (1990). Alterations in cognitive and psychological functioning after organic solvent exposure. *Journal of Occupational Medicine, 32,* 444-450.
Morrow, L. A., Ryan, C. M., Hodgson, M. J., & Robin, N. (1991). Risk factors associated with persistence of neuropsychological deficits in persons with organic solvent exposure. *The Journal of Nervous and Mental Disease, 179,* 540-545.
Morrow, L. A., Robin, M. N., Hodgson, M. J., & Kamis, H. (1992). Assessment of attention and memory efficiency in persons with solvent neurotoxicity. *Neuropsychologia, 30,* 911-922.
Morrow, L. A., Steinhauer, S. R., & Hodgson, M. J. (1992). Delay in P300 latency in patients with organic solvent exposure. *Archives of Neurology, 49,* 315-320.
Morrow, L. A., Kamis, H., & Hodgson, M. J. (1993). Psychiatric symptomatology in persons with organic solvent exposure. *Journal of Consulting and Clinical Psychology, 61,* unpaged preprint.
Morse, D. L., & Baker, E. L., Jr., Kimbrough, R. D., & Wisseman, C. L. (1979). Propanil-chloracne and methomyl toxicity in workers of a pesticides manufacturing plant. *Clinical Toxicology, 15(1),* 13-21.
Morton, W. (1945). Poisoning by barium carbonate. *Lancet, 2,* 738-739.
Morton, W. E., & Caron, G. A. (1989). Encephalopathy: An uncommon manifestation of workplace arsenic poisoning? *American Journal of Industrial Medicine, 15,* 1-15.
Moses, M. (1983). Pesticides. In W. Rom (Ed.), *Environmental and occupational medicine* (pp. 547-571). Boston: Little, Brown.

Muhlendahl, J. E. (1990). Intoxication from mercury spilled on carpets. *Lancet, 336,* 1578.
Muijser, H. Juntunen, J., Matikainen, E., & Seppäläinen, A. M. (1985). Neurophysiological tests of the peripheral nervous system (PNS). In *Neurobehavioral methods in occupational and environmental health* (Document 6, pp. 23–25). Copenhagen: World Health Organization.
Muldoon, S. R., & Hodgson, M. J. (1992). Risk factors for nonoccupational organophosphate pesticide poisoning. *Journal of Occupational Medicine, 34(1),* 38–41.
Muller, K. E., Barton, C. N., & Benignus, V. A. (1984). Recommendations for appropriate statistical practice in toxicologic experiments. *NeuroToxicology, 5,* 113–126.
Mulve, S. J. (1984). The pharmacodynamics of cocaine abuse. *Psychiatric Annals, 14,* 724–727.
Murata, K., Araki, S., Kawakami, N., Saito, Y., & Hino, E. (1991). Ccentral nervous system effects and visual fatigue in VDT workers. *International Archives of Occupational and Environmental Health, 63,* 109–113.
Murata, K., Araki, S., & Yokoyama, K. (1991). Assessment of the peripheral, central and autonomic nervous system function in styrene workers. *American Journal of Industrial Medicine, 20,* 775–784.
Murata, K., Araki, S., Yokoyama, K., Uchida, E., & Fujimura, Y. (1993). Assessment of central, peripheral, and autonomic nervous system functions in lead workers: Neuroelectrophysiological studies. *Environmental Research, 61,* 323–336.
Murphy, L. R., & Colligan, M. J. (1979). Mass psychogenic illness in a shoe factory. A case report. *International Archives of Occupational and Environmental Health, 44,* 133–138.
Murray, J. B. (1986). Marijuana's effects on human cognitive functions, psychomotor functions, and personality. *Journal of General Psychology, 113,* 23–55.
Mushak, P. (1992). Defining lead as the premiere environmental health issue for children in America: Criteria and their quantitative application. *Environmental Research, 59,* 281–309.
Mutti, A., & Franchini, I. (1987). Toxicity of metabolites to dopaminergic systems and the behavioural effects of organic solvents. *British Journal of Industrial Medicine, 44(11),* 721–723.
Mutti, A., Mazzucchi, A., Frigeri, G., Falzoi, M., Arfini, G., & Franchini, I. (1983). Neuropsychological investigation on styrene exposed workers. In A. Gilioli, M. A. Cassitto, & V. Foa (Eds.), *Neurobehavioral methods in occupational health* (pp. 271–281). New York: Pergamon Press.
Mutti, A., Mazzucchi, A., Rustichelli, P., Frigeri, G., Arfini, G., & Franchini, I. (1984). Exposure–effect and exposure–response relationships between occupational exposure to styrene and neuropsychological functions. *American Journal of Industrial Medicine, 5,* 275–286.
Mutti, A., Falzoi, M., Romanelli, A., & Franchini, I. (1985). Regional alterations of brain catecholamines by styrene exposure in rabbits. *Archives of Toxicology, 55,* 173–177.
Mutti, A., Falzoi, M., Romanelli, A., Bocchi, M. C., Ferroni, C., & Franchini, I. (1988). Brain dopamine as a target for solvent toxicity: Effects of some monocyclic aromatic hydrocarbons. *Toxicology, 49(1),* 77–82.
Mutti, A., Folli, D., der Venne, V., Berlin, A., Gerra, G., Caccavari, R., Vescovi, P. P., & Franchini, I. (1992). Long-lasting impairment of neuroendocrine response to psychological stress in heroin addicts. *NeuroToxicology, 13,* 255–260.
Myers, R. A. M., Mitchell, J. T., & Cowley, R. A. (1983). Psychometric testing and carbon monoxide poisoning. *Disaster Medicine, 1,* 279–283.
Myint, R. T. (1990). *Occurrence of abnormal magnetic resonance imaging of brain among employees exposed to mixed solvents, heavy metal and pesticides.* Paper presented at the Eighth International Neurotoxicology Conference, Little Rock, AR, October 1–4.
Naaki, K. (1969). An experimental study on the effect of exposure to organic vapor in human subjects. *Journal of Science and Labour, 50,* 89–96.
Nakajima, H. (1993). An international perspective in neurobehavioral toxicology. *Environmental Research, 63,* 47–53.
Nakatsuka, H., Watanabe, T., Takeuchi, Y., Hisanaga, N., Shibata, E., Suzuki, H., Huang, M.-Y., Chen, Z., Qu, Q.-S., & Ikeda, M. (1992). Absence of blue–yellow color vision loss among workers exposed to toluene or tetrachloroethylene, mostly at levels below occupational exposure limits. *International Archives of Occupational and Environmental Health, 64,* 113–117.
Namba, T. (1971). Cholinesterase inhibition by organophosphorus compounds and its clinical effects. *Bulletin of the World Health Organization, 44,* 289–307.

REFERENCES

Nanson, J. L., & Hiscock, M. (1990). Attention deficits in children exposed to alcohol prenatally. *Alcoholism: Clinical and Experimental Research, 14,* 656–661.
Nasr, A. N. M. (1971). Ozone poisoning in man: Clinical manifestations and differential diagnosis—A review. *Clinical Toxicology, 4,* 461–466.
National Academy of Sciences. (1973). *Medical and biological effects of environmental pollutants: Manganese.* Washington, DC: National Academy Press.
National Academy of Sciences. (1989). *Research on children and adolescents with mental, behavioral, and developmental disorders.* Washington, DC: National Academy Press.
National Association for Perinatal Addiction Research and Education. (1988, August 28). News Release.
National Institute for Occupational Safety and Health. (1973). *Criteria for a recommended standard: Occupational exposure to toluene* (HEW Publication No. (NIOSH) 73-11023). Washington, DC: U.S. Government Printing Office.
National Institute for Occupational Safety and Health. (1975a). *Criteria for a recommended standard: Occupational exposure to arsenic, new criteria* (HEW Publication No. (NIOSH) 75-149). Washington, DC: U.S. Government Printing Office.
National Institute for Occupational Safety and Health. (1975b). *Criteria for a recommended standard: Occupational exposure to xylene* (HEW Publication No. (NIOSH) 75-168). Washington, DC: U.S. Government Printing Office.
National Institute for Occupational Safety and Health. (1977). *Behavioral and neurological effects of methyl chloride* (HEW Publication No. (NIOSH) 77-125). Washington, DC: U.S. Government Printing Office.
National Institute for Occupational Safety and Health. (1978). *Special occupational hazard review of trichloroethylene* (HEW Publication No. (NIOSH) 78-130).
National Institute on Drug Abuse (1988). *National household survey on drug abuse: Population estimates, 1988* Washington, DC: U.S. Department of Health and Human Services.
Needleman, H. L. (1982). The neurobehavioral consequences of low lead exposure in childhood. *Neurobehavioral Toxicology and Teratology, 4,* 729–732.
Needleman, H. L. (1983). Lead at low dose and the behavior of children. *Acta Psychiatrica Scandinavica, 67*(Suppl. 303), 26–37.
Needleman, H. L. (1990). Lessons from the history of childhood plumbism for pediatric neurotoxicology. In B. L. Johnson (Ed.), *Advances in neurobehavioral toxicology: Applications in environmental and occupational health* (pp. 331–338). Chelsea, MI: Lewis Publishers.
Needleman, H. L. (1993). The current status of childhood low-level lead toxicity. *NeuroToxicology, 14,* 161–166.
Needleman, H. L., Gunnoe, C., Leviton, A., Reed, R., Peresie, H., Maher, C., & Barrett, P. (1979). Deficits in psychologic and classroom performance of children with elevated dentine lead levels. *New England Journal of Medicine, 300,* 689–695.
Needleman, H. L., Schell, A., Bellinger, D., Leviton, A., & Allred, E. (1990). The long-term effects of exposure to low doses of lead in childhood. *The New England Journal of Medicine, 322,* 83–88.
Negrotti, A., Calzetti, S., & Sasso, E. (1992). Calcium-entry blockers-induced parkinsonism: Possible role of inherited susceptibility. *Neurotoxicology, 13,* 261–264.
Neil-Dwyer, G., Bartlett, J., McAinsh, J., & Cruickshank, J. M. (1981). Beta-adrenoceptor blockers and the blood-brain barrier. *British Journal of Clinical Pharmacology, 11*(9), 549–553.
Nell, V., Myers, J., Colvin, M., & Rees, D. (1993). Neuropsychological assessment of organic solvent effects in South Africa: Test selection, adaptation, scoring, and validation issues. *Environmental Research, 63,* 301–318.
Nelson, B. K. (1990). Origins of behavioral teratology and distinctions between research on pharmaceutical agents and environmental/industrial chemicals. *Neurotoxicology and Teratology, 12,* 301–305.
Neudorfer, N., & Wolpert, E. (1976). [Neuropsychiatric disturbances after phenol intoxication]. *Muenchener Medizinische Wochenschrift, 118,* 1177–1178.
Neurobehavioral methods in occupational and environmental health. (1985a). Document 3. Copenhagen: World Health Organization, Regional Office for Europe.

Neurobehavioral methods in occupational and environmental health. (1985b). Document 6. Copenhagen: World Health Organization, Regional Office for Europe.

Neurotoxicity. Identifying and controlling poisons of the nervous system. (1990). Office of Technology Assessment. OTA-BA-436, GPO No. 052-003-01184-1.

Neuspiel, D. R., Hamel, S. C., Hochberg, E., Greene, J., & Campbell, D. (1991). Maternal cocaine use and infant behavior. *Neurotoxicology and Teratology, 13,* 229–233.

Newman, P. J., & Sweet, J. J. (1986). The effects of clinical depression on the Luria–Nebraska neuropsychological battery. *Clinical Neuropsychology, 8,* 109–114.

Ng, T. P., Ong, S. G., Lam, W. K., & Jones, G. M. (1990). Neurobehavioral effects of industrial mixed solvent exposure in Chinese printing and paint workers. *Neurotoxicology and Teratology, 12,* 661–664.

Ng, T. P., Foo, S. C., & Young, T. (1992). Risk of spontaneous abortion in workers exposed to toluene. *British Journal of Industrial Medicine, 49,* 804–808.

Ngim, C. H., Foo, S. C., Boey, K. W., & Jeyaratnam, J. (1992). Chronic neurobehavioural effects of elemental mercury in dentists. *British Journal of Industrial Medicine, 49(11),* 782–790.

Nicholi, A. M. (1984). Cocaine use among the college age group: Biological and psychological effects: Clinical and laboratory research findings. *Journal of American College Health, 32,* 258–261.

Nicholson, M. E., Wang, M., Airhihenbuwa, C. O., Mahoney, B. S., Christina, R., & Maney, D. W. (1992). Variability in behavioral impairment involved in the rising and falling BAC curve. *Journal of Studies on Alcohol, 53,* 349–356.

Niklowitz, W. J. (1980). Neurotoxicology of lead. In L. Manzo (ed.), *Advances in neurotoxicology* (pp. 27–34). New York: Pergamon Press.

Niklowitz, W. J., & Mandybur, T. I. (1975). Neurofibrillary changes following childhood lead encephalopathy. *Journal of Neuropathology and Experimental Neurology, 34,* 445–455.

NIOSH. (1978). Current Intelligence Bulletin 20. *Tetrachloroethylene (perchloroethylene).* (NIOSH) Publication 78-112.

NIOSH. (1987). Current Intelligence Bulletin 48. *Organic solvent neurotoxicity.* DHHS (NIOSH) Publication No. 87-104.

NIOSH pocket guide to chemical hazards. (1985). Cincinnati: DHHS (NIOSH) Publication No. 85-114.

Noonberg, A., Goldstein, G., & Page, H. A. (1985). Premature aging in male alcoholics: "Accelerating aging" or "increased vulnerability?" *Alcoholism: Clinical and Experimental Research, 9,* 334–338.

Norback, D., & Edling, C. (1991). Environmental, occupational and personal factors related to the prevalence of sick building syndrome in the general population. *Scandinavian Journal of Work Environment and Health, 48,* 451–462.

Norback, D., & Widstron, M. I. (1990). Indoor air quality and personal factors related to the sick building syndrome. *Scandinavian Journal of Work Environment and Health, 16,* 121–128.

Norback, D., Torgen, M., & Edling, C. (1990). Volatile organic compounds, respirable dust, and personal factors related to prevalence and incidence of sick building syndrome in primary schools. *British Journal of Industrial Medicine, 47,* 733–741.

Nordberg, G. F. (1976). Effects and dose–response levels of toxic metals. *Scandinavian Journal of Work Environment and Health, 2,* 37–43.

Nordberg, G. F. (1979). Neurotoxic effect of metals and their compounds. In L. Manzo (Ed.), *Advances in neurotoxicology* (p. 315). New York: Pergamon Press.

Nordin, C., Rosenqvist, M., & Hollstedt, C. (1988). Sniffing of ethyl chloride—an uncommon form of abuse with serious mental neurological symptoms. *International Journal of Addictions, 23,* 623–627.

Norton, S. (1986). Toxic responses of the central nervous system. In C. D. Klaassen, M. O. Amdur, & J. Doull (Eds.), *Toxicology: The basic science of poisons* (pp. 359–386). New York: Macmillan Co.

Noveske, F. G., Hahn, K. R., & Flynn, R. J. (1989). Possible toxicity of combined fluoxetine and lithium. *American Journal of Psychiatry, 146(11),* 1515.

Nunez, C. M., Klitzman, S., & Goodman, A. (1993). Lead exposure among automobile repair workers and their children in New York City. *American Journal of Industrial Medicine, 23,* 763–777.

Nyfors, A. (1978). Benefits and adverse drug experiences during long-term methotrexate treatment of 248 psoriatics. *Danish Medical Bulletin, 25,* 208–226.

REFERENCES

Nylander, M., Frieberg, L., & Lind, B. (1987). Mercury concentrations in the human brain and kidneys in relation to exposure from dental amalgam fillings. *Swedish Dental Journal, 11,* 179–187.

Oades, R. D. (1987). Attention deficit disorder with hyperactivity (ADDH): The contribution of catecholaminergic activity. *Progress in Neurobiology, 29(4),* 365–391.

Oberg, R. C. E., Udeson, H., Thomsen, A. M., Gade, A., & Mortensen, E. L. (1985). Psychogenic behavioral impairments in patients exposed to neurotoxins. Neuropsychological assessments in differential diagnosis. In *Neurobehavioral methods in occupational and environmental health* (Document 3, pp. 130–135). Copenhagen: World Health Organization.

O'Callaghan, J. P. (1989). Neurotypic and gliotypic proteins as biochemical markers of neurotoxicity. *Neurotoxicology and Teratology, 10,* 445–452.

Occupational exposure to styrene. (1983). DHHS (NIOSH) Publication No. 83-119.

O'Donoghue, J. L. (1985a). Carbon monoxide, inorganic nitrogenous compounds and phosphorus. In J. L. O'Donoghue (Ed.), *Neurotoxicity of industrial and commercial chemicals* (Vol. I, pp. 193–203). Boca Raton: CRC Press.

O'Donoghue, J. L. (1985b). Aromatic hydrocarbons. In J. L. O'Donoghue (Ed.), *Neurotoxicity of industrial and commercial chemicals* (Vol. II, pp. 127–136). Boca Raton: CRC Press.

O'Donoghue, J. L. (1986). *Neurotoxicity risk assessment in industry.* Paper presented at the American Association for the Advancement of Science symposium on Evaluating the Neurotoxic Risk Posed by Chemicals in the Workplace and Environment, May 26.

O'Hanlon, J. F., & Volkerts, E. R. (1986). Hypnotics and actual driving performance. *Acta Psychiatrica Scandinavica, 74*(Suppl. 332), 95–104.

Ohnishi, A., & Murai, Y. (1993). Polyneuropathy due to ethylene oxide, propylene oxide and butylene oxide. *Environmental Research, 60,* 242–247.

Ohnishi, A., & Yoshiyuki, M. (1993). Polyneuropathy due to ethylene oxide, propylene oxide and butylene oxide. *Environmental Research, 60,* 242–247.

Oliver, T. (1901). *Dangerous trades.* London: Murray.

Olney, J. W. (1992). Excitotoxins in foods. *Neurotoxicology, 15,* 535–544.

Olney, J. W., Labruyere, J., & Price, M. T. (1989). Pathological changes induced in cerebrocortical neurons by phencyclidine and related drugs. *Science, 244(4910),* 1360–1362.

Olsen, J. (1983). Risk of exposure to teratogens amongst laboratory staff and painters. *Danish Medical Bulletin, 30(1),* 24–28.

Olsen, J., & Sabroe, S. (1980). A case-referent study of neuropsychiatric disorders among workers exposed to solvents in the Danish wood and furniture industry. *Scandinavian Journal of Social Medicine,* Suppl. 16, 44–49.

Olson, B. A., Gamberale, F., & Iregren, A. (1985). Co-exposure to toluene and p-xylene in man: Central nervous functions. *British Journal of Industrial Medicine, 42,* 117–122.

Olson, B. B. (1982). Effects of organic solvents on behavioral performance of workers in the paint industry. *Neurobehavioral Toxicology and Teratology, 4,* 703–708.

O'Mahony, J. F., & Doherty, B. (1993). Patterns of intellectual performance among recently abstinent alcohol abusers on WAIS-R and WMS-R sub-tests. *Archives of Clinical Neuropsychology, 8,* 373–380.

O'Malley, P. M., Johnston, L. D., & Bachman, J. G. (1985). Cocaine use among American adolescents and young adults. In *Cocaine use in America: Epidemiologic and clinical perspectives* (NIDA Research Monograph 61). Washington, DC: U.S. Government Printing Office.

O'Malley, S. S., & Gawin, F. H. (1990). Abstinence symptomatology and neuropsychological impairment in chronic cocaine abusers. In J. W. Spencer & J. J. Boren (Eds.), *Residual effects of abused drugs on behavior* (NIDA Research Monograph 101, pp. 179–190). Washington, DC: U.S. Government Printing Office.

O'Malley, S. S., Gawin, F. H., Heaton, R., & Kleber, H. D. (1988). *Cognitive deficits associated with cocaine abuse.* Presented at the Annual Meeting of the Society of Psychologists in Addictive Behaviors, Atlanta, GA, August.

Ørbaek, P., & Nise, G. (1989). Neurasthenic complaints and psychometric function of toluene-exposed rotogravure printers. *American Journal of Industrial Medicine, 16(1),* 67–77.

Ørbaek, P., Risberg, J., Rosen, I., Haeger-Aronson, B., Hagstadius, S., Hjortsberg, U., Regnell, G.,

Rehnstrom, S., Svensson, K., & Welinder, H. (1985). Effects of long-term exposure to solvents in the paint industry. *Scandinavian Journal of Work Environment and Health, 11*(Suppl. 2), 128.

O'Reilly, J. P. (1974). Performance decrements under hyperbaric He-O_2. *Undersea Biomedical Research, 1,* 353–361.

Organic solvent neurotoxicity (1987). Cincinnati, OH: U.S. Dept. of Health and Human Services, Public Health Service, Centers for Disease Control, National Institute for Occupational Safety and Health.

Organic solvents and the central nervous system. (1985). Copenhagen: World Health Organization.

Orlando, P., Perdelli, F., Gallelli, G., Reggiani, E., Cristina, M. L., & Oberto, C. (1994). Increased blood levels in runners training in urban areas. *Archives of Environmental Health, 49,* 200–203.

Oscar-Berman, M., Weinstein, A., & Wysocki, D. (1983). Bimanual tactual discrimination in aging alcoholics. *Alcoholism: Clinical and Experimental Research, 7,* 398–403.

Oscar-Berman, M., Clancy, J., & Weber, D. A. (1993). Discrepancies between IQ and memory scores in alcoholism and aging. *The Clinical Neuropsychologist, 7,* 281–296.

O'Shaughnessy, W. B. (1839). On the preparations of the Indian hemp, or gunjah. *Transactions of the Medical and Physical Society of Bengal, 1838–1840,* 421–461.

Osterreith, P. A. (1944). Le test de copie d'une figure complexe. *Archives de Psychologie, 30,* 206–356.

Otto, D. (1983a). Computer based behavioural test batteries. In N. Cherry & H. A. Waldron (Eds.), *The neuropsychological effects of solvent exposure* (pp. 130–135). Havant, Hampshire: Colt Foundation.

Otto, D. (1983b). Event-related brain potentials: An alternative methodology for neurotoxicological research. In N. Cherry & H. A. Waldron (Eds.), *The neuropsychological effects of Solvent Exposure* (pp. 33–40). Havant, Hampshire: Colt Foundation.

Otto, D. A., & Fox, D. A. (1993). Auditory and visual dysfunction following lead exposure. *NeuroToxicology, 14,* 191–208.

Otto, D. A., & Hudnell, H. K. (1990). Electrophysiological systems for neurotoxicity field testing: Pearl II and Alternatives. In B. L. Johnson (Ed.), *Advances in neurobehavioral toxicology: Applications in environmental and occupational health* (pp. 283–296). Chelsea, MI: Lewis Publishers.

Otto, D. A., & Hudnell, H. K. (1993). The use of visual and chemosensory evoked potentials in environmental and occupational health. *Environmental Research, 62,* 159–171.

Otto, D., Benignus, V., Muller, K., Barton, C., & Mushak, P. (1982). Event-related slow brain potential changes in asymptomatic children with secondary exposure to lead. In R. Gilioli, M. G. Cassitto, & V. Foa (Eds.), *Neurobehavioral methods in occupational health* (pp. 295–300). New York: Pergamon Press.

Otto, D., Benignus, V., Muller, K., Barton, C., Seiple, K., Prah, J., & Schroeder, S. (1982). Effects of low to moderate lead exposure on slow cortical potentials in young children: Two year follow-up study. *Neurobehavioral Toxicology and Teratology, 4,* 733–737.

Otto, D., Robinson, G., Baumann, S., Schroeder, S., Mushak, P., Kleinbaum, D., & Boone, L. (1985). Five-year follow-up study of children with low-to-moderate lead absorption: Electrophysiological evaluation. *Environmental Research, 38(1),* 168–186.

Otto, D., Molhave, L., Rose, G., Hudness, H. K., & House, D. (1990a). Neurobehavioral and sensory irritant effects of controlled exposure to a complex mixture of volatile organic compounds. *Neurotoxicology and Teratology, 12,* 649–652.

Otto, D. A., Svendsgaard, D., Soliman, S., Soffar, A., & Ahmed, N. (1990b). Neurobehavioral assessment of workers exposed to organophosphorus pesticides. In B. L. Johnson (Ed.), *Advances in neurobehavioral toxicology* (pp. 305–322). Chelsea, MI: Lewis Publishers.

Otto, D., Hudnell, H. K., House, D. E., Molhave, L., and Counts, W. (1992a). Exposure of humans to a volatile organic mixture. 1. Behavioral assessment. *Archives of Occupational Health, 47,* 23–30.

Otto, D., Mulhare, L., Rose, G., Hudnell, H. K., & House, D. (1992b). Neurobehavioral and sensory irritant effects of controlled exposure to a complex mixture of volatile organic compounds. *Neurotoxicology and Teratology, 12,* 649–652.

Oxford English Dictionary—Compact Edition. (1971). London: Oxford University Press.

Page, J. B., Fletcher, J., & True, W. R. (1988). Psychosociocultural perspectives on chronic cannabis use: The Costa Rican follow-up. *Journal of Psychoactive Drugs, 20,* 57-64.
Palliyath, S. K., Schwartz, B. D., & Gant, L. (1990). Peripheral nerve functions in chronic alcoholic patients on disulfiram: A six month follow up. *Journal of Neurology, Neurosurgery, and Psychiatry, 53(3),* 227-230.
Pankratz, L. (1988). Malingering on intellectual and neuropsychological measures. In R. Rogers (Ed.), *Clinical assessment of malingering and deception* (pp. 169-192). New York: Guilford Press.
Parker, E. S., & Noble, E. P. (1977). Alcohol consumption and cognitive function in social drinkers. *Journal of Studies on Alcohol, 38,* 1224-1232.
Parsons, O. (1986). Overview of the Halstead-Reiton Battery. In T. Incagnoli, G. Goldstein, & C. J. Goldstein (Eds.), *Clinical applications of neuropsychological test batteries* (pp. 155-192). New York: Plenum Press.
Cognitive functioning in sober social drinkers: A review and critique. *Journal of Studies on Alcohol, 47(2),* 101-114.
Parsons, O. A., & Farr, S. P. (1981). The neuropsychology of drug and alcohol abuse. In S. B. Filskov & T. J. Boll (Eds.), *Handbook of clinical neuropsychology* (pp. 320-365). New York: Wiley.
Parsons, O. A., Tarter, R. E., & Edelberg, R. (1972). Altered motor control in chronic alcoholics. *Journal of Abnormal Psychology, 72,* 308-314.
Parsons, O. A., Sinha, R., & Williams, H. L. (1990). Relationships between neuropsychological test performance and event-related potentials in alcoholic and nonalcoholic samples. *Alcohol: Clinical Experiments Research, 14(5),* 746-755.
Pascual-Leone, A., Dhuna, A., & Anderson, D. C. (1991a). Longterm neurological complications of chronic, habitual cocaine abuse. *NeuroToxicology, 12,* 393-400.
Pascual-Leone, A., Dhuna, A., & Anderson, D. C. (1991b). Cerebral atrophy in habitual cocaine abusers: A planimetric CT study. *Neurology, 41,* 34-38.
Patterson, W. B., Craven, D. E., Schwartz, D. A., Nardell, E. A., Kasmer, J., & Noble, J. (1985). Occupational hazards to hospital personnel. *Annals of Internal Medicine, 102,* 658-680.
Paulson, G. W. (1983). Steroid-sensitive dementia. *American Journal of Psychiatry, 140,* 1031-1033.
Pearlson, G. D., Garbacz, D. J., Tomkins, R. H., Ahn, H. S., Gutterman, D. F., Veroff, A. E., & De Paulo J. R. (1984). Clinical correlates of lateral ventricular enlargement in bipolar affective disorder. *American Journal of Psychiatry, 141,* 253-256.
Pelham, R. W., Marquis, J. K., Kugelmann, K., & Munsat, T. L. (1980). Prolonged ethanol consumption produces persistent alterations of cholinergic function in rat brain. *Alcoholism: Clinical and Experimental Research, 4,* 282-287.
Penney, D. G. (1990). Acute carbon monoxide poisoning: Animal models—a review. *Toxicology, 62,* 123-160.
Pennsylvania Department of Labor and Industry. (1938). Survey of carbon disulphide and hydrogen sulphide hazards in the viscose rayon industry, Bulletin No. 46.
Peoples, S. A., Maddy, K. T., & Thomas, W. (1976). *Occupational health hazards of exposure to 1,3-dichloropropene.* California Department of Agriculture, Publication No. ACF 59241.
Peper, M., Klett, M., Frentzel-Beyme, R., & Heller, W.-D. (1993). Neuropsychological effects of chronic exposure to environmental dioxins and furans. *Environmental Research, 60,* 124-135.
Perbellini, L., Brugnone, F., & Gaffuri, E. (1981). Neurotoxic metabolites of "commercial hexane" in the urine of shoe factory workers. *Clinical Toxicology, 18,* 1377-1385.
Perez-Reyes, M., Hicks, R. E., Bumberry, J., Jeffcoat, A. R., & Cook, C. E. (1988). Interaction between marihuana and ethanol: Effects on psychomotor performance. *Alcoholism: Clinical and Experimental Research, 12,* 268-276.
Perlick, D., Stastny, P., Katz, I., Mayer, M., & Mattis, S. (1986). Memory deficits and anticholinergic levels in chronic schizophrenia. *American Journal of Psychiatry, 143,* 230-232.
Peters, J. E., & Steele, W. J. (1982). Differential effect of chronic ethanol administration on rates of protein synthesis on free and membrane-bound polysomes in vivo in rat liver during dependence development. *Biochemical Pharmacology, 31(11),* 2059-2063.
Peterson, C. D. (1981). Seizures induced by acute loxapine overdose. *American Journal of Psychiatry, 138,* 1089-1091.

Peterson, J. B., Finn, P. R., & Pihl, R. O. (1992). Cognitive dysfunction and the inherited predisposition to alcoholism. *Journal of Studies on Alcohol, 53*, 154–160.
Phelan, D. M., Hagley. S. R., & Guerin, M. D. (1984). Is hypokalemia the cause of paralysis in barium poisoning? *British Medical Journal, 289*, 882.
Pierce, C. H., & Tozer, T. N. (1992). Styrene in adipose tissue of nonoccupationally exposed persons. *Environmental Research, 58*, 230–235.
Piikivi, L., & Tolonen, U. (1989). EEG findings in chlor-alkali workers subjected to low long term exposure to mercury vapour. *British Journal of Industrial Medicine, 46(6)*, 370–375.
Piikivi, L., Hänninen, H., Martelin, T., & Mantere, P. (1984). Psychological performance and long-term exposure to mercury vapors. *Scandinavian Journal of Work Environment and Health, 10*, 35–41.
Pishkin, V., Lovallo, W. R., & Bourne, L. E. (1985). Chronic alcoholism in males: Cognitive deficit as a function of age of onset, age, and duration. *Alcoholism: Clinical and Experimental Research, 9*, 400–406.
Platt, J. E., Campbell, M., Green, W. H., & Grega, D. M. (1984). Cognitive effects of lithium carbonate and haloperidol in treatment-resistant aggressive children. *Archives of General Psychiatry, 41*, 657–662.
Pleasure, D. E., Mishler, K. C., & Engel, W. K. (1969). Axonal transport of proteins in experimental neuropathies. *Science, 166(904)*, 524–525.
Poda, G. A. (1966). Hydrogen sulfide can be handled safely. *Archives of Environmental Health, 12(6)*, 795–800.
Podell, R. N. (1988). Regarding the article by Selner et al. (Correspondence). *Journal of Allergy and Clinical Immunology, 82*, 701–702.
Poklis, A., & Burkett, C. D. (1977). Gasoline sniffing: A review. *Clinical Toxicology, 11*, 35–41.
Politis, M. J., Schaumburg, H. H., & Spencer, P. S. (1980). Neurotoxicity of selected chemicals. In P. S. Spencer & H. H. Schaumberg (Eds.), *Experimental and clinical neurotoxicology* (pp. 613–630). Baltimore: Williams & Wilkins.
Pomes, A., Frustace, S., Cattaino, G., De Grandis, D., Bongiovanni, L. G., Tumolo, S., & Quadu, G. (1986). Local neurotoxicity of cisplatin after intra-arterial chemotherapy. *Acta Neurologica Scandinavica, 73*, 302–303.
Pope, H. G., Jonas, J. M., Hudson, J. I., & Kafka, M. P. (1985). Toxic reactions to the combination of monoamine oxidase inhibitors and tryptophan. *American Journal of Psychiatry, 142*, 491–492.
Poulsen, H., Loft, S., Andersen, J. R., & Andersen, M. (1992). Disulfiram therapy—Adverse drug reactions and interactions. *Acta Psychiatrica Scandinavica, 86*, 59–66.
Prakash, R., Kelwala, S., & Ban, T. A. (1982). Neurotoxicity with combined administration of lithium and a neuroleptic. *Comprehensive Psychiatry, 23*, 567–571.
Prejean, J., & Gouvier, W. D. (1994). Neuropsychological sequelae of chronic recreational gasoline inhalation. *Archives of Clinical Neuropsychology, 9*, 173–174.
Preskorn, S. H., & Jerkovich, G. S. (1990). Central nervous system toxicity of tricyclic antidepressants: Phenomenology, course, risk factors, and role of therapeutic drug monitoring. *Journal of Clinical Psychopharmacology, 10*, 88–95.
Press, R. J. (1983). *The neuropsychological effects of chronic cocaine and opiate use*. Ann Arbor: University Microfilms International.
Prigatano, G. P. (1977). Neuropsychological functioning in recidivist alcoholics treated with disulfiram. *Alcoholism: Clinical and Experimental Research, 1*, 81–86.
Princenthal, R. A., Lowman, R., Zeman, R. K., & Burrell, M. (1983). Ureterosigmoidostomy: The development of tumors, diagnosis, and pitfalls. *American Journal of Roentgenology, 141(1)*, 77–81.
Pryor, G. T., Dickinson, J., Howd, R. A., & Rebert, C. S. (1983). Transient cognitive deficits and high frequency hearing loss in weanling rats exposed to toluene. *Neurobehavioral Toxicology and Teratology, 5*, 53–57.
Pulliainen, V., & Jokelainen. M. (1994). Effects of phenytoin and carbamazepine on cognitive functions in newly diagnosed epileptic patients. *Acta Neurologica Scandinavica, 89*, 81–86.

Purisch, A. D., & Sbordone, R. J. (1986). The Luria-Nebraska neuropsychological battery. In G. Goldstein & R. E. Tarter (Eds.), *Advances in clinical neuropsychology* (Vol. 3, pp. 291-316). New York: Plenum Press.
Putz-Anderson, V., Albright, B. E., Lee, S. T., Johnson, B. L., Chrislip, D. W., Taylor, B. J., Brightwell, W. S., Dickerson, N., Culver, M., Zentmeyer, D., & Smith, P. (1983). A behavioral examination of workers exposed to carbon disulfide. *NeuroToxicology, 4*, 67-78.
Rabinowitz, M., & Needleman, H. (1982). Temporal trends in the lead concentrations of umbilical cord blood. *Science, 216*, 1429-1431.
Rabinowitz, M. B., Wang, J.-D, & Soong, W.-T. (1991). Dentine lead and child intelligence in Taiwan. *Archives of Environmental Health, 46*, 351-360.
Raffi, G. B., & Violante, F. S. (1981). Is Freon 113 neurotoxic? *International Archives of Occupational and Environmental Health, 49(2)*, 125-127.
Ramirez, J. A., Arismendi, G., Cedillo, L., Revvelta, A. M., Topete, L. M., & Silver, A. (1990). Advances in a psychodiagnostic test battery for Mexican workers exposed to toxic substances in the petrochemical and other industries. In B. L. Johnson (Ed.), *Advances in neurobehavioral toxicology* (pp. 97-106). Chelsea, MI: Lewis Publishers.
Ramsey, J. M. (1972). Carbon monoxide, tissue hypoxia and sensory psychomotor response in hypoxemia. *Clinical Science, 42*, 619-625.
Randall, C. L., Ekblad, U., & Anton, R. F. (1990). Perspectives on the pathophysiology of fetal alcohol syndrome. *Alcoholism: Clinical and Experimental Research, 14*, 807-812.
Raneletti, A. (1931). Die beruflichen Schwefelkohlenstoffvergiftungen (in Italian). *Archiv fuer Gewerbepathologie und Gewerbehygiene, 2*, 664-676.
Rasmussen, H., Olsen, J., & Lauritsen, J. (1985). Risk of encephalopathia among retired solvent-exposed workers. *Journal of Occupational Medicine, 27*, 561-566.
Rasmussen, K., Jeppesen, H. J., & Sabroe, S. (1993). Solvent-induced chronic toxic encephalopathy. *American Journal of Industrial Medicine, 23*, 779-792.
Ratner, D., Oren, B., & Vigder, K. (1983). Chronic dietary anticholinesterase poisoning. *Israeli Journal of Medical Science, 19*, 810-814.
Rawat, A. K. (1979). Neurotoxic effects of maternal alcoholism on the developing fetus and newborn. In L. Manzo (Ed.), *Advances in neurotoxicology* (pp. 155-164). New York: Pergamon Press.
Rea, W. J. (1977). Environmentally triggered small vessel vasculitis. *Annals of Allergy, 38*, 245-251.
Realmuto, G., Begleiter, H., Odencrantz, J., & Porjesz, B. (1993). Event-related potential evidence of dysfunction in automatic processing in abstinent alcoholics. *Biological Psychiatry, 33*, 594-601.
Rebeta, J. L., McAllister, D. A., Gange, J. J., Jordan, R. C., & Riley, A. L. (1986). *The relationship between alcohol-related variables and Luria-Nebraska Neuropsychological Battery performance in a cross-sectional V.A. population.* Paper presented at the sixth annual meeting of the National Academy of Neuropsychologists, 27-29 October, Las Vegas.
Reed, R., Grant, I., & Rourke, S. (1992). Long-term abstinent alcoholics have normal memory. *Alcoholism: Clinical and Experimental Research, 16*, 677-683.
Rees, D. C., Francis, E. Z., & Kimmel, C. A. (1990). Scientific and regulatory issues relevant to assessing risk for developmental neurotoxicity: An overview. *Neurotoxicology and Teratology, 12*, 175-181.
Reeves, D. L., Winter, K. P., LaCour, S. J., Raynsford, K. M., Vogel, K., & Grissett, J. D. (1991). *The UTC-PAB/AGARD STRES Battery: User's manual and system documentation* (NAMRL Special Report 91-3). Pensacola, FL: Naval Aerospace Medical Research Laboratory.
Reeves, D., Winter, D., LaCour, S., Raynsford, K., Kay, G., Elsmore, T., & Hegge, F. W. (1992). Automated neuropsychological assessment metrics (ANAM). In *ANAM Documentation* (Vol. 1, Test Administrators Guide). Washington, DC: Defense Technical Information Center.
Reggiani, G., & Bruppacher, R. (1985). Symptoms, signs and findings in humans exposed to PCBs and their derivatives. *Environmental Health Perspectives, 60*, 225-232.
Reich, G., & Berner, W. (1968). Aerial application accidents. *Archives of Environmental Health, 17*, 776.
Reichert, E. R., Yauger, W. L., Rashad, M. N., & Klemmer, H. W. (1977). Diazinon poisoning in eight members of related households. *Clinical Toxicology, 11*, 511.

Reitan, R. M. (1966). Problems and prospects in studying the psychological correlates of brain lesions. *Cortex, 2,* 127–154.

Reitan, R.M., & Wolfson, D. (1985). *The Halstead-Reitan Neuropsychological Battery. Theory and clinical interpretation.* Tuscon, AR: Neuropsychology Press.

Reiter, L. W. (1985). Introductory remarks to workshop on neurotoxicity testing in human populations. *Neurobehavioral Toxicology and Teratology, 7,* 287–288.

Reiter, L. W., & Ruppert, P. H. (1984). Behavioral toxicity of trialkyltin compounds: A review. *NeuroToxicology, 5,* 177–186.

Repko, J. D. (1981). Neurotoxicity of methyl chloride. *Neurobehavioral Toxicology and Teratology, 3,* 425–429.

Repko, J. D., Morgan, B. B., & Nicholson, J. A. (1975). *Behavioral effects of occupational exposure to lead* (HEW Publication No. (NIOSH) 75-164). Washington, DC: U.S. Government Printing Office.

Repko, J. D., Jones, P. D., Garcia, L. S., Scheider, E. J., Roseman, E., & Corum, C. R. (1976). *Final report of the behavioral and neurological evaluation of workers exposed to solvents: Methyl chloride* (HEW Publication No. (NIOSH) 77-125). Washington, DC: U.S. Government Printing Office.

Repko, J. D., Corum, C. R. Jones, P. D., & Garcia, L. S. (1978). *The effects of inorganic lead on behavioral and neurological function. Final report* (HEW Publication No. (NIOSH) 78-128). Washington, DC: U.S. Government Printing Office.

Reuhl, K. R. (1991). Delayed expression of neurotoxicity: The problem of silent damage. *NeuroToxicology, 12,* 341–346.

Reuhl, K. R., & Cranmer, J. M. (1984). Developmental neuropathology of organotin compounds. *NeuroToxicology, 5,* 187–204.

Rey, A. (1959). Sollicitation del la memoire de fixation par des mots et des objets presentes simultanement. *Archives de Psychologie, 37,* 126–139.

Reynolds, E. H., & Trimble, M. R. (1985). Adverse neuropsychiatric effects of anticonvulsant drugs. *Drugs, 29,* 570–581.

Richalet, J.-P., Duval-Arnould, G., Darnaud, B., Keromes, A., & Rutgers, V. (1988). Modification of colour vision in the green/red axis in acute and chronic hypoxia explored with a portable anomaloscope. *Aviation, Space, and Environmental Medicine, 59,* 620–623.

Richardson, E. D., Malloy, P. F., Longabaugh, R., Williams, J., Noel, N., & Beattie, M. C. (1991). Liver function tests and neuropsychologic impairment in substance abusers. *Addictive Behaviors, 16,* 51–55.

Richardson, E. P., & Adams, R. D. (1977). Degenerative diseases of the nervous system. In G. W. Thorn, R. D. Adams, E. Braunwald, K. J. Isselbacher, & R. G. Petersdorf (Eds.), *Harrison's principles of internal medicine* (8th ed., pp. 1919–1934). New York: McGraw-Hill.

Richardson, G. A., & Day, N. L. (1991). Maternal and neonatal effects of moderate cocaine use during pregnancy. *Neurotoxicology and Teratology, 13,* 455–460.

Richelle, M. (1983). New approaches and neglected areas. In G. Zbinden, V. Cuomo, G. Racagni, & B. Weiss (Eds.), *Applications of behavioral pharmacology in toxicology* (pp. 127–140). New York: Raven Press.

Rickles, W. H., Jr., Cohen, M. J., Whitaker, C. A., & McIntyre, K. E. (1973). Marijuana induced state-dependent verbal learning. *Psychopharmacologia, 30,* 349–354.

Riederer, P., & St. Wuketich, S. (1976). Time course of nigrostriatal degeneration in Parkinson's disease. *Journal of Neural Transmission, 38,* 277–301.

Riege, W. H., Tomaszewski, R., Lanto, A., & Metter, E. J. (1984). Age and alcoholism: Independent memory decrements. *Alcoholism: Clinical and Experimental Research, 8,* 42–47.

Riley, J. N., & Walker, D. W. (1978). Morphological alterations in hippocampus after long-term alcohol consumption in mice. *Science, 201,* 646–648.

Risberg, J., & Hagstadius, S. (1983). Effects on the regional cerebral blood flow of long-term exposure to organic solvents. *Acta Psychiatrica Scandinavica, 67*(Suppl. 303), 92–99.

Roach, S. A., & Rappaport, S. M. (1990). But they are not thresholds: A critical analysis of the documentations to threshold limit values. *American Journal of Industrial Medicine, 17,* 727–753.

Roache, J. D., Cherek, D. R., Bennett, R. H., Schenkler, J. C., & Cowan, K. A. (1993). Differential effects of triazolam and ethanol on awareness, memory, and psychomotor performance. *Journal of Clinical Psychopharmacology, 13*, 3-15.

Roberts, L. A., & Bauer, L. O. (1993). Reaction time during cocaine versus alcohol withdrawal: Longitudinal measures of visual and auditory suppression. *Psychiatry Research, 46*, 229-237.

Roberts, R. E., & Lee, E. S. (1993). Occupation and the prevalence of major depression, alcohol, and drug abuse in the United States. *Environmental Research, 61*, 266-278.

Rochford, J., Grant, I., & LaVigne, G. (1977). Medical students and drugs. Neuropsychological and use pattern considerations. *International Journal of the Addictions, 12*, 1057-1065.

Rodepierre, J. J., Truhaut, J., Alizon, J., & Champion, Y. (1955). The possible etiological role of methyl chloride in an obscure syndrome. *Society of Medical, Legal and Criminal Police Science Toxicology, 35*, 80.

Rodier, J. (1955). Manganese poisoning in Moroccan miners. *British Journal of Industrial Medicine, 12*, 21-35.

Rodnitzky, R. L. (1973). Neurological and behavioral aspects of occupational exposure to organophosphate pesticides. In C. Xintaras, B. L. Johnson, & I. deGroot (Eds.), *Behavioral toxicology: Early detection of occupational hazards* (HEW Publication No. (NIOSH) 74-126, pp. 165-173). Washington, DC: U.S. Government Printing Office.

Roeleveld, N., Zielhuis, G. A., & Gabreëls, F. (1990). Occupational exposure and effects of the central nervous system in offspring: Review. *British Journal of Industrial Medicine, 47*, 580-588.

Roels, H., Lauwerys, R., Buchet, J. P., Bernard, A., Barthels, A., Oversteyns, M., & Gaussin, J. (1987). Comparison of renal function and psychomotor performance in workers exposed to elemental mercury. *International Archives of Occupational and Environmental Health, 50*, 77-93.

Roels, H., Lauwerys, R., Buchet, J. P., Genet, P., Sarhan, J. S. M., Hanotiau, I., deFays, M., Bernard, A., & Stanescu, D. (1990). Epidemiological survey among workers exposed to manganese: Effects on lung, central nervous system, and some biological indices. *American Journal of Industrial Medicine, 11*, 307-327.

Rogan, W. J., & Gladen, B. C. (1991). PCB's DDE and child development at 18 and 24 months. *Annals of Epidemiology, 1*, 407-413.

Rogan, W. J., & Gladen, B. C. (1992). Neurotoxicology of PCB's and related compounds. *NeuroToxicology, 13*, 27-36.

Rogan, W. J., Gladen, B. C., McKinney, J. D., Carreras, N., Hardy, P., Thullen, J., Tinglestad, J., & Tully, M. (1986). Neonatal effects of transplacental exposure to PCB's and DDE. *Journal of Pediatrics, 109*, 335-341.

Rogan, W. J., Gladen, B. C., McKinley, J. D., et al. (1987). Polychlorinated biphenyls (PCB's) and dichlorophenyl dichloroethene (DDE) in human milk: Effects on growth, morbidity, and duration of lactation. *American Journal of Public Health, 77*, 1294-1297.

Rogers, R., Ed. (1988). *Clinical assessment of malingering and deception.* New York: Guilford Press.

Ron, M. A. (1983). The alcoholic brain: CT scan and psychological findings. *Psychological medicine,* Monograph Suppl. 3, p. 133. London: Cambridge University Press.

Ron, M. A. (1986). Volatile substance abuse: A review of possible long-term neurological, intellectual and psychiatric sequelae. *British Journal of Psychiatry, 148*, 235-246.

Rönnbäck, L., & Hansson, E. (1992). Chronic encephalopathies induced by mercury or lead: Aspects of underlying cellular and molecular mechanisms. *British Journal of Industrial Medicine, 49*, 233-240.

Rosati, G., De Bastiani, P., Gilli, P., & Paolino, E. (1980). Oral aluminum and neuropsychological functioning. *Journal of Neurology, 223*, 251-257.

Rose, C. (1990, January 28). Chemical firms stymie treatment. *Des Moines Sunday Register.*

Rose, C. S., Heywood, P. G., & Costanzo, R. M. (1992). Olfactory impairment after chronic occupational cadmium exposure. *Journal of Occupational Medicine, 34(6)*, 600-605.

Rosen, I., Haeger-Aronsen, B., Rehnstrom, S., and Welinder, H. (1978). Neurophysiological observations after chronic styrene exposure. *Scandinavian Journal of Work Environment and Health, 4*(Suppl. 2), 184-194.

Rosenberg, N. L., Spitz, M. C., Filley, C. M., Davis, K. A., & Schaumberg, H. H. (1988). Central

nervous system effects of chronic toluene abuse—Clinical, brainstem evoked response and magnetic resonance imaging studies. *Neurotoxicology and Teratology, 10,* 489-495.
Rosenberg, S. J., Freedman, M. R., Schmaling, K. B., and Rose, C. (1990). Personality styles of patients asserting environmental illness. *Journal of Occupational Medicine, 32,* 678-681.
Rosengren, L. E., Wronski, A., Briving, C., & Haglid, K. G. (1985). Long lasting changes in gerbil brain after chronic ethanol exposure: A quantitative study of the glial cell marker S-100 and DNA. *Alcoholism, 9,* 109-113.
Rosengren, L. E., Kjellstrand, P., & Haglid, K. G. (1986). Tetrachloroethylene: levels of DNA and S-100 in the gerbil CNS after chronic exposure. *Neurobehavioral Toxicology and Teratology, 8,* 201-206.
Rosenstock, H. A., Simons, D. G., & Meyer, J. S. (1971). Chronic manganism. Neurologic and laboratory studies during treatment with levodopa. *Journal of the American Medical Association, 217,* 1354-1358.
Rosenstock, L., Daniell, W., Barnhart, S., Schwartz, D., & Demers, P. A. (1990). Chronic neuropsychological sequelae of occupational exposure to organophosphate insecticides. *American Journal of Industrial Medicine, 18,* 321-325.
Rosenthal, N. E., & Cameron, C. L. (1991). Exaggerated sensitivity to an organophosphate pesticide. *American Journal of Psychiatry, 148,* 270.
Ross, G. H. (1992). History and clinical presentation of the chemically sensitive patient. *Toxicology and Industrial Health, 8,* 21-28.
Ross, W. D., & Sholiton, M. C. (1983). Specificity of psychiatric manifestations in relation to neurotoxic chemicals. *Acta Psychiatrica Scandinavica, 67*(Suppl. 303), 100-104.
Ross, W. D., Emmett, E. A., Steiner, J., & Tureen, R. (1981). Neurotoxic effects of occupational exposure to organotins. *American Journal of Psychiatry, 138,* 1092-1095.
Rosselli, M., Lorenzana, R., Rosselli, A., & Vergara, I. (1987). Wilson's disease, a reversible dementia: Case report. *Journal of Clinical and Experimental Neuropsychology, 9,* 399-406.
Rostain, J. C., Lemaire, C., Gardette-Chauffour, M. C., Doucet, J., & Naquet, R. (1983). Estimation of human susceptibility to the high-pressure nervous syndrome. *Journal of Applied Physiology, 54,* 1063-1070.
Roth, B. J., Griest, A., Jubilis, P. S., Williams, S. D., & Einhorn, L. H. (1988). Cisplatin-based combination chemotherapy for disseminated germ cell tumors: Long-term follow-up. *Journal of Clinical Oncology, 6,* 1239-1247.
Rothke, S., & Bush, D. (1986). Neuropsychological sequelae of neuroleptic malignant syndrome. *Biological Psychiatry, 21,* 838-841.
Rothwell, P. M., & Grant, R. (1993). Cerebral venous sinus thrombosis induced by 'Ectasy.' *Journal of Neurology, Neurosurgery and Psychiatry, 56(9),* 1035.
Rounsaville, B. J., Jones, C., Novelly, R. A., & Kleber, H. (1982). Neuropsychological functioning in opiate addicts. *Journal of Nervous and Mental Disease, 170,* 209-216.
Roush, G. (1959). The toxicology of the boranes. *Journal of Occupational Medicine, 1,* 46-52.
Roy-Byrne, P., Fleishaker, J., Arnett, C., Dubach, M., Stewart, J., Radant, A., Veith, R., & Graham, M. (1993). Effects of acute and chronic alprazolam treatment on cerebral blood flow, memory, sedation, and plasma catecholamines. *Neuropsychopharmacology, 8,* 161-169.
Ruff, R. M., Marshall, L. F., Klauber, M. R., Blunt, B. A., Grant, I., Foulkes, M. A., Eisenberg, H., Jane, J., & Marmarou, A. (1990). Alcohol abuse and neurological outcome of the severely head injured. *Journal of Head Trauma Rehabilitation, 5,* 21-31.
Ruffin, J. B. (1963). Functional testing for behavioral toxicity: A missing dimension in experimental toxicology. *Journal of Occupational Medicine, 5,* 117-121.
Ruijten, M. W. M. M., Salle, H. J. A., Verberk, M. M., & Muijser, H. (1990). Special nerve functions and colour discrimination in workers with long term low level exposure to carbon disulphide. *British Journal of Industrial Medicine, 47,* 589-595.
Ruitjten, M. W., Verberk, M. M., & Salle, H. J. (1991). Nerve function in workers with long-term exposure to trichloroethene. *British Journal of Industrial Medicine, 48(2),* 87-92.
Ruscak, M., Ruscakova, D., & Hager, H. (1968). The role of the neuronal cell in the metabolism of the rat cerebral cortex. *Physiologica Bohemoslov, 17,* 113-121.

Russell, E. W. (1975). A multiple scoring method for assessment of complex memory functions. *Journal of Consulting and Clinical Psychology, 43*, 800–809.
Russell, E. W., Neuringer, P., & Goldstein, G. (1970). *Assessment of brain damage: A neuropsychological key approach*. New York: Wiley.
Russell, P. R. (1983). *Neuropsychological effects of individuals exposed to Telone (1,3-dichloropropene)*. Presented at the American Psychological Association Convention, Anaheim, CA.
Russell, R. W., Flattau, P. E., & Pope, A. M. (Eds.). (1990). *Behavioral measures of neurotoxicity*, Washington, DC: National Academy Press.
Russo, L. S., Beale, G. B., Sandroni, S., & Ballinger, W. E. (1992). Aluminum intoxication in undialyzed adults with chronic renal failure. *Journal of Neurology, Neurosurgery and Psychiatry, 55*, 697–700.
Rutter, M. (1980). Raised lead levels and impaired cognitive-behavioral functioning: A review of the evidence. *Developmental Medicine and Child Neurology, 22*(Suppl. 1), 126.
Ryan, C. M. (1990). Memory disturbances following chronic, low level carbon monoxide exposure. *Archives of Clinical Neuropsychology, 5*, 59–67.
Ryan, C., & Butters, N. (1980). Learning and memory impairments in young and old alcoholics: Evidence for the premature-aging hypothesis. *Alcoholism: Clinical and Experimental Research, 4*, 288–293.
Ryan, C., Butters, N., & Montgomery, K. (1980). Memory deficits in chronic alcoholics: Continuities between the "intact" alcoholic and the alcoholic Korsakoff patient. In H. Begleiter (Ed.), *Biological effects of alcoholism* (pp. 683–699). New York: Plenum Press.
Ryan, C. M., Morrow, L. A., Bromet, E. J., & Parkinson, D. K. (1987). The assessment of neuropsychological dysfunction in the workplace: Normative data from the Pittsburgh Occupational Exposures Test Battery. *Journal of Clinical and Experimental Neuropsychology, 9*, 665–679.
Ryan, C. M., Morrow, L. A., & Hodgson, M. (1988). Cacosmia and neurobehavioral dysfunction associated with occupational exposure to mixtures of organic solvents. *American Journal of Psychiatry, 145*, 1442–1445.
Ryan, J. J., & Lewis, C. V. (1988). Comparison of normal controls and recently detoxified alcoholics on the Wechsler Memory Scale - Revised. *The Clinical Neuropsychologist, 2*, 173–180.
Ryan, T. V., Sciara, A. D., & Barth, J. T. (1993). Chronic neuropsychological impairment resulting from disulfiram overdose. *Journal of Studies on Alcohol, 54*, 389–392.
Ryback, R. S. (1971). The continuum and specificity of the effects of alcohol on memory: A review. *Quarterly Journal of Studies on Alcohol, 32*, 995–1016.
Rylander, R., & Vesterlund, J. (1981). Carbon monoxide criteria. *Scandinavian Journal of Work Environment and Health*, (Suppl. 1), 20–39.
Sacristan, J. A., Iglesias, C., Arellano, F., & Lequerica, J. (1991). Absence seizures induced by lithium: Possible interaction with fluoxetine. *American Journal of Psychiatry, 148*, 146–147.
Saletu, B., Grunberger, J., & Seighart, W. (1986). Pharmaco-EEG, behavioral methods and blood levels in the comparison of temazepam and flunitrazepam. *Acta Psychiatrica Scandinavica, 74*(Suppl. 332), 67–94.
Salvini, M., Binaschi, S., & Riva, M. (1971a). Evaluation of the psychophysiological functions in humans exposed to the threshold limit value of 1,1,1-trichloroethane. *British Journal of Industrial Medicine, 28*, 256–292.
Salvini, M., Binaschi, S., & Riva, M. (1971b). Evaluation of the psychophysiological functions of humans exposed to trichloroethylene. *British Journal of Industrial Medicine, 28*, 293–295.
Salzman, C. (1992). Behavioral side effects of benzodiazepines. In J. M. Kane & J. A. Lieberman (Eds.), *Adverse effects of psychotropic drugs* (pp. 139–152). New York: The Guilford Press.
Sandyk, R., & Gillman, M. A. (1984). Motor dysfunction following chronic exposure to a fluoroalkane solvent mixture containing nitromethane. *European Neurology, 23(6)*, 479–481.
Sano, M., Wendt, P. E., Wirsen, A., Stenberg, G., Risberg, J., & Ingvar, D. H. (1993). Acute effects of alcohol on regional cerebral blood flow in man. *Journal of Studies on Alcohol, 54*, 369–376.
Sansone, M. E., & Ziegler, D. K. (1985). Lithium, toxicity: A review of neurologic complications. *Clinical Neuropharmacology, 8*, 242–248.
Satz, P., Fletcher, J. M., & Sutker, L. S. (1976). Neuropsychologic, intellectual, and personality

correlates of chronic marijuana use in native Costa Ricans. *Annals of the New York Academy of Sciences, 282,* 266-306.

Saurel-Cubizolles, M. J., Estryn-Behar, M., Maillard, M. F., Mugnier, N., Masson, A., & Monod, G. (1992). Neuropsychological symptoms and occupational exposure to anaesthetics. *British Journal of Industrial Medicine, 49,* 276-281.

Savage, E. P., Keefe, T. J., Mounce, L. M., Lewis, J. A. Heaton, R. K., & Parks, L. H. (1980). *Chronic neurological sequelae of acute organophosphate pesticide poisoning: A case-control study.* (Available from the Epidemiologic Studies Program, Health Effects Branch, Hazard Evaluation Division of the U.S. Environmental Protection Agency, Washington, DC).

Savage, E. P., Keefe, T. J., Mounce, L. M., Lewis, J. A., Heaton, R. K., & Parks, L. H. (1988). Chronic neurological sequelae of acute organophosphate pesticide poisoning. *Archives of Environmental Health, 43,* 38-45.

Savard, R. J., Rey, A. C., & Post, R. M. (1980). Halstead-Reitan Category Test in bipolar and unipolar affective disorders. Relationship to age and phase of illness. *Journal of Nervous and Mental Disease, 118,* 297-304.

Savolainen, H. (1982). Toxicological mechanisms in acute and chronic nervous system degeneration. *Acta Neurologica Scandinavica, 66* (Suppl. 92), 23-35.

Savolainen, H., & Pfaffli, P. (1977). Effects of chronic styrene inhalation on rat brain protein metabolism. *Acta Neuropathologica, 40,* 237-241.

Savolainen, H., Pfaffli, P., Tengen, M. N., & Vainio, H. (1977). Biochemical and behavioral effects on inhalation exposure to tetrachloroethylene and dichloromethane. *Journal of Neuropathology and Experimental Neurology, 36,* 941-949.

Savolainen, K., Riihimaki, V., & Linnoila, M. (1979). Effects of short-term xylene exposure on psychophysiological functions in man. *International Archives of Occupational and Environmental Health, 44,* 201-211.

Savolainen, K., Riihimaki, V., Laine, A., & Kekoni, J. (1981). Short term exposure of human subjects to m-xylene and 1,1,1-trichloroethane. *International Archives of Occupational and Environmental Health, 49,* 89-98.

Scelsi, R., Poggi, P., Fera, L., & Gonella, G. (1981). Industrial neuropathy due to n-hexane. Clinical and morphological findings in three cases. *Clinical Toxicology, 18,* 1387-1393.

Schaeffer, J., Andrysiak, T., & Ungerleider, J. T. (1981). Cognition and long-term use of ganja (cannabis). *Science, 213,* 465-466.

Schaeffer, K. W., Parsons, O. A., & Yohman, J. R. (1984). Neuropsychological differences between male familial and nonfamilial alcoholics and nonalcoholics. *Alcoholism: Clinical and Experimental Research, 8,* 347-351.

Schauben, J. L. (1990). Adulterants and substitutes. *Emergency Medicine Clinics of North America, 8,* 595-611.

Schaumburg, H. H., & Spencer, P. S. (1978). Environmental hydrocarbons produce degeneration in cat hypothalmus and optic tract. *Science, 199,* 199-200.

Schaumburg, H. H., & Spencer, P. S. (1984). Human toxic neuropathy due to industrial agents. In P. J. Dyck, P. K. Thomas, E. H. Lambert, & R. Bunge (Eds.), *Peripheral neuropathy* (Vol. II, pp. 2115-2132). Philadelphia: Saunders.

Schaumburg, H. H., Wisniewski, H. M., & Spencer, P. S. (1974). Ultrastructural studies of the dying-back process. I. Peripheral nerve terminal and axon degeneration in systemic acrylamide intoxication. *Journal of Neuropathology and Experimental Neurology, 33(2),* 260-284.

Schaumburg, H. H., Spencer, P. S., & Thomas, P. (1983). *Disorders of peripheral nerves.* Philadelphia: Davis.

Schecter, A., & Ryan, J. J. (1992). Persistent brominated and chlorinated dioxin levels in a chemist. *Journal of Occupational Medicine, 34,* 702-707.

Schinka, J. A. (1984). *Personal Problems Checklist for Adults.* Odessa, FL: Psychological Assessment Resources, Inc.

Schneider, J. W., & Chasnoff, I. J. (1992). Motor assessment of cocaine/polydrug exposed infants at age 4 months. *Neurotoxicology and Teratology, 14,* 97-101.

REFERENCES

Schneider, J. W., Griffith, D. R., and Chasnoff, I. J. (1989). Infants exposed to cocaine in utero: Implications of developmental assessment and intervention. *IYC, 2,* 25–36.

Schoenberg, B. S. (1982). Analytic, experimental, and theoretical neuroepidemiology: Applications to occupational neurology. *Acta Neurologica Scandinavica, 66*(Suppl. 92), 11–22.

Schottenfeld, R. (1987). Workers with multiple chemical sensitivities: A psychiatric approach to diagnosis and treatment. *State of the Art Reviews in Occupational Medicine, 2,* 739–753.

Schottenfeld, R. S., & Cullen, M. R. (1984). Organic affective illness associated with lead intoxication. *American Journal of Psychiatry, 41,* 1423–1426.

Schuckit, M. A., Greenblatt, D., Gold, E., & Irwin, M. (1991). Reactions to ethanol and diazepam in healthy young men. *Journal of Studies on Alcohol, 52,* 180–187.

Schull, W. J., Norton, S., & Jensh, R. P. (1990). Ionizing radiation and the developing brain. *Neurotoxicology and Teratology, 12,* 249–260.

Schuman, S. H., Whitlock, N. W., Coldwell, S. T., & Horton, P. M. (1989). Update on hospitalized pesticide poisonings in South Carolina, 1983–1987. *Journal of the South Carolina Medical Association, 85,* 62–66.

Schwartz, B. S., Ford, D. P., Bolla, K. I., Agnew, J., Rothman, N., & Bleecker, M. L. (1990). Solvent-associated decrements in olfactory function in pain manufacturing workers. *American Journal of Industrial Medicine 18,* 697–706.

Schwartz, J. (1994a). Low-level lead exposure and children's IQ: A meta-analysis and search for a threshold. *Environmental Research, 65,* 42–55.

Schwartz, J. (1994b). Societal benefits of reducing lead exposure. *Environmental Research, 66,* 105–124.

Schwartz, J., & Otto, D. (1991). Lead and minor hearing impairment. *Archives of Environmental Health, 46,* 300–305.

Schwartz, R. H. (1991). Heavy marijuana use and recent memory impairment. *Psychiatric Annals, 21,* 80–82.

Schwartz, S. P. (1985). Environmental threats, communities and hysteria. *Journal of Public Health, 1,* 58–77.

Scov, P., Valbjorn, O., & Pedersen, B. V. (1989). Influence of personal characteristics, job-related factors and psychosocial factors on the sick building syndrome. *Scandinavian Journal of Work, Environment and Health, 15(4),* 286–295.

Sedman, A. B., Wilkening, G. N., Warady, B. A., Lum, G. M., & Alfrey, A. C. (1984). Clinical and laboratory observations: Encephalopathy in childhood secondary to aluminum toxicity. *Journal of Pediatrics, 105,* 836–838.

Seeber, A. K., Kiesswetter, E., Nerdhart, B., & Blaszkewicz, M. (1990). Neurobehavioral effects of a long-term exposure to tetraalkyllead. *Neurotoxicology and Teratology, 12(6),* 653–665.

Seeber, A., Kiesswetter, E., Vangala, R. R., Blaszkewicz, M. & Golka, K. (1992a). Combined exposure to organic solvents: An experimental approach using acetone and ethyl acetate. *Applied Psychology International Review, 41,* 281–292.

Seeber, A., Kiesswetter, E., & Blaskewicz, M. (1992b). Correlations between subjective disturbances due to acute exposure to organic solvents and internal dose. *NeuroToxicology, 13,* 265–270.

Segal, R., & Sisson, B. V. (1985). Medical complications associated with alcohol use and the assessment of risk of physical damage. In T. E. Bratter & G. G. Forrest (Eds.), *Alcoholism and substance abuse* (pp. 137–175). New York: Free Press.

Seidman, L. J., Pepple, J. R., Faraone, S. V., Kremen, W. S., Green, A. I., Brown, W. A., & Tsuang, M. T. (1993). Neuropsychological performance in chronic schizophrenia in response to neuroleptic dose reduction. *Biological Psychiatry, 33,* 575–584.

Selected Petroleum Products. (1982). Geneva: World Health Organization.

Selvin-Testa, A., Loidl, C. F., Lopez-Costa, J. J., Lopez, E. M., & Pecci-Saavedra, J. (1994). Chronic lead exposure induces astrogliosis in hippocampus and cerebellum. *NeuroToxicology, 15,* 389–402.

Senanayake, N., & Karalliedde, L. (1987). Neurotoxic effects of organophosphorus insecticides. *New England Journal of Medicine, 316,* 761–763.

Seppäläinen, A. M. (1973). Neurotoxic effects of industrial solvents. *Electroencephalography and Clinical Neurophysiology, 34*, 702-703.
Seppäläinen, A. M. (1978). Neurotoxicity of styrene in occupational and experimental exposure. *Scandinavian Journal of Work Environment and Health, 4*(Suppl. 2), 181-183.
Seppäläinen, A. M. (1982a). The use of EMG techniques in solvent exposure. In R. Gilioli, M. G. Cassitto, & V. Foa (Eds.), *Neurobehavioral methods in occupational health* (pp. 177-182). New York: Pergamon Press.
Seppäläinen, A. M. (1982b). Neurophysiological findings among workers exposed to organic solvents. *Acta Neurologica Scandinavica, 66*(Suppl. 92), 106-116.
Seppäläinen, A. M. (1985). Neurophysiological aspects of the toxicity of organic solvents. *Scandinavian Journal of Work Environment and Health, 11*(Suppl. 1), 61-64.
Seppäläinen, A. M., & Antti-Poika, M. (1983). Time course of electrophysiological findings for patients with solvent poisoning: A descriptive study. *Scandinavian Journal of Work Environment and Health, 9*, 15-24.
Seppäläinen, A. M., & Lindstrom, K. (1982). Neurophysiological findings among house painters exposed to solvents. *Scandinavian Journal of Work Environment and Health, 8*(Suppl. 1), 131-135.
Seppäläinen, A. M., Tolonen, M., Karli, P., Hänninen, H., & Hernberg, S. (1972). Neurophysiological findings in chronic carbon disulfide poisoning: A descriptive study. *Scandinavian Journal of Work Environment and Health, 9*, 71-75.
Seppäläinen, A. M., Husman, K., & Martenson, G. (1978). Neurophysiological effects of long-term exposure to a mixture of organic solvents. *Scandinavian Journal of Work Environment and Health, 4*, 304-314.
Seppäläinen, A. M., Lindstrom, K., & Martelin, T. (1980). Neurophysiological and psychological picture of solvent poisoning. *American Journal of Industrial Medicine, 1*, 31-42.
Seppäläinen, A. M., Hernberg, S., Vesanto, R., & Kock, B. (1983). Early neurotoxic effects of occupational lead exposure: A prospective study. *NeuroToxicology, 4*, 181-192.
Seppäläinen, A. M., Laine, A., Salmi, T., Verkkala, E., Riihimaki, V., & Luukkonen, R. (1991). Electroencephalographic findings during experimental human exposure to m-xylene. *Archives of Environmental Health, 46*, 16-24.
Seshia, S. S., Rajani, K. R., Boeckx, R. L., & Chow, P. N. (1978). The neurological manifestations of chronic inhalation of leaded gasoline. *Developmental Medicine and Child Neurology, 20*, 323-334.
Shalev, A., & Munitz, H. (1986). The neuroleptic malignant syndrome: Agent and host interaction. *Acta Psychiatrica Scandinavica, 73*, 337-347.
Shankle, R., & Keane, J. R. (1988). Acute paralysis from inhaled barium carbonate. *Archives of Neurology, 45(15)*, 579-580.
Shapiro, W. R., & Young, D. F. (1984). Neurological complications of antineoplastic therapy. *Acta Neurologica Scandinavica, 70* (suppl. 100), 125-132.
Shenker, S., Becker, H. C., Randall, C. L., Phillips, D. K., Baskin, G. S., & Henderson, G. I. (1990). Fetal alcohol syndrome: Current status of pathogenesis. *Alcoholism: Clinical and Experimental Research, 14*, 635-647.
Shephard, R. J. (1983). *Carbon monoxide.* Springfield, IL: Charles C. Thomas.
Sherman, J. D. (1995). Organophophate pesticides—Neurological and respiratory toxicity. *Toxicology and Industrial Health, 11*, 33-39.
Shield, K. K., Coleman, T. L., & Markesbery, W. R. (1977). Methyl bromide intoxication: Neurologic features, including simulation of Reye's syndrome. *Neurology, 27*, 959-962.
Shindell, S., & Stack, U. (1985). A cohort study of employees of a manufacturing plant using trichloroethylene. *Journal of Occupational Medicine, 27*, 577-579.
Shoemaker, W. J. (1981). The neurotoxicity of alcohols. *Neurobehavioral Toxicology and Teratology, 3*, 431-436.
Shore, D., & Wyatt, R. J. (1983). Aluminum and Alzheimer's disease. *Journal of Nervous and Mental Disease, 171*, 553-558.
Siegel, R. K. (1982). Cocaine smoking. *Journal of Psychoactive Drugs, 14*, 321-337.
Silbergeld, E. K. (1982). Neurochemical and ionic mechanisms of lead neurotoxicity. In K. N. Prasad & A. Vernadakis (Eds.), *Mechanisms of actions of neurotoxic substances.* New York: Raven Press.

REFERENCES

Silbergeld, E. K. (1983). Indirectly acting neurotoxins. *Acta Psychiatrica Scandinavica, 67*(Suppl. 303), 16-25.
Silbergeld, E. (1990). Developing formal risk assessment methods for neurotoxicants: An evaluation of the state of the art. In B. Johnson (Ed.), *Advances in neurobehavioral toxicology* (pp. 133-148). Chelsea, MI: Lewis Publishers.
Silbergeld, E. K. (1993). Neurochemical approaches to developing biochemical markers of neurotoxicity: Review of current status and evaluation of future prospects. *Environmental Research, 63(2),* 274-286.
Silberstein, J. A., & Parsons, O. A. (1979). Neuropsychological impairment in female alcoholics. *Currents in Alcoholism, 7,* 481-495.
Silverman, J. J., Hart, R. P., Garrettson, L. K., Stockman, S. J., Hamer, R. M., Schulz, C., & Narasimhachari, N. (1985). Posttraumatic stress disorder from pentaborane intoxication. *Journal of the American Medical Association, 254,* 2603-2608.
Silverstein, F. S., Parrish, M. A., & Johnston, M. V. (1982). Adverse behavioral reactions in children treated with carbamazepine (Tegretol). *Journal of Pediatrics, 101,* 785-787.
Simon, G. E., Katon, W. J., & Sparks, P. J. (1990). Allergic to life: Psychological factors in environmental illness. *American Journal of Psychiatry, 147,* 901-906.
Simonson, K., & Lund, S. P. (1992). A strategy for delineating risks due to exposure to neurotoxic chemicals. *American Journal of Industrial Medicine, 21,* 773-792.
Singer, R., Valciukas, J. A., & Lilis, R. (1983). Lead exposure and nerve conduction velocity: The differential time course of sensory and motor nerve effects. *NeuroToxicology, 4(2),* 193-202.
Sjogren, B., Lidums, V., Hakansson, M., & Hedstrom, L. (1985). Exposure and urinary excretion of aluminum during welding. *Scandinavian Journal of Work Environment and Health, 11,* 39-43.
Sjogren, B., Elinder, C. G., Lidums, V., & Chang, C. (1988). Uptake and urinary excretion of aluminum among welders. *International Archives of Occupational and Environmental Health, 60,* 77-79.
Sjogren, B., Gustavsson, P., & Hogstedt, C. (1990). Neuropsychiatric symptoms among welders exposed to neurotoxic metals. *British Journal of Industrial Medicine, 47,* 704-707.
Small, I. F., Small, J. G., Milstein, V., & Moore, J. E. (1972). Neuropsychological observations with psychosis and somatic treatment. *Journal of Nervous and Mental Disease, 155,* 613.
Smigan, L., & Perris, C. (1983). Memory functions and prophylactic treatment with lithium. *Psychological Medicine, 13,* 529-536.
Smith, D. E., & Seymour, R. B. (1994). LSD: History and toxicity. *Psychiatric Annals, 24,* 145-147.
Smith, S. J. M., & Kocen, R. S. (1988). A Creutzfeldt-Jakob like syndrome due to lithium toxicity. *Journal of Neurology, Neurosurgery and Psychiatry, 51,* 120-123.
Smith, D. E., Ehrlich, P., & Seymour, R. B. (1991). Heavy marijuana use and recent memory impairment. *Psychiatric Annals, 21,* 80-90.
Smith, E. A., & Oehme, F. M. (1991). A review of selected herbicides and their toxicities. *Veterinary and Human Toxicology, 33,* 596-608.
Smith, G., & Shirley, W. A. (1977). Failure to demonstrate effect of trace concentrations of nitrous oxide and halothane on psychomotor performance. *British Journal of Anaesthesiology, 48,* 65-70.
Smith, J. S., & Brandon, S. (1970). Acute carbon monoxide poisoning—3 years experience in a defined population. *Postgraduate Medical Journal, 46,* 65.
Smith, J. W., Burt, D. W., & Chapman, R. F. (1973). Intelligence and brain damage in alcoholics: A study in patients of middle and upper social class. *Quarterly Journal of Studies on Alcoholism, 34,* 414-422.
Smith, M. J. Colligan, M. J., & Hurrell, J. J. (1978). Three incidents of mass psychogenic illness: A preliminary report. *Journal of Occupational Medicine, 20,* 399-400.
Smith, P. J., & Langolf, G. D. (1981). The use of Sternberg's memory scanning paradigm in assessing effects of chemical exposure. *Human Factors, 23,* 701-708.
Smith, P. J., Langolf, G. D., & Goldberg, J. (1983). Effects of occupational exposure to elemental mercury on short-term memory. *British Journal of Industrial Medicine, 40,* 413-419.
Smith, R. G., Vorwald, A. J., Patil, L. S., & Mooney, T. F. (1970). Effects of exposure to mercury in the manufacture of chlorine. *American Industrial Hygiene Association Journal, 31,* 687-700.

Smitherman, J., & Harber, P. (1991). A case of mistaken identity: Herbal medicine as a cause of lead toxicity. *American Journal of Industrial Medicine, 20,* 795-798.
Soleo, L., Urbano, M. L., Petrera, V., & Ambrosi, L. (1990). Effects of low exposure to inorganic mercury on psychological performance. *British Journal of Industrial Medicine, 47,* 105-109.
Solomon, S., Hotchkiss, E., Saraway, S. M., Bayer, C., Ramsey, P., & Blum, R. S. (1983). Impairment of memory function by antihypertensive medication. *Archives of General Psychiatry, 40,* 1109-1112.
Solvent neurotoxicity. (1985). *British Journal of Industrial Medicine, 42,* 433-434.
Sonne, N. M., & Tonnesen, H. (1992). The influence of alcoholism on outcome after evacuation of subdural haematoma. *British Journal of Neurosurgery, 6,* 125-130.
Spake, A. (1985, March). A new American nightmare? *Ms.* 35-42, 93-95.
Spencer, P. S. (1990a). Are neurotoxins driving us crazy? Planetary observations on the causes of neurodegenerative diseases of old age. In R. W. Russell, P. E. Flattau, & A. M. Pope (Eds.), *Behavioral measures of neurotoxicity* (pp. 11-36). Washington, DC: National Academy Press.
Spencer, P. S. (1990b). Environmental causes of neurodegenerative diseases. In R. W. Russell, P. E. Flattau, & A. M. Pope (Eds.), *Behavioral measures of neurotoxicity* (pp. 268-284). Washington, DC: National Academy Press.
Spencer, P. S. (1990c). *Neurotoxicity. Identifying and controlling poisons of the nervous system.* Washington, DC: Office of Technology Assessment, OTA-BA-436.
Spencer, P. S. (1990d). Neurotoxicological vaticinations (The Hänninen Lecture). In B. L. Johnson (Ed.), *Advances in neurobehavioral toxicology* (pp. 13-25). Chelsea, MI: Lewis Publishers.
Spencer, P. S., & Schaumburg, H. H. (1977). Ultrastructural studies of the dying-back process. IV. Differential vulnerability of PNS and CNS fibers in experimental central-peripheral distal axonopathies. *Journal of Neuropathology and Experimental Neurology, 36,* 300-320.
Spencer, P. S., & Schaumburg, H. H. (1985). Organic solvent neurotoxicity. Facts and research needs. *Scandinavian Journal of Work Environment and Health,* Suppl. 1, 53-60.
Spencer, P. S., Couri, D., & Schaumburg, H. H. (1980). N-hexane and methyl n-butyl ketone. In P. S. Spencer & H. H. Schaumberg (Eds.), *Experimental and clinical neurotoxicology* (pp. 456-475). Baltimore: Williams & Wilkins.
Spencer, P. S., Arezzo, J., & Schaumburg, H. (1985). Chemicals causing disease of neurons and their processes. In J. L. O'Donoghue (Ed.), *Neurotoxicity of industrial and commercial chemicals* (Vol. I, p. 114). Boca Raton: CRC Press.
Spencer, P. S., Ludolph, A., Dwivedi, M. P., Roy, D. N., Hugon, J., & Schaumburg, H. H. (1986). Lathyrism: Evidence for role of the neuroexcitatory aminoacid BOAA. *Lancet, 2,* 1066-1067.
Spencer, P. S., Ludolph, A. C., & Kisby, G. E. (1993). Neurologic diseases associated with use of plant components with toxic potential. *Environmental Research, 62,* 106-113.
Spurgeon, A., Gray, C. N., Sims, J., Calvert, I., Levy, L. S., Harvey, P. G., & Harrington, J. M. (1992). Neurobehavioral effects of long-term occupational exposure to organic solvents: Two comparable studies. *American Journal of Industrial Medicine, 22,* 325-335.
Spyker, D. A., Gallanosa, A. G., & Suratt, P. M. (1982). Health effects of acute carbon disulfide exposure. *Journal of Toxicology—Clinical Toxicology, 19,* 87-93.
Squire, L. R., Lewis, L. L., Janowsky, D. S., & Huey, L. Y. (1980). Effects of lithium carbonate on memory and other cognitive functions. *American Journal of Psychiatry, 137,* 1042-1046.
Stahl, S. M., & Lebedun, M. (1974). Mystery gas: An analysis of epidemic hysteria and social sanction. *Journal of Health and Social Behavior, 15,* 44-50.
Stalberg, E., Hilton-Brown, P., Kolmodin-Hedman, B., Holmstedt, B., & Augustinsson, K.-B. (1978). Effect of occupational exposure to organophosphorus insecticides on neuromuscular function. *Scandinavian Journal of Work Environment and Health, 4,* 255-261.
Staudenmeyer, H., Selner, J. C., & Buhr, M. P. (1993). Double-blind provocation chamber challenges in 20 patients presenting with "multiple chemical sensitivity." *Regulatory Toxicology and Pharmacology, 18,* 44-53.
Steinberg, W. (1981). Residual neuropsychological effects following exposure to trichloroethylene (TCE): A case study. *Clinical Neuropsychology, 2(3),* 14.
Steinhauer, S. R., & Hill, S. Y. (1993). Auditory event-related potentials in children at high risk for alcoholism. *Journal of Studies on Alcohol, 54,* 408-421.

Sterman, A. B., & Schaumburg, H. H. (1980). Neurotoxicity of selected drugs. In P. S. Spencer & H. H. Schaumburg (Eds.), *Experimental and clinical neurotoxicology* (pp. 593-613). Baltimore: Williams & Wilkins.
Sternberg, D. E. (1986). The neuroleptic malignant syndrome: The pendulum swings. *American Journal of Psychiatry, 143*, 1273-1275.
Sternberg, D. E., & Jarvik, M. E. (1976). Memory functions in depression: Improvement with antidepressant medications. *Archives of General Psychiatry, 33*, 219-224.
Steru, L., & Simon, P. (1983). Behavioral toxicology of psychotropic drugs from animal to man. In G. Zbinden, V. Cuomo, G. Racagni, & B. Weiss (Eds.), *Application of behavioral pharmacology in toxicology* (pp. 353-366). New York: Raven Press.
Stewart, D. E., & Raskin, J. (1986). Hypersensitivity disorder and vernacular science. *Canadian Medical Association Journal, 134(12)*, 1344-1346.
Stewart, R. D. (1975). The effect of carbon monoxide on humans. *Annual Review of Pharmacology, 15*, 409-423.
Stewart, R. D., Gay, H. H., Schaffer, A. W., Erley, D. S., & Rowe, V. K. (1969). Experimental human exposure to methyl chloroform vapor. *Archives of Environmental Health, 19*, 467-472.
Stewart, R. D., Peterson, J. E., Bachand, R. T., Baretta, E. D., Hosko, M. J., & Hermann, A. A. (1970). Experimental human exposure to carbon monoxide. *Archives of Environmental Health, 21*, 154-164.
Stiles, K. M., & Bellinger, D. C. (1993). Neuropsychological correlates of low-level lead exposure in school-age children: A prospective study. *Neurotoxicology and Teratology, 15*, 27-35.
Stillman, R. C., Weingartner, H., Wyatt, R. J., Gillin, J. C., & Eich, J. (1974). State-dependent (dissociative) effects on marihuana on human memory. *Archives of General Psychiatry, 31*, 81-85.
Stoller, A., Krupinski, J., Christophers, A. J., & Blansk, G. K. (1965). Organophosphorus insecticides and major mental illness. *Lancet, 1*, 1387-1388.
Stollery, B. T. (1986). Effects of 50 Hz electric currents on mood and verbal reasoning skills. *British Journal of Industrial Medicine, 44*, 339.
Stollery, B. T., Broadbent, D. E., Banks, H. A., & Lee, W. R. (1991). Short term prospective study of cognitive functioning in lead workers. *British Journal of Industrial Medicine, 48*, 739-749.
Stolwijk, J. A. J. (1991) Sick-building syndrome. *Environmental Health Perspectives, 95*, 99-100.
Stoner, G. R., Skirboll, L. R., Werkman, S., & Hommer, D. W. (1988). Preferential effects of caffeine on limbic and cortical dopamine systems. *Biological Psychiatry, 23*, 761-768.
Stoner, H. B., Barnes, J. M., & Dugg, J. I. (1955). Studies on the toxicity of alkyl tin compounds. *British Journal of Pharmacology, 10*, 16-25.
Stopford, W. (1985). The toxic effects of pesticides. In P. L. Williams & J. L. Burson (Eds.), *Industrial toxicology: Safety and health applications in the workplace* (pp. 211-228). New York: Van Nostrand Reinhold.
Strang, J., & Gurling, H. (1989). Computerized tomography and neuropsychological assessment in long-term high-dose heroin addicts. *British Journal of Addiction, 84*, 1011-1019.
Strassman, R. J. (1984). Adverse reactions to psychedelic drugs: A review of the literature. *Journal of Nervous and Mental Disease, 172*, 577-595.
Strauss, M. E., Lew, M. F., Coyle, J. T., & Tune, L. E. (1985). Psychopharmacologic and clinical correlates of attention in chronic schizophrenia. *American Journal of Psychiatry, 142*, 497-499.
Streicher, H. Z., Gabow, P. A., Moss, A. H., Kono, D., & Kaehny, W. D. (1981). Syndromes of toluene sniffing in adults. *Annals of Internal Medicine, 94*, 758-762.
Streissguth, A. P. (1990). Prenatal alcohol-induced brain damage and long-term postnatal consequences: Introduction to the symposium. *Alcoholism: Clinical and Experimental Research, 14*, 648-649.
Streissguth, A. P., Barr, H. M., & Sampson, P. D. (1990). Moderate prenatal exposure: Effects on child IQ and learning problems at age 7 1/2 years. *Alcoholism: Clinical and Experimental Research, 14*, 662-669.
Streissguth, A. P., Aase, J. M., Clarren, S. K., Randels, S. P., LaDue, R. A., & Smith, D. F. (1991). Fetal alcohol syndrome in adolescents and adults. *Journal of the American Medical Association, 265*, 1961-1967.
Strickland, T., Hartman, D. E., & Satz, P. (1991). Neuropsychological consequences of crack cocaine

abuse. Paper presented at the eleventh annual meeting of the National Academy of Neuropsychology, Dallas, Texas.
Strickland, T. L., Mena, I., Villanueva-Meyer, J., Miller, B., Cummings, J., Mehringer, C. M., Satz, P., and Myers, H. (1993). Cerebral perfusion and neuropsychological consequences of chronic cocaine use. *Journal of Neuropsychiatry and Clinical Neurosciences, 5,* 419-427.
Strub, R. L., & Black, W. F. (1981). *Organic brain syndromes.* Philadelphia: Davis.
Struve, F. A., & Willner, A. E. (1983). Cognitive dysfunction and tardive dyskinesia. *British Journal of Psychiatry, 143,* 597-600.
Struve, F., Straumanis, J., Patrick, G., Norris, G., Leavitt, J., & Webb, P. (1991). *Topographic quantitative EEG findings in subjects with 18+ years of cumulative daily THC exposure.* Presented at the 53rd Annual Scientific Meeting of the Committee on Problems of Drug Dependence, Palm Beach, FL, June 16-20.
Struwe, G., & Wennberg, A. (1983). Psychiatric and neurological symptoms in workers occupationally exposed to organic solvents—Results of a differential epidemiological study. *Acta Psychiatrica Scandinavica, 67*(Suppl. 303), 68-80.
Struwe, G., Mindus, P., & Jonsson, B. (1980). Psychiatric ratings in occupational health research: A study of mental symptoms in lacquerers. *American Journal of Industrial Medicine, 1,* 23-30.
Struwe, G., Knave, B., & Mindus, P. (1983). Neuropsychiatric symptoms in workers occupationally exposed tojet fuel—A combined epidemiological and casuistic study. *Acta Psychiatrica Scandinavica, 67*(Suppl. 303), 55-67.
Styrene. (1983). Geneva: World Health Organization.
Sullivan, M. J., Rarey, K. E., & Conolly, R. B. (1989). Ototoxicity of toluene in rats. *Neurobehavioral Toxicology and Teratology, 10,* 525-530.
Summerfield, A. (1978). Behavioral toxicology—The psychology of pollution. *Journal of Biosocial Science, 10,* 335-345.
Summerfield, A. (1983). Psychological viewpoint. In N. Cherry & H. A. Waldron (Eds.), *The neuropsychological effects of solvent exposure* (pp. 168-173). Havant, Hampshire: Colt Foundation.
Suzuki, T., Hongo, T., Yoshinaga, J., Imai, H., Nakazawa, M., Matsuo, N., & Akagi, H. (1993). The hair-organ relationship in mercury concentration in contemporary Japanese. *Archives of Environmental Health, 48,* 221-229.
Svanum, S., & Schladenhauffen, J. (1986). Lifetime and recent alcohol consumption among male alcoholics. *Journal of Nervous and Mental Disease, 174,* 214-220.
Svare, C. W., Peterson, L. C., Reinhardt, J. W., Boyer, D. B., Frank, C. W., Gay, D. D., & Cox, R. D. (1981). The effects of dental amalgams on mercury in expired air. *Journal of Dental Research, 60,* 1668-1671.
Svensson, B. G., Nise, G., Erfurth, E. M., & Olsson, H. (1992). Neuroendocrine effects in printing workers exposed to toluene. *British Journal of Industrial Medicine, 49(6),* 402-408.
Sweet, J. J., & Moberg, P. J. (1990). A survey of practices and beliefs among ABPP and non-ABPP clinical neuropsychologists. *Clinical Neuropsychologist, 4(2),* 101-120.
Sze, P. Y. (1989). Pharmacological effects of phenylalanine in seizure susceptibility: An overview. *Neurochemical Research, 14,* 103-111.
Tabershaw, I. R., & Cooper, W. C. (1966). Sequelae of acute organophosphate poisoning. *Journal of Occupational Medicine, 8,* 5-20.
Taft, R., Jr. (1974). Federal commitment to occupational safety and health. In C. Xintaras, B. L. Johnson, & I. deGroot (Eds.), *Behavioral toxicology: Early detection of occupational hazards* (HEW Publication No. (NIOSH) 74-126). Washington, DC: U.S. Government Printing Office.
Takahata, N., Hayashi, H., Watanabe, S., & Anso, T. (1970). Accumulation of mercury in the brains of two autopsy cases with chronic inorganic mercury poisoning. *Folio Psychiatrica Neurologica, Japan, 24,* 59-69.
Takeuchi, T., Eto, N., & Eto, K. (1979). Neuropathology of childhood cases of methylmercury poisoning (Minamata disease) with prolonged symptoms, with particular reference to the decortication syndrome. *NeuroToxicology, 1,* 120.
Tallent, N. (1987). Computer-generated psychological reports: A look at the modern psychometric machine. *Journal of Personality Assessment, 51,* 95-108.

REFERENCES

Tarbox, A. R., Connors, G. J., & McLaughlin, E. J. (1986). Effects of drinking pattern on neuropsychological performance among alcohol misusers. *Journal of Studies on Alcohol, 47,* 176–179.
Tarcher, A. B. (1992). Environmental and biological monitoring. In A. B. Tarcher (Ed.), *Principles and practice of environmental medicine* (pp. 495–529). New York: Plenum Press.
Tarter, R. E., & Edwards, K. L. (1986). Multifactorial etiology of neuropsychological impairment in alcoholics. *Alcohol: Clinical and Experimental Research, 10,* 128–135.
Tarter, R. E., & Van Thiel, D. H. (1985). *Alcohol and the brain: Chronic Effects.* New York: Plenum Press.
Tarter, R. E., Jones, B. M., Simpson, C. D., & Vega, A. (1971). Effects of task complexity and practice on performance during acute alcohol intoxication. *Perceptual and Motor Skills, 33,* 307–318.
Tarter, R. E., Hegedus, A. M., Gavaler, J. S. J., Schade, R. R., Van Thiel, D. H., & Starzl, T. E. (1983). Cognitive and psychiatric impairments associated with alcoholic and nonalcoholic cirrhosis: Compared and contrasted. (Abstract) *Hepatology, 3,* 830.
Tarter, R. E., Hegedus, A. M., Goldstein, G., Shelly, C., & Alterman, A. (1984). Adolescent sons of alcoholics: Neuropsychological and personality characteristics. *Alcohol: Clinical and Experimental Research, 8,* 216–222.
Tarter, R. E., Hegedus, A. M., Van Thiel, D. H., Gavaler, J. S. J., & Schade, R. R. (1986). Hepatic dysfunction and neuropsychological test performance in alcoholics with cirrhosis. *Journal of Studies on Alcohol, 47,* 74–77.
Tarter, R. E., Edwards, K. L., & Van Thiel, D. H. (1988). Perspective and rationale for neuropsychological assessment of medical disease. In R. E. Tarter, D. H. Van Thiel & K. L. Edwards (Eds.), *Medical neuropsychology. The impact of disease on behavior* (pp. 1–10). New York: Plenum Press.
Tarter, R. E., Jacob, T., & Bremer, D. L. (1989). Specific cognitive impairment in sons of early onset alcoholics. *Alcoholism: Clinical and Experimental Research, 13,* 786–789.
Tarter, R. E., Moss, H., Arria, A., & Van Thiel, D. (1990). Hepatic, nutritional, and genetic influences on cognitive process in alcoholics. In J. W. Spencer & J. J. Boren (Eds.), *Residual effects of abused drugs on behavior,* Research Monograph 101, pp. 124–135. U. S. Department of Health and Human Services.
Tashkin, D. P., Wu, T.-C., & Djahed, B. (1988). Acute and chronic effects of marijuana smoking compared with tobacco smoking on blood carboxyhemoglobin levels. *Journal of Psychoactive Drugs, 20,* 27–31.
Tato, S., Hendler, N., & Brodie, J. (1989). Effects of active and completed litigation on treatment results: Worker's compensation patients compared with other litigation patients. *Journal of Occupational Medicine, 31,* 265–269.1
Taueg, C., Sanfilippo, D. J., Rowens, B., Szejda, J., & Hesse, J. L. (1992). Acute and chronic poisoning from residential exposures to elemental mercury—Michigan, 1989-1990. *Clinical Toxicology, 30,* 63–67.
Taxay, E. P. (1966). Vikane inhalation. *Journal of Occupational Medicine, 8,* 425–426.
Taylor, D., & Ho, B. T. (1977). Neurochemical effects of cocaine following acute and repeated injection. *Journal of Neuroscience Research, 3,* 95–101.
Taylor, D., & Ho, B. T. (1978). Comparison of inhibition of monoamine uptake by cocaine, methylphenidate and amphetamine. *Research Communications in Chemical Pathology and Pharmacology, 21,* 67–75.
Taylor, J. R. (1982). Neurological manifestations in humans exposed to chlordecone and follow-up results. *NeuroToxicology, 3,* 916.
Taylor, J. R., Selhorst, J. B., & Calabrese, V. P. (1980). Chlordecone. In P. S. Spencer & H. H. Schaumburg (Eds.), *Experimental and clinical neurotoxicology* (pp. 407–421). Baltimore: Williams & Wilkins.
Taylor, P. R., & Lawrence, C. E. (1992). Polychlorinated biphenyls: Estimated serum half lives. *British Journal of Industrial Medicine, 49,* 527–528.
Tell, I., Somervaille, L. J., Nilsson, U., Bensryd, I., Schutz, A., Chettle, D. R., Scott, M. C., & Skerfving, A. (1992). Chelated lead and bone lead. *Scandinavian Journal of Work Environment and Health, 18,* 113–119.

Terr, A. I. (1986). Environmental illness: A clinical review of 50 cases. *Archives of Internal Medicine, 146*, 145-149.
Terr, A. I. (1989). Clinical ecology. *Annals of Internal Medicine, 111*, 168-178.
Tham, R., Bunnfors, I., Eriksson, B., Lindren, S., & Odkvist, L. M. (1984). Vestibulo-ocular disturbances in rats exposed to organic solvents. *Acta Pharmacologica et Toxicologica, 54*, 58-63.
Thompson, P. J. & Trimble, M. R. (1982). Anticonvulsant drugs and cognitive functions. *Epilepsia, 23*, 531-544.
Thompson, P. J., & Trimble, M. R. (1983). Anticonvulsant serum levels: Relationship to impairments of cognitive function. *Journal of Neurology, Neurosurgery and Psychiatry, 46*, 227-233.
Thompson, P. J., Huppert, F. A., & Trimble, M. R. (1981). Phenytoin and cognitive functions: Effects on normal volunteers and implications for epilepsy. *British Journal of Clinical Pharmacology, 20*, 155-161.
Thorne, D. R., Genser, S. G., Sing, H. C., & Hegge, F. W. (1985). The Walter Reed performance assessment battery. *Neurobehavioral Toxicology and Teratology, 7*, 415-418.
Thrasher, J. D., Madison, R., & Broughton, A. (1993). Immunologic abnormalities in humans exposed to chlorpyrifos: Preliminary observations. *Archives of Environmental Health, 48*, 89-93.
Tiller, J. R., Schilling, R. S. F., & Morris, J. N. (1968). Occupational toxic factors in mortality from coronary heart disease. *British Medical Journal, 4*, 407-411.
Tilson, H. A. (1981). The neurotoxicity of acrylamide: An overview. *Neurobehavioral Toxicology and Teratology, 3*, 445-461.
Tilson, H. A. (1990). Neurotoxicology in the 1990's. *Neurotoxicology and Teratology, 12*, 293-300.
Ting, R., Keller, A., Berman, P., & Finnegan, L. P. (1974). Follow-up studies of infants born to methadone-dependent mothers. *Pediatric Research, 8*, 346.
Tinklenberg, J. R., Melges, F. T., Hollister, L. E., & Gillespie, H. K. (1970). Marijuana and immediate memory. *Nature, 226*, 1171-1172.
Tiplady, B., Sinclair, W. A., & Morrison, L. M. M. (1992). Effects of nitrous oxide on psychological performance. *Psychopharmacology Bulletin, 28*, 207-211.
Tollefson, L. (1993). Multiple chemical sensitivity: Controlled scientific studies as proof of causation. *Regulatory Toxicology and Pharmacology, 18*, 32-43.
Tolonen, M. (1975). Vascular effects of carbon disulfide. A review. *Scandinavian Journal of Work Environment and Health, 1*, 63-77.
Tolonen, M., Nurminen, M., & Hernberg, S. (1975). A follow up study of coronary heart disease in viscose rayon workers exposed to carbon disulphide. *British Journal of Industrial Medicine, 32*, 1-10.
Tonge, J. I., Burry, A. F., & Saal, J. R. (1977). Cerebellar calcification: A possible marker of lead poisoning. *Pathology, 9*, 289-300.
Torvick, A., Lindboe, C. F., & Rodge, S. (1982). Brain lesions in alcoholics. A neuropathological study with clinical correlations. *Journal of Neurological Science, 56(2-3)*, 233-248.
Trape, A. Z. (1990). Exposure to pesticides—Situation in Brazil. In B. L. Johnson (Ed.), *Advances in neurobehavioral toxicology* (pp. 59-74). Chelsea, MI: Lewis Publishers.
Triebig, G., & Lang, C. (1993). Brain imaging techniques applied to chronically solvent-exposed workers: Current results and clinical evaluation. *Environmental Research, 61*, 239-250.
Triebig, G., & Schaller, K.-H. (1982). Neurotoxic effects in mercury exposed workers. *Neurobehavioral Toxicology and Teratology, 4*, 717-720.
Triebig, G., Schaller, K. H., Erzigkeit, H., & Valentin, H. (1977). Biochemical investigations and psychological studies of persons chronically exposed to trichloroethylene with regard to non-exposure intervals [original in German]. *International Archives of Environmental Health, 38*, 149-162.
Triebig, G., Lehrl, S., Weltle, D., Schaller, K. H., & Valentin, H. (1989). Clinical and neurobehavioral study of the acute and chronic neurotoxicity of styrene. *British Journal of Industrial Medicine, 46*, 799-804.
Trzepacz, P. T., DiMartini, A., & Tringali, R. (1993). Psychopharmacologic issues in organ transplantation. *Psychosomatics, 34*, 199-207.
Tsushima, W. T., & Towne, W. S. (1977). Effects of paint sniffing on neuropsychological test performance. *Journal of Abnormal Psychology, 86*, 402-407.

REFERENCES

Tune, L. E., Strauss, M. E., Lew, M. F., Breitlinger, E., & Coyle, J. T. (1982). Serum levels of anticholinergic drugs and impaired recent memory in chronic schizophrenic patients. *American Journal of Psychiatry, 139,* 1460–1461.
Tunving, KI., Thulin, S. O., Risberg, J., & Warkentin, S. (1986). Regional cerebral blood flow in long-term heavy cannabis use. *Psychiatry Research, 17,* 15–21.
Turner, A. J. (1987). *Lead poisoning among Queensland children.* Sydney: Australas.
Tuttle, T. C., Wood, G. D., & Grether, C. B. (1976). *Behavioral and neurological evaluation of workers exposed to carbon disulfide (CS_2)* (HEW Publication No. (NIOSH) 77-128). Washington, DC: U.S. Government Printing Office.
Tuttle, T. C., Wood, G. D., & Grether, C. B. (1977). *A behavioral and neurological evaluation of dry cleaners exposed to perchlorethylene.* Cincinnati: NIOSH.
Tvedt, B., Skyberg, K., Aaserud, O., Hobbesland, A., & Mathiesen, T. (1991). Brain damage caused by hydrogen sulfide: A follow-up study of six patients. *American Journal of Industrial Medicine, 20,* 91–101.
Uitti, R. J., Rajput, A. H., Ashenhurst, E. M., & Rizdilsky, B. (1985). Cyanide-induced parkinsonism: A clinicopathologic report. *Neurology, 35,* 291–295.
Ukai, H., Watanabe, T., Nakatsuka, H., Satoh, T., Liu, S.-J., Qiao, X., Yin, H., Jin, C., Li, G.-L., & Ikeda, M. (1993). Dose-dependent increase in subjective symptoms among toluene-exposed workers. *Environmental Research, 60,* 274–289.
United Nations (1990). Drug abuse extent, patterns and trends. Report of the Secretary General, UN Economic and Social Council, Commission on Narcotic Drugs, 11th Special Session, Vienna, January 29–February 2, 1990.
United States Department of Health and Human Services. (1985). *Preventing lead poisoning in young children, a statement by the Centers for Disease Control.* Atlanta: Public Health Service, Centers for Disease Control.
United States Department of Health and Human Services. (1988). *The nature and extent of lead poisoning in children in the United States: A report to Congress.* Washington, DC: Public Health Service, Agency for Toxic Substances and Disease Registry.
Unrug-Neervoort, A., Luijtelaar, G. V., & Coenen, A. (1992). Cognition and vigilance: Differential effects of diazepam and buspirone on memory and psychomotor performance. *Neuropsychobiology, 26,* 146–150.
Urban, P., & Lukas, E. (1990). Visual evoked potentials in rotogravure printers exposed to toluene. *British Journal of Industrial Medicine, 47,* 819–823.
Uzzell, B., & Oler, J. (1986). Chronic low-level mercury exposure and neuropsychological functioning. *Journal of Clinical and Experimental Neuropsychology, 8,* 581–593.
Vainio, H. (1982). Inhalational anesthetics, anticancer drugs and sterilants as chemical hazards in hospitals. *Scandinavian Journal of Work Environment and Health, 8,* 94–107.
Valciukas, J. A. (1984). A decade of behavioral toxicology: Impressions of a NIOSH/WHO workshop in Cincinnati, May 1983. *American Journal of Industrial Medicine, 5,* 405–406.
Valciukas, J. A., & Lilis, R. (1980). Psychometric techniques in environmental research. *Environmental Research, 21,* 275–297.
Valciukas, J. A., Lilis, R., Singer, R. Fischbein, A., & Anderson, H. A. (1980). Lead exposure and behavioral changes: Comparisons of four occupational groups with different levels of lead absorption. *American Journal of Industrial Medicine, 1,* 421–426.
Valciukas, J. A., Lilis, R., Singer, R. M., Glickman, L., & Nicholson, W. J. (1985). Neurobehavioral changes among shipyard painters exposed to solvents. *Archives of Environmental Health, 40,* 47–52.
Vandekar, M., Plestina, R., & Wilhelm, K. (1971). Toxicity of carbamates for mammals. *Bulletin of the World Health Organization, 44,* 241–249.
van Vliet, C., Swaen, G. M. H., Volovics, A., Tweehuysen, M., Meijers, J. M. M., de Boorder, T., & Sturmans, F. (1990). Neuropsychiatric disorders among solvent-exposed workers. *International Archives of Occupational and Environmental Health, 62,* 127–132.
Varga, J., Uitto, J., & Jiminez, S. A. (1992). The cause and pathogenesis of the eosinophilia-myalgia syndrome (review). *Annals of Internal Medicine, 116(2),* 140–147.
Varma, V. K., Malhotra, A. K., Dang, R., Das, K., & Nehra, R. (1988). Cannabis and cognitive functions: A prospective study. *Drug and Alcohol Dependence, 21,* 147–152.

Varney, N. R., Alexander, B., & MacIndoe, J. H. (1984). Reversible steroid dementia in patients without steroid psychosis. *American Journal of Psychiatry, 141*, 369-372.

Vaughan, E. (1993). Chronic exposure to an environmental hazard: Risk perceptions and self-protective behavior. *Health Psychology, 12*, 74-85.

Veil, C. L., Bartoli, D., Baume, S., et al. (1970). Bilan psychologique, socioprofessionnel, psychopathologique et physiopathologique a un an de distrance de l'intoxication oxycarbone aigue. *Annales Medico-Psychologie (Paris), 2*, 343-398.

Vernon, R. J., & Ferguson, R. K. (1969). Effects of trichloroethylene on visual-motor performance. *Archives of Environment and Health, 18*, 895-900.

Vicente, P. J. (1980). Neuropsychological assessment and management of a carbon monoxide intoxication patient with consequent sleep apnea: A longitudinal case report. *Clinical Neuropsychology, 2*, 91-94.

Victor, M., Adams, R. D., & Collins, G. H. (1971). *The Wernicke-Korsakoff syndrome.* Philadelphia: Davis.

Vining, E. P. G., Mellits, E. D., Cataldo, M. F., Dorsen, M. M., Spielberg, S. P., & Freeman, J. M. (1983). Effects of phenobarbital and sodium valproate on neuropsychological function and behavior. *Annals of Neurology, 14*, 360.

Vogt, R. F., Jr. (1991). Use of laboratory tests for immune biomarkers in environmental health studies concerned with exposure to indoor air pollutants. *Environmental Health Perspectives, 95*, 85-91.

Volkow, N. D., Mullani, N., Gould, K. L., Adler, S., & Krajewski, K. (1988). Cerebral blood flow in chronic cocaine users: A study with positron emission tomography. *British Journal of Psychiatry, 152*, 641-648.

Volkow, N. D., Hitzemann, R., Wolf, A. P., Logan, J., Fowler, J. S., Christman, D., Dewey, S. L., Schlyer, D., Burr, G., Vitkon, S., & Hirschowitz, J. (1990). Acute effects of ethanol on regional brain glucose metabolism and transport. *Psychiatry Research: Neuroimaging, 35*, 39-48.

Volkow, N. D., Gillespie, H., Mullani, N., Tancredi, L., Grant, C., Ivanovic, M., & Hollister, L. (1991). Cerebellar metabolic activation by delta-9-tetrahydrocannabinol in human brain: A study with positron emission tomography and ^{18}F-2-fluoro-2-deoxyglucose. *Psychiatry Research: Neuroimaging, 40*, 69-78.

Volkow, N. D., Wang, G.-J., Hitzemann, R., Fowler, J. S., Overall, J. E., Burr, G., & Wolf, A. P. (1994). Recovery of brain glucose metabolism in detoxified alcoholics. *American Journal of Psychiatry, 151*, 178-183.

Vorhees, C. V. (1994). Developmental neurotoxicity induced by therapeutic and illicit drugs. *Environmental Health Perspectives, 102*(Suppl. 2), 145-153.

Walker, D. W., Hunter, B. E., & Abraham, W. C. (1981). Neuroanatomical and functional deficits subsequent to chronic ethanol administration in animals. *Alcoholism: Clinical and Experimental Research, 5*, 267-282.

Walker, D. W., Heaton, M. B., Lee, N., King, M. A., & Hunter, B. E. (1994). Effect of chronic ethanol on the septohippocampal system: A role for neurotrophic factors. *Alcoholism: Clinical and Experimental Research, 17*, 12-18.

Walsh, K. W. (1978). *Neuropsychology: A clinical approach.* Edinburgh: Churchill Livingstone.

Walsh, K. W. (1985). *Understanding brain damage: A primer of neuropsychological evaluation.* Edinburgh: Churchill Livingstone.

Walsh, T. J., Clark, A. W., Parhad, I. M., & Green, W. R. (1982). Neurotoxic effects of cisplatin therapy. *Archives of Neurology, 39*, 719-720.

Wang, J.-D., & Chen, J.-D. (1993). Acute and chronic neurological symptoms among paint workers exposed to mixtures of organic solvents. *Environmental Research, 61*, 107-116.

Wang, J.-D., Huang, C.-C., Hwang, Y.-H., Chiang, J.-R., Lin, J.-M., & Chen, J.-S. (1989). Manganese induced parkinsonism: An outbreak due to an unrepaired ventilation control system in a ferromanganese smelter. *British Journal of Industrial Medicine, 46(12)*, 856-859.

Wang, J.-D., Shy, W.-Y., Chen, J.-S., Yang, K.-H., & Hwang, Y.-H. (1989). Parental occupational lead exposure and lead concentration of newborn cord blood. *American Journal of Industrial Medicine, 15*, 111-115.

Ward, A. A. (1972). Topical convulsant metals. In D. Purpura, J. Penny, D. Tower, D. Woodbury, & R. Walter (Eds.), *Experimental models of epilepsy—A manual for laboratory workers* (p. 13). New York: Raven Press.
Washton, A. M., & Gold, M. S. (1984). Chronic cocaine abuse: Evidence for adverse effects on health and functioning. *Psychiatric Annals, 14,* 733-743.
Washton, A. M., & Tatarsky, A. (1984). Adverse effects of cocaine abuse. *National Institute on Drug Abuse: Research Monograph Series, 49,* 247-254.
Watanabe, I. (1980). Organotins (triethyltin). In P. S. Spencer & H. H. Schaumburg (Eds.), *Experimental and clinical neurotoxicology.* Baltimore: Williams & Wilkins.
Wechsler, L. S., Checkoway, H., & Franklin, G. H. (1991). A pilot study of occupational and environmental risk factors for Parkinson's disease. *NeuroToxicology, 12,* 387-392.
Wedeen, R. P. (1984). *Poison in the pot: The legacy of lead.* Carbondale: Southern Illinois University Press.
Wedeen, R. P. (1989). Were the hatters of New Jersey "mad"? *American Journal of Industrial Medicine, 16,* 225-223.
Wedeen, R. P. (1992). Removing lead from bone. *NeuroToxicology, 13,* 843-852.
Weingartner, H., Rudorfer, M. V., & Linnoila, M. (1985). Cognitive effects of lithium treatment in normal volunteers. *Psychopharmacology, 86,* 472-474.
Weiss, B. (1978). The behavioral toxicology of metals. *Federation Proceedings, 37,* 22-27.
Weiss, B. (1983). Behavioral toxicology and environmental health science. *American Psychologist,* November, 1174-1187.
Weiss, B., & Clarkson, T. W. (1986). Toxic chemical disasters and the implications of Bhopal for technology transfer. *Milbank Quarterly, 64,* 216-240.
Weiss, B., & Laties, V. (Eds.). (1975). *Behavioral toxicology.* New York: Plenum Press.
Welch, L. S. (1991). Severity of health effects associated with building-related illness. *Environmental Health Perspectives, 95,* 67-69.
Welch, L., Kirshner, H., Heath, A., Gilliland, R., & Broyles, S. (1991). Chronic neuropsychological and neurological impairment following acute exposure to a solvent mixture of toluene and methyl ethyl ketone (MEK). *Clinical Toxicology, 29,* 435-445.
Wendroff, A. P. (1990). Domestic mercury pollution. *Nature, 347,* 623.
Wennberg, A., & Otto, D. (1985). Neurophysiological tests of the central nervous system (CNS). In P. Grandjean (Ed.), *Neurobehavioral methods in occupational and environmental health* (Document 6). Copenhagen: World Health Organization.
Wennberg, A., Hagman, M., & Johansson, L. (1992). Preclinical neurophysiological signs of parkinsonism in occupational manganese exposure. *NeuroToxicology, 13,* 271-274.
Wernick, R., & Smith, R. (1989). Central nervous system toxicity associated with weekly low-dose methotrexate treatment. *Arthritis and Rheumatism, 32,* 770-775.
Weschler, C. J., Shields, H. C., & Naik, D. V. (1989). Indoor ozone exposures. *Journal of the Air Pollution Control Association, 32,* 1568-1662.
Wesson, D. R., Grant, I., Carlin, A. A., Adams, K., & Harris, C. (1978). Neuropsychological impairment and psychopathology. In D.R. Wesson, A. S. Carlin, K. M. Adams, & G. Beschner (Eds.), *Polydrug abuse: The results of a national collaborative study* (pp. 263-272). New York: Academic Press.
West, J. R., Goodlett, C. R., & Brandt, J. P. (1990). New approaches to research on the long-term consequences of prenatal exposure to alcohol. *Alcoholism: Clinical and Experimental Research, 14,* 684-689.
Wetherill, S. F., Guarino, M. J., & Cox, R. W. (1981). Acute renal failure associated with barium chloride poisoning. *Annals of Internal Medicine, 95,* 187-188.
Wheeler, M. G., Rozycki, A. A., & Smith, R. P. (1992). Recreational propane inhalation in an adolescent male. *Clinical Toxicology, 30,* 135-139.
Whipple, S. C., Berman, S. B., & Noble, E. P. (1987). Event-related potentials in alcoholic fathers and their sons. *Alcohol, 8,* 321-327.
Whipple, S. C., Parker, E. S., & Noble, E. P. (1988) An atypical neurocognitive profile in alcoholic fathers and their sons. *Journal of Studies on Alcohol, 49,* 240-244.

White, D. M., Daniell, W. E., Maxwell, J. K., & Townes, B. D. (1990). Psychosis following styrene exposure: A case report of neuropsychological sequelae. *Journal of Clinical and Experimental Neuropsychology, 12(5),* 798-806.

White, R. F., Feldman, R. G., Moss, M. B., & Proctor, S. P. (1993). Magnetic resonance imaging (MRI), neurobehavioral testing, and toxic encephalopathy: Two cases. *Environmental Research, 61,* 117-123.

White, R. F., Robines, T. G., Proctor, S., Echeverria, D., & Rocskay, A. S. (1994). Neuropsychological effects of exposure to naptha among automotive workers. *Occupational and Environmental Medicine, 51,* 102-112.

Whorton, M. D., & Obrinsky, D. L. (1983). Persistence of symptoms after mild to moderate acute organophosphate poisoning among 19 farm field workers. *Journal of Toxicology and Environmental Health, 11,* 347-354.

Wilkinson, D. A., & Carlen, P. L. (1980a). Neuropsychological and neurological assessment of alcoholism. *Journal of Studies on Alcohol, 41,* 129-139.

Wilkinson, D. A., & Carlen, P. L. (1980b). Relation of neuropsychological test performance in alcoholics to brain morphology measured by computed tomography. In H. Begleiter (Ed.), *Biological effects of alcoholism* (pp. 683-699). New York: Plenum Press.

Wilkinson, D. A., & Carlen, L. (1980c). Relationship of neuropsychological test performance to brain morphology in amnesic and non-amnesic chronic alcoholics. *Acta Psychiatrica Scandinavica, 62*(Suppl. 286), 89-101.

Williams, P. L., & Burson, J. L. (Eds.). (1985). *Industrial toxicology: Safety and health applications in the workplace.* New York: Van Nostrand Reinhold.

Williamson, A. M. (1990a). Current status of test development in neurobehavioral toxicology. In R. W. Russell, P. E. Flattau, & A. M. Pope (Eds.), *Behavioral measures of neurotoxicity* (pp. 56-68). Washington, DC: National Academy Press.

Williamson, A. M. (1990b). The development of a neurobehavioral test battery for use in health evaluations in occupational settings. *Neurotoxicology and Teratology, 12,* 509-514.

Williamson, A. M., & Teo, R. K. C. (1986). Neurobehavioral effects of occupational exposure to lead. *British Journal of Industrial Medicine, 43,* 374-380.

Williamson, A. M., & Winder, C. (1993). A prospective cohort study of the chronic effects of solvent exposure. *Environmental Research, 62,* 256-271.

Williamson, A. M., Teo, R. K. C., & Sanderson, J. (1982). Occupational mercury exposure and its consequences for behaviour. *International Archives of Occupational and Environmental Health, 50,* 273-286.

Wills, M. R., & Savory, J. (1983). Aluminum poisoning: Dialysis encephalopathy, osteomalacia and anaemia. *Lancet, 2,* 29-34.

Wilson, G. S., Desmond, M. M., & Vernlaud, W. M. (1973). Early development of infants of heroin-addicted mothers. *American Journal of Diseases of Children, 126,* 457-462.

Windebank, A. J., McCall, J. T., & Dyck, P. J. (1984). Metal neuropathy. In P. J. Dyck, P. K. Thomas, E. H. Lambert, & R. Bunge (Eds.), *Peripheral neuropathy* (Vol. II, pp. 2133-2161). Philadelphia: Saunders.

Winneke, G. (1982). Acute behavioral effects of exposure to some organic solvents—Psychophysiological aspects. *Acta Neurologica Scandinavica, 66*(Suppl. 92), 117-129.

Winneke, G. (1992). Cross species extrapolation in neurotoxicology: Neurophysiological and neurobehavioral aspects. *NeuroToxicology, 13,* 15-26.

Winneke, G., & Collet, W. (1985). Components of test batteries for the detection of neuropsychological dysfunction in children. In *Neurobehavioral methods in occupational and environmental health* (Document 3, pp. 44-48). Copenhagen: World Health Organization.

Winneke, G., & Kraemer, U. (1984). Neuropsychological effects of lead in children: Interactions with social background variables. *Neuropsychobiology, 11,* 195-202.

Winneke, G., Hrdina, K.-G., & Brockhaus, A. (1982). Neuropsychological studies in children with elevated tooth-lead concentrations. 1. Pilot study. *International Archives of Occupational and Environmental Health, 51,* 169-183.

Winneke, G., Kraemer, U., Brockhaus, A., Ewers, U., Kujanek, G., Lechner, H., & Janke, W.

(1983). Neuropsychological studies in children with elevated tooth-lead concentrations. II. Extended study. *International Archives of Occupational and Environmental Health, 51,* 231-252.
Winneke, G., Brockhaus, A., Ewers, U., Kramer, U., & Nef, M. (1990). Results from the European Multicenter Study on lead neurotoxicity in children: Implications for risk assessment. *Neurotoxicity and Teratology, 12,* 553-559.
Winter, P. M., & Miller, J. N. (1976). Carbon monoxide poisoning. *Journal of the American Medical Association, 236,* 1502-1504.
Wise, R. A. (1984). Neural mechanisms of the reinforcing action of cocaine. In J. Grabowski (Ed.), *Cocaine: Pharmacology, effects, and treatment of abuse* (NIDA Research Monograph 50, pp. 15-33). Washington, DC: Department of Health and Human Services.
Wittenborn, J. R. (1980). Behavioral toxicity of psychotropic drugs. *Journal of Nervous and Mental Disease, 168,* 171-176.
Wolff, M., Osborne, J. W., & Hanson, A. L. (1983). Mercury toxicity and dental amalgam. *Neurotoxicology, 4,* 201-204.
Wolkowitz, O. W., Weingartner, H., Thompson, K., Pickar, D., Paul, S. M., & Hommer, D. W. (1987). Diazepam-induced amnesia: A neuropharmacological model of an "organic amnestic syndrome." *American Journal of Psychiatry, 144,* 25-29.
Wolkowitz, O. M., Reus, V. I., Weingartner, H., Thompson, K., et al. (1990a). Cognitive effects of corticosteroids. *American Journal of Psychiatry, 147(10),* 1297-1303.
Wolkowitz, O. M., Rubinow, D., Doran, A. R., Breier, A., Berretini, W. H., Kling, M. A., & Pickar, D. (1990b). Prednisone effects on neurochemistry and behavior. *Archives of General Psychiatry, 47,* 963-968.
Wolkowitz, O. M., Weingartner, H., Rubinow, D. R., Jimerson, D., Kling, M., Berretini, W., Thompson, K., Breier, A., Doran, A., Reus, V. I., & Pickar, D. (1993). Steroid modulation of human memory: Biochemical correlates. *Biological Psychiatry, 33(10),* 744-746.
Wood, R. (1981). Neurobehavioral toxicity of carbon disulfide. *Neurobehavioral Toxicology and Teratololgy, 3,* 397-405.
Woolley, H. C., & Hall, R. W. (1926). Non-medical workers and the mental hospital. *The American Journal of Psychiatry, 5,* 411-414.
World Health Organization. (1976). *Mercury.* Geneva: WHO.
World Health Organization. (1977). *Selenium.* Geneva: WHO.
World Health Organization. (1979a). *Carbon disulfide.* Geneva: WHO.
World Health Organization. (1979b). *Carbon monoxide.* Geneva: WHO.
World Health Organization. (1980). Recommended health-based limits in occupational exposure to heavy metals. *World Health Organization Technical Report Series, 647,* 102-116.
World Health Organization (1985). *Chronic effects of organic solvents on the central nervous system and diagnostic criteria.* Copenhagen: WHO.
World Health Organization. (1990). *Public health impact of pesticides used in agriculture.* Geneva: WHO.
Wright, M., & Hogan, T. P. (1972). Repeated LSD ingestion and performance on neuropsychological tests. *Journal of Nervous and Mental Disease, 154,* 432-438.
Wyse, D. G. (1973). Deliberate inhalation of volatile hydrocarbons: A review. *Canadian Medical Association Journal, 108,* 71-74.
Wysocki, J. J., & Sweet, J. J. (1985). Identification of brain damage, schizophrenic, and normal medical patients using a brief neuropsychological screening battery. *International Journal of Clinical Neuropsychology, 7,* 40-44.
Yant, W. P. (1929). Hydrogen sulphide industry occurrence, effect treatment. *American Journal of Public Health, 19,* 598-608.
Yasui, M., Yase, Y., Ota, K., Mukoyama, M., & Adachi, K. (1991). High aluminum deposition in the central nervous system of patients with amyotrophic lateral sclerosis from the Kii Peninsula, Japan: Two case reports. *NeuroToxicology, 12,* 277-284.
Yerkes, R. M., Bridges, J. W., & Hardwick, R. S. (1915) *A point scale for measuring mental ability.* Baltimore, MD: Warwick & York.

Yesavage, J. A., Leirer, V. O., Denari, M., & Hollister, L. E. (1985). Carry-over effects of marijuana intoxication on aircraft pilot performance: A preliminary report. *American Journal of Psychiatry, 142,* 1325–1329.

Yohman, J. R., Parsons, O. A., & Leber, W. R. (1985). Lack of recovery in male alcoholics' neuropsychological performance one year after treatment. *Alcoholism: Clinical and Experimental Research, 9,* 114–117.

Yokoyama, K., Araki, S., & Murata, K. (1992). Effects of low level styrene exposure on psychological performance in FRP boat laminating workers. *NeuroToxicology, 13,* 551–556.

Yoshida, S. H., Chang, C. C., Teuber, S. S., & Gershwin, M. E. (1993). Silicon and silicone: Theoretical and clinical implications of breast implants. *Regulatory Toxicology and Pharmacology, 17,* 3–18.

Yoshida, S. H., German, J. B., Fletcher, M. P., & Gershwin, M. E. (1994). The toxic oil syndrome: A perspective on immunontoxicological mechanism. *Regulatory Toxicology and Pharmacology, 19,* 60–79.

Yule, W., Lansdown, R., Millar, I. B., & Urbanowicz, M. A. (1981). The relationship between blood lead concentrations, intelligence and attainment in a school population: A pilot study. *Developmental Medicine and Child Neurology, 23,* 567–576.

Zagami, A., Lethlean, A. K., & Mellick, R. (1993). Delayed neurological deterioration following carbon monoxide poisoning: MRI findings. *Journal of Neurology, 240,* 113–116.

Zalewski, C., & Thompson, W. (1994). Comparison of neuropsychological test performance in PTSD, generalized anxiety disorder and control Vietnam veterans. *Assessment, 1,* 133–142.

Zbinden, G. (1983). Definition of adverse behavioral effects. In G. Zbinden, V. Cuomo, G. Racagni, & B. Weiss (Eds.), *Applications of behavioral pharmacology in toxicology* (p. 114). New York: Raven Press.

Zeigler, E. E., Edwards, B. B., Jensen, R. K. Mahaffey, K. R., & Fomon, S. J. (1978). Absorption and retention of lead by infants. *Pediatric Research, 12,* 29–34.

Zhang, J. (1993). Investigation and evaluation of zinc protoporphyrin as a diagnostic indicator of lead intoxication. *American Journal of Industrial Medicine, 24,* 707–712.

Ziem, G. E., & Castleman, B. I. (1989). Threshold limit values: Historical perspectives and current practice. *Journal of Occupational Medicine, 31,* 910–918.

Zillmer, E. A., Lucci, K.-A., Barth, J. T., Peake, T. H., & Spyker, D. A. (1986). Neurobehavioral sequelae of subcutaneous injection with metallic mercury. *Clinical Toxicology, 24,* 91–110.

Ziporyn, T. (1986). A growing industry and menace: Makeshift laboratory's designer drugs. *Journal of the American Medical Association, 256,* 3061–3063.

Zorick, T. R., Sicklesteel, J., & Stepanski, E. (1981). Effects of benzodiazepines on sleep and wakefulness. *British Journal of Clinical Pharmacology,* Suppl. *11,* 31s–35s.

Zorumski, C. F., & Olney, J. W. (1992). Acute and chronic neurodegenerative disorders produced by dietary excitotoxins. In R. L. Isaacson & K. F. Jensen (Eds.), *The vulnerable brain and environmental risks* (pp. 273–292). New York: Plenum Press.

Index

Abortion, spontaneous
  anesthetic gases-related, 353
  toluene-related, 152
Abstract reasoning tests, use in alcohol abuse, 248–249
Acetone, 189, 192
Acetone methyl ethyl ketone, 150
Acetylcholine, interaction with alcohol, 231
Acetylcholinesterase
  carbamates-related inhibition, 17
  organophosphate-related inhibition, 17, 331
Acetylene tetrabomide: see Tetrabromethane
Acetylethyl tetramethyl tetralin, 1
Acquired immune deficiency syndrome (AIDS), 321
Acrylamide, 16, 349–350
Acyclovir, 296
"Adam": see 3,4-Methylenedioxymethamphetamine
Adolescents
  cocaine use, 300
  hallucinogen use, 311
  illicit drug use, 297
  marijuana use, 312
  solvent abuse, 38, 150, 186–187
Adoxycarb, 339
Adrenocorticotropin, heroin-related decrease, 320
Advocacy, by neuropsychologists, 32
Affect, relationship to neuropsychological performance, 19
Agent Orange, 344–345
Aging
  accelerated, in alcoholics, 254
  premature, in alcoholics, 253–255
Aiming test, as WHO/NIOSH Neurobehavioral Core Test Battery component, 60
Airline pilots, marijuana use by, 315
Air pollutants
  carbon monoxide as, 357
  children's exposure to, 37
  immunologic effects, 410
  in sick building syndrome, 415
Alcohol(s), 150
  metabolism of, 226
    gender-related differences, 39, 226–227, 259
    rate, 227

Alcohol abuse, 225–266
  acute effects, 226–228
  blood alcohol level (BAL), calculating, 226
  chronic effects, 229–248
    central pontine myelinosis, 236
    disease pathology of, 234–235
    hepatic dysfunction, 235–236
    imaging and electrophysiological studies of, 236–242
    Marchiafava–Bignami syndrome, 236
    prenatal exposure-related, 37, 38, 229–231, 234–235
    with Wernicke–Korsakoff syndrome, 242–245
    without Wernicke–Korsakoff syndrome, 231–233, 245–248
  disulfiram therapy for, 283–285
  Fetal Alcohol Syndrome, 229–231
  imaging and electrophysiological studies, 236–242
  of neuropsychological function recovery, 260–262
  interaction with cocaine use, 306
  interaction with marijuana use, 316
  interaction with neurotoxin exposure, 41
  interaction with solvent exposure, 233–234
  interaction with styrene exposure, 209
  neurological disorders associated with, 235
  neuropsychological effects, 225–226
    acute effects, 226–228
    adaptation to, 227
    alcohol threshold in, 228, 229
    blood alcohol level in, 226
    comparison with benzodiazepine effects, 281
    drinking history and, 256
    drinking patterns and, 256
    familial factors in, 256–257
    gender factors in, 226–227, 259–260
    handedness and, 258
    head trauma and, 233, 259, 264–265
    locus of damage hypothesis, 246–248
    nutritional factors in, 258
    onset of, 255
    psychopathology of, 258
    of social drinking, 228–229

503

Alcohol abuse (*cont.*)
  neuropsychological function recovery in, 260–265
  neuropsychological testing of, 248–260
    factors affecting performance, 252–260
    individual tests for, 248–250
    relationship to computed tomographic scan results, 239–240
    relationship to liver function variables, 235–236, 237
    test batteries for, 250–252
  prenatal effects, 229–231
  sites of damage by, 16, 20
  social drinking, 228–229
Alcohol consumption, annual per capita rate, 225
Alcohol, Drug Abuse, and Mental Health Administration, neurotoxicity research funding by, 22
Alcoholics, number of, 225
Alcohol inhalation, 186
Aldehydes, 150
Aldicarb, 339
Aldrin, 340, 341, 342–343
Alkyl halides, 188–189
Alkyl nitrites, 189
Allyxycarb, 339
Alopecia, in thallium poisoning, 143
Alprazolam, 276, 277, 279
Aluminum compounds, neurotoxicity, 81
Aluminum exposure, 82–85
  action mechanism of, 19
  antacids, 83
  as dialysis dementia cause, 83–84
  diseases associated with, 84
  individuals at risk for, 82–83
  occupational, 84–85
  synergistic effects with lead, 83
Alzheimer-like symptoms, lead-related, 99–100
Alzheimer's disease
  aluminum accumulation in, 82
  dementia associated with, 8
  memory deficit associated with, 315
American Conference of Governmental Industrial Hygienists
  neurotoxic symptoms, list of, 41, 42
  threshold limit values, recommendations of, 23–24, 32–33
Aminocarb, 339
Aminoglycosides, 296
Amotivational syndrome, 319
Amphetamines, 297–298
  developmental neurotoxicity of, 298
  neurological consequences of abuse, 297
Amyl nitrite, 298
Amyotrophic lateral sclerosis, neurotoxic exposure-related, 8
  aluminum-related, 19

Amyotrophic lateral sclerosis, neurotoxic exposure-related (*cont.*)
  cycads-related, 378
Anabolic steroids, 296–297
Analogue Enforcement Act, 308
ANAM (Automated Neuropsychological Assessment Metrics), 70
Anesthetic gases, 350–354
Anglicus Bartholomaeus, 11
Animal models, of neurotoxic exposure, limitations of, 27–28
Anoxia, as neurotoxic injury mechanism, 13–14
  in carbon monoxide exposure, 357–358
Antabuse (disulfiram), 209, 283–285
Anticholinergics, 17–18, 267–268, 288–289
Anticonvulsants, 268–273
Antidepressants, 273–274
Antihypertensives, 275
Antimony, 85
Antineoplastic drugs, 282–283
Antiparkinsonian drugs, 17–18, 289
Anxiety
  alcoholism-related, 258
  oranophosphates-related, 335
Aphasia Screening Test: *see* Halstead–Reitan Battery, Aphasia Screening Test
Arsenic, 86–89
  criminal use, 86
  exposure case study, 88–89
  exposure limits, 86
  neurological effects, 87–88
  neuropsychological effects, 25, 87–89
  white matter damage by, 16
Art conservators, lead poisoning risk of, 101
Arthrinium, 354
Aspartame, 355, 391, 392
Aspartate, 19, 391, 392
Aspartic acid, as aspartame metabolite, 355
Assyrians, arsenic use by, 86
Astrocytes
  lead-related damage, 15, 99
  mercury-related damage, 15
Ativan: *see* Lorazepam
Attention deficit disorder, 18
Auditory evoked potentials, 76
Autoimmune diseases
  multiple chemical sensitivity as, 409–410
  silicone/silicon-related, 142–143, 404–405
Automated Neuropsychological Assessment Metrics (ANAM), 70
Automobile driving
  benzodiazepine-related impairment of, 278–279
  while intoxicated, 226, 276, 278–279
Automobile exhaust
  as cyanide source, 376

Automobile exhaust (*cont.*)
  as lead source, 101, 103
Autonomic nervous system, examination of, 47

Babylonians, metallurgy development by, 79
Bacillus amylolique, 380
Baclofen, 294–295
Baker, Feldman, White, Harley, Dinse, and Berkey Battery, 61–62
Bard, Samuel, 128
Barium, 89–90, 148
Bayley Mental Development Index, use in lead poisoning, 105
Bayley Scales of Infant Development
  use in fetal alcohol syndrome, 230
  use in maternal marijuana use, 319–320
  use in phenobarbital administration, 271
Beck Depression Scale, 51, 161
Behavioral toxicology, 2
Bender–Gestalt test
  use in DDT exposure, 342
  use in organophosphate exposure, 334
Bender Visual Motor Gestalt
  use in lead exposure, 110
  use in marijuana abuse, 318
  use in mercury exposure, 139
Bendicarb, 339
Benomyl, 340
Benton, Arthur, 59
Benton's tests, use in steroid administration, 295
Benton Visual Retention Test, 53, 59–60
  use in alcohol abuse
    during recovery, 263
    with women, 259
  use in anticholinergic neuroleptic administration, 288–289
  for children, 71
  use in marijuana use, 317
  use in organophosphate exposure, 334
  use in phenytoin administration, 270
  use in solvent exposure, 168
  use in tricyclic antidepressant administration, 273
Benzene(s), 189
  alkylated, 162
  as aromatic hydrocarbon, 149
  as jet fuel component, 394
Benzodiazepines, 275–281
  neurophysiological effects, 275
  neuropsychological effects
    acute, 276–279
    chronic, 279–281
    comparison with alcohol effects, 281
  psychological and psychiatric effects, 281
Beta blockers, 275

Bexley–Maudsley Category Scoring test, use in benzodiazepine administration, 280
Birth defects: *see* Congenital anomalies
Bismuth, 90
Bisphenol A, 401
Block Design test: *see* Wechsler Adult Intelligence Scales-Revised (WAIS-R), Block Design subtest
Blood, lead content, 98
  in children, 103–104, 108, 109, 112, 121
Blood alcohol content (BAC), of automobile drivers, 226
Blood alcohol level (BAL)
  acute effects, 226, 227
  formula to calculate, 226
Blood–brain barrier
  in Alzheimer's disease, 82
  cocaine passage across, 300
  intravenous drug passage across, 321
  lead-related damage, 99
  mercury-related damage, 135
  neurotoxic agents' passage across, 14
Bone, lead accullmation in, 97, 98
Borderline personality disorder, multiple chemical sensitivity as, 413
Borel v. Fibreboard, 421
Boron, 148
Botanicas, 1
Bourdon–Wiersma Vigilance Test, use in jet fuel exposure, 394–395
BPMC, 339
Brain
  fetal, neurotoxicity vulnerability of, 37
  lead accumulation in, 99–101
    chelation therapy for, 108
    in children, 104
  lipid content, 14, 37
  mercury accumulation in, 135
  neurotoxin-induced damage, 15–20
  See also Central nervous system, neurotoxin-related damage of
Brain Age Quotient, use in alcohol abuse, 255
Brain damage
  lesion localization in, 6
  See also Head injury
Brain electrical activity mapping, 76–77
Brain function, behavioral neurology theories of, 5
Brain size, alcohol-related reduction, 232, 245–246
Brain stem auditory evoked potentials, 75–76
  in carbon monoxide poisoning, 361
  in cyanide poisoning, 378
Brain tumors, in children, parental neurotoxin exposure-related, 38
Brassavola, Antonio Musa, 128
Brazier's disease, 147

Breast implants, silicone, 142, 404–405
Breast milk
  mercury content, 140
  solvent content, 152–153
  tetrahydrocannabinol content, 319–320
British Anti-Lewisite, as lead exposure chelation therapy, 108
Bromazepam, 279
Bromine, 355
Brown–Peterson Interference test, use with recovering alcoholics, 264
Bruninks–Oseretsky test, 71
Bufencarb, 339
Building-related illness, 415
Buschke Selective Reminding Test
  use in benzodiazepine administration, 277
  use in lithium administration, 294
Buspirone, 279
Butacarb, 339
Butane, 189, 356
Butylene oxide, 388

Cabamazepine, 272
Cadmium, 80, 91
  olfactory impairment, as cause of, 91
Caffeine, effect on cognitive performance, 298–299
Calcium disodium ethylenediaminetetraacetic acid (EDTA)
  for lead body burden assessment, 98
  as lead chelation therapy, 108
California Neuropsychological Screening-Revised, 61
California Verbal Learning Test
  Associative Learning subtest, 308
  use in eosinophilia myalgia syndrome, 380, 381
  use in multiple chemical sensitivity, 412
Calomel, 128
  George Washington's use of, 129
Cancellation tasks
  use in benzodiazepine administration, 280, 281
  use in carbamazepine administration, 268
  for children, 71
  use in solvent exposure, 171
Cancer, electromagnetic radiation-related, 378
Cannabis sativa, 312; *see also* Marijuana use
Caprolactam, 401
Carbamates, 331, 340
  acetylcholinesterase-inhibiting action, 17
Carbamazepine, 268–269
Carbanolate, 339
Carbaryl, 339, 340
Carbendazim, 340
Carbofuran, 339

Carbon disulfide, 192–197
  Hanninen's study of, 6
  historical background, 11
  litigation-related case critique, 425–429
  neuropathology, 194–195
  neuropsychological effects, 195–196
    testing for, 196
  neurotoxicity, 152, 191, 193–194
  occupational exposure, 36
  white matter damage by, 16
Carbon monoxide, 356–375
  as air pollutant, 357
  elimination, 357–358
  epidemiology, 357
  exposure limits, 356, 362
  historical background, 356–357
  neurological and neuropsychological effects, 25, 358–375
    case reports, 363–375
    chronic low-level exposure-related, 362–363
    delayed exposure-related, 361–363
    severe acute exposure-related, 360–361
    severe poisoning-related, 360
    short-term and low-exposure-related, 358–360
  recovery from exposure to, 372–373
  toxicological effects, 357
Carbon monoxide hemoglobin (COHb)
  dose-effects reponse to, 359–360
  elimination, 357–358
  in marijuana users, 313
  relationship to carbon monoxide poisoning symptom severity, 358
  relationship to inhaled carbon monoxide, 359
  in severe carbon monoxide poisoning, 360
  in smokers, 357, 358
Carbon tetrachloride, 149, 189, 196–197
  as chlorinated hydrogen, 149–150
Carroll, Lewis, 129
Cassavism, 375
Category Test: *see* Halstead–Reitan Battery, Category Test
Causation, in toxic exposure, criteria for, 423–424
Cellular damage, by neurotoxins, 15–16
Central nervous system, neurotoxin-related damage of
  by acrylamide, 349
  by alcohol abuse, 231–233
  by antihypertensives, 275
  by arsenic, 87, 88
  by cadmium, 91
  by carbon disulfide, 193, 194–195
  by cisplatin, 141
  by cocaine, 301
  by ethylene oxide, 388
  by *n*-hexane, 199–200

Central nervous system, neurotoxin-related damage of (*cont.*)
  by lead, 100, 121
  by lithium, 291–292
  by mercury, 133
  by metals, 79
  by methyl butyl ketone, 201
  by neuroleptics, 285
  by perchloroethylene, 202
  by phenytoin, 269
  by solvents, 152, 160, 162
  by steroids, 295
  by styrene, 207–208
  by thallium, 143
  by tricyclic antidepressants, 273
Central pontine myelinosis, 236
Cerebellar atrophy, alcoholism-related, 232–233, 238
Cerebellar examination, 75
Cerebral blood flow studies, 77
  in alcoholism, 242
    during recovery, 261–262
  of benzodiazepine activity, 276
  in cocaine use, 302–303
  in marijuana use, 313
  in solvent-related syndromes, 154, 160–161
Chemicals
  industrial, number of, 7
  OSHA exposure standards for, 32
  See also names of specific chemicals
Chemotherapy, 282–283
Children
  of alcoholics
    evoked-related potentials in, 242, 256–257
    neuropsychological testing of, 257
  aluminum neurotoxicity in, 83
  Arthrinium poisoning in, 354
  aspartame use by, 355
  behavioral toxicology test batteries for, 71
  carbamazepine use by, 269
  electromagnetic radiation-related cancer in, 378
  excitotoxin-containing food intake of, 391, 392
  fetal alcohol syndrome in, 231
  illicit drug use by, 297
  lead exposure in, 102–118
    blood levels, 103–104, 108, 109, 112, 121
    case study, 112–118
    economic costs, 38
    incidence, 102–104
    longitudinal and chronic effects, 105–107
    neurophysiological effects, 107
    neuropsychological effects, 108–112
    neuropsychological testing for, 71
    through parental occupational exposure, 38
    relative risk, 104–105
    treatment, 108

Children (*cont.*)
  mercury poisoning in, 129, 132
  mercury "therapy" for, 128
  metal exposure of, 80
  neuroleptics use by, 290
  phenobarbital use by, 271–272
  polychlorinated biphenyl exposure of, 402
  solvent abuse by, 186–187
  thallium ingestion by, 143
  toxic exposure-related mental disorders in, 37
Chloramphenicol, 296
Chlorbufam, 340
Chlordane, 341, 342, 343–344
Chlordecone, 345
Chlordiazepoxide, 276
Chlorinated cyclodienes, 343–346
Chlorinated hydrocarbons, 149–150, 340–343
Chlorine, 375–376
Chloroform, 189, 219–220, 350–351
  abuse of, 186
Chloromethane: *see* Methyl chloride
Chloropicrin, 347
Chlorothane: *see* 1,1,1-Trichloroethane
Chlorpromazine, 288
Chlorpyrifos, 331
  case study, 336–340
  neuropsychological effects, 334
  office workers' exposure to, 329
Cholinesterase, in organophosphate-exposed individuals, 332, 335
Chorea, 4
  in Wilson's disease, 92
Chromium, 91
Chronic technologial disaster, 420
Cicero, 356
Cinnabar (mercury), 101, 128
Ciprofloxacin, 296
Cirrhosis
  alcohol abuse-related, 235–236
  primary biliary, vitamin E deficiency-related, 258
Cis-platinum, 141
Citrate, as organic brain syndrome risk factor, 83
Clean Air Act of 1970, 22
Clean Water Act, 22
Clinical cases, research data applied to, 428–429
Clinical psychology, 4
Clioquinol, 20, 296
Clobazam, 276, 278, 279
Cloethocarb, 339
Clorazepate dipotassium, 279–280
Clozapine, 290
Cobalt, 146
Coca, 299
Cocaine, 297, 299–308
  chronic effects, 302–304

Cocaine (cont.)
"crack," 41, 297, 299, 300, 308
freebase, 300, 303, 306–307
Freud's experience with, 299
historical background, 299–300
interaction with alcohol, 306
neurochemical effects, 300–301
neurophysiological effects, 301–302
neuropsychological effects, 306–308
prenatal effects, 304
psychological effects, 305
seizures, as cause of, 301–302
withdrawal, 305
Coffee, effect on cognitive performance, 298
Color perception impairment
alcohol abuse-related, 250
LSD-related, 310
solvent-related, 171–172
toluene-related, 214
Color vision discrimination tests, 54
Columbus, Christopher, 186
Comprehensive Employment Training Act, 320
Computed tomographic (CT) scans
of alcoholics, 236, 238–240
of recovering alcoholics, 260–261
of cocaine abusers, 302
of solvent-related syndrome patients, 154–155
of toluene abusers, 214
Conditioning, in neurotoxic syndromes, 50
Congenital anomalies
alcohol-related, 229–230
anesthetic gases-related, 353
carbamazepine-related, 269
cocaine-related, 304
lead-related, 96
phenytoin-related, 270–271
solvents-related, 152–153
toluene-related, 213
Continuous Performance Test, 54, 66
use in ethylene oxide exposure, 389
use in nitrous oxide exposure, 354
Continuous Visual Memory Test, use in eosinophilia myalgia syndrome, 380
Contraceptives, lead as, 96
Controlled Oral Word Association test, use in benzodiazepine administration, 280
Controlled Substances Act, 309
Cook County Hospital, neurotoxic syndrome diagnostic decision tree of, 45, 46
Copper, 92–94
Babylonians' use of, 79
exposure limits, 92
as Wilson's disease cause, 92–94
Cornell Medical Index, 48
Corticosteroids, 295–296

Cosmetics, neurotoxic potential, 1
"Crack" cocaine, 41, 297, 299, 300, 308
Cranial nerves, neurological examination of, 47
Creatine kinase-B, 15
Creutzfeldt–Jacob-like syndrome, 292
Critical Flicker Fusion test
use in alcohol use, 227–228
use in benzodiazepine administration, 280, 281
Cumene, 189
Cyanide, 376–378
Cyanocobalamin, 376
Cyas circinalis, 378
Cycads, 378
Cycasin, 378
Cyclodienes, 340
chlorinated, 342–346
Cyclohexane, 150
Cyclosporin, 282
Cysteine, 391
Cytochrome oxidase-cyanide complex, 377

Dangerous Trades (Oliver), 192
Dantrolene, 295
Dapsone, 296
Datura metel, 267
Datura stramonium, 268
DDT: *see* Dichlorodiphenyltrichloroethane
Delphic oracle, 186
Dementia
Alzheimer's, 8
dialysis, 83–84
lead in, 123
solvent exposure in, 179
Wilson's disease in, 92
Dementia praecox, 4
Denmark
chronic toxic encephalopathy definition use in, 156
solvent syndrome-related disabilities in, 172
Dental amalgam fillings, mercury content, 131–132
Dental offices, mercury exposure risk in, 130–131
Depakote (sodium valproate), 271, 272
Depression
alcoholism-related, 258
cocaine-related, 305
effect on memory, 274
lead-related, 123
methyl bromide-related, 347
organophosphates-related, 333, 335
phencyclidine-related, 322
steroids-related, 295, 297
Designer drugs, 308–309
Desmedipham, 340
Detroit Learning and Aptitude Test, use with children of alcoholics, 257

INDEX 509

Dialysis patients
  aluminum neurotoxicity in, 82–85
  formaldehyde exposure, 393
Diazepam, 275, 276
  neuropsychological effects, 277–278, 279, 281
Dichlorodiphenyltrichloroethane (DDT), 20, 340, 341, 342
Dichloromethane, 189
1,3-Dichloropropene, 345–346
Dichotic Listening task, use in alcohol abuse, 227, 247
Dickens, Charles, 96
Dieldrin, 340, 341, 342–343
Diethylene oxide, 210
Diethyl ether, 191, 350
Diethylin diiodide, 144
Differential diagnosis, in neurotoxicity evaluation, 29–30
Digit cancellation test, use in carbamazepine administration, 268
Digits Backwards test, use in tricyclic antidepressant administration, 273
Digit Span test: see Wechsler Adult Intelligence Scales-Revised (WAIS-R), Digit Span subtest
Digit Symbol test: see Wechsler Adult Intelligence Scales-Revised (WAIS-R), Digit Symbol subtest
Dilantin: see Phenytoin
Dimercaptosuccinic acid, as lead chelation therapy, 108
Dimethalan, 339
Dioscorides, 96
Dioxacarb, 339
Dioxane, 150
Dioxin, 344–345, 402
Diphtheria toxin, 20
Disability, occupational neurotoxin exposure-related, 7–8
Disulfiram, 283–285
  interaction with styrene, 209
Divalproex sodium, as eosinophilia myalgia syndrome therapy, 381
Dopamine
  depletion with MPTP, 308–309
  interaction with cocaine, 300–301, 302
  interaction with styrene, 206
  in parkinsonism, 16–17, 39
  in solvent neurotoxicity, 17
Dow Chemical Company, 345
Down syndrome, brain aluminum accumulation-related, 84
Doxorubicin, 20
Drinking water
  lead content, 103–104
  pesticide contamination, 329

Driving while intoxicated, 226
  comparison with benzodiazepines' effects, 278–279
  under influence of benzodiazepines, 276
Drug abuse
  economic costs, 38
  maternal, fetal risks, 21, 37, 38
  by workers, 41
  See also Drugs, abused and nonprescription; names of specific drugs
Drugs, 267–325
  abused and nonprescription, 297–325
    amphetamines, 297–298
    amyl nitrate, 298
    caffeine, 298–299
    cocaine, 299–308
    designer drugs, 308–309
    LSD and psychedelic drugs, 309–312
    marijuana, 312–320
    opiates, 320–324
    polydrug abuse, 324
    sedative-hypnotics, 324–325
  prescription drugs, 267–297
    anticholinergics, 267–268
    anticonvulsants, 268–273
    antidepressants, 273–274
    antihypertensives, 275
    benzodiazepines, 275–281
    chemotherapy and antineoplastic drugs, 282–283
    disulfiram (Antabuse), 209, 283–285
    lithium, 290–294
    muscle relaxants, 294–295
    naloxone, 285
    neuroleptics, 285–290
    neurotoxicity of, in elderly persons, 39
    steroids, 295–297
Dursban: see Chlorpyrifos

Ecological illness, 407
"Ecstasy": see 3,4-Methylenedioxymethamphetamine
Egyptians, ancient, lead use by, 79, 95
Elderly persons
  antidepressant use, 273–274
  metals exposure, 80
  neurotoxicity in, 39
Electricity, "allergic" reactions to, 411–412
Electroencephalogram (EEG), 77
  use in carbon disulfide exposure, 194
  use in cocaine exposure, 301
  use in lithium exposure, 291
  use in mercury exposure, 133
  quantitative (QEEG), 77
  use in solvent exposure, 154, 160, 222
  use in styrene exposure, 206–207, 208
  use in toluene exposure, 212

Electromagnetic fields
 "allergic" reactions to, 411–412
 extremely low-frequency (ELF), 378–379
Embedded Figures Test, use in alcohol abuse, 249
 during recovery, 263–264
Emotional effects
 of arsenic exposure, 88
 of lead exposure, 118, 123
 of mercury exposure, 131, 137–138
Encephalitis, herpes simplex, 315
Encephalopathy
 alcohol abuse-related, 235
 aluminum-related, 84
 arsenic-related, 86
 Arthrinium-related, 354
 baclofen-related, 294
 bismuth-related, 90
 carbon disulfide-related, 193
 carbon monoxide-related, 361, 362
 hepatic, alcohol abuse-related, 235
 hydrogen sulfide-related, 396
 lead-related, 120
  case study, 123–125
 lithium-related, 291
 mercury-related, 133
 nickel-related, 140–141
 phenytoin-related, 269
 solvents-related, 152
  diagnostic criteria, 156–157
  neuropsychological recovery from, 222–224
  physiological effects, 161
  neuropsychological impairment in, 159
 tetrabromomethane-related, 77
 1,1,1-trichloroethane-related, 217
 Wernicke's, 243
  brain weight reduction in, 232
  computed tomographic scan of, 238
  relationship to Korsakoff's syndrome, 244
Endosulfan, 342–343
Endrin, 340, 341, 342–343
Enflurane, 351
Environmental Protection Agency (EPA)
 neurotoxicity testing guidelines of, 9
 political restraints on, 31
 registry of toxic chemicals of, 7
 regulatory legislation of, 21–22
Eosinophil-derived neurotoxin, as eosinophilia myalgia syndrome causal agent, 380
Eosinophilia myalgia syndrome (EMS), 379–386
 case report, 381–386
 causal agent, 380
 comparison with silicone-related disorders, 404–405
Epidemiological studies, neuropsychological testing in, 52–53

Epilepsy, 4
1,4-Epoxybutane: *see* Tetrahydrofuran
Epstein-Barr virus, 410
Equanil, 281
Erethism, 129, 137–138
Ergot, as lysergic acid source, 310
Erythroxylon coca, 299
Esters, 150
Ethane, 189
Ethanol, 150
Ethenyl benzene: *see* Styrene
Ether, 149, 186, 189
Ethiofencarb, 339
Ethionamide, 296
Ethosuximide, 271
Ethyl chloride, 189, 386–387
Ethylene, 150
[1,1-Ethylenebis(tryptophan)], as eosinophilia myalgia syndrome causal agent, 380
Ethylene chloride, 191
Ethylene oxide, 387–391
 exposure limits, 387
 neuropsychological effects, 388–390
  case report, 390–391
"Ethyl Four Star" (ethyl chloride), 386
"Ethyl Gaz" (ethyl chloride), 386
Europe, neuropsychological research in, 8
European Economic Community, pharmaceutical developmental neurotoxicity screening in, 8
Event-related potentials
 in alcoholics, 241–242
 P300, 76
 in solvent-related syndromes, 155–156
Event threshold, 13
Evoked potentials, 75
 in electromagnetic radiation exposure, 379
EVOSTIK (inhalant), 190
Excitotoxicity, 392
Excitotoxins, food content, 391, 392
Expert witnesses, neuropsychologists as, 421–428
Exposure limits: *see under specific chemicals and compounds*
Eysenck Personality Inventory
 use in lead exposure, 121–122
 use in mercury exposure, 138

Face Learning test, use in alcoholism recovery, 264
Face powder, white lead content, 1
False sago palm (*Cyas circinalis*), 378
Federal Insecticide, Fungicide and Rodenticide Act, 22
Fetal alcohol syndrome (FAS), 37, 38, 229–231, 234–235
Fetal hydantoin syndrome (FHS), 270
Fetotoxic environments, regulatory legislation for, 21

Fetus, neurotoxin exposure: *see* Prenatal exposure
Finger Oscillation (Finger Tapping) Test
 of Halstead–Reitan Battery: *see* Halstead–Reitan Battery, Finger Oscillation Test
 of Neurobehavioral Evaluation System, 66
Finland, number of occupational health researchers in, 8
Fish/seafood, methylmercury contamination of, 129, 132, 134, 135, 140
Flunitrazepam, 278
Fluoroalkanes: *see* Freon(s)
Fluoxetine, 274
 interaction with MDMA, 311–312
Flupenthixol, 290
Fluphenazine, 288
Flurazepam, 275, 276, 278, 279–280
Folk medicine, neurotoxic compound use in, 1, 132–133, 267–268
Food
 aluminum content, 82
 arsenic content, 86
 cyanide content, 376
 excitotoxin content, 391
 neurotoxin cotent, 1
 vanadium content, 146
Food additives, 1, 392
Forensic issues, 421–429
Formaldehyde, 392–394
 exposure limits, 392
 neuropsychological effects, 393–394
Formaldehyde-marijuana abuse, 316
Formethanate, 339
Formication, cocaine-related, 305
Four Word STM test, use in alcohol abuse recovery, 263
Franklin, Benjamin, 96
Franz, Shepherd Ivory, 4
Freebase cocaine, 300, 303, 306–307
Freon(s), 189, 197
Freon 10: *see* Carbon tetrachloride
Freud, Sigmund, 299
Frontal-limbic-diencephalic injury hypothesis, of alcohol-related neuropsychological effects, 247–248
Frontal lobe impairment, fluoxetine-related, 274
5-FU, 282
Fugu, 1
Fumigants, 347–348
Fungicides, 339, 340, 347–348
Furan, 345, 402

Gasoline, lead content, 97, 103, 118, 119–120
Gasoline sniffing, 119–120, 187
Generally Recognized As Safe (GRAS) list, 353
"Ginger Jake" paralysis, 330

Glasgow Outcome Scale (GOS), 233
Glutamate, 19, 391, 392
γ-Glutamate transferase, 236
Glutethimide, 296
Glycols, 150
Gold, 94
 Babylonians' use of, 79
 neurotoxicity, 94
 therapeutic applications, 94
Gordon Personal Profile, 138
Grain alcohol, 150
Grip Strength (Dynamometer) Test: *see* Halstead–Reitan Battery, Grip Strength (Dynamometer) Test
Grooved Pegboard test
 use in benzodiazepine administration, 279
 use in toluene exposure, 214, 215
Guillain–Barre syndrome, gold therapy-related, 94

Hair loss, in thallium poisoning, 143
Haldol, 290
Hallervorden–Spatz disease, 95
Hallucinogens, 309–312
Halogenated compounds, 149–150
Halon 104: *see* Carbon tetrachloride
Haloperidol, 288
Halothane, 350–351, 352, 353
Halstead, Ward, 6, 55–56
Halstead–Reitan Battery, 55–56
 use in alcohol abuse, 236, 239, 245, 248, 249, 250–251, 252, 259
  with and without Wernicke–Korsakoff syndrome, 245
 during recovery, 262, 263, 264
 Aphasia Screening Test, 56, 311
 use in benzodiazepine administration, 281
 Category Test, 56
  use in alcohol abuse, 236, 239, 245, 248, 251, 252, 259
  use in disulfiram administration, 284–285
  use in LSD abuse, 311
  use in neuroleptics administration, 287
  use in phencyclidine abuse, 323
 Finger Oscillation Test, 56
  use in alcohol abuse, 247–248
  use in benzodiazepine administration, 279
  use in eosinophilia myalgia syndrome, 381
  use in lithium administration, 293
  use in mercury exposure, 139, 140
  use in nitrous oxide exposure, 354
  use in formaldehyde exposure, 393–394
 Grip Strength Test, 56, 248–249
 lateral Discrimination Test, 56
 limitations of, 55
 use in lithium administration, 293

Halstead–Reitan Battery (*cont.*)
  use in LSD abuse, 310–311
  use in organophosphate exposure, 333
  use in polydrug abuse, 324
  Reitan Klove Sensory–Perceptual Examination, 56
  Seashore Rhythm Test, 56
    use in disulfuram administration, 284
    use in lead exposure, 109
    use in solvent exposure, 172
  Speech Sounds Perception Test, 56
    use in disulfiram administration, 284, 285
  Tactual Performance Test, 56
    use in alcohol abuse, 227, 245, 246, 251, 252, 259
    use in disulfiram administration, 284, 285
    use in Korsakoff's syndrome, 244
    use in lithium administration, 293–294
    use in solvent abuse, 191
    use in Wilson's disease, 93
  Trail-Making Tests A & B, 56
    use in benzodiazepine administration, 279, 280
    use in carbamazepine administration, 268
    use in eosinophilia myalgia syndrome, 380
    use in lead exposure, 110
    use in marijuana use, 317
    use in solvents exposure, 171, 172
  Trail-Making Test A
    use in alcohol abuse, 239, 240
    use in cocaine abuse, 307
    use in dioxin exposure, 345
    use in opiate abuse, 320
    use in solvents exposure, 172
    use in toluene exposure, 214
  Trail-Making Test B
    use in alcohol abuse, 239, 240, 245, 251, 252
    use in cocaine abuse, 307
    use in lithium administration, 293
    use in LSD abuse, 311
    use in opiates abuse, 320
    use in organophosphate exposure, 334
    use in phencyclidine abuse, 323
    use in toluene exposure, 214
    use in 1,1,1-trichloroethane exposure, 216
    use in 1,1,1-trichloromethane exposure, 219–220
  use in Wilson's disease, 93
Halstead–Reitan Impairment Index
  use in alcohol abuse, 239
  use in cocaine abuse, 307
  use in disulfiram administration, 284, 285
  use in mercury exposure, 138
  use in solvent exposure, 172, 191
Handbook of Mental Examination Methods (Franz), 4
Hand Dynamometer
  use in benzodiazepine administration, 279
  use in solvent abuse, 191

Handedness, relationship to alcoholism, 258
Hand–Eye Coordination test, use in ethylene oxide exposure, 389
Hanninen, Helena, 6
"Hard metals," 146
Hatters, mercury poisoning in, 129
Hatter's shakes, 129
Hazardous waste sites, 7
  phenol component, 401
Head injury, relationship to alcoholism, 233, 259, 264–265
Health care costs, neurotoxicity-related, 38
Health questionnaires, design and administration of, 47–51
  medical history component, 48–49
  occupational history component, 49
  psychological history component, 49–51
Hearing loss, lead exposure-related, 107, 121
Heavy metals, accumulation in children, 37
Hemachromatosis, iron-related, 95
Heptachlor, 341, 342–343
Herbicides, 339, 340, 346, 348
  annual poisonings with, 348
  as Parkinson's disease cause, 309
Heroin, 16, 300
Heroin addiction, 16–17
Hexacarbons, 19
Hexachlorophene, 20
Hexane(s), 149
n-Hexane, 189, 199–200, 201
  exposure limits, 199
  neurotoxicity, 152, 191
  white matter damage by, 16
Hexanedione, 16
Hippocrates, mercuric compounds use by, 128
Hoffman, Albert, 310
Homicide, arsenic-related, 86, 87
Homovanillic acid, as lead exposure indicator, 104
Hoppcide, 339
Horizontal Addition test, use in ethylene oxide exposure, 389
Hospital staff, pesticide exposure, 329
Hydrocarbons
  aliphatic, 149, 188, 189
  aromatic, 149, 188, 189
  chromogenic, 192
  chlorinated, 149–150, 340–342
  fluorinated, 187
  as jet fuel component, 394
Hydrogen cyanide, 377, 378
Hydrogen selenium, 142
Hydrogen sulfide, 395–397
b-Hydroxylase, 17
Hyperbaric nitrogen, 397–398
Hyperbaric oxygen, in cotreatment, 357–358

Hypertension, lead-related, 99
Hypopituitarism, 4
Hypothalamic-pituitary-gonadal injury hypothesis, of alcohol-related neuropsychological effects, 248
Hypothalamus, effect of excitotoxins on, 391, 392
Hysteria, 4

"Ice," 297
Iminodipropionitrile, 19
Imipramine, 273
Immune system, conditioning of, 414
IMT (Instituto de Medicina del Trabajo) Battery, 65
Incas, cocaine use by, 299
Inhalants: *see* Solvents
Institute of Occupational Health Battery, 62
Intelligence–memory discrepancy, in alcohol abuse, 249–250
Intelligence quotient (IQ)
  decreasing, social impact of, 22–23
  of fetal alcohol syndrome children, 231
  lead exposure-related decrease, 105, 111
  toluene exposure-related decrease, 215
  in Wilson's disease, 93
Intelligence tests, 58–59
  use in alcohol abuse, 249
  See also Wechsler Adult Intelligence Scales-Revised (WAIS-R); Wechsler Intelligence Scales for Children-Revised (WISC-R)
Ionizing radiation, 398
Iowa Tests of Educational Development, use in marijuana use, 317
Iraq methylmercury grain poisoning incident, 129–130, 134
Iron, 19, 94–95
Isobutane, 189
Isonazid, 296
Isophorone, 150
Isoprocarb, 339
Isopropyl alcohol abuse, 265

Jackson, J. Hughlings, 5
"Jamaica Ginger" beverage, tri-*o*-cresyl phosphate contamination of, 330
Japan, pharmaceutical developmental neurotoxicity screening in, 8, 38
Jet fuel, 189, 394–395
Johns Hopkins Hospital, Phipps Psychiatric Clinic, 3
Journals, neuropsychological research articles in, 9

Kaiser Steel Corporation, 7
Kepone, 346
Kerosene, 189
Ketones, 150, 188, 189
Kindling, 301–302

Kleist, Karl, 5
Knob Turning test, use in alcohol abuse, 248
Knox Cube Test, use in benzodiazepine administration, 279
Korea, occupational solvent exposure in, 8
Korsakoff, S. S., 244
Korsakoff-like syndrome, benzodiazepines-related, 281
Korsakoff's psychosis, 244
  as alcoholism end-point, 245
  IQ–memory discrepancy in, 250
  memory deficit associated with, 315
  PET scan of, 240

Lampodius, 192
Lanthony D-15 color panel test, 54
Lathyrus sativus (neurolathyrism), 398–399
Lead, 95–125
  adult exposure, 118–125
    case study, 123–125
    cognitive effects, 122–123
    emotional effects, 123
    with inorganic lead, 120–121
    neuropsychological effects, 121–122
    neuropsychological recovery from, 123
    with organic lead, 118–120
  ancient Egyptians' use of, 79
  ancient Romans' use of, 11, 95–96
  childhood exposure: *see* Lead, prenatal and childhood exposure
  current uses, 96–97
  effect on brain astrocytes, 15
  exposure limits, 95, 98
    in children, 102, 104–105, 112
  gasoline content, 97, 103, 118, 119–120
  historical background, 11, 79, 95–96
  inorganic, 120–121
  intake, 97–98
  lack of neurotoxicology research on, 34
  neuropathology of, 98–101
  neuropsychological effects, 25, 81
  neurotoxicity, 80
  non-occupational sources, 101
  occupational exposure, 36, 101
    historical background, 12
    industries at risk for, 97
  organic, 118–120
  paint content, 96, 101, 102–103
  prenatal and childhood exposure, 1, 102–118
    blood levels in, 103–104, 108, 109, 112, 121
    case study, 112–118
    economic costs, 38
    incidence, 102–104
    longitudinal and chronic effects, 105–107
    neurophysiological effects, 107

Lead (*cont.*)
　prenatal and childhood exposure (*cont.*)
　　neuropsychological effects, 108–112
　　neuropsychological tests for, 71
　　relative risk, 104–105
　　through parental occupational exposure, 38
　　treatment, 108
　regulatory control of, 79–80
　renal toxicity, 18–19
　subclinical toxicity, 102
　synaptic action effects, 15
Lead metabolism, age-specific kinetic model of, 100–101
Lead paint, 96, 101, 102–103
Legislative issues, in neurotoxic injury, 21–24
Le secret, 129
Librium: *see* Chlordiazepoxide
Life Sciences Products Company, 345
Lincoln–Oseretsky test, 71
Lindane, 341
Lipophilicity, of neurotoxic agents, 14
Lithium, 290–294
　in combination with neuroleptics, 292
　gender-related response differences, 40
　neurological effects, 291–292
　neuropsychological effects, 292–294
Litigation, toxic exposure-related, 31, 421–428
Liver function tests, in alcohol abuse, 235–236, 237
Livestock, pesticide treatment of, 327
"Locked-in" syndrome, 236
London School of Hygiene Battery, 62
Lorazepam, 276, 277, 278, 279–281
　gender-related response differences, 40
Love Canal, 420
LSD: *see* Lysergic acid diethylamide
Luria, Alexander, 5, 54
Luria–Nebraska Neuropsychological Battery, 55, 56
　use in alcohol abuse, 251–252
　use in cocaine abuse, 306
　use in 1,1,1-trichloroethane exposure, 216
　use in 1,1,1-trichloromethane exposure, 219–220
Lysergic acid diethylamide (LSD), 297, 309–311

Madame Bovary (Flaubert), 408
Maddox Wing Test, use in xylene exposure, 221
Mad Hatter (literary character), 129
Magnetic resonance imaging (MRI)
　use in cocaine abuse, 303
　use in solvent-related syndromes, 155
Magnetophosphenes, 378
Malathion, 328, 329, 331
Malingering, 29, 50, 424
Mandelic acid, as styrene metabolite, 205

Manganese, 19, 125–127
　exposure limits, 125
　as parkinsonism cause, 17
Manganism, 126
"Manikin" spatial orientation task, use in benzodiazepine administration, 280
Manipulation tests, use in benzodiazepine administration, 281
MANS (Milan Automated Neurobehavioral System) Battery, 67
Marchiafava–Bignami syndrome, 236
Marijuana use, 297, 312–320
　combined with formaldehyde, 316
　interaction with alcohol, 316
Marine Protection Research and Sanctuaries Act, 22
Mass psychogenic illness, 416–417
Mattis–Kovner Memory Inventory, use in anticholinergic neuroleptic administration, 288–289
Maudsley Automated Psychological Screening Tests, use in alcohol abuse, 246
Maudsley Personality Inventory
　use in marijuana use, 318
　use in styrene exposure, 207
Maxacarbate, 339
Maze Coordination Errors, use in solvent abuse, 191
McCarthy Scales of Children's Abilities
　use in lead exposure, 107–108
　use in maternal polychlorinated bipheny exposure, 403
　use in phenobarbital administration, 271, 272
MDE, 309
MDMA (3,4-methylenedioxymethamphetamine), 308, 309, 311–312
Medical examination, for neurotoxic syndrome evaluation, 45–47, 74–77
Medical history, of neurotoxic symptoms, 48–49
Memory for Designs test, use in toluene exposure, 215
Mental retardation
　fetal alcohol syndrome-related, 230
　lead-related, 107, 118
Mental testing, 3–4
Meprobamate, 281
Mercury, 127–140
　amalgam, levels in, 131–132
　asymptomatic poisoning effects, 12
　brain astrocyte effects, 15
　exposure limits, 127, 139
　folk medicine containing, 132–133
　historical background, 11, 128–130
　inorganic, 135–136
　methylated, 80
　neuropathology, 133–136
　neuropsychological effects, 25, 81, 136–140
　neurotoxicity, 79, 80, 99

INDEX 515

Mercury (cont.)
  nonoccupational exposure, 131–133
  occupational exposure, 130–131
  oral ingestion, 138
  protein-inhibiting action, 15
  regulatory control, 79–80
  renal toxicity, 18–19
  RNA synthesis-inhibiting action, 15
  sources, 79–80
  teratogenicity, 37, 38, 132, 140
Mercury vapor, 130–131
  neuropathlogy, 135–136
  sources, 132–133
  teratogenicity, 140
Mesoridazine, 288
Metallurgy, historical background, 79
Metals exposure, 79–148
  acute versus chronic, 8–81
  aluminum, 82–85
  antimony, 85–89
  barium, 89–90
  cadmium, 91
  chromium, 91
  copper, 92–94
  gold, 94–95
  lead, 95–125
    adult, 118–125
    exposure limits, 95, 98, 102, 104–105, 112
    historical background, 11, 79, 95–96
    intake methods, 97–98
    neuropathology, 98–101
    occupational, 36, 97, 101
    prenatal and childhood, 1, 102–118
    regulatory control, 79–80
    renal toxicity, 18–19
    subclinical toxicity, 102
    synaptic action, 15
  lithium, 125–127
  mercury, 127–140
  neuropsychological testing of, 81–82
  nickel, 140–141
  occupational exposure, historical background, 12, 79
  platinum, 141
  selenium, 141–142
  silicon and silicone, 142–143
  tellurium, 143
  thallium, 143
  tin and organotins, 144–146
  tungsten, 146
  vanadium, 146–147
  variables affecting neurotoxicity of, 80–81
  zinc, 147
Methamidophos, 331
Methamphetamine, 297, 311

Methane, 189
Methane trichloride: see 1,1,1-Trichloromethane
Methanol, 150
Methomyl, 339
Methotrexate, 282
Methoxyflurane, 350–351
Methoxymethylmercury acetate poisoning, 130
b-N-Methyl amino-L-alanine, 378
Methyl-D-aspartate receptors, binding with phencyclidine, 322
Methyl benzene: see Toluene
Methyl benzol: see Toluene
Methyl butyl ketone, 200–201
  exposure limits, 199
  neuropsychological effects, 190
  neurotoxicity, 152, 191
Methylcarbamic acid, 329
Methyl chloride, 189, 197–201
  exposure limits, 197
  neurological effects, 198
  neuropsychological effects, 198–199
Methyl chloroform, 189
Methyldopa, 275
Methylene chloride, 149–150, 188–189
3,4-Methylenedioxymethamphetamine (MDMA), 308, 309, 311–312
Methyl ethyl ketone, 152, 189
  co-exposure with toluene, case study, 181–186
3-Methyl fentanyl, 308
Methyl isobutyl ketone, 201
Methylmercury
  conversion from mercury, by oral bacteria, 131
  Iraqi poisoning incident, 129–130, 134
  Minamata poisoning incident, 12, 37, 129, 134, 140
  neuropathology, 133
  effect on nissl substance, 16
  prenatal exposure, 37, 38, 133–135, 140
Methyl n-butyl ketone, 16, 150, 189
Methylparathion, 331
Methyl-4-phenyl-4-propinoxy-piperidine (MPPP), 308
1-Methyl-4-phenyl-1,2,5-tetrahydropyridine (MPTP), 16–17, 308–309
  aging-related neurotoxicity, 39
  damage sites, 20
  MPP metabolite, 18
Metolcarb, 339
Metronidazole, 296
Mevinphos, 332
Mexico
  occupational exposure in, 8
  solvent abuse in, 8, 38, 186–187
Meyer, Adolf, 5
Michigan Eye–Hand Coordination test, use in methyl chloride exposure, 198

Microcomputer-based neuropsychological batteries, 57–58, 68
Microcomputer-Based Testing System (MTS) Battery, 68
Milan Automated Neurobehavioral System (MANS) Battery, 67
Miltown, 281
Minamata methylmercury poisoning incident, 12, 37, 129, 134, 140
Minamata syndrome, 37
Mineral spirits, 149, 189
Minnesota Multiphasic Personality Inventory (MMPI), 51
  use in Agent Orange exposure, 344
  use in formaldehyde exposure, 393–394
  as Halstead–Reitan Battery component, 56
  use in multiple chemical sensitivity, 412
  use neuroleptic administration, 287
  use in solvent exposure, 154
  use in telone exposure, 346
Minnesota Paper Form Board Test, 293
Minnesota Rate of Manual Dexterity Test, 53
Mira Test, 169
Misonidazole, 296
Monochloromethane, 189
Monoethanolamine foam, 162
Moses (biblical figure), 94
Motor system, examination of, 47
MPPP (methyl-4-phenyl-4-propionoxy-piperidine), 308
MPTP (1-methyl-4-phenyl-1,2,5-tetrahydropyridine), 16–17, 308–309
  aging-related neurotoxicity, 39
  damage sites, 20
  MPP metabolite, 18
MTS (Microcomputer-Based Testing System) Battery, 68
Multiple Affect Adjective Checklist, 51
Multiple chemical sensitivity (MCS), 407–414, 416, 418
  etiology, 409–414
  patient profile, 408–409
  treatment, 414
Muscle relaxants, 294–295

Naloxone, 285
Naphtha, 189
National Adult Reading Test, use in benzodiazepine administration, 280
National Institute of Mental Health, neurotoxicity research funding by, 22
National Institute of Occupational Safety and Health (NIOSH)
  Pocket Guide, computer disk version, 44
  neurobehavioral core test battery of, 56–57, 59–60, 65

National Institute of Occupational Safety and Health (NIOSH) (*cont.*)
  neurotoxic agents list, 34
  research mandate of, 22
National Institute on Drug Abuse, 318–319
  neurotoxicity research funding by, 22
National Toxicology Program, 7
Naval Biodynamics Laboratory Battery, 68–69
Neisser vigilance task, 195
Nerve conduction studies, 76–77
"Nerve gas," 330
Nervous system, neurotoxic damage of
  assessment, 24–30
  selective vulnerability of, 14–20
    cellular damage, 15–16
    damage loci, 20
    neurochemical damage, 16–20
    synaptic damage, 15
  See also Central nervous system, neurotoxin-related damage; Peripheral nervous system, neurotoxin-related damage
Neurasthenia, 408
Neurasthenic syndrome, 157
Neurobehavioral Evaluation System (NES) Battery, 52, 58, 66–67
  Chinese version, 66–67
  use in ethylene oxide exposure, 389–390
  use in lead exposure, 106
  use in mercury exposure, 140
  use in sick building syndorme, 416
  use in solvents exposure, 168, 173
  use in styrene exposure, 209
  Visual Retention test of, 66
Neurobehavioral toxicology, 2
  historical background, 3–6
  modern, 7–9
Neurochemical damage, neurotoxin-related, 16–20
Neurochemical markers, of neurotoxic exposure, 27
Neurodegeneration, as neurotoxic injury mechanism, 13, 19
Neurolathyrism, 398–399
Neuroleptic malignant syndrome, 286–287
Neuroleptics, 285–290
  anticholinergic, 17–18
  chronic effects, 287
  in combination with lithium, 292
  as neuroleptic malignant syndrome cause, 286–287
  neurological effects, 285–286
  as tardive dyskinesia cause, 287, 288
Neurological examination, 75
  of cranial nerves, 47
Neurology, behavioral, 5
Neuron-specific enolase, 15

Neuropathy, peripheral
    alcohol-related, 233
    disulfiram-related, 283
    eosinophilia myalgia syndrome-related, 379–380
    lead-related, 100
    methyl butyl ketone-related, 201
    organophosphate-related, 331
    phenytoin-related, 269
    solvents-related, 172
    trichloroethylene-related, 217
    tri-*o*-cresyl phosphate-related, 330
Neuropsychological effects: *see under specific neurotoxins*
Neuropsychological monitoring, in the workplace, obstacles to, 31–32
Neuropsychological Screening Test battery, 64–65
Neuropsychological testing
    historical background, 4, 5, 6
    implications for neurotoxicity-related litigation, 423–424
    strategies of, 9–11
        flexible batteries, 9
        performance testing, 10–11
        process approaches, 10
        standardized (fixed) batteries, 9
    See also under specific neurotoxins; names of specific tests
Neuropsychological testing batteries, 51–59
    construction of, 51–59
        clinical testing approaches, 54–59
        epidemiological versus clinical batteries, 52–53
        test structure and content, 53–54
    flexible batteries, 52–53, 56–57
    intelligence tests, 58–59
    microcomputer-based batteries, 57–58
    standard batteries, 55–56
Neuropsychological toxicology
    assessment techniques, 24–30
        advantages, 24–27
        difficulties in, 27–30
    obstacles to, 30–32
Neuropsychological toxicology researchers, limited number of, 31
Neuropsychologists
    as expert witnesses, 421–428
    involvement in neurotoxic exposure issues, 2–3
Neuropsychology
    clinical, 3
    definition, 3
    historical precedents, 3–6
        behavioral neurology, 5
        clinical and experimental psychology, 4
        mental testing, 3–4
Neurotoxic damage
    legislative issues, 21–24

Neurotoxic damage (*cont.*)
    principles of, 12–20
        cellular damage, 15–16
        direct damage, 12
        indirect damage, 12–13
        neurochemcial damage, 16–20
        subclinical, 13
        synaptic damage, 15
Neurotoxicity
    causation criteria, 423–424
    developmental, 37–39
    health care costs attributed to, 38
    historical views of, 11–12
    linear without threshold, 19
    of metals, 79
        factors affecting, 80–81
    neuropsychological symptoms, 41, 42, 43
    nonmonotonic threshold model, 19
    sex factors, 39–40
    "silent" (asymptomatic), models of, 19
    subclinical, 12, 13
    testing of, 24
Neurotoxicity testing, developmental, 38
Neurotoxicology, definition, 2
Neurotoxic symptoms, 41–42
Neurotoxic syndromes
    evaluation of, 45–77
        ambiguity of neuropsychological testing in, 73–74
        children's tests for, 71
        complementary approaches to, 74–77
        health questionnaire construction for, 47–51
        individual tests for, 70–71
        initial medical evaluation, 45–47
        interpretation of neuropsychological test results, 71
        neuropsychological test batteries for, 51–70
Neurotoxins
    "all or none" hypothesis of, 12
    definition, 2
    developmental, 38
    lack of research regarding, 34
Neurotransmitters
    acrylamide's interaction with, 349
    alcohol's interaction with, 231–232
New Learning test, use in benzodiazepine administration, 280
Nickel, 80, 140–141
Nikander of Colophon, 96
Nissl substance, 16
Nitrazepam, 278
Nitrobenzene, 191–192
Nitrofurantoin, 296
Nitrogen, hyperbaric, 397–398
Nitrogen narcosis, 397–398

Nitromethane: see Freon(s)
Nitrous oxide, 296, 350–351, 352, 353–354
    abuse of, 186
Norepinephrine
    interaction with cocaine, 301
    in selective attention, 17
Nortriptyline, 273
    gender-related response differences to, 40
Nuclear power generators, as ionizing radiation source, 398

Object Assembly test, use in alcohol abuse, 249
Occupational history, 49
Occupational neurotoxic agents, number of, 34
Occupational neurotoxin exposure
    international comparison, 35
    litigation related to, 31
    as mortality cause, 7
    number of workers at risk for, 35
    occupations at risk for, 36–37, 40
    premorbid function assessment in, 28–29
Occupational researchers, number of, U.S./Finnish comparison, 8
Occupational Safety and Health Act, 22
Occupational Safety and Health Administration (OSHA)
    chemical exposure standards of, 32
    political restraints on, 31
Occupations, at risk for neurotoxic exposure, 36–37, 40
Octanes, as aliphatic hydrocarbons, 149
Ocular fundi, examination of, 47
Odors, as multiple chemical sensitivity "triggers," 413, 414
Office of Management and Budget, cost–benefit criteria of, 31
Old Testament, 94
Oleic anilide, 405
Olfactory sensitivity, solvent-related impairment of, 154
O'Conner Dexterity Test, use in anesthetic gas exposure, 352
Operating rooms, anesthetic gas exposure in, 351, 353
Opiates, 320–324
Optic nerve, neurological examination of, 47
Organic affective syndrome, 157
Organic brain syndrome
    carbon monoxide-related, 368–372
    jet fuel-related, 395
    secondary to aluminum absorption, 83
    solvent-related, 8, 160
Organophosphates, 329–336
    acetylcholinesterase-inhibiting action, 17
    active ingredients, 330

Organophosphates (cont.)
    annual poisonings from, 327–328
    emotional effects, 335–336
    exposure limits, 329
    neurochemical pathway damage from, 17
    neurological effects, 332–333
    neuropsychological effects, 25, 333–335
    personality effects, 335
Organotin, 144–145
Osler, William, 3
b-N-Oxalyl-L-a,b-diaminopropionic acid, 399
Oxamyl, 340
Oxazepam, 40, 276, 279
Ozone, 399–400

Paced Auditory Serial Addition Task
    use in cocaine abuse, 307
    use in mercury exposure, 139
    use in solvent abuse, 190
Paint
    aliphatic hydrocarbon content, 149
    lead content, 96, 101, 102–103
Painters' naphtha, 149
Paired Associate Learning test
    use in alcohol abuse, 239
    use in tricyclic antidepressants administration, 273
Panax ginseng, 267
Pao Pu Tzu, 94, 128
Paracelsus, 94
Parallel processing, 54
Paraquat, 18, 347
Parasitosis, cocaine-related, 305
Parathion, 329, 330, 331
Parental exposure, 38
Parkinson dementia complex of Guam, 84
Parkinson's disease
    anticholinergic therapy, 267, 289
    pathophysiology, 16
    relationship to neurotoxin exposure, 16–17
    in elderly persons, 39
Parkinsonism
    cocaine-related, 302
    lithium-related, 291
    MPPP-related, 308
    MPTP-related, 308–309
    neuroleptics-related, 286
    neurotoxic exposure as risk factor for, 8
    event threshold in, 13
Pattern Memory test, use in ethylene oxide exposure, 389
Pattern reversal evoked potentials, 75
Peabody Picture Vocabulary test, use in toluene exposure, 215
"Peak E," as eosinophilia myalgia syndrome causal agent, 380

Pentaborane, 400
Pentanes, as aliphatic hydrocarbons, 149
People's Republic of China, acrylamide exposure in, 350
Perceptual disorders, LSD-related, 310
Perchloroethylene, 189, 201–205
    as chlorinated hydrogen, 149–150
    chronic exposure, case study, 203–205
    exposure limits, 201
    neuropsychological effects, 162
    neuropsychological recovery following exposure to, 222
    neurotoxicity, 191–192
Performance testing, 10–11
Perhexiline maleate, 296
Peripheral nervous system, neurotoxin-related damage, 15
    acrylamide-related, 349
    arsenic-related, 87, 88
    cadmium-related, 91
    cisplatin-related, 141
    cocaine-related, 301
    ethylene oxide-related, 388
    gold-related, 94
    hexane-related, 199
    lead-related, 121
    lithium-related, 291
    metals-related, 79
    phenytoin-related, 269
    styrene-related, 207
    thallium-related, 143
Personality
    effect of tin exposure on, 146
    relationship to neuropsychological performance, 19
Personality Assessment Inventory, 51
    use in 1,1,1-trichloroethane exposure, 216
    use in 1,1,1-trichloromethane exposure, 219–220
Pesticides, 327–348
    carbamates, 331, 339–340
    children's exposure to, through parental exposure, 38
    chlorinated cyclodienes, 342–346
    chlorinated hydrocarbons, 340–342
    combined exposure to, 348
    Dursban (chlorpyrifos), 329, 331, 334, 336–339
    fungicides and fumigants, 346–348
    herbicides, 339, 340, 346, 348
    individuals at risk for, 327–328, 329
    as mortality cause, 8
    neurotoxic potential of, 7
    neurotransmitter degradation-inhibiting action, 17
    number of, 327
    occupational exposure, 41
    organophosphates, 329–336

Pesticides (cont.)
    paraquat, 18, 346
    pyrethroids, 346
    routes of exposure, 329
    worldwide number of annual poisionings from, 8
PET scans, 77
    in studies of
        alcohol, 240
        cocaine, 302–303
        Tetrabromoethane, 210
    See also Positron emission tomographic (PET) scans
Petroleum, 189
Petroleum ether, 189
Phencyclidine (PCP), 321–324
Phenmedipham, 340
Phenobarbital, 271–272
Phenol, 400–401
Phenylalanine, as aspartame metabolite, 355
Phenylethylene: see Styrene
Phenylglyoxylic acid, as styrene metabolite, 205
Phenyl methane: see Toluene
Phenytoin, 269–271, 272
Phosphamidon, 332
Physicians, lack of neurotoxicity knowledge, 1–2
Pica, as lead ingestion cause, 102
Piperazine derivatives, 286
Piperazine phenothiazines, 288
Pirimicarb, 339
Pittsburgh Occupational Exposures Test (POET) Battery, 60–61
Placenta
    mercury accumulation in, 140
    neurotoxin passage across
        cocaine, 304
        lead, 104, 107
        mercury, 132
        phencyclidine, 322
        solvents, 152–153
    See also Prenatal exposure
Platinum, 141
Poison gas, arsenic content, 86
Political factors, in neurotoxin exposure regulation, 31
Polvinyl chloride stablizers, 144
Polychlorinated biphenols (PCBs), 401–404
    developmental effects, 402–403
    exposure limits, 401
    neurological manifestations, 403–404
Polychlorinated dibenzofurans (PCDFs), 402
Polydrug abuse, 324
    with phencyclidine, 323
Polyneuritis, gold therapy-related, 94
Polyneuropathy
    hexane-related, 200
    solvent-related, 161

Poppelreuter, Walther, 5
Porteus Maze Test
  use in Agent Orange exposure, 344
  use in neuroleptic administration, 290
Positron emission tomographic (PET) scans, 77
  of alcoholics, 240, 247
    following detoxification, 261
  of benzodiazepine activity, 276
  of cocaine users, 302
  of marijuana users, 313
  See also PET scans
Posthallucinogen perceptual disorder, 310
Posttraumatic stress disorder, 418
  diagnostic criteria, 419
  nuclear power plant accident-related, 398
  in pentaborane-exposed individuals, 400
  relationship to multiple chemical sensitivity, 411
  relationship to workpalce stress, 418
  in World War II prisoners-of-war, 154
Prednisolone, 296
Pregnancy, drug abuse during, mother's legal responsiblity for, 21
Prenatal exposure, to neurotoxins, 37–39
  alcohol, 37, 38, 229–231, 234–235
  cocaine, 300, 304
  lead, 104
  marijuana, 319–320
  mercury, 37, 38, 132, 140
  methymercury, 37, 38, 133–135
  opiates, 321
  phencyclidine, 322–323
  polychlorinated biphenyls, 402–403
  solvents, 152–153
  toluene, 213
Prescription drugs: *see* Drugs, prescription
Problem drinkers, number of, 225
Profile of Mood States, 51
  use in dioxin exposure, 345
  use in lead exposure, 123
  use in mercury exposure, 140
  use in naloxone administration, 285
  use in solvent exposure, 170
  use in in toluene exposure, 211–212
  as WHO/NIOSH Neurobehavioral Core Test Battery component, 60
Project for a Scientific Psychology (Freud), 299
Promacyl, 339
Promecarb, 339
Propane, 189, 404
Propoxur, 339
Propranolol, 275
Propylene glycol, 150
Propylene oxide, 388
Prozac: *see* Fluoxetine
Psychedelic drugs, 309–312

Psychiatric disorders, solvent-related, 153, 161–162
Psychogenic illness: *see* Psychosomatic disorders
Psychological history, 49–51
Psychological toxicology, 2
Psychology, experimental, 4
Psychometric Assessment/Dementia Screening Battery, 67–68
Psychometrics, 32
Psycho-organic syndrome, 157
Psychosis
  alcoholic, 4; *see also* Korsakoff's psychosis
  arsenic-related, 88
  cocaine-related, 305
Psychosomatic disorders, 29, 407–420
  chronic technological disaster, 418, 420
  mass psychogenic illness, 416–417
  multiple chemical sensitivity, 407–414
  posttraumatic stress disorder, 418, 419
  sick building syndrome, 415–416
  workplace stress, 417–418
Public health researchers, neurotoxicity research involvement, 22–23
Purdue Pegboard test
  use in alcohol abuse, 227, 248–249
  use in carbon monoxide exposure, 358–359
  use in in phenytoin administration, 270
Purkinje cells, aluminum-related degeneration, 84
Putz–Anderson *et al.* Battery, 63
Pyrethroids, 346
Pyridine, 19, 150, 191–192
Pyridoxine, 296
Pythrethroids, 20

Questionnaires, for neurotoxic syndrome evaluation: *see* Health questionnaires, design and administration of
Quinolinic acid, as eosinophilia myalgia syndrome causal agent, 380

Radiation
  electromagnetic, 378–379, 411–412
  ionizing, 398
Raven's Colored Progressive Matrices IQ
  use in alcohol abuse, 249
  use in anesthetic gas exposure, 352
  use in lead exposure, 109–110
  use in LSD abuse, 311
  use in mercury exposure, 137
"Raves," 311
Reaction Time Test, 53–54
  use in benzodiazepine administration, 280
Reagan administration, opposition to dioxin toxicity research, 345
Recurrent Figures Test, use in mercury exposure, 139
Reflex examination, 47

INDEX 521

Regulation, of neurotoxins, political obstacles to, 31, 345
Reitan, Ralph, 6, 56
Renal failure patients: *see* Dialysis patients
Research data, applied to clinical cases, 428–429
Reserpine, 267
Resource and Conservation and Recovery Act, 22
Rey Auditory Verbal Learning Test
　use in arsenic exposure, 87
　use in lead exposure, 122
　use in marijuana use, 317
　use in mercury exposure, 139
Rey's Embedded Figures test, use in styrene exposure, 209
Rey–Osterreith test, use with children of alcoholics, 257
Right hemisphere hypothesis, of alcohol-related neuropsychological damage, 247
Romans, ancient
　aluminum use by, 82
　lead poisioning in, 11, 95–96
Rubber solvent, as aliphatic hydrocarbon, 149
"Rubbing alcohol" abuse, 265
Ruffin, Joseph, 6, 7

Safe Drinking Water Act, 22
St. Basil, 226
Santa Ana Dexterity Test, 53, 59–60
　use in carbon disulfide exposure, 195
　use in lead exposure, 123
　use in mercury exposure, 139
　use in organophosphate exposure, 334
　use in solvent exposure, 168, 170, 171, 174
　use in xylene exposure, 221
Santeria practitioners, mercury exposure of, 132–133
Sarin, 330, 332
Saxitotoxin, 20
Scandinavia
　neuropsychological research in, 8
　solvent-related syndrome diagnostic and classification criteria in, 156, 157
　solvent syndrome-related disabilities in, 172
Schizophrenics, neuroleptic therapy, 288–290
Scopolamine, 267
Scorpion toxin, 20
Scribes, lead poisoning risk of, 101
Seafood, mercury-contaminated, 129, 132, 134, 135, 140
Seashore Rhythm Test: *see* Halstead–Reitan Battery, Seashore Rhythm Test
Secondary gain, 424
Secretage, 129
Sedative-hypnotics, 324–325
Seizures
　aspartame-related, 355
　baclofen-related, 294–295

Seizures (*cont.*)
　cocaine-related, 300, 301–302
　formaldehyde-related, 394
　sulfuryl fluoride-related, 347
Selective serotonin reuptake inhibitors (SSRIs), 274
Selenium, 80, 141–142
Self-report questionnaires
　implication for toxic exposure litigation, 423
　See also Health questionnaires
Sensory system, examination of, 47
Sensory testing, quantitative, 77
Serax: *see* Oxazepam
Serotonin syndrome, 274
Shipley–Hartford test, use in alcohol abuse, 248, 249, 256
Sick building syndrome, 415–416
Silicon and silicone, 142–143, 404–405
Silicon Valley, electronics industry, 217
Silver, Babylonians' use of, 79
Simple Reaction Time test, 59–60, 66
　use in ethylene oxide exposure, 389
Single photon computed tomography scans (SPECT), 77
　of cocaine users, 303
　of solvent-related syndrome patients, 155
Smith and Langolf Battery, 63
Smokers, carbon monoxide hemoglobin (COHb) levels in, 357, 358
Social drinking, 228–229
Sodium divalproex, 381
Sodium valproate, 271, 272
Soil, lead content, 104
Solvent-containing products, constituents of, 188
Solvent mixtures, 159–186
　acute effects, 159
　chronic effects, 160–186
　　case studies, 162–168, 174–186
　　of long-term exposure, 168–174
　　neuropsychological effects, 162, 165–168, 174–186
　　physiological effects, 160–161
　　psychiatric/emotional effects, 161–162
Solvents, 34, 41, 149–224
　abuse, prevalence of, 34
　acute effects, 187–190
　adolescents' abuse of, 38, 150, 187–188
　central nervous system toxicity, 17
　characteristics, 150
　chronic exposure, 190–191
　　prognosis, 221–224
　classification, 149–150
　complex, 150
　exposure, 150–154
　　general toxic effects, 151
　　industrial, 150

Solvents (cont.)
  exposure (cont.)
    intake, 151
    neurophysiological effects, 151
    neuropsychological effects, 153–154
    neurotoxicity, 152
    as psychiatric disorder risk factor, 153
    teratogenicity, 152–153
  historical background, 186
  interaction with alcohol, 41, 233–234
  as neurochemical toxins, 18
  neuropsychological effects, 25
  organic
    definition, 149
    neuropsychological effects, 25
  oxygenated (alcohols), 150
  visual evoked potential response, 75
Solvents-related syndromes, 191, 203
  diagnostic procedures for, 154–159
    diagnostic process, 158–159
    international diagnostic criteria, 156–158
    neuroradiological and neurophysiological techniques, 154–156
Soman, 330
Somatoform disorders, multiple chemical sensitivity as, 411, 413
Somatosensory evoked potentials, in carbon monoxide poisoning, 361, 362
Speech Sounds Perception Test: see Halstead–Reitan Battery, Speech Sounds Perception Test
SPES (Swedish Performance Evaluation System) Battery, 65, 126–127
Stalinon, 145
Standard Progressive Matrices, use in marijuana use, 317
Sternberg task, use in mercury exposure, 138–139, 140
Steroids, 295–297
Stoddard solvent, 149, 159, 170–171
Story Memory Task, use in solvent abuse, 191
Stress, workplace, 417–418
Striatonigral syndrome, aluminum accumulation-related, 84
Stroke, cocaine-related, 303–304
Stroop Color–Word Interference test
  use in alcohol abuse, 227
  use in carbamazepine administration, 268
  use in dioxin exposure, 345
  use in lead exposure, 106, 110
  use in phenytoin administration, 269
  use in solvent exposure, 171
  use in toluene exposure, 215
Styrene, 189, 205–209
  acute effects, 206
  as aromatic hydrocarbon, 149

Styrene (cont.)
  chronic effects, 206–208
  exposure limits, 205
  neuropsychological effects, 208–209
  neurotoxicity, 191
Styrene monomer: see Styrene
Styrol: see Styrene
Sugar, arsenic contamination of, 86
Suicide
  with arsenic, 86, 87
  with carbon monoxide, 356, 357
  with cyanide, 377
  with herbicides, 348
  with mercury injection, 136
  with organophosphates, 331
  with paraquat, 18
  with pesticides, 8, 329
Sulfonamide, 296
Sulfuryl fluoride, 348
Swedish Performance Evaluation System, 65, 126–127
Symbol Copying Test, use in benzodiazepine administration, 280
Symbol Digit Paired Associate Learning Test, use with recovering alcoholics, 263–264
Symbol Digit Substitution Test
  use in ethylene oxide exposure, 389
  use in nitrous oxide exposure, 354
  use in phenytoin administration, 270
  use with recovering alcoholics, 263
Symmetry Drawing Test, use in solvent exposure, 169
Symptom Checklist-90 Revised
  Global Severity Index, use in solvents exposure, 161–162
  use in mercury exposure, 137, 139
Synapses, neurotoxicant-induced damage, 15
Syphilis, 4
  arsenic treatment, 86
  mercury treatment, 128
Systemic lupus erythematosus
  silicone-related, 404
  trichloroethylene-related, 218

Tabun, 330
Tachistoscopic Perception test, use in alcohol abuse, 227
Tactual Performance Test: see Halstead–Reitan Battery, Tactual Performance Test
Tapping Preferred Hand Test, use in disulfiram administration, 284
Tardive dyskinesia, neuroleptics-related, 287, 288
Taurine, solvents-related decrease, 152
Taylor Manifest Anxiety Scale, use in organophosphate exposure, 335

Tea, effect on cognitive performance, 298–299
Teeth, lead content, 106, 109–110
Tegretol: see Carbamazepine
Tellurium, 143
Telone, 346
Temazepam, 278
Teratology, behavioral, 37; see also Prenatal exposure
Tetrabromoethane, 209–210
2,3,7,8-Tetrachlorodibenzo-p-dioxin, 344–345
Tetrachloroethane, 191–192
Tetrachloroethylene: see Perchloroethylene
Tetrachloromethane: see Carbon tetrachloride
Tetradotoxin, 20
Tetrahydrocannabinol (THC), 312, 313–314, 315, 317, 318, 319
Tetrahydrofuran, 210
Tetramethylene oxide: see Tetrahydrofuran
Thallium, 143
Thiamine deficiency, alcoholism-related, 233, 258
  as Wernicke–Korsakoff syndrome cause, 242–243
Thiodicarb, 340
Thiofonax, 340
Thiophanate-ethyl, 340
Thiophanate-methyl, 340
Thioridazine, 288
Thorn apple tea, 268
Threshold limit values, 23–24, 32–33
Tight building syndrome, 415
Time estimation tests, use in benzodiazepine administration, 281
Times Beach toxic contamination incident, 420
Tin, 144–146
  elemental, 80
  exposure limits, 144
  inorganic, 144
  organotin, 144–145
Tobacco smoke
  as carbon monoxide source, 357
  as cyanide source, 376–377
Toluene, 189, 210–216
  abuse of, 213, 214–215
  acute exposure, 211–212
  as aromatic hydrocarbon, 149
  cerebral damage by, 15
  chronic exposure, 212–214, 215–216
  co-exposure with methyl ethyl ketone, case study, 181–186
  exposure limits, 210
  as jet fuel component, 394
  neuropsychological effects, 172–173, 189
  neurotoxicity, 152, 191
  teratoxigenicity, 213
Toluol: see Toluene
Total allergy syndrome, 407
Toxaphene, 340, 341

Toxic chemicals, EPA registry of, 7
Toxic contamination, as chronic technological disaster cause, 418, 420
Toxic oil syndrome, 142, 404–405
Toxicology screens, 45, 47
Toxic Substances Control Act of 1976, 22
Toxic torts, 421–428
Trail-Making Tests: see Halstead–Reitan Battery, Trail-Making Tests A & B
Tranquilizers, prevalence of use, 275
Tranxene: see Clorazepate dipotassium
Tremometry, 77
Tremor, mercury-related, 136
Triazolam, 277, 279–280
Tri-c-cresyl phosphate, 330
Tri-ortho-cresyl phosphate, 16
Trichlorfon, 331
1,1,1-Trichloroethane, 149–150, 162, 187, 189, 216–217
1,1,2-Trichloroethane, 189
Trichloroethylene, 217–219, 350–351
  as chlorinated hydrocarbon, 150
  neuropsychological effects, 172–173
  neuropsychological recovery following exposure to, 222
  neurotoxicity, 152, 191
1,1,1-Trichloromethane, 186, 189, 219–220, 350–351
Trichloronate, 331
Trichlorotrifluoroethane, 197
Tricyclic antidepressants, 273–274
Triethyltin, 16, 144, 145
Trigeminal nerve, neurological examination of, 47
Trimethacarb, 339
Trimethyltin, 15, 144, 145
Tri-ortho-cresyl phosphate, 16
L-Tryptophan, as eosinophilia myalgia syndrome (EMS) cause, 1, 379, 380, 404–405
  case report, 382–386
TUFF Battery, 64
Tungsten carbide, 146
Turpentine, 150
Tuttle, Wood, and Grether Battery, 62–63
Twentieth-century syndrome, 407
Twin studies
  of alcohol abuse, 246
  of solvent exposure, 174

United Kingdom, pharmaceutical developmental neurotoxicity screening in, 8, 38
University of Pennsylvania Smell Identification Kit, 154
Uranium, as ionizing radiation source, 398

Valcuikas and Lilis Battery, 63
Valentine, Basil, 386
Valium: see Diazepam

Valproate, 271, 272
Vanadium, 146–148
Vanillylmandelic acid, as lead exposure indicator, 104
Varnish makers' naphtha, 149
Verbal Memory Recall test, use in benzodiazepine administration, 280
Verbal Pair Associate Learning test, use with recovering alcoholics, 264
Video display terminals (VDTs), 378
Vienna Reaction Device, 110
Vincristine, 283
Vinyl benzene: see Styrene
Visual Afterimages, 227
Visual evoked potentials, 75
　in carbon monoxide poisoning, 361, 362
　in cyanide poisoning, 378
　in solvent-related syndrome, 156
Visual Memory test, use in solvents exposure, 170
Visual pursuit rotor test, use in mercury exposure, 138, 139
Visual Retention Test, of Neurobehavioral Evaluation System, 66
Vitamin B_6, sites of damage by, 20
Vitamin deficiencies, alcoholism-related, 232, 233, 258
Vitamin E deficiency
　alcoholism-related, 258
　paraquat-related, 346
Vitamin megadoses, 1
Voodoo practitioners, mercury exposure of, 132–133

Walter Reed Performance Assessment Battery, 69–70
War of the Ghosts (Bartlett), 314
Washington, George, 129
Water, perchloroethylene-contaminated, 202; *see also* Drinking water
Wechsler Adult Intelligence Scale-Revised (WAIS-R)
　use in alcohol abuse, 228, 240, 245
　use in alcohol abuse recovery, 264
　use in aluminum exposure, 85
　Arithmetic subtest, 303
　Block Design subtest, 59
　　use in alcohol abuse, 239, 249
　　use in anticholinergic neuroleptic administration, 288–289
　　use in benzodiazepine administration, 280
　　use in carbon disulfide exposure, 195, 196
　　use in cocaine use, 307
　　use in lithium administration, 293
　　use in mercury exposure, 137
　　use in organophosphate exposure, 334
　　use in sodium valproate exposure, 271
　　use in solvent exposure, 168, 169, 170, 172, 174, 190
　　use in styrene exposure, 209

Wechsler Adult Intelligence Scale-Revised (WAIS-R) (*cont.*)
　Digit Span subtest, 59
　　use in alcohol abuse, 227, 259
　　use in anesthetic gas exposure, 351, 352
　　use in carbon disulfide exposure, 196
　　use in carbon monoxide exposure, 372–373
　　use in cocaine use, 303
　　use in ethylene oxide exposure, 389
　　use in mercury exposure, 139
　　use in solvent exposure, 168, 169, 170, 171
　　use in toluene exposure, 214
　Digit Symbol subtest, 59–60
　　use in alcohol abuse, 236, 239, 245, 263
　　use in benzodiazepine administration, 280, 281
　　use in carbon disulfide exposure, 194, 195, 196
　　use in fetal alcohol syndrome, 231
　　use in lithium exposure, 293
　　use in mercury exposure, 137
　　use in nitrous oxide exposure, 354
　　use in opiates use, 320
　　use in solvent exposure, 168, 169, 170, 171, 172, 173, 174
　　use in styrene exposure, 207
　　use in toluene exposure, 214
　　use in tricyclic antidepressant administration, 273
　Full Scale IQ
　　use in sodium valproate administration, 271
　　use in solvent exposure, 191
　as Halstead–Reitan Battery component, 56
　Information subtest, use with children of alcoholics, 257
　use in marijuana use, 317
　use in mercury exposure, 139
　Object Assembly subtest, use in alcohol abuse, 249
　Performance IQ, use in tricyclic antidepressant administration, 273
　Picture Arrangement subtest, use in alcohol abuse, 249
　Picture Completion subtest
　　use in anticholinergic administration, 288–289
　　use in dioxin exposure, 345
　　use in sodium valporate administration, 271
　　use in styrene exposure, 207
　as Pittsburgh Occupational Exposures Test component, 60
　Similarities subtest, 59
　　use in anticholinergic neuroleptic administration, 288–289
　Verbal IQ, 191
　Vocabulary subtest
　　in anticholinergic neuroleptic administration, 288–289
　　use in sodium valporate administration, 271
　　use in solvent exposure, 190

INDEX

Wechsler Intelligence Scale for Children-Revised (WISC-R)
　use in fetal alcohol syndrome, 231
　use in lead exposure, 106, 109, 110
　use in toluene exposure, 215
Wechsler Memory Scale-Revised (WMS-R)
　use in Agent Orange exposure, 344
　use in alcohol abuse, 238, 249
　　with and without Wernicke–Korsakoff syndrome, 245
　　during recovery, 264
　use in aluminum exposure, 85
　use in antihypertensives administration, 275
　Associate Learning subtest
　　use in anticholinergic neuroleptic administration, 288–289
　　use in cocaine use, 307
　　use in solvent exposure, 169
　Difficult Paired Associates subtest, 257
　Figural Memory subtest, 72
　　use in benzodiazepine administration, 279
　　use in clozapine administration, 290
　　use in solvent exposure, 170, 171, 190
　Logical Memory subtest
　　use in antihypertensive administration, 275
　　use in arsenic exposure, 87
　　use in benzodiazepine administration, 279
　　use in cocaine abuse, 307
　　use in Korsakoff's psychosis, 244
　　use in solvent exposure, 190
　　use in styrene exposure, 209
　　use in tricyclic antidepressant administration, 273
　Mental Control subtest, use in alcohol abuse, 256
　use in mercury exposure, 138
　use in neuroleptics administration, 287
　Russell's modification, use in alcohol abuse, 236, 248
　use in solvent exposure, 169, 171, 174
　use in steroids administration, 295
　Visual Reproduction subtest, 139, 389
　use in Wilson's disease, 93
Welders, aluminum exposure, 84
Wernicke, Carel, 243, 244
Wernicke–Korsakoff syndrome, 229, 242–243
　cerebral blood flow in, 262

Wernicke–Korsakoff syndrome (*cont.*)
　in chronic alcoholism, 244–245
　neuropsychological investigations of, 244
　Wernicke's encephalopathy phase of: *see* Encephalopathy, Wernicke's
Wernicke's triad, 243
Whipped-cream cans, nitrous oxide propellant, 353–354
White Lead Convention, 96
White matter, neurotoxin-related damage, 16
"White spirit": *see* Stoddard solvent
Williamson, Teo, and Sanderson Battery, 63–64
Wilson's disease, 92, 92–94
Wine, lead contamination, 95–96, 101
Wisconsin Card Sorting Test
　use in alcohol abuse, 247, 249
　use in lead exposure, 106, 110
　use in Wilson's disease, 93
Women
　alcohol abuse vulnerability of, 259–260
　alcohol metabolism in, 226–227, 259
　MDMA metabolism in, 311
Wood alcohol, 150
Workplace stress, 417–418
World Health Organization (WHO), solvent intoxication effects classification scheme, 156–158
World Health Organization/National Institute of Occupational Safety and Health Neuorbehavioral core test battery (NCTB), 56–57, 59–60, 65
　automated version, 67
"Wrist drop," lead-related, 100, 121

"X": *see* 3,4-Methylenedioxymethamphetamine
Xanax: *see* Alprazolam
XMC, 339
X-ray fluorescence, *in vivo* tibial X-ray-induced, 98
X rays, as ionizing radiation source, 398
Xylene, 189, 220–221
　as aromatic hydrocarbon, 149
　neurotoxicity, 191–192
Xylycarb, 339

Zinc, 147–148
Zinc protoporphyrin screening, for lead exposure, 98

Printed in the United States
100081LV00001BA/1-10/A